Y0-BYJ-964

ALGEBRA AND TRIGONOMETRY

For College Students

ALGEBRA AND TRIGONOMETRY

For College Students

Richard S. Paul

Ernest F. Haeussler, Jr.

Department of Mathematics
The Pennsylvania State University

RESTON PUBLISHING COMPANY, INC.

Reston, Virginia

A Prentice-Hall Company

Library of Congress Cataloging in Publication Data

Paul, Richard S
Algebra and trigonometry: for college students.

Includes index.
1. Algebra. 2. Trigonometry. I. Haeussler, Ernest F., joint author. II. Title.
QA152.2.P38 512'.13 77-13393
ISBN 0-87909-031-6

Reston Publishing Company, Inc.
A Prentice-Hall Company
Reston, Virginia 22090

10 9 8 7 6 5

Printed in the United States of America.

Contents

Preface

This text is designed for students who need to develop manipulative skills in algebra and trigonometry. It can be used either by students taking a terminal course in mathematics or by those who need a background for more advanced work in mathematics. Included are functions and their applications as well as other topics that we feel are most basic to students' needs. We have purposely omitted certain topics so as not to forfeit our emphasis on basics.

Every attempt has been made to write in "down-to-earth" language with short paragraphs, simple vocabulary, and conversational style. Over 800 examples, in which all steps are shown, have been included. Where appropriate, diagrams appear in order to convey and reinforce ideas.

After most sections, two sets of exercises appear. First, there are Completion Sets, which include programmed-type problems. These are followed by Problem Sets, which include traditional-type problems. In each Problem Set the exercises are given in order of increasing difficulty. However, there are a sufficient number of problems that students can do rather easily. This allows students to gain confidence and skills and to fix firmly in their minds the basic concepts studied.

At the end of each chapter there is a Review section that contains a list of important terms and symbols, together with page numbers that indicate where these items first appear. Following the list are numerous review problems. In total, there are more than 4100 exercises in the text. At the back of the book are answers to all Completion Sets and all odd-numbered exercises in the Problem Sets and Review Problems.

Numerous verbal problems (requiring no prior knowledge) are scattered throughout. Thus, students can see relevant ways in which the mathematics they are learning can be used.

A skull and crossbones symbol is employed to highlight errors commonly made by students. It is intended to alert them and cause them to "beware." We feel that it is just as important for students to know what they cannot do as to know what they can do.

The solution of basic equations occurs early in the book. Using equations as a springboard, the text proceeds to manipulative aspects of algebra that enable students to solve more complicated problems.

Chapter 1 contains selected topics in arithmetic. Although not necessarily part of a formal course, it nevertheless provides a convenient reference for students. In Chapter 3, positive integral exponents and radicals are introduced. A more complete treatment is given in Chapter 13 where zero, fractional, and negative exponents are considered along with further properties of radicals. By dividing exponents and radicals in this way, students are not overburdened at the beginning of the course with laborious calculations involving negative and fractional exponents.

Available from the publisher is an extensive instructor's manual that contains answers to all exercises in the Problem Sets and all Review Problems. Worked-out solutions to many of the problems, including all verbal problems, are also included. In addition, the manual offers examination questions and their answers.

We wish to express our sincere appreciation to the following people who contributed comments and suggestions for the improvement of the manuscript: Elmar Zemgalis of Highline Community College, Irene Buttery of Long Beach City College, Shirley Sorenson of the University of Maryland, Marie A. Ritten of the National College of Business, Harold Harutunian of Salem State College, and George W. Schultz of St. Petersburg Junior College.

Finally, we express our thanks to our typist Mary Walko who never failed us in meeting deadlines.

Richard S. Paul
Ernest F. Haeussler, Jr.

CHAPTER 1

Arithmetic Refresher

Many things in algebra are done in the same way as in arithmetic. That's why a good understanding of arithmetic may be important to you in learning algebra. For this reason we'll begin with a brief review of some basic topics in arithmetic.[†]

1-1 BASIC OPERATIONS

There are four rules to keep in mind when you work with numbers. Two are:

1. You can add numbers in any order.
2. You can multiply numbers in any order.

For example, $5 + 4 = 9$ and $4 + 5 = 9$. So,

$$5 + 4 = 4 + 5 = 9.$$

[†]Note to instructor: In this chapter, only nonnegative rational numbers are considered.

Rule 2 involves multiplication. Some ways to show a multiplication, say 6 times 2, are (6)(2) and 6(2). Thus by Rule 2,

$$(6)(2) = (2)(6) = 12.$$

EXAMPLE 1

Order is not important in addition or multiplication.

a. $2 + 3 + 4 = 3 + 2 + 4 = 3 + 4 + 2 = 9.$

b. $(5)(2)(3) = (5)(3)(2) = (3)(5)(2) = 30.$

When you multiply numbers, the answer is called their **product**. Each of the numbers is called a **factor** of the product. For example,

$$(3)(2) = 6.$$

↑ ↑ ↑

These are *factors* of 6. This is the *product* of 3 and 2.

The next rules are:

3. You must subtract numbers in the given order.
4. You must divide numbers in the given order.

$5 - 2$ is **not** the same as $2 - 5$. Also, $6 \div 3$ is **not** the same as $3 \div 6$. When you subtract or divide, order **is** important. We point out that in division there are some words you should know:

$$6 \div 3 = 2$$

↑ ↑ ↑

dividend *divided by* **divisor** *equals* **quotient**

In some problems with more than one operation, it may not be clear to you as to what operation to do first. *In these cases do any multiplications and divisions first.* Then do any additions and subtractions.

EXAMPLE 2

The Right Way	*The Wrong Way*
a. $2 + (3)(4) = 2 + 12 = 14.$	**Don't** add 2 to 3 and then multiply by 4. (This gives 20.)

b. $10 - (3)(2) = 10 - 6 = 4.$ **Don't** subtract 3 from 10 and then multiply by 2. (This gives 14.)

c. $(5)(7) - 6 = 35 - 6 = 29.$ **Don't** subtract 6 from 7 and then multiply by 5. (This gives 5.)

EXAMPLE 3

$$\begin{aligned} 16 + 4(5) \div 2 - 3(2) &= 16 + 20 \div 2 - 6 \\ &= 16 + 10 - 6 \\ &= 26 - 6 = 20. \end{aligned}$$

When operations are inside parentheses, do those first, as the next example shows.

EXAMPLE 4

a. $14 + \overbrace{(1 + 4)} - 3\overbrace{(6 - 3)}$ ← Do these first.

$= 14 + 5 - 3(3)$ [Next, do the multiplication.]

$= 14 + 5 - 9$ [Do the addition next.]

$= 19 - 9$ [Finally, do the subtraction.]

$= 10.$

b. $12 + \overbrace{(15 \div 3)}\overbrace{(4 - 2)}$ ← Do these first.

$= 12 + (5)(2)$ [Next, do the multiplication.]

$= 12 + 10$

$= 22.$

c. $(2 - 2)(5) = (0)(5) = 0.$ *Remember*: **0 times any number is 0.**

Don't mix up $2 + 5(6)$ *with* $(2 + 5)6$. *We have*

$$2 + 5(6) = 2 + 30 = 32,$$

$$\textit{but} \quad (2 + 5)(6) = (7)(6) = 42.$$

Problem Set 1–1

In Problems **1–32**, *compute the numbers.*

1. $5 + (2)(8)$ **2.** $17 - (3)(4)$ **3.** $(2)(4) - 7$

4. $(3)(2) - (2)(2)$ **5.** $(8 - 3)(4)$ **6.** $2(5 + 1)$

7. $15 - 3(4)$ **8.** $(6 - 4) + 3$ **9.** $3(4 + 6)$

10. $(6 - 6)8$ **11.** $(7 - 2)(3 + 1)$ **12.** $(4 + 5)(2 - 0)$

13. $0(8 - 5)$ **14.** $(3 - 3)(4 - 4)$ **15.** $(6 - 1) - (10 - 7)$

16. $(6 + 7) - (8 - 4)$ **17.** $(8 \div 4) + 3$ **18.** $(7 - 2)(3) + 2(3 + 1)$

19. $10 + (9 \div 3)$ **20.** $(3 \div 3) + (2)(3 - 1)$

21. $18 \div (6 + 3)$ **22.** $55 \div (17 - 6)$

23. $(6 + 2 - 3) \div (7 - 2)$ **24.** $(8 - 4)(7 - 3) + (16 \div 4)$

25. $13 - 2(5 - 2) + (2 + 3)$ **26.** $8 + (15 \div 3)(3 - 1)$

27. $(36 \div 6)(6 \div 2) - 6$ **28.** $16 + (12 \div 3)(4 - 3)$

29. $4(4 - 1) - 5(4 \div 2) + 9$ **30.** $2(3 + 4) + 5 - (2 - 1)(2 + 1)$

31. $3(6 \div 3) + 2 - 4(2 \div 2)$ **32.** $(4 + 3)(3) + 5(5 - 2)(2 - 2)$

33. In $8 \div 2 = 4$, what is the dividend? What is the divisor?

34. In $16 \div 8 = 2$, what is the quotient?

35. In $30 \div (3 + 2) = 6$, what number is the divisor?

36. The product of three factors is 30. Two of the factors are 2 and 3. What is the other factor?

37. The product of five factors is 32. Three of the factors are each 2. What is the *product* of the other factors?

38. The numbers 2 and 3 are factors of 6. Find two other whole numbers that are also factors of 6. (Whole numbers are the numbers 0, 1, 2, 3, etc.)

1-2 FRACTIONS

You can write the division $6 \div 3$ as the fraction $\frac{6}{3}$. In this fraction, 6 is the **numerator** and 3 is the **denominator**. The fraction is also called the **quotient** of 6 divided by 3, or simply the quotient $\frac{6}{3}$. Since $6 \div 3 = 2$, then $\frac{6}{3} = 2$. Here is another way of looking at $\frac{6}{3}$:

$$\frac{6}{3} \text{ is the number which when multiplied by 3 gives 6.}$$

Since $(3)(2) = 6$, then $\frac{6}{3} = 2$.

In some fractions the numerator is *less* than the denominator, as in $\frac{3}{8}$. We call this a **proper fraction**. Its value is less than 1. Fractions where the numerator is *equal to or larger than* the denominator, as in $\frac{9}{2}$, are called **improper**. The value is at least 1.

$$\text{Proper fractions:} \quad \frac{1}{2}, \frac{12}{17}, \frac{100}{102}.$$

$$\text{Improper fractions:} \quad \frac{15}{8}, \frac{3}{3}, \frac{102}{100}.$$

It's possible to write an improper fraction as a whole number† plus a proper fraction. We call this a **mixed number**. To write $\frac{19}{5}$ as a mixed number, we divide 19 by 5.

$$\begin{array}{r} 3 \\ 5\overline{)19} \\ \underline{15} \\ 4 \end{array}$$

This gives 3 with a remainder of 4. So $\frac{19}{5}$ is 3 plus the fraction $\frac{4}{5}$.

$$\underset{\substack{\uparrow \\ \text{improper} \\ \text{fraction}}}{\frac{19}{5}} = 3 + \frac{4}{5} = \underset{\substack{\uparrow \\ \text{mixed} \\ \text{number}}}{3\frac{4}{5}}.$$

EXAMPLE 1

Writing improper fractions as mixed numbers.

a. $\frac{15}{11} = 1 + \frac{4}{11} = 1\frac{4}{11}$. b. $\frac{17}{3} = 5\frac{2}{3}$. c. $\frac{27}{4} = 6\frac{3}{4}$.

---

†Whole numbers are the numbers 0, 1, 2, 3, etc.

Fractions that have the same value are called **equivalent fractions**. Here's an important rule for getting equivalent fractions. It is called the **fundamental principle of fractions**.

> **You can multiply or divide both the numerator and denominator of a fraction by the same number (except 0). The result is equivalent to the original fraction.**

EXAMPLE 2

Using the fundamental principle of fractions.

a. *Write* $\frac{5}{8}$ *as an equivalent fraction whose denominator is* 24.

24 can be written as a product of 8 and 3. So we must multiply the denominator by 3 to get 24. Thus we must also multiply the numerator by 3.

$$\frac{5}{8} = \frac{(5)(3)}{(8)(3)} = \frac{15}{24}.$$

b. *Write* 7 *as an equivalent fraction with denominator* 2.

Since 7 is the same as the fraction $\frac{7}{1}$, we have

$$7 = \frac{7}{1} = \frac{(7)(2)}{(1)(2)} = \frac{14}{2}.$$

c. *Write* $\frac{12}{20}$ *as an equivalent fraction with numerator* 3.

The numerator will be 3 if we divide it by 4. So we also divide the denominator by 4.

$$\frac{12}{20} = \frac{12 \div 4}{20 \div 4} = \frac{3}{5}.$$

A fraction is in **lowest terms**, or is (completely) **reduced**, when its numerator and denominator have no whole-number factors in common except 1. For example, $\frac{18}{12}$ is not in lowest terms because 18 and 12 have a common factor of 6:

$$18 = (6)(3) \quad \text{and} \quad 12 = (6)(2).$$

To reduce $\frac{18}{12}$, let's use the fundamental principle by dividing the numerator and denominator by this common factor 6.

$$\frac{18}{12} = \frac{\frac{18}{6}}{\frac{12}{6}}.$$

But $\frac{18}{6} = 3$ and $\frac{12}{6} = 2$. Thus,

$$\frac{18}{12} = \frac{\frac{18}{6}}{\frac{12}{6}} = \frac{3}{2}.$$

Since 3 and 2 have no factors in common except 1, then $\frac{3}{2}$ is the reduced form of $\frac{18}{12}$. More simply, to reduce $\frac{18}{12}$ we usually write

$$\frac{18}{12} = \frac{\overset{3}{\cancel{18}}}{\underset{2}{\cancel{12}}} = \frac{3}{2},$$

where the slashes indicate the division by a common factor. Sometimes this division is called **cancellation**. You can think of this problem in another way by showing the common factor 6:

$$\frac{18}{12} = \frac{\overset{1}{\cancel{(6)}}(3)}{\underset{1}{\cancel{(6)}}(2)} = \frac{3}{2}.$$

The fraction $\frac{18}{12}$ can also be reduced by repeated cancellation. Watch how we cancel a factor of 2 and then 3:

$$\frac{18}{12} = \frac{\overset{9}{\cancel{18}}}{\underset{6}{\cancel{12}}} = \frac{\overset{3}{\cancel{9}}}{\underset{2}{\cancel{6}}} = \frac{3}{2}.$$

EXAMPLE 3

Write $\frac{(16)(5)}{(15)(24)}$ *in lowest terms.*

$$\frac{(16)(5)}{(15)(24)} = \frac{(16)\overset{1}{\cancel{(5)}}}{\underset{3}{\cancel{(15)}}(24)} = \frac{\overset{2}{\cancel{16}}}{(3)\underset{3}{\cancel{(24)}}} = \frac{2}{(3)(3)} = \frac{2}{9}.$$

We can save some space by writing

$$\frac{\overset{2}{\cancel{(16)}}\overset{1}{\cancel{(5)}}}{\underset{3}{\cancel{(15)}}\underset{3}{\cancel{(24)}}} = \frac{2}{9}.$$

In Example 3, by cancellation we reduced a fraction whose numerator and denominator were written as a product of numbers. In a fraction whose numerator or denominator is a sum or difference of numbers, this type of cancellation cannot be done. For example, $\frac{6}{2}$*, which is 3, can be written* $\frac{4+2}{2}$. *But*

$$\frac{4+\cancel{2}^{1}}{\cancel{2}_{1}} = \frac{5}{1} \text{ is } \textbf{\textit{false}}.$$

Problem Set 1-2

1. *Fill in the blanks:* $\frac{24}{8}$ is the number which when multiplied by ____ gives ______.
2. In the fraction $\frac{6}{11}$, what is the numerator?
3. In the fraction $\frac{5}{6}$, what is the denominator?
4. Which of the following are proper fractions?

$$\frac{3}{5}, \frac{8}{7}, \frac{2}{9}, \frac{7}{11}, \frac{8}{8}$$

5. Which of the following are improper fractions?

$$\frac{8}{7}, \frac{12}{13}, \frac{7}{7}, \frac{8}{1}$$

6. Which of the following fractions are in lowest terms?

$$\frac{9}{27}, \frac{4}{7}, \frac{13}{3}, \frac{14}{21}$$

In Problems **7–14**, *write the fractions as mixed numbers.*

7. $\frac{23}{3}$ **8.** $\frac{37}{6}$ **9.** $\frac{18}{5}$ **10.** $\frac{74}{9}$

11. $\frac{3}{2}$ **12.** $\frac{251}{24}$ **13.** $\frac{13}{4}$ **14.** $\frac{5003}{1000}$

15. Write $\frac{2}{3}$ as an equivalent fraction with denominator 18.

16. Write $\frac{3}{7}$ as an equivalent fraction with numerator 21.

In Problems **17–28**, *fill in the missing numbers.*

17. $\frac{4}{5} = \frac{\quad}{25}$ **18.** $\frac{15}{24} = \frac{5}{\quad}$ **19.** $\frac{2}{3} = \frac{22}{\quad}$

20. $\frac{12}{32} = \frac{\quad}{8}$ 21. $\frac{8}{24} = \frac{1}{\quad}$ 22. $\frac{100}{5} = \frac{\quad}{1}$

23. $\frac{18}{12} = \frac{\quad}{2}$ 24. $7 = \frac{\quad}{3}$ 25. $5 = \frac{\quad}{3}$

26. $3 = \frac{\quad}{12}$ 27. $\frac{1}{7} = \frac{7}{\quad}$ 28. $\frac{6}{6} = \frac{\quad}{12}$

In Problems **29–50**, *completely reduce the fractions.*

29. $\frac{28}{12}$ 30. $\frac{210}{26}$ 31. $\frac{60}{45}$

32. $\frac{18}{99}$ 33. $\frac{24}{30}$ 34. $\frac{275}{1000}$

35. $\frac{66}{144}$ 36. $\frac{7+9}{2}$ 37. $\frac{4}{8+8}$

38. $\frac{(10)(3)}{5}$ 39. $\frac{6}{(12)(5)}$ 40. $\frac{(9)(10)}{(6)(8)}$

41. $\frac{(3)(16)}{(4)(9)}$ 42. $\frac{(4)(14)}{(21)(2)}$ 43. $\frac{(21)(15)}{(5)(14)}$

44. $\frac{(4)(9)(6)}{(6)(18)}$ 45. $\frac{(8)(12)(15)}{(3)(4)(6)}$ 46. $\frac{(100)(12)(5)}{(6)(15)(25)}$

47. $\frac{(12)(6)(3)(14)}{(22)(14)(21)(18)}$ 48. $\frac{(4)(12)(10)(7)}{(14)(8)(5)(6)}$

49. $\frac{(81)(5)(3)(7)}{(27)(18)(4)}$ 50. $\frac{(24)(50)(3)}{(75)(6)(16)(5)}$

1-3 ADDITION AND SUBTRACTION OF FRACTIONS

It's easy to add or subtract fractions that have the same denominator. Just add or subtract the numerators, but keep the same (common) denominator. For example,

$$\frac{2}{9} + \frac{5}{9} = \frac{2+5}{9}$$

(add numerators; keep same denominator)

$$= \frac{7}{9}.$$

EXAMPLE 1

a. $\frac{3}{7} + \frac{5}{7} = \frac{3+5}{7} = \frac{8}{7}$.

b. $\frac{4}{5} - \frac{2}{5} = \frac{4-2}{5} = \frac{2}{5}$.

c. $\frac{8}{11} + \frac{6}{11} - \frac{3}{11} = \frac{8+6-3}{11} = \frac{11}{11} = 1$. **Always reduce your answer if possible.**

Don't write $\frac{1}{2} + \frac{1}{2} = \frac{1+1}{2+2}$, *which is **false.***

$$\frac{1}{2} + \frac{1}{2} = \frac{1+1}{2} = \frac{2}{2} = 1, \quad \textit{but} \quad \frac{1+1}{2+2} = \frac{2}{4} = \frac{1}{2}.$$

It takes a little more work to add or subtract fractions when the denominators aren't the same. The first step is to rewrite the fractions so that they *do* have the same denominator. There is a special denominator we use. It is the smallest number that can be divided "exactly" by each of the given denominators. We call this the **least common denominator** (L.C.D.).

Sometimes the L.C.D. of fractions can be found by "inspection." For example, the L.C.D. of $\frac{1}{2}$ and $\frac{1}{3}$ is 6, since 6 is obviously the smallest number that can be divided exactly by 2 and 3.

The denominators of

$$\frac{1}{3}, \frac{5}{8}, \text{ and } \frac{7}{12}$$

are 3, 8, and 12. All these denominators divide 24. Also, 24 is the smallest number divisible by 3, 8, and 12. Thus, 24 is the L.C.D.

Another way to find an L.C.D. involves *prime numbers*.

A *prime number* is a whole number greater than 1 whose only whole number factors are itself and 1.

For example, 2, 3, 5, 7, 11, 13, and 17 are prime numbers, but 14 is *not* prime because it has 2 and 7 as factors. It turns out that

Every whole number greater than 1 can be written as a product of prime numbers.

For example, 30 can be written as (2)(15). The 2 is prime, but 15 isn't. So, we write 15 as a product of factors: 15 = (3)(5). Both 3 and 5 are primes. Thus,

$$30 = (2)(3)(5), \text{ a product of primes}.$$

EXAMPLE 2

Write each number as a product of primes.

a. 20 = (2)(10) = (2)(2)(5) [since 10 = (2)(5)].

b. 35 = (5)(7).

c. 42 = (2)(21) = (2)(3)(7).

d. 24 = (2)(12) = (2)(2)(6) = (2)(2)(2)(3).

Here's how to find the L.C.D. of a group of fractions.

First, write each denominator as a product of prime factors. Next, write down each of the different prime factors the greatest number of times it occurs in any single denominator. Multiply these factors together. *This is the* L.C.D.

The next two examples should make the explanation clearer.

EXAMPLE 3

Find the L.C.D. *of the fractions* $\frac{1}{3}$, $\frac{5}{12}$, *and* $\frac{3}{10}$.

First write each denominator as a product of primes.

$$3 = 3.$$

$$12 = (2)(6) = (2)(2)(3).$$

$$10 = (2)(5).$$

There are three different prime factors involved: 3, 2, and 5. The greatest number of times that 3 occurs in any *single* denominator is once. For the 2, it is twice. For 5, it is once. Thus,

$$\text{L.C.D.} = (3)(2)(2)(5) = 60.$$

EXAMPLE 4

Find the L.C.D. *of* $\frac{1}{6}$, $\frac{2}{9}$, *and* $\frac{5}{12}$.

$$6 = (3)(2).$$
$$9 = (3)(3).$$
$$12 = (2)(2)(3).$$

There are two different prime factors involved: 3 and 2. The most often that the 3 occurs is twice. The most often that the 2 occurs is also twice. Thus,

$$\text{L.C.D.} = (3)(3)(2)(2) = 36.$$

Now we're ready to add fractions with different denominators. Let's try

$$\frac{1}{6} + \frac{2}{9} + \frac{5}{12}.$$

First, find the L.C.D. From Example 4, it is 36. Next, write each fraction as an equivalent fraction with a denominator of 36. To do this for $\frac{1}{6}$, multiply its numerator and denominator by 6:

$$\frac{1}{6} = \frac{(1)(6)}{(6)(6)} = \frac{6}{36}.$$

Similarly,

$$\frac{2}{9} = \frac{(2)(4)}{(9)(4)} = \frac{8}{36},$$

$$\frac{5}{12} = \frac{(5)(3)}{(12)(3)} = \frac{15}{36}.$$

Now that all of the denominators are the same, we add the way we did in Example 1 on p. 10.

$$\frac{1}{6} + \frac{2}{9} + \frac{5}{12} = \frac{6}{36} + \frac{8}{36} + \frac{15}{36}$$
$$= \frac{6 + 8 + 15}{36} = \frac{29}{36}.$$

EXAMPLE 5

Add $\frac{2}{3} + \frac{7}{8} + \frac{3}{16}$.

Since $3 = 3$, $8 = (2)(2)(2)$, and $16 = (2)(2)(2)(2)$,

$$\text{L.C.D.} = (3)(2)(2)(2)(2) = 48.$$

Thus,

$$\frac{2}{3} + \frac{7}{8} + \frac{3}{16} = \frac{(2)(16)}{(3)(16)} + \frac{(7)(6)}{(8)(6)} + \frac{(3)(3)}{(16)(3)}$$

$$= \frac{32}{48} + \frac{42}{48} + \frac{9}{48} = \frac{32 + 42 + 9}{48} = \frac{83}{48}.$$

EXAMPLE 6

Combine and simplify.

a. $\frac{3}{5} - \frac{1}{2} + \frac{1}{4} = \frac{12}{20} - \frac{10}{20} + \frac{5}{20}$

$$= \frac{12 - 10 + 5}{20} = \frac{7}{20}.$$

b. $2 + \frac{1}{6} + \frac{1}{3} = \frac{2}{1} + \frac{1}{6} + \frac{1}{3}$

$$= \frac{12}{6} + \frac{1}{6} + \frac{2}{6} = \frac{15}{6} = \frac{5}{2}.$$

Just as you can write an improper fraction as a mixed number, you can also write a mixed number as an improper fraction. To do this, write the whole-number part as a fraction and add it to the fractional part.

EXAMPLE 7

a. $3\frac{3}{4} = 3 + \frac{3}{4} = \frac{3}{1} + \frac{3}{4} = \frac{12}{4} + \frac{3}{4} = \frac{15}{4}.$

b. $5\frac{2}{7} = \frac{5}{1} + \frac{2}{7} = \frac{35}{7} + \frac{2}{7} = \frac{37}{7}.$

Here's how to add or subtract mixed numbers. Write each number as a fraction and then combine as we did before.

EXAMPLE 8

a. $$3\frac{3}{4} + 4\frac{2}{3} = \left(3 + \frac{3}{4}\right) + \left(4 + \frac{2}{3}\right)$$
$$= \left(\frac{12}{4} + \frac{3}{4}\right) + \left(\frac{12}{3} + \frac{2}{3}\right)$$
$$= \frac{15}{4} + \frac{14}{3} = \frac{45}{12} + \frac{56}{12} = \frac{101}{12}.$$

b. $$4\frac{3}{4} - 2\frac{7}{8} = \left(4 + \frac{3}{4}\right) - \left(2 + \frac{7}{8}\right)$$
$$= \left(\frac{16}{4} + \frac{3}{4}\right) - \left(\frac{16}{8} + \frac{7}{8}\right)$$
$$= \frac{19}{4} - \frac{23}{8} = \frac{38}{8} - \frac{23}{8} = \frac{15}{8}.$$

Problem Set 1–3

In Problems **1–8**, *combine and simplify.*

1. $\frac{7}{8} + \frac{17}{8}$ **2.** $\frac{8}{9} - \frac{6}{9}$ **3.** $\frac{7}{12} - \frac{5}{12}$

4. $\frac{7}{18} + \frac{3}{18} - \frac{5}{18}$ **5.** $\frac{5}{13} + \frac{11}{13} - \frac{4}{13}$ **6.** $\frac{12}{27} - \frac{6}{27} + \frac{4}{27} - \frac{1}{27}$

7. $\frac{2}{8 + 8}$ **8.** $\frac{6 + 6}{3 + 3}$

In Problems **9–16**, *find the* L.C.D. *of the fractions.*

9. $\frac{5}{6}, \frac{2}{9}$ **10.** $\frac{4}{21}, \frac{3}{14}$ **11.** $\frac{7}{4}, \frac{3}{2}, \frac{1}{5}$

12. $\frac{1}{15}, \frac{3}{8}, \frac{7}{10}$ **13.** $\frac{1}{3}, \frac{7}{18}, \frac{5}{12}$ **14.** $\frac{5}{6}, \frac{9}{20}, \frac{2}{15}$

15. $\frac{7}{30}, \frac{5}{12}, \frac{11}{20}$ **16.** $\frac{3}{20}, \frac{14}{25}, \frac{27}{50}$

In Problems **17–20**, *write the mixed numbers as fractions.*

17. $4\frac{3}{5}$ **18.** $3\frac{2}{3}$ **19.** $7\frac{2}{7}$ **20.** $10\frac{1}{10}$

In Problems **21–38**, *combine and simplify.*

21. $\frac{3}{4} + \frac{5}{6}$ **22.** $\frac{3}{5} + \frac{2}{3}$ **23.** $\frac{3}{8} - \frac{5}{14}$

24. $\frac{5}{6} - \frac{2}{21}$ **25.** $\frac{1}{3} + \frac{8}{9} + \frac{5}{12}$ **26.** $\frac{2}{3} + \frac{3}{4} + \frac{5}{8}$

27. $\frac{7}{4} - \frac{3}{2} + \frac{1}{5}$ **28.** $\frac{1}{5} + \frac{1}{8} + \frac{1}{12}$ **29.** $\frac{5}{6} + \frac{2}{15} - \frac{1}{3}$

30. $\frac{7}{12} - \frac{5}{18} + \frac{3}{8}$ **31.** $7\frac{2}{3} - 4\frac{3}{4}$ **32.** $2\frac{1}{12} - 1\frac{7}{8}$

33. $3\frac{2}{3} + 4\frac{1}{8} - 6\frac{1}{4}$ **34.** $4\frac{2}{7} - 2\frac{2}{5} + 1\frac{2}{35}$ **35.** $3 - \frac{7}{15} + \frac{3}{10}$

36. $2 + \frac{3}{8} - \frac{1}{6}$ **37.** $\frac{2}{5} + 2 - \frac{3}{8}$ **38.** $\frac{5}{12} + \frac{11}{18} - 1$

1-4 MULTIPLICATION AND DIVISION OF FRACTIONS

To multiply fractions, multiply their numerators and multiply their denominators, giving you a new fraction.

$$\frac{2}{3} \cdot \frac{5}{7} = \frac{(2)(5)}{(3)(7)} = \frac{10}{21}.$$

The dot stands for multiplication.

EXAMPLE 1

a. $$\frac{3}{5} \cdot \frac{8}{4} \cdot \frac{2}{9} = \frac{(3)(8)(2)}{(5)(4)(9)} = \frac{\overset{1}{\cancel{(3)}}\,\overset{2}{\cancel{(8)}}\,(2)}{(5)\,\underset{1}{\cancel{(4)}}\,\underset{3}{\cancel{(9)}}} = \frac{4}{15}.$$

We could have cancelled right away, as the following shows:

$$\frac{\overset{1}{\cancel{3}}}{5} \cdot \frac{\overset{2}{\cancel{8}}}{\underset{1}{\cancel{4}}} \cdot \frac{2}{\underset{3}{\cancel{9}}} = \frac{4}{15}.$$

b. $$\frac{17}{3} \cdot \frac{11}{4} \cdot \frac{2}{17} = \frac{\overset{1}{\cancel{17}}}{3} \cdot \frac{11}{\underset{2}{\cancel{4}}} \cdot \frac{\overset{1}{\cancel{2}}}{\underset{1}{\cancel{17}}} = \frac{11}{6}.$$

c. $$3\left(\frac{1}{2}\right) = \left(\frac{3}{1}\right)\left(\frac{1}{2}\right) = \frac{3}{2}.$$

d. $$5 \cdot \frac{2}{5} = \frac{\overset{1}{\cancel{5}}}{1} \cdot \frac{2}{\underset{1}{\cancel{5}}} = \frac{2}{1} = 2.$$

Usually we just write

$$\not{5} \cdot \frac{2}{\not{5}} = 2.$$

e. $\frac{4}{9} \cdot 18 = \frac{4}{\underset{1}{\not{9}}} \cdot \overset{2}{\not{18}} = 8.$

f. $\left(2\frac{1}{2}\right)\left(\frac{4}{5}\right) = \frac{\overset{1}{\not{5}}}{\underset{1}{\not{2}}} \cdot \frac{\overset{2}{\not{4}}}{\underset{1}{\not{5}}} = \frac{2}{1} = 2.$

Note that we first wrote the mixed number $2\frac{1}{2}$ as the improper fraction $\frac{5}{2}$.

Let's look at division. Suppose we want to divide $\frac{2}{3}$ by $\frac{3}{4}$.

$$\frac{2}{3} \div \frac{3}{4} = \frac{\frac{2}{3}}{\frac{3}{4}}.$$

Watch what happens when we use the fundamental principle by multiplying the numerator and denominator by $\frac{4}{3}$:

$$\frac{\frac{2}{3}}{\frac{3}{4}} = \frac{\frac{2}{3} \cdot \frac{4}{3}}{\frac{\not{3}}{\not{4}} \cdot \frac{\not{4}}{\not{3}}} = \frac{\frac{2}{3} \cdot \frac{4}{3}}{1} = \frac{2}{3} \cdot \frac{4}{3} = \frac{8}{9}.$$

Notice that to *divide* $\frac{2}{3}$ by $\frac{3}{4}$, all we have to do is (look at the next-to-last step) interchange the numerator and denominator of $\frac{3}{4}$ and *multiply* $\frac{2}{3}$ by the result.

$$\frac{2}{3} \div \frac{3}{4} = \frac{2}{3} \cdot \frac{4}{3} = \frac{8}{9}.$$

We say that we *inverted* the divisor and multiplied. Thus,

> ***to divide two fractions,* invert the fraction following the division sign and proceed as in multiplication.**

EXAMPLE 2

a. $\frac{7}{8} \div \frac{5}{4} = \frac{7}{\underset{2}{\not{8}}} \cdot \frac{\overset{1}{\not{4}}}{5} = \frac{7}{10}.$

b. $$\frac{\frac{3}{5}}{\frac{9}{10}} = \frac{3}{5} \div \frac{9}{10} = \frac{\overset{1}{\cancel{3}}}{\underset{1}{\cancel{5}}} \cdot \frac{\overset{2}{\cancel{10}}}{\underset{3}{\cancel{9}}} = \frac{2}{3}.$$

c. $$\frac{4}{\frac{3}{2}} = 4 \div \frac{3}{2} = 4 \cdot \frac{2}{3} = \frac{4}{1} \cdot \frac{2}{3} = \frac{8}{3}.$$

d. $$\frac{\frac{3}{8}}{4} = \frac{3}{8} \div 4 = \frac{3}{8} \div \frac{4}{1} = \frac{3}{8} \cdot \frac{1}{4} = \frac{3}{32}.$$

e. $$4\frac{1}{3} \div \frac{5}{3} = \frac{13}{\underset{1}{\cancel{3}}} \cdot \frac{\overset{1}{\cancel{3}}}{5} = \frac{13}{5}.$$

EXAMPLE 3

Simplify $$\frac{2 - \frac{3}{4}}{3 + \frac{1}{8}}.$$

We first separately simplify the numerator $2 - \frac{3}{4}$ and the denominator $3 + \frac{1}{8}$.

$$\frac{2 - \frac{3}{4}}{3 + \frac{1}{8}} = \frac{\frac{8}{4} - \frac{3}{4}}{\frac{24}{8} + \frac{1}{8}} = \frac{\frac{5}{4}}{\frac{25}{8}} = \frac{\overset{1}{\cancel{5}}}{\underset{1}{\cancel{4}}} \cdot \frac{\overset{2}{\cancel{8}}}{\underset{5}{\cancel{25}}} = \frac{2}{5}.$$

EXAMPLE 4

a. $$\frac{\frac{1}{5} + \frac{3}{4}}{\frac{1}{8}} = \frac{\frac{4}{20} + \frac{15}{20}}{\frac{1}{8}} = \frac{\frac{19}{20}}{\frac{1}{8}} = \frac{19}{\underset{5}{\cancel{20}}} \cdot \frac{\overset{2}{\cancel{8}}}{1} = \frac{38}{5}.$$

b. $$\frac{\left(\frac{3}{5}\right)\left(\frac{4}{7}\right)}{\frac{6}{11}} = \frac{\overset{1}{\cancel{3}}}{5} \cdot \frac{\overset{2}{\cancel{4}}}{7} \cdot \frac{11}{\underset{1}{\underset{\cancel{2}}{\cancel{6}}}} = \frac{22}{35}.$$

Sometimes we speak of the *reciprocal* of a number.

The *reciprocal* of a number (except 0) is 1 divided by the number.

EXAMPLE 5

a. The reciprocal of 2 is $\frac{1}{2}$.

b. The reciprocal of $\frac{2}{3}$ is $\dfrac{1}{\frac{2}{3}} = 1 \div \frac{2}{3} = 1 \cdot \frac{3}{2} = \frac{3}{2}$.

In Example 5(b) notice that to find the reciprocal of a fraction, just interchange its numerator and denominator. Thus the reciprocal of $\frac{3}{8}$ is $\frac{8}{3}$.

The reciprocal is "tied in" to division. For example,

$$\frac{2}{3} \div \frac{5}{7} = \frac{2}{3} \cdot \frac{7}{5} = \frac{14}{15}.$$

This is the same as multiplying $\frac{2}{3}$ by the *reciprocal* of $\frac{5}{7}$.

There is no reciprocal of 0. One thing to fix in your mind is that

division by 0 has no meaning.

Thus it's completely ridiculous to write $1 \div 0$ or $\frac{1}{0}$, because $\frac{1}{0}$ has no value. To see why, suppose $\frac{1}{0}$ were a number. Then, by the meaning of division, this number when multiplied by 0 should be 1. But there is no such number. Sometimes students write $\frac{1}{0} = 0$, which is **false**. But you *can* write

$$\frac{0}{1} = 0,$$

because $\frac{0}{1}$ is the number which when multiplied by 1 gives 0. That number is 0. Similarly,

$$\frac{0}{5} = 0,$$

because $0 = (5)(0)$.

Problem Set 1-4

In Problems **1–42**, *compute and give answers in reduced form.*

1. $\left(\frac{3}{5}\right)\left(\frac{25}{9}\right)$ **2.** $\left(\frac{8}{3}\right)\left(\frac{15}{4}\right)$ **3.** $\frac{14}{15} \cdot \frac{25}{24}$

4. $\frac{7}{12} \cdot 9$ **5.** $\frac{6}{7} \cdot \frac{0}{3}$ **6.** $0 \cdot \frac{2}{15}$

7. $(7)\left(\frac{6}{21}\right)$ **8.** $\left(\frac{2}{3}\right)(5)$ **9.** $\frac{1}{9} \cdot \frac{10}{3}$

10. $\frac{1}{4} \cdot \frac{3}{1}$

11. $\frac{3}{4} \cdot \frac{8}{5} \cdot \frac{4}{9}$

12. $\frac{2}{5} \cdot \frac{3}{6} \cdot \frac{7}{5}$

13. $\frac{3}{5} \cdot \frac{4}{11} \cdot \frac{7}{3} \cdot \frac{25}{4}$

14. $7 \cdot \frac{8}{5} \cdot \frac{6}{12} \cdot \frac{10}{49}$

15. $\left(2\frac{2}{5}\right)\left(1\frac{1}{8}\right)$

16. $(8)\left(3\frac{1}{8}\right)$

17. $\left(5\frac{2}{3}\right)\left(2\frac{3}{4}\right)\left(\frac{2}{17}\right)$

18. $\left(2\frac{2}{3}\right)\left(3\frac{1}{2}\right)(4)$

19. $\frac{8}{3} \div \frac{5}{4}$

20. $\frac{7}{5} \div \frac{3}{10}$

21. $16 \div \frac{12}{5}$

22. $0 \div \frac{8}{4}$

23. $\dfrac{\frac{7}{10}}{\frac{21}{5}}$

24. $\dfrac{\frac{14}{3}}{\frac{6}{15}}$

25. $\dfrac{\frac{18}{11}}{\frac{8}{33}}$

26. $\dfrac{\frac{2}{5}}{\frac{2}{5}}$

27. $\dfrac{\frac{3}{5}}{2}$

28. $\dfrac{7}{\frac{1}{4}}$

29. $\dfrac{4}{\frac{1}{5}}$

30. $\dfrac{\frac{3}{5}}{6}$

31. $\dfrac{4}{\frac{8}{9}}$

32. $\dfrac{1}{\frac{2}{3}}$

33. $\dfrac{\frac{12}{25} \cdot \frac{15}{7}}{20}$

34. $\dfrac{\frac{4}{9}}{\frac{2}{3} \cdot 8}$

35. $\dfrac{6 + \frac{1}{3}}{7}$

36. $\dfrac{\frac{3}{4} - \frac{3}{16}}{\frac{1}{3}}$

37. $\dfrac{7 - \frac{2}{3}}{15 - \frac{1}{3}}$

38. $\dfrac{1 + \frac{1}{2}}{1 - \frac{1}{2}}$

39. $\dfrac{\frac{8}{5} + \frac{2}{3}}{2 + \frac{4}{7}}$

40. $\dfrac{\frac{6}{7} - \frac{6}{7}}{\frac{4}{3} + \frac{5}{3}}$

41. $\dfrac{\frac{1}{2} - \frac{1}{3}}{\frac{1}{4} + \frac{1}{5}}$

42. $\dfrac{\left(\frac{2}{3}\right)\left(\frac{4}{5}\right) + 1}{2 + \frac{1}{15}}$

In Problems **43–48**, *find the reciprocals.*

43. 6

44. $\frac{3}{2}$

45. $\frac{3}{5}$

46. 1

47. $\frac{1}{9}$

48. $\frac{1}{2}$

49. Prove to yourself that $\frac{5}{0}$ has no meaning. *Hint*: If $\frac{5}{0}$ were to have meaning, it would be that number which when multiplied by ____ gives ____. But there is no such number.

1-5 PERCENTAGE

At times we use *percent* to write a fraction whose denominator is 100. For example, $\frac{2}{100}$ is written 2% (read "2 percent"). The percent symbol replaces the denominator 100. Of course, you may write $\frac{2}{100}$ as the decimal .02. So,

$$2\% = \frac{2}{100} = .02$$

There is a quick way to change a percent to a decimal. Just move the decimal point *two places to the left* and remove the percent symbol. For example, to write 2% as a decimal, first write 2% as 2.%. Then move the decimal point and remove the percent symbol.

$$2\%, \qquad 2.\%, \qquad .02$$

EXAMPLE 1

Changing percents to fractions and decimals.

a. $30\% = \frac{30}{100} = \frac{3}{10}$ $\qquad 30\% = .30 = .3$

b. $75\% = \frac{75}{100} = \frac{3}{4}$ $\qquad 75\% = .75$

c. $100\% = \frac{100}{100} = 1$ $\qquad 100\% = 1.00 = 1$

d. $125\% = \frac{125}{100} = \frac{5}{4}$ $\qquad 125\% = 1.25$

EXAMPLE 2

Changing percents to fractions and decimals.

a. $\frac{1}{2}\% = \frac{\frac{1}{2}}{100} = \frac{\frac{1}{2}}{\frac{100}{1}} = \frac{1}{2} \cdot \frac{1}{100} = \frac{1}{200}$ $\qquad \frac{1}{2}\% = .5\% = .005$

b. $10\frac{1}{4}\% = \frac{10\frac{1}{4}}{100} = \frac{\frac{41}{4}}{\frac{100}{1}} = \frac{41}{4} \cdot \frac{1}{100} = \frac{41}{400}$ $\qquad 10\frac{1}{4}\% = 10.25\% = .1025$

c. $.4\% = \frac{.4}{100} = \frac{\frac{4}{10}}{\frac{100}{1}} = \frac{\overset{1}{\cancel{4}}}{10} \cdot \frac{1}{\underset{25}{\cancel{100}}} = \frac{1}{250}$ $\qquad .4\% = .004$

To change a decimal to a percent, move the decimal point *two places to the right*. Then attach the percent symbol.

EXAMPLE 3

Changing decimals to percents.

	Decimal	*Percent*
a.	.06	06.% = 6%
b.	.8	80.% = 80%
c.	1.1	110.% = 110%
d.	.045	04.5% = 4.5%
e.	.0001	00.01% = .01%
f.	2 (= 2.)	200.% = 200%

To change a fraction to a percent, you may either

write the fraction as a decimal and then write the decimal as a percent (as above)

or you may

write the fraction as one whose denominator is 100. The numerator then gives the percent.

EXAMPLE 4

Changing fractions to percents.

	Fraction	*Percent*
a.	$\frac{1}{5} = .2$	20.% = 20%
	$\frac{1}{5} = \frac{(1)(20)}{(5)(20)} = \frac{\textcircled{20}}{100}$	20%
b.	$\frac{3}{4} = .75$	75.% = 75%
	$\frac{3}{4} = \frac{(3)(25)}{(4)(25)} = \frac{\textcircled{75}}{100}$	75%

c. $\frac{1}{8} = .125$ $\qquad 12.5\% = 12.5\%$

d. $\frac{1}{3} = \frac{1 \cdot \frac{100}{3}}{3 \cdot \frac{100}{3}} = \frac{33\frac{1}{3}}{100}$ $\qquad 33\frac{1}{3}\%$

To find a percentage of a number (say 10% of 50), first change the percent to a decimal or a fraction (10% is .1). Then multiply that result by the number. Thus, 10% of 50 is (.1)(50) = 5. In general, the words "percent of" imply the operation of multiplication.

EXAMPLE 5

a. *Find* 60% *of* 80.

$$60\% = .60$$
$$(.60)(80) = 48$$

b. *Find* $2\frac{1}{2}$% *of* 5.

$$2\frac{1}{2}\% = 2.5\% = .025$$
$$(.025)(5) = .125$$

c. *Find* 150% *of* 20.

$$150\% = \frac{150}{100} = \frac{3}{2}$$
$$\frac{3}{2} \cdot 20 = 30$$

60% of 80 ***is not*** $60 \cdot 80 = 4800$. *Make sure that you convert 60% to a decimal or fraction as in Example* 5(a).

Percent is important in *interest problems*. For example, suppose that you deposit \$20 in a savings account that pays interest at the rate of 5% per year. At the end of a year you would receive interest:

$$\text{Interest} = 5\% \text{ of } \$20 = (.05)(20) = \$1.00$$

The total amount in your account would then be \$20 + \$1 = \$21. The amount on which interest is computed (\$20) is called the *principal*.

EXAMPLE 6

Suppose that you deposit a principal of \$2000 in a savings account that pays interest at the rate of $6\frac{1}{2}\%$ per year. How much money will be in the account at the end of one year?

$$\text{Interest} = 6\tfrac{1}{2}\% \text{ of } \$2000 = (.065)(2000) = \$130.$$

$$\left.\begin{array}{c}\text{Total amount in account}\\ \text{at end of one year}\end{array}\right\} = \text{principal} + \text{interest}$$

$$= 2000 + 130$$

$$= \$2130.$$

Problem Set 1-5

In Problems **1–15**, *write each percent as a reduced fraction and as a decimal.*

1. 25%

2. 10%

3. 4%

4. 65%

5. 1%

6. $\frac{1}{4}\%$

7. $\frac{1}{10}\%$

8. .5%

9. 8.5%

10. .01%

11. .001%

12. $12\frac{1}{2}\%$

13. $50\frac{1}{2}\%$

14. 130%

15. 210%

In Problems **16–24**, *write each decimal as a percent.*

16. .8

17. .03

18. .97

19. .00001

20. 1.2

21. 2.65

22. 3.625

23. 6

24. 10

In Problems **25–33**, *write each fraction as a percent.*

25. $\frac{7}{100}$

26. $\frac{24}{100}$

27. $\frac{1}{1000}$

28. $\frac{15}{4}$

29. $\frac{3}{2}$

30. $\frac{13}{50}$

31. $\frac{5}{8}$

32. $\frac{200}{500}$

33. $\frac{15}{200}$

In Problems **34–42**, *find the given percent of the number.*

34. 20% of 60

35. 40% of 72

36. 4.25% of 50

37. 10.3% of 31

38. $5\frac{1}{4}\%$ of 200

39. $1\frac{1}{2}\%$ of 3

40. .02% of 8.5

41. 145% of 60

42. 120% of 100

43. Suppose you deposit \$500 in a savings account. The bank pays interest at the rate of 6% per year. How much money will be in the account at the end of one year?

44. Suppose you deposit \$1200 in a savings account. The bank pays interest at the rate of $7\frac{1}{4}\%$ per year. How much money will be in the account at the end of one year?

45. Suppose that you borrow \$850 for one year and must pay interest on that loan at the rate of 9.5% per year. How much interest do you owe at the end of one year?

46. Suppose that you borrow \$600 for one year. The bank charges interest on that loan at the rate of 8% per year. How much interest do you owe at the end of one year?

47. You are filling out your federal income tax form. According to instructions, you must pay a tax of \$2630 plus 29% of your taxable income over \$12,000. If your total taxable income is \$13,360, what is your federal tax?

48. The income statement of a company shows that operating expenses were 22.84% of net sales. Net sales were \$800,000. Find the operating expenses.

49. Sales last year for a certain store were \$205,000. This year it hopes to increase sales by 14%. For this to occur, what must be the sales this year?

1-6 REVIEW

IMPORTANT TERMS

product *(p. 2)*
factor *(p. 2)*
dividend *(p. 2)*
divisor *(p. 2)*
quotient *(p. 2)*
numerator *(p. 4)*
denominator *(p. 4)*
proper fraction *(p. 5)*
improper fraction *(p. 5)*
mixed number *(p. 5)*
equivalent fractions *(p. 6)*
fundamental principle of fractions *(p. 6)*
reduced fraction *(p. 6)*
lowest terms *(p. 6)*
least common denominator *(p. 10)*
prime number *(p. 10)*
reciprocal *(p. 17)*
percent *(p. 20)*

REVIEW PROBLEMS

In Problems **1–10**, *compute the numbers.*

1. $(8)(5) - 2$

2. $6(8 - 2)$

3. $6 + (9 + 2)$

4. $(2 - 1)(8 + 1 + 2)$

5. $(2 + 3 - 1)(7 - 2 - 5)$

6. $(9 - 5) - (12 \div 4)$

7. $(5 + 4 - 3) \div (5 - 3)$

8. $(20 \div 2)(11 - 6)$

9. $10 - (3)(6 - 4) - 3$

10. $2(8 + 2) + 3(2 + 4)$

In Problems **11–14**, *write the fractions as mixed numbers.*

11. $\frac{12}{5}$

12. $\frac{38}{7}$

13. $\frac{325}{100}$

14. $\frac{51}{50}$

In Problems **15–38**, *compute and give answers in reduced form.*

15. $\frac{2}{3} + \frac{3}{5}$

16. $\frac{8}{9} - \frac{1}{6}$

17. $\frac{4}{15} - \frac{3}{20}$

18. $2\frac{3}{4} + 3\frac{1}{3}$

19. $\frac{19}{3} + \frac{1}{9} - 2$

20. $5 + \frac{1}{2} - \frac{2}{3}$

21. $3\frac{1}{2} - 1\frac{2}{3} + 2\frac{2}{4}$

22. $4\frac{2}{3} - \frac{1}{6} + \frac{1}{9}$

23. $\frac{5}{16} \cdot \frac{24}{25}$

24. $8 \cdot \frac{5}{12}$

25. $\left(2\frac{1}{2}\right)(8)$

26. $\left(1\frac{1}{3}\right)\left(2\frac{1}{3}\right)$

27. $\frac{2}{3} \div \frac{5}{8}$

28. $\frac{4}{5} \div \frac{8}{25}$

29. $\dfrac{\frac{3}{50}}{\frac{27}{5}}$

30. $\dfrac{\frac{1}{3}}{\frac{7}{4}}$

31. $\dfrac{\frac{4}{3}}{4}$

32. $\dfrac{\frac{11}{7}}{7}$

33. $\dfrac{15}{\frac{25}{3}}$

34. $\dfrac{5}{\frac{1}{5}}$

35. $\dfrac{2 - \frac{1}{5}}{\frac{1}{2} + 3}$

36. $\dfrac{3 - \frac{9}{3}}{\frac{1}{4} + \frac{2}{3}}$

37. $\dfrac{\frac{1}{2} - \frac{1}{3}}{\frac{4}{9} \cdot \frac{3}{6}}$

38. $\dfrac{\frac{3}{4} + \frac{1}{8}}{5 - \frac{7}{2}}$

In Problems **39–42**, *write each percent as a fraction and as a decimal.*

39. 80%

40. 7.3%

41. $6\frac{1}{4}\%$

42. .02%

In Problems **43–46**, *write each decimal as a percent.*

43. .06

44. .005

45. 1.025

46. 3

In Problems **47–50**, *write each fraction as a percent.*

47. $\frac{7}{25}$

48. $\frac{1}{6}$

49. $\frac{75}{300}$

50. $\frac{5}{1000}$

In Problems **51–53**, *find the given percent of the number.*

51. 15% of 62

52. $2\frac{1}{2}$ % of 20

53. .04% of 500

54. John Smith pays 23% of his salary for taxes. His salary is $12,000 this year. How much must he pay for taxes?

55. Suppose that you deposit $700 in a savings account. The bank pays interest at the rate of $5\frac{1}{4}$% per year. How much money will be in the account at the end of one year?

56. Grace Wells pays 4.2% of her salary into a retirement fund. Her salary this year is $13,750. How much will she put into the fund this year?

57. Three-eighths of the volume of a certain concrete mixture is sand. What percentage of the mixture is sand?

CHAPTER 2

Real Numbers

2-1 REAL NUMBERS

In mathematics there are different types of numbers. To describe them we'll use the idea of a *set*.

> A **set** is a collection of objects or numbers, called **members** or **elements**.

We shall use braces, { }, to enclose the elements of a set.

EXAMPLE 1

The set of numbers 1, 3, and 12 may be written

$$\{1, 3, 12\}.$$

Thus 1, 3, and 12 are members of this set. The order in which we list the elements is not important. We can also write this set as {3, 12, 1}. Can you think of other ways?

Probably the numbers most familiar to you are 1, 2, 3, 4, 5, etc., which are used in counting. They're called **positive integers.**

SET OF POSITIVE INTEGERS
{1, 2, 3, 4, 5, . . . }

The three dots above mean "and so on." This set has infinitely many members.

The numbers $-1, -2, -3, \ldots$ are called **negative integers**. Putting together the positive integers, negative integers, and zero, we get the set of **integers**.

SET OF INTEGERS
{ . . . ,−2, −1, 0, 1, 2, . . . }

Numbers that can be written as an integer divided by an integer are called **rational numbers.**†

EXAMPLE 2

Some rational numbers.

a. $\frac{2}{3}$ b. $\frac{9}{5}$ c. $6\left(\text{since } 6 = \frac{6}{1}\right)$

d. $-6\left(\text{since } -6 = \frac{-6}{1}\right)$ e. $0\left(\text{since } 0 = \frac{0}{1}\right)$ f. $3\frac{1}{8}\ \left(\text{since } 3\frac{1}{8} = \frac{25}{8}\right)$

In c, d, and e you can see that integers are rational numbers because they can be written with a denominator of 1.

All rational numbers can be represented by decimal numbers that *terminate*, such as $\frac{3}{4} = .75$ and $\frac{3}{2} = 1.5$, or by *nonterminating repeating* decimals (a group

† $\frac{1}{0}$ is an integer divided by an integer, but division by 0 is not defined. So $\frac{1}{0}$ isn't a number at all.

of digits repeats without end), such as $\frac{2}{3} = .666\ldots$ and $\frac{-4}{11} = -.3636\ldots$. Numbers represented by *nonterminating nonrepeating* decimals are called **irrational numbers**. An irrational number cannot be written as an integer divided by an integer. The numbers π (pi) and $\sqrt{2}$ are irrational. By the way, sometimes π is replaced by $\frac{22}{7}$ or 3.14 in a calculation, but these are just approximations of π.

Putting the rational numbers and irrational numbers together, we get the set of **real numbers**. See Fig. 2-1. Thus 2, -5, 0, $\frac{2}{3}$, $16\frac{1}{2}$, $-\frac{12}{5}$, π, and $-\sqrt{2}$ are

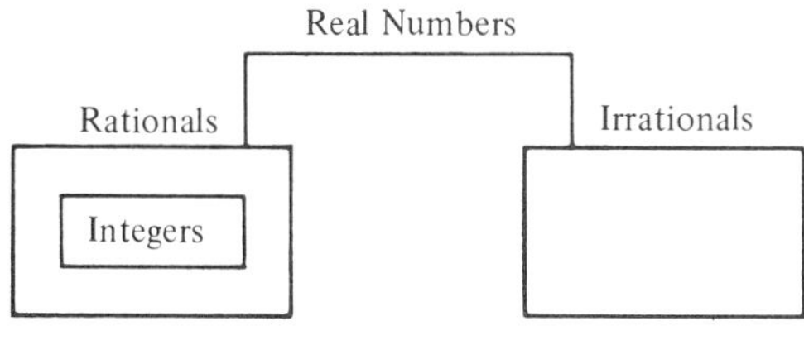

FIG. 2-1

examples of real numbers.

Real numbers can be represented by points on a line, like the markings on a thermometer. On a line we choose a point. This point is called the **origin** and is used to represent the number 0. See Fig. 2-2(a).

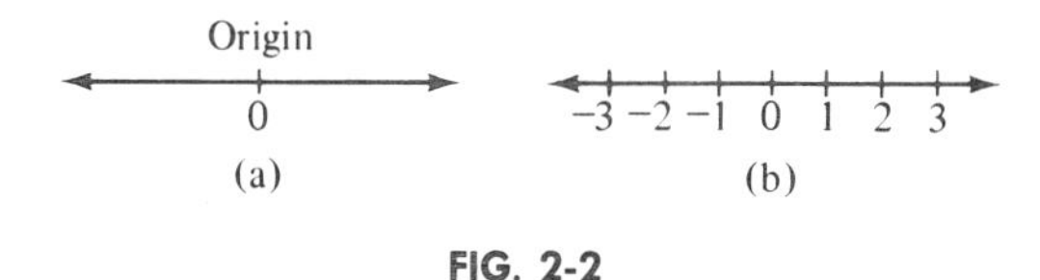

FIG. 2-2

Next, we mark off equal distances to the right and left of the origin. These positions represent the positive and negative integers. See Fig. 2-2(b).

The rest of the positions represent all other rational and irrational numbers. You can see some in Fig. 2-3. For obvious reasons we call this line the (real) **number line**.

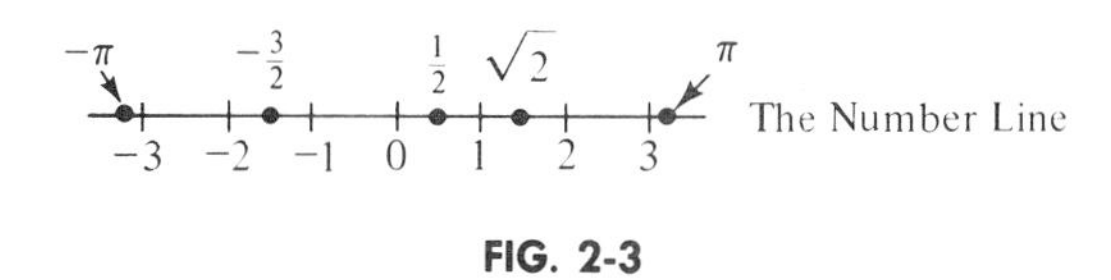

FIG. 2-3

Figure 2-3 shows that all positive numbers lie to the *right* of 0 and all negative numbers lie to the *left*. In this way real numbers are involved with *direction*. In fact, we sometimes write 3 as $+3$ to stress that it is to the *right* of 0.

Thus you may think of real numbers as *signed numbers*: that is, either positive $(+)$, negative $(-)$, or 0.

In the future, whenever we use the word *number* we shall mean *real number*.

Completion Set 2-1

In Problems **1–8**, *fill in the blanks.*

1. A collection of objects is called a ________.

2. The objects in a set are called its members or ____________________.

3. The numbers 1, 2, 3, . . . are called ____________________ integers.

4. The number of members in $\{1, 3, 0, -6\}$ is equal to ____.

5. The set of integers consists of the positive integers, the negative integers, and ___________.

6. Every real number is either positive or negative except for the number ____.

7. Real numbers that are not rational numbers must be ________________________ numbers.

8. Insert *rational* or *irrational*: 5 is a(n) ____________________ number.

In Problems **9–12**, *you are given two numbers. Circle the number that is farther from* 0 *on the number line.*

9. 3, 4 **10.** 7, -8 **11.** -3, 1 **12.** -5, -8

In Problems **13–16**, *you are given two numbers. Circle the number that is to the right of the other number on the number line.*

13. 6, -5 **14.** -3, -4 **15.** -13, -12 **16.** 0, -2

In Problems **17–22**, *mark the statement either* T (= *true*) *or* F (= *false*).

17. 56 is a positive number. ____

18. $\frac{2}{3}$ is an integer. ____

19. $\frac{9}{2}$ is a rational number. ____

20. -123 is an integer. ____

21. The sets $\{1, 2, 3\}$ and $\{2, 3, 1\}$ are not the same. ____

22. All real numbers are irrational. ____

2-2 INEQUALITIES AND ABSOLUTE VALUE

We'll now begin to represent numbers by letters, such as a, b, x, y, and so on. Letters used in this way are called **literal numbers.**

Sometimes we want a way to say that one number is "smaller" than another. If a lies to the *left* of b on the number line, we say that a is **less than** b. In symbols we write $a < b$, where "$<$" is read "is less than." See Fig. 2-4.

$a < b$

a b

$a < b$, a is less than b

FIG. 2-4

EXAMPLE 1

In Fig. 2-5 you can see that $4 < 7$, since 4 is to the left of 7 on the number line.

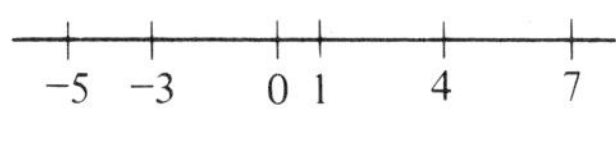

FIG. 2-5

Similarly,

$$-3 < 1, \quad 0 < 4, \quad -5 < -3.$$

If a is less than b, we also say that b is **greater than** a and write $b > a$.

$b > a$; b is greater than a.

This means that b is to the *right* of a on the number line (see Fig. 2-6). Thus,

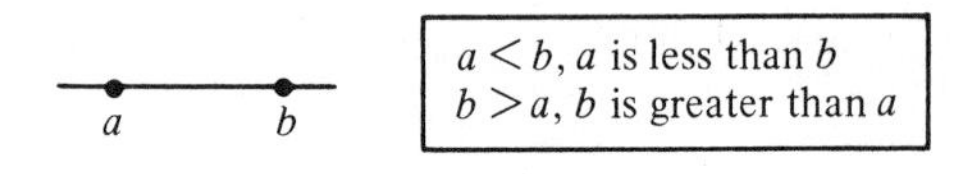

FIG. 2-6

from Fig. 2-5, $4 < 7$ and $7 > 4$. The statements $a < b$ and $b > a$ are called **inequalities**. If you think of the inequality symbols "$<$" and "$>$" as pointers, then they should always point to the "smaller" number. Keep in mind that

$a < b$ and $b > a$ have the same meaning.

EXAMPLE 2

In Fig. 2-5 we see that

$$7 > 0, \quad 0 > -3, \quad 4 > -5, \quad -3 > -5.$$

There's a way to relate positive and negative numbers to inequalities. In Fig. 2-5 all positive numbers are to the right of 0. Thus,

a is positive and $a > 0$ mean the same.

All negative numbers are to the left of 0 and so

a is negative and $a < 0$ mean the same.

There are two more inequality symbols. Each plays a double role:

$a \leqslant b$, a is less than or equal to b;
$a \geqslant b$, a is greater than or equal to b.

Look at the double roles:

$a \leqslant b$ means that ***either*** $a < b$ ***or*** $a = b$;
$a \geqslant b$ means that ***either*** $a > b$ ***or*** $a = b$.

EXAMPLE 3

a. $7 \geqslant 6$, since $7 > 6$. Similarly, $5 \geqslant 2$.

b. $3 \leqslant 3$, since $3 = 3$. Similarly, $0 \leqslant 0$.

c. $-4 \leqslant 3$, since $-4 < 3$. Similarly, $-3 \leqslant -2$.

Sometimes we're interested in the distance of a number from 0, but *not its direction from* 0. This distance is called the number's **absolute value**. We use two vertical bars to write it.

$$|a|, \quad \text{the absolute value of } a.$$

For example, both 3 and -3 are three units from 0. See Fig. 2-7. Thus,

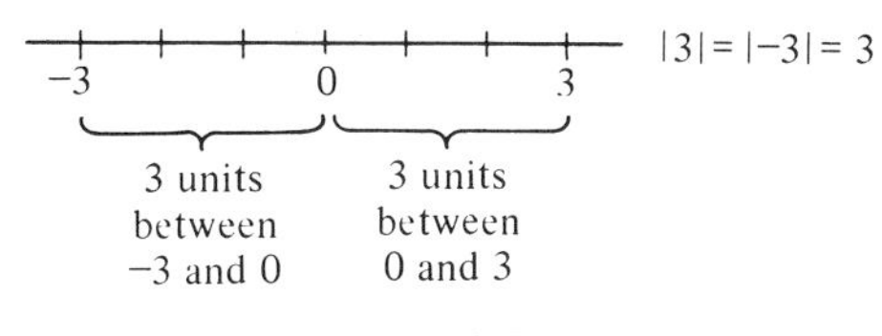

FIG. 2-7

$|3| = 3$ and $|-3| = 3$.

EXAMPLE 4

$$|5| = 5, \qquad |-8| = 8, \qquad |0| = 0,$$

$$\left|-\frac{2}{3}\right| = \frac{2}{3}, \qquad |7-4| = |3| = 3, \qquad |-\pi| = \pi.$$

Since the absolute value of a number is a distance, it is always *positive* or 0. It is *never* negative:

$$|a| \geqslant 0.$$

Look at Example 4 and check this result.

In the next examples we combine inequalities and absolute value.

EXAMPLE 5

a. $|-8| > -7$ because $|-8| = 8$ and $8 > -7$.

b. $|-5| < |8|$ because $|-5| = 5$, $|8| = 8$, and $5 < 8$.

c. $-3 < 2$, but $|-3| > |2|$ because $|-3| = 3$, $|2| = 2$, and $3 > 2$.

Completion Set 2-2

In Problems **1–7**, *fill in the blanks.*

1. The statement $a > b$ is read "a is ________________ than b."

2. The statement $x < y$ means that x is to the __________ of y on the number line.

3. Using the symbol $>$, we can write $3 < 4$ as ____________.

4. If $a > 0$, then a is a ________________ number.

5. Since 4 and -4 are four units from 0, they have the same ________________

____________.

6. $a \geqslant 6$ means that either $a = 6$ or ____________.

7. The absolute value of a number must be either positive or __________. It is never ________________.

In Problems **8–13**, *put "$<$" or "$>$" in the blank so that the inequality is true.*

8. 7 ____ 16 **9.** 8 ____ 0 **10.** -5 ____ -7

11. -3 ____ 2 **12.** 0 ____ -2 **13.** -4 ____ -1

In Problems **14–19**, *fill in the blanks.*

14. $|-6| =$ ____ **15.** $|270| =$ ______ **16.** $\left|-\frac{7}{2}\right| =$ ____

17. $|0| =$ ___ **18.** $\left|3 \cdot \frac{1}{3}\right| =$ ___ **19.** $|3 - 1| =$ ___

In Problems **20–22**, *you are given a pair of numbers. Circle the one that has the greater absolute value.*

20. 5, 3 **21.** -5, -3 **22.** -4, 2

23. Arrange the numbers -6, -3, 0, 4, 5 in a list so that each number has a smaller absolute value than the numbers which follow it in the list.____________

24. Circle those inequalities that are *false*.

a. $|-2| > |-1|$ b. $|-1| \leqslant 0$ c. $|-9| < -8$ d. $3 \geqslant -1$

2-3 OPERATIONS WITH SIGNED NUMBERS

The sum of the numbers a and b is written $a + b$, and a and b are called **terms** of the sum. The product of the numbers a and b is written $a \cdot b$, and a and b are called **factors** of the product. Other ways to write the product $a \cdot b$ are ab, $a(b)$, $(a)b$, and $(a)(b)$. Thus we may write the product of 4 and x as $4x$. The subtraction of b from a is written $a - b$. The quotient of a divided by b is written $\frac{a}{b}$.

Let's go over how to add, subtract, multiply, and divide signed numbers. Don't rush through this section. Go over each example carefully.

ADDITION OF POSITIVE NUMBERS

To add *positive* numbers, add them as in arithmetic.

For example,

$$2 + 3 = 5 \quad \text{and} \quad (+6) + (+7) = 13.$$

ADDITION OF NEGATIVE NUMBERS

To add *negative* numbers, forget signs, add as in arithmetic, and then put a minus sign in front of the result.

Let's add -6 and -8.

$(-6) + (-8)$

Forget signs: $6 + 8 = 14$.

Put minus sign to result and get -14.

Thus, $(-6) + (-8) = -14$.

More simply we write

$$(-6) + (-8) = -(6 + 8) = -14.$$

Note that *the sum of positives is positive* and *the sum of negatives is negative*.

EXAMPLE 1

a. $(-4) + (-3) = -(4 + 3) = -7.$

b. $(+9) + (+7) = 16.$

c. $\left(-\frac{1}{2}\right) + \left(-\frac{3}{2}\right) = -\left(\frac{1}{2} + \frac{3}{2}\right) = -\frac{4}{2} = -2.$

d. $(-2) + (-9) + (-3) = -(2 + 9 + 3) = -14.$

Adding a positive number and a negative number involves absolute value.

ADDITION WITH UNLIKE SIGNS

To add a positive number and a negative number, subtract the smaller absolute value *from* the larger absolute value. To this result put the sign of the number with the larger absolute value. When two numbers with unlike signs have the same absolute value, their sum is zero.

EXAMPLE 2

Add: $(+2) + (-9)$.

The absolute values are 2 and 9.

$$(\text{larger absolute value}) - (\text{smaller absolute value}) = 9 - 2 = 7.$$

Of $+2$ and -9, the sign of the number with the larger absolute value is "$-$".
Put this sign in front of 7: -7.
Thus,

$$(+2) + (-9) = -7.$$

More simply, we write

$$(+2) + (-9) = -(9 - 2) = -7.$$

After some practice you should be able to do a problem like Example 2 in your head.

EXAMPLE 3

a. $(-7) + (10) = +(10 - 7) = 3$.

b. $(-15) + (5) = -(15 - 5) = -10$.

c. $4 + (-4) = 0$, because 4 and -4 have unlike signs but have the same absolute value.

d. $4 + (-2) = 2$.

e. $6 + (-9) = -3$.

Every number has an *opposite*. The **opposite of *a***, written $-a$, is that number which when added to a gives 0. Thus $a + (-a) = 0$. For example,

the opposite of 4 is -4,

because $4 + (-4) = 0$. Similarly,

the opposite of -11 is 11,

because $(-11) + 11 = 0$. Since the opposite of -11 is represented by $-(-11)$, we have $-(-11) = 11$. In general, the opposite of the opposite of a number is that number:

$$-(-a) = a.$$

Also,

the opposite of 0 is 0,

because $0 + 0 = 0$.

We do subtraction by using addition and an opposite.

SUBTRACTION

$$a - b = a + (-b).$$

To subtract b from a, add the opposite of b to a.

For example,

opposite

$$7 - (9) = 7 + (-9) = -2.$$

change to addition

EXAMPLE 4

a. $5 - (3) = 5 + (-3) = 2.$

b. $5 - 8 = 5 + (-8) = -3.$

c. $-10 - 8 = -10 + (-8) = -18.$

d. $7 - (-3) = 7 + (+3) = 10.$

e. $-6 - (-6) = -6 + (6) = 0.$

EXAMPLE 5

To subtract -4 *from* 9, we have

$$9 - (-4) = 9 + (4) = 13.$$

Every number except 0 has a *reciprocal*. The **reciprocal** of a is 1 divided by a.

$$\boxed{\text{Reciprocal of } a \text{ is } \frac{1}{a}.}$$

EXAMPLE 6

	Number	*Reciprocal*
a.	2	$\frac{1}{2}$
b.	$\frac{3}{2}$	$\frac{1}{\frac{3}{2}} = 1 \div \frac{3}{2} = 1 \cdot \frac{2}{3} = \frac{2}{3}$
c.	$\frac{5}{x}$	$\frac{1}{\frac{5}{x}} = \frac{x}{5}$

Zero does not have a reciprocal, since division by zero has no meaning. ***Never divide by 0.*** $\frac{1}{0}$ *is not a number.*

Look at b and c in Example 6. Notice that to find the reciprocal of a fraction, you can just reverse the numerator and denominator.

When we multiply 2 by its reciprocal, $\frac{1}{2}$, we get $2 \cdot \frac{1}{2} = 1$. In fact, *any number times its reciprocal is* 1.

$$\boxed{a \cdot \frac{1}{a} = 1.}$$

Thus,

$$\left(\frac{3}{2}\right)\left(\frac{2}{3}\right) = 1.$$

Before giving rules for multiplication and division, we point out that division is a form of multiplication because of the following definition of division:

$$\frac{a}{b} = a \cdot \frac{1}{b}.$$

Thus, dividing by b is the same as multiplying by the reciprocal of b.

EXAMPLE 7

$$\frac{3}{4} = 3 \cdot \frac{1}{4}; \qquad 6 \cdot \frac{1}{8} = \frac{6}{8} = \frac{3}{4}; \qquad \frac{x}{3} = \frac{1}{3}x.$$

Here is the rule for handling signs in multiplication and division.

> **MULTIPLICATION AND DIVISION**
> The product or quotient of two positive or two negative numbers is positive. The product or quotient of a positive number and a negative number is negative.

EXAMPLE 8

a. $(+2)(+6) = +12 = 12.$

b. $(-3)(-5) = +15 = 15.$

c. $(-2)(4) = -8.$

d. $3\left(-\frac{1}{3}\right) = -\left(3 \cdot \frac{1}{3}\right) = -1.$

EXAMPLE 9

a. $\frac{+14}{+2} = +7 = 7.$

b. $\frac{-6}{-3} = +\frac{6}{3} = +2 = 2.$

c. $\frac{-4}{5} = -\frac{4}{5}.$

d. $\frac{18}{-6} = -\frac{18}{6} = -3.$

e. $(-2)\left(\frac{1}{-4}\right) = (-2)\left(-\frac{1}{4}\right) = +\left(2\cdot\frac{1}{4}\right) = \frac{2}{4} = \frac{1}{2}.$

All of the results in Examples 8 and 9 can be matched with the following forms.

MULTIPLICATION	DIVISION
$(a)(b) = (-a)(-b) = ab$	$\frac{-a}{-b} = \frac{a}{b}$
$(a)(-b) = (-a)(b) = -ab$	$\frac{-a}{b} = -\frac{a}{b} = \frac{a}{-b}$

For example, $\frac{-4}{5} = -\frac{4}{5}$ matches $\frac{-a}{b} = -\frac{a}{b}$. Here are more examples of all the basic operations.

EXAMPLE 10

a. $(-2)(-7+4) = (-2)(-3) = +6 = 6.$

b. $(-2)(-x) = 2\cdot x = 2x.$

c. $\frac{6(-7)}{9} = \frac{\overset{2}{\cancel{6}}(-7)}{\underset{3}{\cancel{9}}} = \frac{2(-7)}{3} = \frac{-14}{3} = -\frac{14}{3}.$

d. $\frac{6(-2)}{-3} = \frac{-12}{-3} = \frac{12}{3} = 4.$

e. $(-2)x = -2x.$

f. $\frac{(-x)y}{z} = \frac{-xy}{z} = -\frac{xy}{z}.$

EXAMPLE 11

a. $\dfrac{-6-2}{-4} = \dfrac{-8}{-4} = \dfrac{8}{4} = 2.$

b. $\dfrac{-3-6}{-2-1} = \dfrac{-9}{-3} = \dfrac{9}{3} = 3.$

c. $\left(-\dfrac{4}{9}\right)(18) = -\left(\dfrac{4}{\underset{1}{\cancel{9}}} \cdot \overset{2}{\cancel{18}}\right) = -(4 \cdot 2) = -8.$

d. $\dfrac{-6}{(-5)(-7)} = \dfrac{-6}{35} = -\dfrac{6}{35}.$

e. $\dfrac{-\dfrac{7}{2}}{4} = -\dfrac{\dfrac{7}{2}}{4} = -\left(\dfrac{7}{2} \cdot \dfrac{1}{4}\right) = -\dfrac{7}{8}.$

f. $-\dfrac{-3}{-8} = -\left(\dfrac{-3}{-8}\right) = -\left(\dfrac{3}{8}\right) = -\dfrac{3}{8}.$

Here are two more rules that you should know.

$$(-1)a = -a;$$
$$-(-a) = a.$$

EXAMPLE 12

$$(-1)x = -x; \qquad -(-3) = 3; \qquad -6 = (-1)(6).$$

EXAMPLE 13

To simplify $-(-xy)$, match it with the form $-(-a) = a$. Here xy plays the role of a.

$$-(-xy) = xy.$$

EXAMPLE 14

a. $-\frac{-3}{6} = -\left(\frac{-3}{6}\right) = -\left(-\frac{3}{6}\right) = \frac{3}{6} = \frac{1}{2}.$

b. $-\frac{x}{y(-z)} = -\frac{x}{-(yz)} = -\left(-\frac{x}{yz}\right) = \frac{x}{yz}.$

When you work with the numbers 0 and 1, here are the rules to use.

$0 + a = a.$	$0 - a = -a.$
$a - 0 = a.$	$a \cdot 0 = 0.$
$\frac{0}{a} = 0$ if $a \neq 0.$	$\frac{a}{0}$ has no meaning.
$1 \cdot a = a.$	$\frac{a}{1} = a.$

EXAMPLE 15

a. $0 + (-9) = -9.$

b. $0 - 7 = -7.$

c. $xy - 0 = xy.$

d. $(-6)(0) = 0.$

e. $\frac{0}{12} = 0.$

f. $\frac{9}{0}$ has no meaning.

g. $x = 1 \cdot x,$

h. $3xy = \frac{3xy}{1}.$

When the order of doing operations is not clear to you, do the multiplications and divisions first. Then do the additions and subtractions. For example,

$$-4 + 2(6) = -4 + 12 = 8,$$
$$-3 - (4)(-2) = -3 - (-8) = -3 + 8 = 5.$$

But in a fraction such as $\frac{6+2}{7-3}$, you must first simplify both the numerator and the denominator.

$$\frac{6+2}{7-3} = \frac{8}{4} = 2.$$

Note that $-4 + 2(6) \neq -2(6)$. *Here the symbol "*$\neq$*" means "is not equal to."* *Also,* $-3 - 4(-2) \neq -7(-2)$.

When operations are inside parentheses, do those first.

EXAMPLE 16

a. $7 - (4 - 9) = 7 - (-5) = 7 + (5) = 12.$

b. $-2(1 - 5) = -2(-4) = 8.$

c. $\dfrac{8(4 + 3)}{-3} = \dfrac{8(7)}{-3} = \dfrac{56}{-3} = -\dfrac{56}{3}.$

Completion Set 2-3

In Problems **1–14**, *fill in the blanks.*

1. The sum of four negative numbers is a ____________________ number.

2. $\frac{8}{5}$ means ____ times $\frac{1}{5}$.

3. If -6 is subtracted from 6, the result is _____.

4. $\frac{-7}{1}$ in a simpler form is just _____.

5. The opposite of 5 is _____, and the reciprocal of 5 is _____.

6. The sum of a number and its opposite is ____.

7. The product of a nonzero number and its reciprocal is ____.

8. The reciprocal of $\frac{3}{5}$ is _____.

9. The opposite of -4 is ____.

10. All numbers have a reciprocal except ____.

11. The product of $\frac{x}{2}$ and ______ equals 1.

12. $\frac{7}{0}$ has no meaning, but $\frac{0}{7}$ = ____.

13. $8 - 5$ means $8 +$ ______.

14. $5 +$ ______ $= 0$.

Problem Set 2-3

In Problems **1–71**, *simplify.*

1. $(+4) + (+8)$

2. $(-2) + (-3)$

3. $(-7) + (-2)$

4. $-5 + 0$

5. $6 + (-4)$

6. $-2 + 4$

7. $5 - 7$

8. $6 - 9$

9. $-5 + 2$

10. $8 + (-9)$

11. $-2 - 8$

12. $3 - (-4)$

13. $-6 - (-5)$

14. $(-8) - 7$

15. $-14 - (-14)$

16. $(+3)(+4)$

17. $(-3)(-4)$

18. $(-5)(-2)$

19. $7\left(-\frac{2}{7}\right)$

20. $(-3)(7)$

21. $\left(-\frac{4}{9}\right)(-18)$

22. $\frac{+4}{+2}$

23. $\frac{-6}{-3}$

24. $\frac{10}{-2}$

25. $\frac{-\frac{1}{2}}{4}$

26. $\frac{-5}{25}$

27. $\frac{10}{-1}$

28. $\frac{-4}{-2}$

29. $(-2)\left(\frac{3}{-2}\right)$

30. $-(-4)$

31. $(-1)a$

32. $\frac{2}{3}(-6)$

33. $\left(-\frac{2}{3}\right)\cdot 0$

34. $\left(-\frac{4}{5}\right)\left(\frac{-25}{8}\right)$

35. $(5)\left(\frac{-3}{5}\right)$

36. $(-6 - 7) + 8$

37. $\frac{-3}{6(-2)}$

38. $(-1)(-abc)$

39. $\frac{6 - 6}{3}$

40. $(-6)(-7) - 8$

41. $-2(-2 + 3)$

42. $\frac{12}{-6 + 7}$

43. $\frac{3 - 5}{4 - 6}$

44. $3(5 + 1)$

45. $(8 + 1)\cdot 7$

46. $\dfrac{-7-5}{4}$ **47.** $-8+(-2-5)$ **48.** $-7-(-3-2)$

49. $\dfrac{(-3)(3-5)}{(2-5)-6}$ **50.** $\dfrac{7-12}{-3}$ **51.** $\dfrac{2(-2+1)(3-4)}{12}$

52. $(-2)(-3)-(-4)(2)$ **53.** $\dfrac{(-3)(2)-(-2)(4)}{-4-(2-2)}$

54. $(-1)(-x)$ **55.** $-\dfrac{-6}{-8}$ **56.** $-\dfrac{7}{-14}$

57. $-\dfrac{-8}{24}$ **58.** $-\dfrac{-10}{-2}$ **59.** $\dfrac{\frac{8}{3}}{-2}$

60. $\dfrac{-\frac{7}{3}}{7}$ **61.** $-(-5xy)$ **62.** $-\dfrac{x}{(-y)z}$

63. $-\dfrac{-x}{(-y)(-z)}$ **64.** $(-7)(x)$ **65.** $(-6)(-y)$

66. $0-xy$ **67.** $\dfrac{x}{(-y)(z)}$ **68.** $\dfrac{(-x)(-y)}{(-w)(-z)}$

69. $\dfrac{1}{\frac{x}{6}}$ **70.** $\dfrac{1}{\frac{6}{y}}$ **71.** $\dfrac{0}{3-2}$

In Problems **72–75**, *find the reciprocals of the numbers.*

72. $\frac{1}{7}$ **73.** $\frac{5}{x}$ **74.** $\frac{x}{9}$ **75.** y

2-4 PROPERTIES OF REAL NUMBERS

Numbers† or numbers involved with operations are called **algebraic expressions.** Thus 12, $4x$, $x+7$, $4-y$, and $\dfrac{x+1}{4}$ are algebraic expressions.

There are special laws of real numbers that are handy when you are working with (algebraic) expressions. You're probably familiar with some, but may not be familiar with their names.

For example, you know that

$$\left.\begin{aligned} &3+2=5\\ \text{and}\quad &2+3=5.\end{aligned}\right\}\quad \text{Thus, } 3+2=2+3.$$

†Including literal numbers.

That is, *you can add numbers in any order*. This is called the **commutative law of addition.**

There's also a commutative law for *multiplication*. For example,

$$\left.\begin{array}{rr} & 7\cdot 6 = 42 \\ \text{and} & 6\cdot 7 = 42. \end{array}\right\} \quad \text{Thus, } 7\cdot 6 = 6\cdot 7.$$

That is, *you can multiply numbers in any order*. In symbols,

$a + b = b + a,$	commutative law of addition .
$a\cdot b = b\cdot a,$	commutative law of multiplication .

EXAMPLE 1

Use of commutative laws.

a. $3\cdot 9 = 9\cdot 3.$

b. $2(-6) = (-6)2.$

c. $2x + y = y + 2x.$

d. $y(2x) = (2x)y.$

e. $x\cdot\frac{3}{4} = \frac{3}{4}x.$

f. $2 + (x + 3) = 2 + (3 + x).$

EXAMPLE 2

$$\begin{aligned} -y + 2x &= (-y) + 2x \\ &= 2x + (-y) && \text{[commutative]} \\ &= 2x - y && [\text{since } a - b = a + (-b)]. \end{aligned}$$

Another law involves grouping. To find $7 + 2 + 3$, the law says that we may add two numbers first: either the first two, or the last two.

$$\left.\begin{array}{rr} & (7 + 2) + 3 = 9 + 3 = 12 \\ \text{and} & 7 + (2 + 3) = 7 + 5 = 12. \end{array}\right\} \quad \text{Thus, } (7 + 2) + 3 = 7 + (2 + 3).$$

This shows that *when adding, you can group numbers in any way*. This is called the **associative law of addition.**

There's also an associative law of *multiplication*. It too involves grouping. To find $2\cdot 3\cdot 4$, the law says that we may multiply two numbers first: either the first two, or the last two.

$$\left.\begin{array}{ll} & (2\cdot3)\cdot4 = 6\cdot4 \;\;= 24 \\ \text{and} & 2\cdot(3\cdot4) = 2\cdot12 = 24. \end{array}\right\} \quad \text{Thus, } (2\cdot3)\cdot4 = 2\cdot(3\cdot4).$$

In symbols,

$a + (b + c) = (a + b) + c,$	associative law of addition.
$a(bc) = (ab)c,$	associative law of multiplication.

EXAMPLE 3

Use of associative laws.

a. $3x + (y + 5) = (3x + y) + 5.$

b. $2(6x) = (2\cdot6)x = 12x.$

c. $\frac{1}{2}\left(\frac{2}{3}p\right) = \left(\frac{1}{2}\cdot\frac{2}{3}\right)p = \frac{1}{3}p.$

d. $4(-3x) = 4[(-3)x] = [4(-3)]x = -12x$. Note the use of brackets for grouping. Thus, $4(-3x) = -12x$.

EXAMPLE 4

a. To find $(-6)(-2)(-1)$, we shall begin by multiplying the first two numbers. This grouping will be indicated by brackets:

$$[(-6)(-2)](-1) = [12](-1) = -12.$$

In general, *the product of an* ***odd*** *number of negative numbers is negative.*

b. $(-2)(-5)(-3)(-4) = [(-2)(-5)][(-3)(-4)] = [10][12] = 120.$

The product of an ***even*** *number of negative numbers is positive.*

c. $(-2)(6)(-4) = 2\cdot6\cdot4 = 48.$

d. $2(-6)(4) = -(2\cdot6\cdot4) = -48.$

EXAMPLE 5

Use of the commutative and associative laws.

a. $(5x)\cdot7 = 7(5x)$ [commutative]

$= (7\cdot5)x$ [associative]

$= 35x.$

b. $(2 + x) + 4 = (x + 2) + 4$ [commutative]

$= x + (2 + 4)$ [associative]

$= x + 6.$

c. $(3x)(4y) = 3 \cdot x \cdot 4 \cdot y$

$= 3 \cdot 4 \cdot x \cdot y$ [commutative; we usually write the numerical product in front of the literal product]

$= 12xy.$

d. $(4x)(3y)(2z) = 4 \cdot x \cdot 3 \cdot y \cdot 2 \cdot z = 4 \cdot 3 \cdot 2 \cdot x \cdot y \cdot z = 24xyz.$

When we write

$$7 - 2 + 6 - 4,$$

what we mean is the *sum*

$$7 + (-2) + 6 + (-4).$$

By the commutative and associative laws you may "shuffle" and group these numbers to get the positive numbers together and the negative numbers together.

$$\underbrace{7 + 6} + \underbrace{(-2) + (-4)} = 13 + (-6) = 7.$$

More simply, we write

$$7 - 2 + 6 - 4 = 7 + 6 - 2 - 4 = 13 - 6 = 7.$$

That is, "get the pluses together and the minuses together."

EXAMPLE 6

a. $7 - 6 - 11 = 7 - 17 = -10.$

b. $-12 + 4 - 6 + 3 = 4 + 3 - 6 - 12 = 7 - 18 = -11.$

There are two laws that deal with both multiplication and addition. They are called the *distributive laws*. To illustrate, look at

$$2(3 + 5) = 2 \cdot 8 = 16$$

and $$2 \cdot 3 + 2 \cdot 5 = 6 + 10 = 16.$$

Notice that

$$2(3 + 5) = 2 \cdot 3 + 2 \cdot 5.$$

We say that multiplication *distributes* over addition. Here's what this means. To find $2(3 + 5)$ you can multiply the first factor, 2, by *each* term in the second factor, $(3 + 5)$, and add the results. Let's see it again.

$$2(3 + 5) = 2 \cdot 3 + 2 \cdot 5.$$

first factor

Let's do another one.

$$5(7 + 3) = 5 \cdot 7 + 5 \cdot 3 = 35 + 15 = 50.$$

first factor

By reversing the order of all the factors, we have

$$(7 + 3)(5) = 7 \cdot 5 + 3 \cdot 5 = 35 + 15 = 50.$$

We can look at this as multiplying *each* term in the first factor, $(7 + 3)$, by the second factor, (5), and then adding the results. In symbols,

DISTRIBUTIVE LAWS

$$a(b + c) = ab + ac;$$

$$(a + b)c = ac + bc.$$

EXAMPLE 7

Use of distributive laws.

a. $2(x + 3) = 2 \cdot x + 2 \cdot 3 = 2x + 6.$

b. $(7 + x)y = 7 \cdot y + x \cdot y = 7y + xy.$

Note that $2(x + 3) \neq 2x + 3$. *You must multiply both* x *and* 3 *by* 2 [*see Example* 7(a)].

The distributive laws can be extended. We can write

$$2(3 + 1 + 4) = 2\cdot 3 + 2\cdot 1 + 2\cdot 4 = 6 + 2 + 8 = 16.$$

EXAMPLE 8

Simplify.

a. $a(b + c + d) = ab + ac + ad.$

b. $(2x)(y + 3z + 4w) = (2x)(y) + (2x)(3z) + (2x)(4w)$

$$= 2xy + 6xz + 8xw.$$

Other useful properties related to the distributive laws are

$$a(b - c) = ab - ac;$$

$$(a - b)c = ac - bc.$$

We also refer to these as distributive laws.

EXAMPLE 9

Simplify.

a. $(x - y)z = xz - yz.$

b. $x(y - 6) = xy - x(6)$

$= xy - 6x.$ [commutative]

c. $(2 - 3a)b = 2b - 3ab.$

We can combine the distributive laws and write things like

$$x(y + z - w) = xy + xz - xw.$$

Now that you are familiar with the commutative, associative, and distributive laws, we end this section with three rules that may be useful when you are working with fractions.

1. $\boxed{\dfrac{ab}{c} = \dfrac{a}{c} \cdot b.}$ **2.** $\boxed{\dfrac{a}{bc} = \dfrac{a}{b} \cdot \dfrac{1}{c}.}$ **3.** $\boxed{\dfrac{ab}{cd} = \dfrac{a}{c} \cdot \dfrac{b}{d}.}$

EXAMPLE 10

a. By Rule 1, $\dfrac{3x}{3} = \dfrac{3}{3}x = 1 \cdot x = x.$

b. By Rule 1, $\dfrac{4x}{8} = \dfrac{4}{8}x = \dfrac{\overset{1}{\cancel{4}}}{\underset{2}{\cancel{8}}}x = \dfrac{1}{2}x.$ Using Rule 1 again, we can write $\dfrac{1}{2}x = \dfrac{1 \cdot x}{2} = \dfrac{x}{2}$. Thus, $\dfrac{4x}{8} = \dfrac{x}{2}$. Usually we simply write

$$\frac{\overset{1}{\cancel{4}}x}{\underset{2}{\cancel{8}}} = \frac{x}{2}.$$

c. By Rule 1, $\dfrac{-3t}{-3} = \dfrac{(-3)t}{-3} = \left(\dfrac{-3}{-3}\right)t = (1)t = t.$ Usually we simply write

$$\frac{-3t}{3} = \frac{(\overset{1}{\cancel{-3}})t}{\underset{1}{\cancel{-3}}} = t.$$

d. By Rule 2, $\dfrac{3}{9y} = \dfrac{3}{9} \cdot \dfrac{1}{y} = \dfrac{\overset{1}{\cancel{3}}}{\underset{3}{\cancel{9}}} \cdot \dfrac{1}{y} = \dfrac{1}{3} \cdot \dfrac{1}{y}$. Using Rule 2 again, we have $\dfrac{1}{3} \cdot \dfrac{1}{y} = \dfrac{1}{3y}$. Thus, $\dfrac{3}{9y} = \dfrac{1}{3y}$. Usually we simply write

$$\frac{\overset{1}{\cancel{3}}}{\underset{3}{\cancel{9}}y} = \frac{1}{3y}.$$

e. $\dfrac{-2x}{4} = -\dfrac{2x}{4} = -\dfrac{\overset{1}{\cancel{2}}x}{\underset{2}{\cancel{4}}} = -\dfrac{x}{2}.$

f. By Rule 1, $25\left(\dfrac{x}{5}\right) = \dfrac{25x}{5}$. But $\dfrac{25x}{5} = \dfrac{\overset{5}{\cancel{25}}x}{\underset{1}{\cancel{5}}} = \dfrac{5x}{1} = 5x$. Thus,

$$25\left(\frac{x}{5}\right) = \frac{\overset{5}{\cancel{25}}x}{\underset{1}{\cancel{5}}} = \frac{5x}{1} = 5x.$$

g. By Rule 3, $\frac{12x}{16y} = \frac{12}{16} \cdot \frac{x}{y} = \frac{\overset{3}{\cancel{12}}}{\underset{4}{\cancel{16}}} \cdot \frac{x}{y} = \frac{3}{4} \cdot \frac{x}{y} = \frac{3x}{4y}$. Usually we simply write

$$\frac{\overset{3}{\cancel{12}}x}{\underset{4}{\cancel{16}}y} = \frac{3x}{4y}.$$

Completion Set 2-4

In Problems **1–12**, *fill in the blanks.*

1. The commutative law says that $a + b =$ ____________.

2. The associative law says that $a(bc) =$ ____________.

3. By the distributive law, $2(b + c) = 2b +$ _____.

4. By the commutative law of addition, $2 + 3y =$ ____________.

5. $(x + y) + z = (y + x) + z$ by the ______________________ law.

6. $7 + (4 + y) = (7 + 4) + y$ by the ______________________ law.

7. $4(x + a) = 4x + 4a$ by the ______________________ law.

8. The sign of the product $(-800)(752)(-48)$ is ____.

9. The sign of the product $(-21)(-25)(-87)$ is ____.

10. $(3a)(b + 2c + 3d) =$ ______ + ______ + ______.

11. $(x - 2)y =$ _____ $-2y$.

12. $a(x +$ ____ $) = ax + ay$.

In Problems **13–18**, *mark the statements either* T (= *true*) *or* F (= *false*).

13. $3(x - 4) = 3x - 4.$ ____

14. $(2 + x)y = 2 + xy.$ ____

15. $(3 + 2) + 2 = 3 + (2 + 2).$ ____

16. $2(3 \cdot 4 \cdot x) = (2 \cdot 3) + (2 \cdot 4) + (2 \cdot x).$ ____

17. $x + 3y = 3y + x.$ ____

18. $ab + xy = yx + ba.$ ____

Problem Set 2-4

In Problems **1–50**, *perform the indicated operations.*

1. $5(2x)$

2. $(3x)(6)$

3. $(3 + x) + 4$

4. $(-2)(2)(-1)(-2)(-1)$

5. $(-2)(-1)(-2)(-1)(-1)$

6. $-2 - 5 + 9 - 1$

7. $(-2) + (-1) - 5$

8. $5 - 6 + 1$

9. $3 - 5 + 6 - 4$

10. $(-2)(4)(-3) + 6(-2)$

11. $4 - 8 + 6(-2)$

12. $(x + 3) + 5$

13. $4(2 + x)$

14. $(5a)(x + 3)$

15. $2(3x - 2)$

16. $(x + 2)y$

17. $(3x + 4y)z$

18. $(2x)(y - 6)$

19. $x(y + 3z)$

20. $w(2a + 3b)$

21. $(4x - 3)y$

22. $(5 - 2a)b$

23. $(3x)(4y)$

24. $x(7 - y)$

25. $(3x)(2 - 3y)$

26. $(3 - 5x) \cdot 7$

27. $x(y - 2z + w)$

28. $(8x)\left(\frac{1}{4}t\right)$

29. $(5x)(2 - y - 3z)$

30. $x(2y)(2z)$

31. $4\left(\frac{3}{4}x\right)$

32. $\frac{1}{5}(5a)$

33. $6\left(\frac{3}{8}x\right)$

34. $3\left(\frac{1}{3}x\right)$

35. $(2x)(3y)(4z)$

36. $5\left(\frac{2}{9}x\right)$

37. $\frac{-12x}{-6}$

38. $\frac{-6x}{24}$

39. $\frac{5x}{-10}$

40. $(-8x)(-2)$

41. $5(-3x)$

42. $4 - \frac{2(-4)}{2} + \frac{12}{-6}$

43. $\frac{7x}{14}$

44. $6\left(\frac{x}{3}\right)$

45. $\dfrac{24}{18z}$ 46. $\dfrac{-4x}{2}$ 47. $\dfrac{27y}{6}$

48. $\dfrac{25y}{15x}$ 49. $\dfrac{18}{12p}$ 50. $\left(\frac{6}{5}\right)\left(\frac{10}{9}x\right)$

2-5 EVALUATING FORMULAS

Knowing how to handle signed numbers is important in evaluating formulas.

EXAMPLE 1

The formula $F = 32 + \frac{9}{5}C$ is used to convert a celsius temperature C to a fahrenheit temperature F. To what does $-20°$ C correspond?

In the formula, replace C by -20 and simplify.

$$F = 32 + \frac{9}{5}C$$

$$F = 32 + \frac{9}{5}(-20)$$

$$F = 32 + \left[-\frac{9 \cdot \cancelto{4}{20}}{\cancelto{1}{5}}\right] = 32 - 36$$

$$F = -4.$$

Thus, -20°C corresponds to -4°F.

EXAMPLE 2

The formula

$$d = |a - b|$$

gives the distance d between the numbers a and b on the number line. Find d when $a = 7$ and $b = -9$.

$$d = |a - b| = |7 - (-9)| = |7 + 9| = |16| = 16.$$

Thus the numbers 7 and -9 are 16 units apart.

Note that $|7 - (-9)| \neq |7| - |-9| = 7 - 9 = -2$. Moral: Don't invent your own shortcuts and rules.

EXAMPLE 3

Given the formula

$$m = \frac{y_2 - y_1}{x_2 - x_1},$$

find m when $x_1 = 5$, $x_2 = 3$, $y_1 = -5$, *and* $y_2 = 7$. *The small numbers to the right and just below the letters are called* ***subscripts***. *We read* x_1 *as "x sub one," and so on.*

$$m = \frac{y_2 - y_1}{x_2 - x_1} = \frac{7 - (-5)}{3 - 5} = \frac{12}{-2} = -6.$$

You will see and use this formula again when we study straight lines. It's called the *slope formula*.

Problem Set 2-5

In Problems **1** *and* **2**, *use the formula* $F = 32 + \frac{9}{5}C$, *as given in Example* 1, *to convert the given celsius temperature to a fahrenheit temperature.*

1. $-35°C$

2. $-15°C$

The formula $C = \frac{5}{9}(F - 32)$ *is used to convert a fahrenheit temperature F to a celsius temperature C. In Problems* **3** *and* **4**, *use this formula to convert the given fahrenheit temperature to a celsius temperature.*

3. $-40°F$

4. $32°F$

In Problems **5** *and* **6**, *use the distance formula* $d = |a - b|$, *as given in Example* 2, *to find the distance d between the given numbers.*

5. $a = -3, b = 5$

6. $a = -4, b = -9$

In Problems **7** *and* **8**, *use the slope formula* $m = \frac{y_2 - y_1}{x_2 - x_1}$, *as given in Example* 3, *to find m from the given information.*

7. $x_1 = 5, x_2 = -8, y_1 = -1, y_2 = -2$

8. $x_1 = 3, x_2 = 5, y_1 = 4, y_2 = 1$

If your grades in two exams were 70 *and* 80, *then the average grade for both exams is* $\frac{70 + 80}{2} = \frac{150}{2} = 75$. *In statistics, we say the average or* ***mean*** *of the numbers* x_1, x_2, x_3, *and* x_4 *is given by the formula*

$$\bar{x} = \frac{x_1 + x_2 + x_3 + x_4}{4},$$

where the symbol $\bar{x}$ *(read "x bar") represents the mean. In Problems* **9** *and* **10**, *find* $\bar{x}$ *from the given information.*

9. $x_1 = -8, x_2 = 7, x_3 = 16, x_4 = -3$

10. $x_1 = -3, x_2 = -5, x_3 = 12, x_4 = -4$

11. A math student taking calculus finds that he must compute y' (read "y prime"), where

$$y' = \frac{(x)(1) - (x-5)(1)}{(x)(x)}$$

and $x = -2$. What answer should he get?

12. Repeat Problem 11 if $x = 3$.

For an arrangement of three particles on a number line, the location $\bar{x}$ (read "x bar") of the so-called center of mass is given by the formula

$$\bar{x} = \frac{m_1x_1 + m_2x_2 + m_3x_3}{m_1 + m_2 + m_3}.$$

Here m_1, m_2, and m_3 are the masses of the particles, and x_1, x_2, and x_3 are their locations, respectively. In Problems **13** *and* **14**, *find $\bar{x}$ from the given information.*

13. $m_1 = 2, m_2 = 3, m_3 = 4, x_1 = -2, x_2 = -3, x_3 = 8$

14. $x_1 = -4, x_2 = 4, x_3 = -4, m_1 = 2, m_2 = 2, m_3 = 2$

2-6 SUPPLEMENT ON ABSOLUTE VALUE†

In this chapter you looked at absolute value in terms of the number line. Here's a brief note on defining it in another way.

You know that $|-2| = 2$. It might have occurred to you that all we did was drop the negative sign so that we don't have a negative answer. This idea can lead to problems if we don't know the sign of the number.

For example, we can't always write $|x| = x$, because we don't know whether x is positive or negative. Here's a way out.

If $x = -2$, then

$$\begin{aligned}|x| &= |-2| = 2\\ &= -(-2)\\ &= -x.\end{aligned}$$

Thus, if x is negative, then $|x| = -x$, which is positive. But if $x = 3$, then

$$|x| = |3| = 3 = x.$$

---

†This material is not needed in the rest of the book.

Thus, if x is positive, then $|x| = x$. Of course, if $x = 0$, then $|x| = 0$. Now we can define $|x|$ as

$$|x| = x \text{ if } x \geqslant 0;$$
$$|x| = -x \text{ if } x < 0.$$

2-7 REVIEW

IMPORTANT TERMS AND SYMBOLS

set *(p. 27)*
positive integer *(p. 28)*
integer *(p. 28)*
rational number *(p. 28)*
irrational number *(p. 29)*
real number *(p. 29)*
number line *(p. 29)*
literal number *(p. 31)*
$a < b, a > b, a \leqslant b, a \geqslant b$ *(p. 31)*
inequality *(p. 32)*
absolute value *(p. 33)*
$|a|$ *(p. 33)*
term *(p. 35)*
opposite of a number *(p. 37)*
reciprocal *(p. 38)*
algebraic expression *(p. 45)*
commutative laws *(p. 46)*
associative laws *(p. 46)*
distributive laws *(p. 49)*
subscript *(p. 55)*

REVIEW PROBLEMS

In Problems **1–8**, *name the law used in the given statement.*

1. $8 + y = y + 8$

2. $2(x + 3y) = 2x + 6y$

3. $2x + (x + y) = (2x + x) + y$

4. $2(4x) = (2 \cdot 4)x$

5. $5(x + 4) = (x + 4)5$

6. $5(x + 4) = 5(4 + x)$

7. $(a - 3)b = ab - 3b$

8. $(3x)(y) = 3(xy)$

In Problems **9–16**, *determine whether each statement is true or false.*

9. $-3 < -2$

10. $3 > 0$

11. $|-6| = 6$

12. $-5 > 1$

13. $|-3| < |-2|$

14. $-|-4| = 4$

15. $2(3x) = (2 \cdot 3)(2 \cdot x)$

16. $2(3 - x) = 6 - x$

In Problems **17–52**, *simplify.*

17. $\frac{(-2)(-4)}{-16}$

18. $(-3)(-4+7)$

19. $(3x)(y+5)$

20. $2(x-y)$

21. $2(-5y)$

22. $\frac{7}{2-2}$

23. $-6-(-5)$

24. $\frac{-2}{6}$

25. $3(-8+4)$

26. $(-2)(-4)(-1)$

27. $\frac{14}{(-2)(-7)}$

28. $\frac{6(-3)}{-2}$

29. $4(5x-3)$

30. $8\left(\frac{5}{12}x\right)$

31. $(7+x)-8$

32. $\frac{6-8}{-4}$

33. $2(-6)+(-4)(-1)$

34. $\frac{8-9}{7-6}$

35. $-(-3xz)$

36. $-6+4-9(2)$

37. $\left(-\frac{3}{4}\right)(-20)$

38. $-9(-12-8)$

39. $(-8)\left(\frac{-5}{8}\right)$

40. $(a+b)c$

41. $(4x)(7y)$

42. $-\frac{-8}{-64}$

43. $\frac{7-9}{(7)(-9)}$

44. $(8-8)(8x)$

45. $\frac{6(-4)}{-2}$

46. $\frac{(-2)(4)(-6)}{0-3}$

47. $\frac{(-2)-(-1)(0)}{-2}$

48. $\frac{8(-2)-(-2)(-8)}{3(-2)-2}$

49. $-4(3x)$

50. $(3x)(1-y+z)$

51. $(8-x)y$

52. $(x+7)-(7-4)$

53. Compute $\frac{a-2b+c}{ad}$ if

a. $a=2,\ b=-3,\ c=4,\ d=-1$

b. $a=-3,\ b=4,\ c=-2,\ d=-2$.

In Problems **54** *and* **55**, *for the given values of x and y, find* (a) $|x+y|$, (b) $|x-y|$, (c) $|x|+|y|$, and (d) $|x|-|y|$.

54. $x=2, y=5$

55. $x=-3, y=-8$

CHAPTER 3

A First Look at Exponents and Radicals

3-1 EXPONENTS

Recall that when we multiply numbers to form a product, each number is called a **factor** of the product. For example, in $2ab$ we can say that 2, a, and b are factors. Sometimes the letters a and b are called *literal factors*.

As a bit of shorthand, the product

$$a \cdot a \cdot a \cdot a \text{ is written } a^4.$$

It is read, "the fourth power of a" or "a raised to the fourth." We call a the **base** and 4 the **exponent**.

$$\text{base} \rightarrow a^{4} \leftarrow \text{exponent}$$

In general, if a is used as a factor n times, we have

> The nth power of a:
>
> $$a^n = \underbrace{a \cdot a \cdot \ldots \cdot a}_{n \text{ factors of } a}$$

$x^2 = x \cdot x$. ***Don't write*** $x^2 = x + x$.

EXAMPLE 1

a. $x^3 = x \cdot x \cdot x$. We read x^3 as x *cubed* or x *to the third.*

b. $6^2 = 6 \cdot 6 = 36$. We read 6^2 as 6 *squared.*

c. $2x^3 = 2 \cdot x \cdot x \cdot x$. Only x is cubed here.

d. $(2x)^3 = 2x \cdot 2x \cdot 2x$.

e. $4^2x^4 = 4 \cdot 4 \cdot x \cdot x \cdot x \cdot x$.

f. $(2x - 3)^2 = (2x - 3)(2x - 3)$.

g. $(1 + 3)^2 = (1 + 3)(1 + 3) = 4 \cdot 4 = 16$. Note that

$$(1 + 3)^2 \neq 1^2 + 3^2,$$

since $(1 + 3)^2 = 16$ and $1^2 + 3^2 = 1 + 9 = 10$.

EXAMPLE 2

a. $(-2)^3 = (-2)(-2)(-2) = -8$. *An odd power of a negative number is negative.*

b. $(-1)^4 = (-1)(-1)(-1)(-1) = 1$. *An even power of a negative number is positive.*

An exponent applies only to the quantity immediately to the left and below it. For example:

In $(-2)^2$ *the base is* -2. $(-2)^2 = (-2)(-2) = 4$.

In -2^2 *the base is* 2. $-2^2 = -(2 \cdot 2) = -4$.

In $(2x)^2$ *the base is* $2x$. $(2x)^2 = (2x)(2x)$.

In $2x^2$ *the base is* x. $2x^2 = 2 \cdot x \cdot x$.

In $2(x + 1)^2$ *the base is* $x + 1$. $2(x + 1)^2 = 2(x + 1)(x + 1)$.

This is what an exponent of 1 means:

$$\boxed{a^1 = a.}$$

EXAMPLE 3

a. $5^1 = 5.$

b. $x - 3 = (x - 3)^1.$

Notice that

$$a^2 \cdot a^3 = (a \cdot a)(a \cdot a \cdot a) = a \cdot a \cdot a \cdot a \cdot a$$
$$= a^5 = a^{2+3}.$$

So, $a^2 \cdot a^3 = a^{2+3}$. In general,

to multiply numbers with the same base, add the exponents and keep the base the same.

$$\boxed{a^m a^n = a^{m+n}.}$$

EXAMPLE 4

a. $4^3 \cdot 4^8 = 4^{3+8} = 4^{11}.$

b. $x^5 \cdot x^2 = x^{5+2} = x^7.$

c. $(-1)^2(-1) = (-1)^2(-1)^1 = (-1)^{2+1} = (-1)^3 = -1.$

d. $(3y - 1)^7(3y - 1)^3 = (3y - 1)^{10}.$

e. $x^3x^4x^6 = x^{3+4+6} = x^{13}.$

f. $3 \cdot 3^2 = 3^1 \cdot 3^2 = 3^3 = 27.$

g. $x(x^n) = x^1x^n = x^{1+n} = x^{n+1}.$

EXAMPLE 5

a. $\dfrac{1}{x^2x^6} = \dfrac{1}{x^{2+6}} = \dfrac{1}{x^8}.$

b. $\dfrac{x^2x^3}{y^3y^4} = \dfrac{x^5}{y^7}.$

c. $(2x^3)(x^2) = 2 \cdot x^3 \cdot x^2 = 2x^5.$

d. $(3x^5)(9x^2) = 3 \cdot 9 \cdot x^5x^2 = 27x^7.$

Remember that in the above rule, the bases must be the same before you add exponents.

$$2^2 \cdot x^3 \neq (2x)^5.$$

$$(-2)^4(-2^2) \neq (-2)^6, \quad \textit{but} \quad (-2)^4(-2^2) = (16)(-4) = -64.$$

Watch what happens when we raise a power of a number to a power.

$$(a^2)^3 = a^2 \cdot a^2 \cdot a^2 = a^6 = a^{2 \cdot 3}.$$

In general,

> *to find a "power of a power," multiply the exponents and keep the base the same.*

$$\boxed{(a^m)^n = a^{mn}.}$$

EXAMPLE 6

a. $(3^2)^4 = 3^{2 \cdot 4} = 3^8.$

b. $(x^3)^5 = x^{3 \cdot 5} = x^{15}.$

c. $(t^4)^9 = t^{36}.$

d. $[(y - 2)^3]^2 = (y - 2)^6.$

e. $(r^n)^p = r^{np}.$

f. $(2^n)^n = 2^{n \cdot n} = 2^{n^2}.$

Don't confuse $(x^2)^3$ with $x^2 \cdot x^3$.

$$(x^2)^3 = x^6 \quad \textit{but} \quad x^2 \cdot x^3 = x^5.$$

Let's look at division.† In $\frac{a^5}{a^3}$ there are more a's in the numerator than in the denominator:

$$\frac{a^5}{a^3} = \frac{\not{a} \cdot \not{a} \cdot \not{a} \cdot a \cdot a}{\not{a} \cdot \not{a} \cdot \not{a}} = a^2 = a^{5-3}.$$

†Here, as elsewhere in this book, we assume no division by 0.

But in $\frac{a^3}{a^5}$ the opposite is true:

$$\frac{a^3}{a^5} = \frac{\not{a}\cdot\not{a}\cdot\not{a}}{\not{a}\cdot\not{a}\cdot\not{a}\cdot a\cdot a} = \frac{1}{a^2} = \frac{1}{a^{5-3}}.$$

Thus,

$$\frac{a^m}{a^n} = a^{m-n} \quad \text{for } m > n,$$

$$\frac{a^m}{a^n} = \frac{1}{a^{n-m}} \quad \text{for } n > m,$$

$$\frac{a^n}{a^n} = 1.$$

EXAMPLE 7

a. $\frac{2^8}{2^3} = 2^{8-3} = 2^5 = 32.$

b. $\frac{6^{14}}{6^{16}} = \frac{1}{6^{16-14}} = \frac{1}{6^2} = \frac{1}{36}.$

c. $\frac{x^{11}}{x^7} = x^{11-7} = x^4.$

d. $\frac{(x^2+1)^4}{(x^2+1)^{12}} = \frac{1}{(x^2+1)^{12-4}} = \frac{1}{(x^2+1)^8}.$

e. $\frac{-y^6}{y^4} = -\frac{y^6}{y^4} = -y^2.$

f. $\frac{(-2)^3}{(-2)^4} = \frac{1}{(-2)^{4-3}} = \frac{1}{(-2)^1} = \frac{1}{-2} = -\frac{1}{2}.$

g. $\frac{x^{1+n}}{x^n} = x^{1+n-n} = x^1 = x.$

EXAMPLE 8

a. $\frac{(x^6)^4}{(x^4)^5} = \frac{x^{24}}{x^{20}} = x^4.$

b. $\frac{x^2(x^4)^7}{x^{90}} = \frac{x^2x^{28}}{x^{90}} = \frac{x^{30}}{x^{90}} = \frac{1}{x^{60}}.$

Let's look at $(3 \cdot 5)^2$:

$$(3 \cdot 5)^2 = (3 \cdot 5)(3 \cdot 5) = 3 \cdot 3 \cdot 5 \cdot 5 = 3^2 5^2.$$

Notice that to raise a product to a power, you raise each factor to that power.

$$\boxed{(ab)^n = a^n b^n.}$$

Now look at $\left(\frac{3}{5}\right)^2$:

$$\left(\frac{3}{5}\right)^2 = \frac{3}{5} \cdot \frac{3}{5} = \frac{3 \cdot 3}{5 \cdot 5} = \frac{3^2}{5^2}.$$

To raise a quotient to a power, you raise both the numerator and denominator to that power.

$$\boxed{\left(\frac{a}{b}\right)^n = \frac{a^n}{b^n}.}$$

EXAMPLE 9

a. $(xy)^8 = x^8 y^8$.

b. $\left(\frac{x}{y}\right)^{22} = \frac{x^{22}}{y^{22}}$.

c. $\left(\frac{2}{3}\right)^3 = \frac{2^3}{3^3} = \frac{8}{27}$.

d. $\left(\frac{1}{x}\right)^4 = \frac{1^4}{x^4} = \frac{1}{x^4}$.

e. $(2x)^4 = 2^4 x^4 = 16x^4$. Note that $(2x)^4 \neq 2x^4$.

f. $(abc)^5 = a^5 b^5 c^5$.

EXAMPLE 10

a. $(2x^2y^3)^3 = 2^3(x^2)^3(y^3)^3 = 8x^6y^9$.

b. $\left(\frac{x}{y^4}\right)^2 = \frac{x^2}{(y^4)^2} = \frac{x^2}{y^8}$.

c. $\left(\frac{2x^2}{y^3}\right)^3 = \frac{(2x^2)^3}{(y^3)^3} = \frac{2^3(x^2)^3}{y^9} = \frac{8x^6}{y^9}$.

d. $\left(\dfrac{ab^2}{c^3d^4}\right)^6 = \dfrac{(ab^2)^6}{(c^3d^4)^6} = \dfrac{a^6(b^2)^6}{(c^3)^6(d^4)^6} = \dfrac{a^6b^{12}}{c^{18}d^{24}}.$

e. $\dfrac{(20)^3}{5^3} = \left(\dfrac{20}{5}\right)^3 = 4^3 = 64.$

f. $\left(\dfrac{x^{2a}}{y^b}\right)^c = \dfrac{(x^{2a})^c}{(y^b)^c} = \dfrac{x^{2ac}}{y^{bc}}.$

The next example shows you how to handle minus signs involved with powers.

EXAMPLE 11

a. $(-x)^9 = [(-1)x]^9 = (-1)^9x^9 = (-1)x^9 = -x^9.$

b. $(-x^2y)^8 = (-1 \cdot x^2 \cdot y)^8 = (-1)^8(x^2)^8y^8 = 1 \cdot x^{16}y^8 = x^{16}y^8.$

Completion Set 3-1

In Problems **1–12**, *fill in the blanks.*

1. In 4^5, the number 4 occurs as a factor __________ times.

2. In 6^9, the base is ____ and the exponent is ____.

3. In $(3x)^2$ the base is _____, but in $3x^2$ it is ____.

4. The value of 1^5 is ____.

5. The value of $(-1)^5$ is _____.

6. The value of $(-1)^8$ is ____.

7. The value of $(18)^1$ is _____.

8. (*Insert* + *or* ·) $x^4(x^5)$ is x to the 4 ____ 5 power.

9. (*Insert* + *or* $\cdot$) $(x^4)^5$ is x to the 4 ____ 5 power.

10. (*Insert* + *or* $-$) $\dfrac{x^8}{x^3}$ is x to the 8 ____ 3 power.

11. In raising a power of a number to a power, we ____________ the exponents.

12. The cube of $2x$ is equal to ________.

Problem Set 3-1

In Problems **1–12**, *compute the given numbers.*

1. 2^3
2. $(-3)^2$
3. $2^4 - 2^5$
4. $(-7)^1$
5. $-(-2)^4$
6. $\dfrac{-2^2}{(-2)^3}$
7. $(-2^3)(-3)^2$
8. $(3 - 5)^2$
9. $\dfrac{-(-3)^2}{(-3)^3}$
10. $2^3 \cdot 3^2 - 6^2$
11. $(-2^2)^3$
12. $(-2 + 3^2)^2$

In Problems **13–75**, *simplify.*

13. x^3x^8
14. x^4x^4
15. y^5y^4
16. t^2t
17. x^2x^4x
18. x^ax^b
19. $(x - 2)^5(x - 2)^3$
20. $y^9y^{91}y^2$
21. $\dfrac{x^5x^2}{y^2y^3y^4}$
22. $\dfrac{x(x^2)}{y^2y^3}$
23. $(2x^2)(7x^6)$
24. $x^4(3x^2)$
25. $(-3x)(4x^3)$
26. $(-2x^5)(-3x^4)$
27. $(x^8)^2$
28. $(x^4)^3$
29. $(x^3)^3$
30. $(x^5)^7$
31. $(t^2)^n$
32. $(x^b)^c$
33. $(x^4)^2(x^3)^7$
34. $x^6(x^4)^2$
35. $\dfrac{x^7}{x^3}$
36. $\dfrac{x^8}{x^{12}}$
37. $\dfrac{x^{21}}{x^{22}}$
38. $\dfrac{(a + b)^{16}}{(a + b)^{12}}$
39. $\dfrac{y^{14}}{-y^8}$
40. $\dfrac{-y^2}{-y^5}$
41. $\dfrac{x^2x^8}{x^{16}}$
42. $\dfrac{x^{18}}{x^{10}x^{10}}$

43. $\dfrac{(x^5)^3}{x^2}$ **44.** $\dfrac{(x^4)^2}{(x^5)^3}$ **45.** $\dfrac{(x^4)^2}{x(x^6)}$

46. $\dfrac{t^{12}(t^6)}{(w^5)^3}$ **47.** $\dfrac{(x^2)^4(x^4)^2}{(x^3)^7}$ **48.** $\dfrac{1}{(x^4)^5}(x^4)^5$

49. $(ab)^6$ **50.** $(xy)^4$ **51.** $(2x)^4$

52. $(3x)^3$ **53.** $(3y)^2(5y^4)$ **54.** $(3y^2)(2y^2)^4$

55. $\left(\dfrac{a}{b}\right)^3$ **56.** $\left(\dfrac{x}{2}\right)^4$ **57.** $\left(\dfrac{3}{x}\right)^4$

58. $\left(\dfrac{x}{y}\right)^2$ **59.** $(xy^2)^4$ **60.** $(3x^2)^3$

61. $\left(\dfrac{2y}{z}\right)^3$ **62.** $\left(\dfrac{1}{x^2y^3}\right)^5$ **63.** $\left(\dfrac{2}{3}a^2b^3c^6\right)^2$

64. $(-4)(2x^2)^2$ **65.** $\left(\dfrac{x^2}{y^5}\right)^3$ **66.** $\left(\dfrac{2x^2}{x^2}\right)^4$

67. $\left(\dfrac{x^2y^3}{2z^4}\right)^4$ **68.** $\left(\dfrac{2x^4}{5x^2}\right)^3$ **69.** $(-x)^{13}$

70. $(-3x)^4$ **71.** $(-2x^2y)^4$ **72.** $(-3)^2(-x)^3$

73. $\dfrac{(-xy)^5}{(-t)^4}$ **74.** $\dfrac{-y^3}{(-z)^2}$ **75.** $(-x)^3(-x)^7$

In Problems **76–79**, *find the values in an easy way.*

76. $\dfrac{5^{100}}{5^{99}}$ **77.** $\dfrac{2^6 2^{11}}{(2^5)^3}$ **78.** $\dfrac{(80)^6}{(40)^6}$ **79.** $(14)^5\left(\dfrac{1}{28}\right)^5$

80. Here's a problem you might find in statistics. Compute

$$\frac{(x_1-\bar{x})^2+(x_2-\bar{x})^2+(x_3-\bar{x})^2+(x_4-\bar{x})^2+(x_5-\bar{x})^2}{4},$$

if $x_1 = 6$, $x_2 = 8$, $x_3 = 5$, $x_4 = 3$, $x_5 = 3$, and $\bar{x} = 5$.

81. You might find a problem like this in calculus. Compute y' (read "y prime") if

$$y' = (x^2 - 2x)(2x) + (x^2 + 4)(2x - 2)$$

and $x = -1$.

3-2 RADICALS

Since $2^4 = 16$, we say 2 is a *fourth root* of 16. More generally,

r is an nth root of a means $r^n = a$.†

Thus, 5 is a *cube* (or third) *root* of 125 because $5^3 = 125$.

Similarly, since $3^2 = 9$ and $(-3)^2 = 9$, both 3 and -3 are *square* (or second) *roots* of 9. However, *when there is a positive* nth *root, it is customary to call that one the* ***principal* nth *root***. For this reason we say that 3 is the principal square root of 9. Sometimes the principal nth root is simply called *the* nth root. Thus,

3 is *the* square root of 9.

When there is only a negative nth *root, that root is the principal* nth *root.* Thus, *the* (principal) cube root of -8 is -2 because $(-2)^3 = -8$ and there are no positive cube roots of -8. By the way, we define any root of 0 to be 0.

To refer to the principal nth root of a we use the symbol $\sqrt[n]{a}$ (called a **radical**).

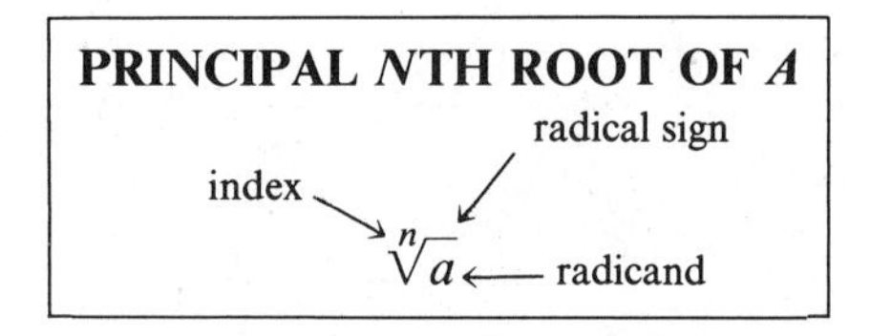

Thus, $\sqrt[3]{-8} = -2$ and $\sqrt[2]{9} = 3$ (**not** -3). In $\sqrt[3]{-8}$, the index is 3 and the radicand is -8. For a principal square root we drop the index 2; thus, $\sqrt{9} = 3$. In the back of the book there is a table of square roots and cube roots.

There are some troubles with even roots of negative numbers. For example, there is no real number whose square is -4. This means that $\sqrt{-4}$ does not represent a real number. Right now we won't bother with problems involving such roots.

---

†n is a positive integer.

EXAMPLE 1

Some principal roots.

a. $\sqrt{25} = 5$; 5 is *the* square root of 25, since $5^2 = 25$.

b. $\sqrt[4]{16} = 2$; 2 is *the* fourth root of 16, since $2^4 = 16$.

c. $\sqrt[3]{-1} = -1$, because $(-1)^3 = -1$ and no positive number has its cube equal to -1.

d. $\sqrt[5]{0} = 0$, because any root of 0 is 0.

EXAMPLE 2

a. $\sqrt[3]{\frac{1}{125}} = \frac{1}{5}$, because $\left(\frac{1}{5}\right)^3 = \frac{1}{125}$.

b. $\sqrt{.01} = .1$, because $(.1)^2 = .01$.

c. $-\sqrt{81} = -(\sqrt{81}) = -(9) = -9$.

d. $-\sqrt[3]{-8} = -(-2) = 2$.

Not all roots are integers or rational numbers. Some may be irrational. For example, $\sqrt{2}$ is irrational. Although an approximate value of $\sqrt{2}$ is 1.414, we'll usually keep $\sqrt{2}$ in radical form. Don't think you always have to go to a decimal. There is nothing wrong in writing $\sqrt{2}$; it is just a number whose square is 2. Similarly, we'll keep $\sqrt{3}$ in radical form.

Since $\sqrt{9}$ is a number whose square is 9, we have $\sqrt{9} \cdot \sqrt{9} = 9$. Similarly,

$$\sqrt{5} \cdot \sqrt{5} = 5$$

and

$$\sqrt[3]{4} \cdot \sqrt[3]{4} \cdot \sqrt[3]{4} = 4.$$

In general,

$$\boxed{(\sqrt[n]{a})^n = a.}$$

Here are two important properties of radicals that involve products or quotients.

$$\sqrt[n]{ab} = \sqrt[n]{a} \cdot \sqrt[n]{b}$$

$$\sqrt[n]{\frac{a}{b}} = \frac{\sqrt[n]{a}}{\sqrt[n]{b}}.\dagger$$

These properties let us simplify many radicals by writing them as a product or quotient of radicals so that one of them is easy to compute.

EXAMPLE 3

Simplifying radicals.

a. $\sqrt{20}$.

We may factor 20 so that one factor is the square of an integer.

$$\sqrt{20} = \sqrt{4 \cdot 5} = \sqrt{4} \cdot \sqrt{5} = 2\sqrt{5}.$$

Here we wrote $\sqrt{20}$ as a product of two radicals so that we could easily compute one of them. It would not have helped us to write something like $\sqrt{20} = \sqrt{10} \cdot \sqrt{2}$.

b. $\sqrt{18} = \sqrt{9 \cdot 2} = \sqrt{9} \cdot \sqrt{2} = 3\sqrt{2}$.

c. $\sqrt{\frac{7}{4}} = \frac{\sqrt{7}}{\sqrt{4}} = \frac{\sqrt{7}}{2}$. Here we wrote $\sqrt{\frac{7}{4}}$ as a quotient of two radicals, since we could compute $\sqrt{4}$ easily.

d. $\sqrt[3]{54} = \sqrt[3]{27 \cdot 2} = \sqrt[3]{27} \cdot \sqrt[3]{2} = 3\sqrt[3]{2}$.

---

†Let's see why $\sqrt[n]{ab} = \sqrt[n]{a} \cdot \sqrt[n]{b}$. Look at $(\sqrt[n]{a} \cdot \sqrt[n]{b})^n$:

$$(\sqrt[n]{a} \cdot \sqrt[n]{b})^n = (\sqrt[n]{a})^n (\sqrt[n]{b})^n = ab.$$

Since $(\sqrt[n]{a} \cdot \sqrt[n]{b})^n = ab$, then $\sqrt[n]{a} \cdot \sqrt[n]{b}$ is the nth root of ab. That is, $\sqrt[n]{ab} = \sqrt[n]{a} \cdot \sqrt[n]{b}$. In a similar way we could show that $\sqrt[n]{\frac{a}{b}} = \frac{\sqrt[n]{a}}{\sqrt[n]{b}}$.

EXAMPLE 4

Simplify $\dfrac{\sqrt{20}}{\sqrt{5}}$.

$$\frac{\sqrt{20}}{\sqrt{5}} = \sqrt{\frac{20}{5}} = \sqrt{4} = 2.$$

EXAMPLE 5

Given the formula

$$d = \sqrt{(x_2 - x_1)^2 + (y_2 - y_1)^2},$$

find d *if* $x_1 = 1$, $x_2 = -7$, $y_1 = 2$, *and* $y_2 = 8$.

$$\begin{aligned} d &= \sqrt{(x_2 - x_1)^2 + (y_2 - y_1)^2} \\ &= \sqrt{(-7-1)^2 + (8-2)^2} \\ &= \sqrt{(-8)^2 + (6)^2} \\ &= \sqrt{64 + 36} = \sqrt{100} \\ d &= 10. \end{aligned}$$

This formula, called the *distance formula*, is used in analytic geometry.

In Example 5, note that $\sqrt{64+36} = \sqrt{100} = 10$. ***Don't write*** $\sqrt{64+36}$ *as* $\sqrt{64} + \sqrt{36} = 8 + 6 = 14$. *In general,* $\sqrt{a+b}$ *is* ***not*** $\sqrt{a} + \sqrt{b}$.
Moral: Don't try fancy tricks of your own. Stick to basics.

Completion Set 3-2

In Problems **1–10**, *fill in the blanks.*

1. Since $2^4 = 16$, we say that 2 is a ______________ root of 16.

2. Although $2^4 = 16$ and $(-2)^4 = 16$, by the symbol $\sqrt[4]{16}$ we mean the number ____.

3. The principal square root of 16 is ____.

4. Since $\sqrt{7}$ is a number whose square is 7, we have $(\sqrt{7})^2 =$ ____.

5. In the radical $\sqrt[4]{16}$, the 4 is called the ____________.

6. In the radical $\sqrt[5]{32}$, the 32 is called the ____________________.

7. Any root of 0 is equal to ____.

8. $\sqrt{28} = \sqrt{4 \cdot 7} = \sqrt{\_\_\_} \cdot \sqrt{7} = \_\_\_ \cdot \sqrt{7}$.

9. $\dfrac{\sqrt{48}}{\sqrt{3}} = \sqrt{\dfrac{48}{3}} = \sqrt{\_\_\_} = \_\_\_$.

10. $\sqrt{\dfrac{3}{25}} = \dfrac{\sqrt{3}}{\sqrt{\_\_\_}} = \dfrac{\sqrt{3}}{\_\_\_}$.

Problem Set 3-2

In Problems **1–32**, *compute the numbers.*

1. $\sqrt{49}$ **2.** $\sqrt[3]{125}$ **3.** $\sqrt[3]{8}$

4. $\sqrt{81}$ **5.** $\sqrt{36}$ **6.** $\sqrt{100}$

7. $\sqrt[3]{-27}$ **8.** $\sqrt[3]{-8}$ **9.** $\sqrt[3]{-64}$

10. $\sqrt[4]{1}$ **11.** $\sqrt[4]{16}$ **12.** $\sqrt[5]{-1}$

13. $\sqrt[5]{0}$ **14.** $\sqrt[3]{64}$ **15.** $\sqrt[3]{-125}$

16. $\sqrt[5]{32}$ **17.** $\sqrt[6]{64}$ **18.** $\sqrt{12 \cdot 3}$

19. $\sqrt{.04}$ **20.** $\sqrt{.25}$ **21.** $\sqrt{\dfrac{1}{16}}$

22. $\sqrt{\dfrac{1}{100}}$ **23.** $-\sqrt{25}$ **24.** $-\sqrt[3]{-1}$

25. $\sqrt{81} - \sqrt[3]{-8}$ **26.** $\sqrt{64} - \sqrt[3]{-8}$

27. $\dfrac{\sqrt{64} + \sqrt[3]{-64}}{\sqrt{81} + \sqrt[4]{81}}$ **28.** $\sqrt[3]{|-8|}$

29. $\sqrt{5} \cdot \sqrt{5}$

30. $\sqrt[3]{7} \cdot \sqrt[3]{7} \cdot \sqrt[3]{7}$

31. $(\sqrt[4]{4})^4$

32. $(\sqrt{3})^2$

In Problems **33–59**, *simplify the numbers.*

33. $\sqrt{50}$

34. $\sqrt{75}$

35. $\sqrt{12}$

36. $\sqrt{32}$

37. $\sqrt{8}$

38. $\sqrt{18}$

39. $\sqrt{54}$

40. $\sqrt[3]{24}$

41. $\sqrt{48}$

42. $\sqrt[4]{162}$

43. $\sqrt[5]{64}$

44. $\sqrt[3]{-54}$

45. $\sqrt[3]{-500}$

46. $\sqrt{\frac{3}{2}} \cdot \sqrt{6}$

47. $\sqrt{\frac{3}{5}} \cdot \sqrt{\frac{1}{15}}$

48. $\sqrt{\frac{5}{4}}$

49. $\sqrt{\frac{14}{9}}$

50. $\sqrt{\frac{2}{25}}$

51. $\sqrt[3]{\frac{10}{27}}$

52. $\frac{\sqrt{90}}{\sqrt{10}}$

53. $\frac{\sqrt{50}}{\sqrt{2}}$

54. $\frac{\sqrt[4]{64}}{\sqrt[4]{4}}$

55. $\frac{\sqrt[3]{-2}}{\sqrt[3]{16}}$

56. $\frac{\sqrt[5]{-4}}{\sqrt[5]{-128}}$

57. $\sqrt{8} \cdot \sqrt{2}$

58. $\sqrt{27} \cdot \sqrt{3}$

59. $\sqrt[3]{4} \cdot \sqrt[3]{16}$

In Problems **60** *and* **61**, *use the distance formula* $d = \sqrt{(x_2 - x_1)^2 + (y_2 - y_1)^2}$, *as given in Example* 5, *to find* d *with the given information.*

60. $x_1 = 10, y_1 = 2, x_2 = 7, y_2 = -2$

61. $x_1 = -1, y_1 = 1, x_2 = -6, y_2 = 13$

In the right triangle shown in Fig. 3-1, *the length of the hypotenuse* c *is given by*

$$c = \sqrt{a^2 + b^2}.$$

In Problems **62** *and* **63**, *find* c *from the given information.*

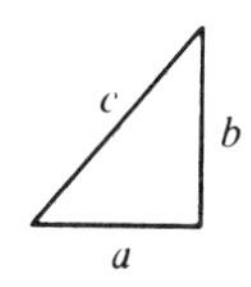

FIG. 3-1

62. $a = 5, b = 12$

63. $a = 7, b = 24$

The expression $\frac{-b + \sqrt{b^2 - 4ac}}{2a}$ *is used in solving a certain type of equation. In*

Problems **64–67**, *evaluate this expression for the given values of a, b, and c.*

64. $a = 4, b = 5, c = 0$

65. $a = 1, b = 2, c = -15$

66. $a = 1, b = -9, c = 14$

67. $a = 4, b = -12, c = 9$

68. Compute y' (read "y prime") where

$$y' = \frac{\dfrac{x + 5}{2\sqrt{x + 1}} - \sqrt{x + 1}}{(x + 5)^2}$$

and $x = 3$.

69. In statistics, the standard deviation of the numbers x_1 and x_2 is given by

$$\sqrt{\frac{(\bar{x} - x_1)^2 + (\bar{x} - x_2)^2}{2}},$$

where $\bar{x} = \dfrac{x_1 + x_2}{2}$. Find the standard deviation of $x_1 = 0$ and $x_2 = 4$.

3-3 REVIEW

IMPORTANT TERMS AND SYMBOLS

base *(p. 59)*

exponent *(p. 59)*

principal root *(p. 68)*

nth root *(p. 68)*

index *(p. 68)*

radical *(p. 68)*

$\sqrt[n]{a}$ *(p. 68)*

radicand *(p. 68)*

REVIEW PROBLEMS

In Problems **1–24**, *determine whether each statement is true or false.*

1. $a^3a^2 = a^6$

2. $\left(\dfrac{1}{x}\right)^3 = \dfrac{1}{x^3}$

3. $2^2 + 3^2 = 5^2$

4. $(-a)^4 = a^4$

5. $3 + 3 + 3 = 3^3$

6. $(x^2y)^2 = x^4y$

7. $(2x^2)^3 = 2^3x^5$

8. $\dfrac{x^5}{x^3} = x^2$

9. $(-y)^3 = -y^3$

10. $2^2 \cdot 2^4 = 4^8$

11. $3x = xxx$

12. $(-3)^3 = -3^3$

13. $(-2)^2 = -2^2$

14. $(2 + 5)^2 = 2^2 + 5^2$

15. $a^3a^7 = a^7a^3$

16. $(x^2)^y = x^{2y}$

17. $(2x)^2 = 2x^2$

18. $\sqrt{16} = -4$

19. $\sqrt{5}\,\sqrt{5} = \sqrt{25}$

20. $\sqrt{4 \cdot 2} = 4\sqrt{2}$

21. $\sqrt[3]{-\frac{1}{8}} = -\frac{1}{2}$

22. $\sqrt[3]{0} = 0$

23. $\sqrt{1} + \sqrt{1} = \sqrt{2}$

24. $(\sqrt[3]{-1})^2 = 1$

In Problems **25–56**, *simplify*.

25. $x^6x^4x^3$

26. $\dfrac{x^2x^{20}}{y^5y^2}$

27. $\dfrac{(x^2)^5}{(y^5)^{10}}$

28. $(5x^2)(2x^3)$

29. $(-2xy^4)^5$

30. $\left(\dfrac{2xy^3}{z^2}\right)^2$

31. $\dfrac{(x^3)^6}{x(x^3)}$

32. $\dfrac{(xy^2)^2}{(t^2w)^3}$

33. $(-x)^2(-x)^3$

34. $\dfrac{(x^2)^3(x^4)^5}{(x^3)^8}$

35. $\dfrac{(5^7)^2}{(5^4)^4}$

36. $\dfrac{2^6 2^{11}}{(2^5)^3}$

37. $-\sqrt{36} + \sqrt{81}$

38. $\dfrac{-\sqrt{144}}{|-12|}$

39. $|-2| - \sqrt[4]{16}$

40. $(\sqrt[3]{2})^3$

41. $\sqrt{13} \cdot \sqrt{13}$

42. $\dfrac{(-1)^2 - 3\sqrt{9}}{|-4| + \sqrt{16}}$

43. $\dfrac{1}{3}\sqrt[3]{\dfrac{1}{27}}$

44. $\dfrac{\sqrt[5]{-32}}{\sqrt{9}}$

45. $\sqrt{.01} + \sqrt{.0025}$

46. $\dfrac{\sqrt{4} - \sqrt[3]{8}}{\sqrt{6}}$

47. $\sqrt{45}$

48. $\sqrt{72}$

49. $\sqrt[4]{810}$

50. $\sqrt[3]{40}$

51. $\sqrt{\dfrac{7}{16}}$

52. $\sqrt{\dfrac{10}{9}}$

53. $\dfrac{\sqrt[3]{24}}{\sqrt[3]{3}}$

54. $\dfrac{\sqrt{72}}{\sqrt{2}}$

55. $\sqrt{18} \cdot \sqrt{2}$

56. $\sqrt[3]{8} \cdot \sqrt[3]{16}$

57. The area of the right triangle in Fig. 3-2 is given by

$$\frac{1}{2}a\sqrt{c^2 - a^2}.$$

If $a = 3$ and $c = 5$, find the area of the triangle.

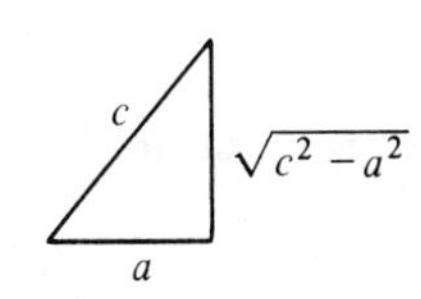

FIG. 3-2

CHAPTER 4

Basic Equations

4-1 SOLVING EQUATIONS

One important use of mathematics is to solve equations. *An equation is a statement that two expressions are equal.* For example,

$$3x - 1 = 14 \text{ is an equation.}$$

Here we say that

$3x - 1$ is the *left side*,
14 is the *right side*,
$3x$, -1, and 14 are *terms*, and
x is a *variable* or *unknown*.

Variables are letters, such as x, that stand for unspecified numbers. There may be values of x that make the above equation a true statement. Other values may make it false. For example, if x is 0, then $3x - 1 = 14$ becomes

$$3(0) - 1 = 14,$$

or simply

$$-1 = 14, \text{ which is a } \textbf{false} \text{ statement}.$$

We want to find values of x for which this equation is **true**. These values are called **solutions** or **roots**. *Solving* the equation means to find all of its solutions.

To solve an equation, we apply certain rules to it until the value of x is obvious. These rules are

I.	You can add (or subtract) the same number to (or from) **both** sides of an equation.
II.	You can multiply (or divide) **both** sides of an equation by the same number. *But that number must **not** be* 0.

Let's use these rules to solve

$$3x - 1 = 14.$$

First, we look for the term involving x:

This term involves x $\longrightarrow$ $\textcircled{3x} - 1 = 14.$

Next, we use the rules to get this term, $3x$, by itself. To get rid of the 1 that's *subtracted* from $3x$, we *add* 1 to ***both*** sides (Rule I).

$$3x - 1 + 1 = 14 + 1$$
$$3x + 0 = 15$$
$$3x = 15. \qquad (4\text{-}1)$$

Now $3x$ is by itself. But we want to find x. Since x is *multiplied* by 3, we'll *divide **both*** sides of Eq. (4-1) by 3 (Rule II).

$$\frac{3x}{3} = \frac{15}{3}.$$

Finally, we simplify both sides.

$$\frac{\overset{1}{\cancel{3}}x}{\underset{1}{\cancel{3}}} = \frac{\overset{5}{\cancel{15}}}{\underset{1}{\cancel{3}}}$$

$$\boxed{x = 5.}$$

Thus the solution of $3x - 1 = 14$ is 5. Let's check it to be sure. In the original equation we replace x by 5:

Check:

$$3x - 1 = 14$$
$$3(5) - 1 = 14$$
$$14 = 14 \checkmark$$

We wrote our solution as $x = 5$. But some people write it as a set, $\{5\}$. They call $\{5\}$ the *solution set*.

By the way, it's a good idea to put a box (or circle) around a solution as we did. It shows at a glance the result of all your hard work.

EXAMPLE 1

Solve $-2x = 8$.

Here the term involving x is by itself. Since $-2x$ is $(-2)x$, then to get x by itself we divide both sides of the equation by -2.

$$-2x = 8$$
$$(-2)x = 8$$
$$\frac{\overset{1}{\cancel{(-2)}}x}{\underset{1}{\cancel{(-2)}}} = \frac{8}{-2}$$
$$\boxed{x = -4.}$$

Be sure you do the same thing to ***both*** *sides of an equation. If*

$$3x = 9,$$

don't write $x = 9 - 3$. *Here the left side was* ***divided*** *by 3, but on the right side 3 was* ***subtracted*** *from 9. You should divide both sides of* $3x = 9$ *by 3:*

$$\frac{3x}{3} = \frac{9}{3}$$
$$\boxed{x = 3.}$$

EXAMPLE 2

Solve $-x = 7$.

Here we just multiply both sides by -1.

$$-x = 7$$
$$(-1)(-x) = (-1)(7)$$
$$\boxed{x = -7.}$$

EXAMPLE 3

Solve $8 + 4y = 10$.

Here the unknown is y.

$$8 + 4y = 10$$
$$8 + 4y - 8 = 10 - 8 \qquad \text{[subtracting 8 from both sides].}$$

On the left side the sum of the terms 8 and -8 is 0.

$$4y = 2$$
$$\frac{\overset{1}{\cancel{4}}y}{\underset{1}{\cancel{4}}} = \frac{2}{4} \qquad \text{[dividing both sides by 4]}$$
$$\boxed{y = \frac{1}{2}.}$$

EXAMPLE 4

Solve $2 = -6 - 3t$.

Here the unknown is t. The term involving t is $-3t$.

$$2 = -6 - 3t$$
$$2 + 6 = -6 - 3t + 6 \qquad \text{[adding 6 to both sides]}$$
$$8 = -3t$$
$$\frac{8}{-3} = \frac{(\cancel{-3})t}{(\cancel{-3})} \qquad \text{[dividing both sides by } -3\text{]}$$
$$\frac{8}{-3} = t$$

or

$$\boxed{t = -\frac{8}{3}.}$$

Completion Set 4-1

In Problems **1–6**, *mark the statement either* T (= *true*) *or* F (= *false*).

1. If $4x = 5$, then $x = 5$. ____

2. If $3x = 7$, then $x = 7 - 3$. ____

3. If $-x = -3$, then $x = 3$. ____

4. If $3x - 7 = 2$, then $3x = \frac{2}{7}$. ____

5. If $2x + 1 = 5$, then $2x = 6$. ____

6. If $S + 2 = 0$, then $S = 2$. ____

In Problems **7–12**, *fill in the blanks.*

7. In the equation $3y + 7 = 2$, the unknown is ____.

8. In the equation $6 + 5z = 4$, the term involving the unknown is ______.

9. If $4x - 5 = 15$, then $4x =$ ______, and so $x =$ ____.

10. If $4 - 2S = 8$, then $-2S =$ ____, and so $S =$ ______.

11. If $-t - 5 = 0$, then $-t =$ ____, and so $t =$ ______.

12. If $-3y + 1 = -5$, then $-3y =$ ______, and so $y =$ ____.

Problem Set 4-1

Solve the following equations. Check your solutions.

1. $x + 3 = 0$

2. $x - 6 = 0$

3. $x + 4 = 7$

4. $3 + x = 7$

5. $y - 6 = 8$

6. $y - 1 = -3$

7. $1 + x = -5$

8. $x + 10 = -4$

9. $6x = 18$

10. $3x = 14$

11. $-4t = 12$

12. $7t = -21$

13. $-x = -15$

14. $-4x = 8$

15. $2x - 8 = 12$

16. $3x - 8 = 7$

17. $6x + 7 = 7$

18. $10x + 15 = 5$

19. $5 + 3r = -1$

20. $-9 = 6r - 1$

21. $2 = 4t - 5$

22. $-8 + 7t = 13$

23. $13 - 2x = 19$

24. $8 - x = 4$

25. $4 - y = 16$

26. $1 - 3y = -8$

27. $-2x - 3 = -4$

28. $-5x - 6 = -7$

29. $-5 = 7 - 8u$

30. $-1 = 4 + 2u$

4-2 SOME EQUATIONS WITH FRACTIONS

Let's solve some equations that have fractions.

EXAMPLE 1

Solve $\frac{x}{4} = -3$.

It's a good idea to clear this equation of fractions. Let's multiply both sides by the denominator 4.

$$\frac{x}{4} = -3$$

$$4 \cdot \frac{x}{4} = 4(-3)$$

$$\cancel{4} \cdot \frac{x}{\cancel{4}} = 4(-3)$$

$$\boxed{x = -12.}$$

EXAMPLE 2

Solve $3x + 6 = \frac{5}{2}$.

$$3x + 6 = \frac{5}{2}$$

$$2(3x + 6) = 2 \cdot \frac{5}{2} \quad \text{[multiplying both sides by denominator 2]}$$

$$2 \cdot 3x + 2 \cdot 6 = \cancel{2} \cdot \frac{5}{\cancel{2}} \quad \text{[distributive law]}$$

$$6x + 12 = 5$$

$$6x + 12 - 12 = 5 - 12 \quad \text{[subtracting 12 from both sides]}$$

$$6x = -7$$

$$\frac{\cancel{6}x}{\cancel{6}} = \frac{-7}{6} \quad \text{[dividing both sides by 6]}$$

$$\boxed{x = -\frac{7}{6}.}$$

When you multiply both sides of the equation in Example 2 by 2, you ***do not*** *get* $2 \cdot 3x + 6$ *on the left side. What you do get is* $2(3x + 6)$. *The parentheses mean that you multiply the* ***entire*** *left side by 2.*

EXAMPLE 3

Solve $-\frac{3}{8}u + 1 = -2.$

We could first clear of fractions by multiplying both sides by 8. But instead, let's get the term involving u by itself.

$$-\frac{3}{8}u + 1 = -2$$

$$-\frac{3}{8}u + 1 - 1 = -2 - 1 \qquad \text{[subtracting 1 from both sides]}$$

$$-\frac{3}{8}u = -3$$

$$8(-\frac{3}{8}u) = 8(-3) \qquad \text{[multiplying both sides by 8]}$$

$$-3u = -24 \qquad [\text{since } 8(-\frac{3}{8}u) = -(\cancel{8} \cdot \frac{3}{\cancel{8}})u = -3u]$$

$$\frac{(\cancel{-3})u}{\cancel{-3}} = \frac{-24}{-3} \qquad \text{[dividing both sides by } -3]$$

$$\boxed{u = 8.}$$

EXAMPLE 4

Solve $\frac{2x}{3} + \frac{1}{2} = 5.$

When two or more terms of an equation have fractions, you may multiply both sides by the least common denominator (L.C.D.). Here the L.C.D. is 6.

$$\frac{2x}{3} + \frac{1}{2} = 5$$

$$6\left(\frac{2x}{3} + \frac{1}{2}\right) = 6 \cdot 5 \qquad \text{[multiplying both sides by L.C.D.]}$$

$$6 \cdot \frac{2x}{3} + 6 \cdot \frac{1}{2} = 30 \qquad \text{[distributive law]}$$

$$\overset{2}{\cancel{6}} \cdot \frac{2x}{\cancel{3}} + \overset{3}{\cancel{6}} \cdot \frac{1}{\cancel{2}} = 30$$

$$4x + 3 = 30$$

$$4x + 3 - 3 = 30 - 3 \quad \text{[subtracting 3 from both sides]}$$

$$4x = 27$$

$$\frac{\not{4}x}{\not{4}} = \frac{27}{4} \quad \text{[dividing both sides by 4]}$$

$$\boxed{x = \frac{27}{4}.}$$

EXAMPLE 5

Solve $\dfrac{1 + 2x}{2} = \dfrac{1}{4}$.

Here the L.C.D. is 4.

$$\frac{1 + 2x}{2} = \frac{1}{4}$$

$$\overset{2}{\not{4}} \cdot \frac{1 + 2x}{\not{2}} = 4 \cdot \frac{1}{4} \quad \text{[multiplying both sides by L.C.D.]}$$

$$2(1 + 2x) = 1 \quad \text{[simplifying]}$$

$$2 + 4x = 1 \quad \text{[distributive law]}$$

$$4x = -1 \quad \text{[subtracting 2 from both sides]}$$

$$\boxed{x = -\frac{1}{4}.} \quad \text{[dividing both sides by 4]}$$

Completion Set 4-2

In Problems **1–8**, *fill in the blanks.*

1. If $\dfrac{x}{5} = 7$, then $x =$ ____ $\cdot 7$, and so $x =$ ______.

2. If $-\dfrac{x}{3} = 2$, then $-x =$ ____ $\cdot 2$, and so $x =$ ______.

3. If $\dfrac{4}{5}x = 3$, then $4x =$ ____ $\cdot 3$, and so $x =$ ______.

4. If $-\dfrac{2x}{3} = 5$, then $-2x =$ ____ $\cdot 5$, and so $x =$ _______.

5. If $-\frac{1}{4}x + 3 = 7$, then $-\frac{1}{4}x =$ ____, $-x =$ ______, and so $x =$ ________.

6. If both sides of $x + 2 = \frac{1}{5}$ are multiplied by 5, then $5x +$ ______ $= 1$.

7. To clear $\frac{x}{3} - \frac{5}{4} = 10$ of fractions, multiply both sides by the L.C.D., which is ______.

8. To clear $\frac{5x}{2} + 6 = \frac{7}{8}$ of fractions, multiply both sides by the L.C.D., which is ____.

Problem Set 4-2

In Problems **1–38**, *solve the equations.*

1. $\frac{x}{8} = 3$ **2.** $\frac{x}{5} = -6$ **3.** $\frac{x}{7} = 0$

4. $\frac{y}{3} = 1$ **5.** $\frac{3}{4}x = 2$ **6.** $\frac{x}{-7} = 2$

7. $-\frac{y}{6} = 3$ **8.** $-\frac{2y}{3} = \frac{5}{2}$ **9.** $\frac{x}{3} = \frac{7}{2}$

10. $\frac{1}{8} = \frac{y}{6}$ **11.** $\frac{2}{5} = \frac{y}{12}$ **12.** $y + \frac{1}{2} = \frac{1}{2}$

13. $\frac{2x}{5} = -\frac{3}{2}$ **14.** $\frac{9}{8}x = \frac{3}{2}$ **15.** $2t = \frac{4}{9}$

16. $\frac{7}{6}t = 0$ **17.** $4x + 1 = \frac{7}{3}$ **18.** $-\frac{1}{3}x = \frac{1}{3}$

19. $3x - \frac{1}{5} = 4$ **20.** $3x - \frac{9}{4} = 2$ **21.** $\frac{u}{4} - \frac{4}{3} = 8$

22. $\frac{u}{2} + 4 = \frac{3}{7}$ **23.** $\frac{2w}{3} - \frac{4}{5} = 0$ **24.** $1 - \frac{3w}{5} = 0$

25. $2 = 7 - \frac{x}{3}$ **26.** $4 = 6 + \frac{4}{3}x$ **27.** $\frac{5}{3}S - 6 = -1$

28. $\frac{2S}{15} + 3 = -5$ **29.** $\frac{3x}{7} + 5 = \frac{2}{3}$ **30.** $8 = \frac{6}{5}x - \frac{1}{2}$

31. $3 - \frac{2}{5}x = 2$ **32.** $4 - \frac{9x}{8} = -6$ **33.** $\frac{13}{9} = 1 - \frac{4}{3}x$

34. $-\frac{1}{12} = -2 - \frac{3}{4}x$ **35.** $\frac{x-3}{4} = 5$ **36.** $\frac{1+x}{6} = \frac{1}{3}$

37. $\frac{3}{4} = \frac{2-x}{3}$ **38.** $\frac{3}{8} = \frac{2x-2}{6}$

39. A gas has a volume of 200 cubic centimeters (cm^3) at a temperature of 273°K (the Kelvin scale). Under certain conditions, its volume (in cm^3) at 373°K is given by v, where

$$\frac{v}{200} = \frac{373}{273}.$$

Find v. You may round off your answer to the nearest cm^3.

4-3 FORMULAS, WORD PROBLEMS, PERCENTAGES

Formulas are equations that involve literal numbers. Let's look at some formulas.

EXAMPLE 1

The formula for the circumference of a circle is

$$C = 2\pi r,$$

where C is the circumference and r is the radius. If the circumference is 16 centimeters (cm), find the radius.

Letting $C = 16$, we have

$$16 = 2\pi r.$$

But $2\pi r$ is just $(2\pi)r$. So we divide both sides by 2π.

$$\frac{16}{2\pi} = \frac{2\pi r}{2\pi}$$

$$\boxed{\frac{8}{\pi} = r.}$$

Thus the radius is $\frac{8}{\pi}$ cm.

EXAMPLE 2

The formula for the perimeter[†] *P of a rectangle is $P = 2l + 2w$. Here l is the length and w is the width. Solve for w.*

[†]The perimeter of a rectangle is the total length of all its sides.

Since we want to solve for w, think of w as the unknown.

$$P = 2l + 2w$$

$$P - 2l = 2l + 2w - 2l \quad \text{[subtracting } 2l \text{ from both sides].}$$

On the right side the sum of the terms $2l$ and $-2l$ is 0.

$$P - 2l = 2w$$

$$\frac{P - 2l}{2} = \frac{2w}{2} \quad \text{[dividing both sides by 2]}$$

$$\frac{P - 2l}{2} = w.$$

Our answer is

$$\boxed{w = \frac{P - 2l}{2}.}$$

*In the **answer** above, **don't** cancel the 2's. That is,* $w \neq \frac{P - \cancel{2}l}{\cancel{2}}$. *In a fraction we can cancel only **factors** of the numerator with **factors** of the denominator. Now, **if** it turned out (which it doesn't) that w were equal to* $\frac{2(P - 2l)}{2}$, *then we could cancel the factor 2 of the numerator with the factor 2 of the denominator:* $\frac{\cancel{2}(P - 2l)}{\cancel{2}}$.

EXAMPLE 3

Solve $A = \frac{1}{2}bh$ *for b. This is the formula for the area A of a triangle with base b and height h.*

$$A = \frac{1}{2}bh$$

$$2A = 2 \cdot \frac{1}{2}bh \quad \text{[multiplying both sides by 2]}$$

$$2A = bh$$

$$\frac{2A}{h} = \frac{b\cancel{h}}{\cancel{h}} \quad \text{[dividing both sides by } h\text{]}$$

$$\frac{2A}{h} = b.$$

Thus,

$$\boxed{b = \frac{2A}{h}.}$$

Now let's turn to "word" problems. With word problems, an equation is not handed to you. You have to set it up by translating verbal statements into an equation. This is called *mathematical modeling*. After modeling, solve the equation.

It takes a lot of practice to get a good feeling for word problems. Read the following examples carefully.

EXAMPLE 4

A student installed an 8-*track tape player in his car. He has* \$34 *and wants to spend it all on tapes of his favorite country group. If tapes sell for* \$4.25 *each, how many tapes can he buy*?

You might be able to solve this problem in your head. But let's model it. First, read the problem again so that you understand it. Then say to yourself, "What is the problem asking?"

How many tapes can the student buy?

Next, choose a letter to represent a quantity that you want to find. Now, write down exactly what this letter stands for. We'll let

$$n = \text{number of tapes the student can buy}.$$

We chose the letter n as the unknown because it reminds us of what we want to find.

Now let's set up an equation. Since each tape costs \$4.25, then n tapes cost $4.25n$.

$$4.25n = \text{total cost of } n \text{ tapes}.$$

But the total cost of these n tapes equals the amount he will spend, \$34. So the equation to solve is

$$4.25n = 34.$$

Solving, we obtain

$$\frac{4.25n}{4.25} = \frac{34}{4.25}$$

$$n = \frac{34}{4.25} = \boxed{8.}$$

Thus, the student can buy 8 tapes.

EXAMPLE 5

The Ipsy-Wipsy Company makes the famous Ipsy-Wipsy electric fly swatter. The company has ***fixed costs*** *that total* \$20,000 *per year. This includes rent, light, heat, and the boss's salary. There is also the manufacturing cost (labor, material) of* \$3.50 *for each electric fly swatter produced. Next year the company wants the* **total cost** *of making fly swatters to be* \$48,000. *How many fly swatters should be made?*

Did you reread the problem? If not, do it now. Here the question is "How many fly swatters should be made?" Let

$$n = \text{number of fly swatters that should be made}\,.$$

The total cost is made up of two parts:

$$\text{Total cost} = (\text{fixed costs}) + \begin{pmatrix}\text{manufacturing cost}\\ \text{for } n \text{ fly swatters}\end{pmatrix}.$$

But we know

$$\text{total cost} = 48{,}000$$

and

$$\text{fixed costs} = 20{,}000.$$

Since the manufacturing cost for each fly swatter is \$3.50, then

$$\begin{pmatrix}\text{manufacturing cost}\\ \text{for } n \text{ fly swatters}\end{pmatrix} = 3.50n.$$

The equation to solve is

$$48{,}000 = 20{,}000 + 3.50n.$$

Solving, we obtain

$$\begin{aligned} 48{,}000 - 20{,}000 &= 20{,}000 + 3.50n - 20{,}000 \\ 28{,}000 &= 3.50n \\ \frac{28{,}000}{3.50} &= \frac{3.50n}{3.50} \\ \boxed{8000 = n.} \end{aligned}$$

Thus, 8000 fly swatters should be made next year.

In Section 1-5, we worked percentage problems. For example, to find 20%

of 50, we first write 20% as a decimal, .20. Then we multiply 50 by .20:

$$(.20)\cdot 50 = 10.$$

Thus, 20% of 50 is 10.

Notice the connection between the phrase "20% *of* 50 *is* 10" and the equation $(.20)\cdot 50 = 10$:

$$\begin{array}{ccccc} 20\% & \text{of} & 50 & \text{is} & 10 \\ \updownarrow & \updownarrow & & \updownarrow & \\ (.20) & \cdot & 50 & = & 10 \end{array}$$

The 20% is replaced by its decimal form. The word *of* is replaced by the multiplication symbol. And the word *is* is replaced by "=".

This connection will help you set up different types of percentage problems.

EXAMPLE 6

70 is 20% of what number?

Let n be the number we're looking for. The decimal form of 20% is .20.

$$\begin{array}{ccccc} 70 & \text{is} & 20\% & \text{of} & \underbrace{\text{what number}} \\ \downarrow & \downarrow & \downarrow & \downarrow & \downarrow \\ 70 & = & (.20) & \cdot & n \end{array}$$

The equation to solve is $70 = (.20)\cdot n$ or

$$.20n = 70$$

$$\frac{.20n}{.20} = \frac{70}{.20}$$

$$\boxed{n = 350.}$$

Thus, 70 is 20% of 350.

EXAMPLE 7

At Sheepskin College there are 2100 freshmen, and 1365 of them take a math course. What percentage of the freshmen class is taking math?

Here we want to find a percent. Let P be the percentage of the freshmen class taking math. Since P will occur in an equation, P will be expressed in decimal form.

$$P = \begin{cases} \text{percentage of freshmen class} \\ \text{taking math (expressed as a decimal)} \end{cases}$$

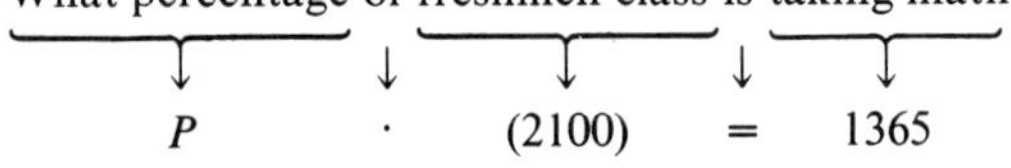

$$P \quad \cdot \quad (2100) \quad = \quad 1365$$

The equation we want to solve is $P \cdot (2100) = 1365$ or

$$2100P = 1365$$

$$\frac{2100P}{2100} = \frac{1365}{2100}$$

$$\boxed{P = .65}$$

But $P = .65$ is the decimal form of percentage. So, 65% of the freshmen take math.

Completion Set 4-3

Fill in the blanks.

1. To solve $ab = c$ for b, divide both sides of the equation by ____.
2. To find how many widgets (w) costing \$3.50 each can be bought for \$84, you may solve $(3.50) \cdot$ ____ = ______.
3. 20 is 12% of some number n. To find n you may solve ________ $\cdot n =$ ______.
4. 15 is some percentage of 90. If P is the decimal form of this percent, to find P you may solve the equation $P \cdot$ ______ = ______.

Problem Set 4-3

1. Use the formula $P = 2l + 2w$ to find the width w of a rectangle whose perimeter P is 960 meters (m) and whose length l is 360 m.
2. Use the formula $A = \frac{1}{2}bh$ to find the height h of a triangle whose area A is 75 square centimeters (cm^2) and whose base b is 15 cm.
3. Solve $C = 2\pi r$ for r.

4. $F = ma$ is a formula found in physics. It is one of Newton's laws. Here F is force, m is mass, and a is acceleration. Solve for m.

5. The formula for simple interest I is $I = Prt$. Here P is the principal, r is the rate of interest, and t is time. Solve for r. *Hint:* First write Prt as $(Pt)r$.

6. Solve $I = Prt$ for P.

7. The amount A to which a principal P grows at a simple rate of interest r after t years is given by $A = P + Prt$. Solve for t.

8. Solve $A = P + Prt$ for r.

9. The total surface area A of a right circular cylinder (think of a can of soda with a top and bottom) is given by $A = 2\pi r^2 + 2\pi rh$. Here r is the radius of the cylinder and h is the height. Solve for h.

10. The beam in Fig. 4-1 with forces F_1 and F_2 acting on it will balance on the pivot if

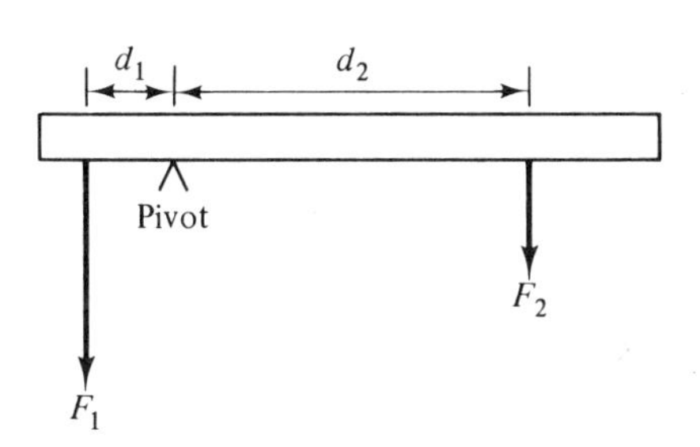

FIG. 4-1

$F_1d_1 = F_2d_2$. Solve this equation for F_1.

11. In Problem 10, suppose $F_1 = 20$ lb, $d_1 = 2$ ft, and $F_2 = 8$ lb. Find d_2 (in ft).

12. Use the formula in Problem 10 to find the force F (in lb) in Fig. 4-2 so that the beam

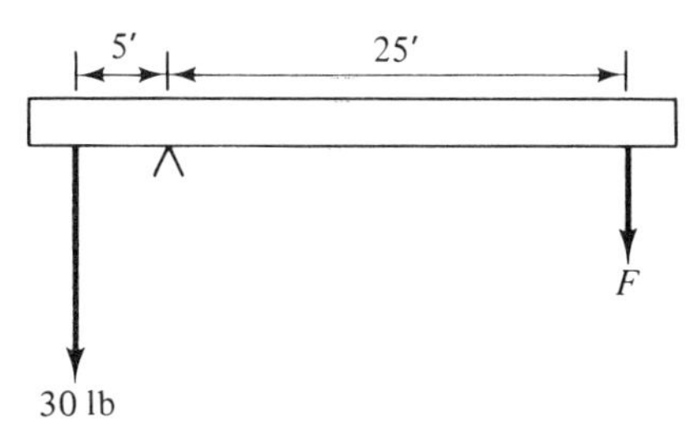

FIG. 4-2

balances.

13. At time t, the height s of an object thrown up from the ground with a velocity v_0 is given by

$$s = v_0t - 16t^2.$$

If an object is to have a height of 16 (ft) in 2 (sec), find v_0 (in ft per sec).

14. Boyle's law for gases is given by $V_1P_1 = V_2P_2$. Here the V's are volumes and the P's are pressures. If $V_1 = 100$ cubic centimeters (cm^3), $P_1 = 756$ millimeters (mm) of mercury, and $P_2 = 720$ mm of mercury, find V_2 (in cubic centimeters).

In Problems **15–22**, *model each of the problems and solve.*

15. A city set aside $26,000 to purchase parking meters. If each meter costs $65, how many meters can be bought?

16. Yankee Doodle Dummy Co. makes manikins. The manufacturing cost for each manikin is $45. The company has fixed costs of $4150 per month. How many manikins can it make next month for a total cost of $10,000?

17. The Piddle Paddle Co. has fixed costs of $50,000 per year. The manufacturing cost for each paddle is $5. How many paddles can the company make next year for a total cost of $125,000?

18. A farmer can harvest 45 bushels of corn per acre. How many acres should he plant so that he can harvest a total of 3600 bushels?

19. A hospital has only private and semiprivate rooms. It has 90 private rooms (one bed to a room). However, the total number of beds in the hospital is 318. How many semiprivate rooms (two beds to a room) does it have?

20. A certain machine can perform 34 chemical analyses per day. But a lab technician can perform only 7. A laboratory must make 110 analyses tomorrow, and it has only two machines. How many technicians will be needed to complete the job?

21. The *current ratio* of a company is the value of its current assets divided by its current liabilities. The J. P. Jellybelly Jellybean Co. has a current ratio of 2.5 and current liabilities of $80,000. What are its current assets?

22. The IQ (intelligence quotient) of a person is found by dividing his mental age by his chronological age and then multiplying that result by 100. For example, a person with a mental age of 11 and a chronological age of 10 has an IQ of $\frac{11}{10} \cdot 100 = 110$. Find the mental age of a person with chronological age 12 if his IQ is 125.

23. 90 is 30% of what number?

24. 72 is 45% of what number?

25. 75 is 12% of what number?

26. 34 is 85% of what number?

27. What percentage of 200 is 8?

28. What percentage of 60 is 9?

29. What percentage of 50 is 55?

30. What percentage of 1400 is 70?

31. A group of people were polled by Snoopy Surveys and 20%, or 700, of them favored a new product over the best selling brand. How many people were polled?

32. Approximately 21% of the air we breathe is oxygen. To the nearest milliliter, how many milliliters of air contain one milliliter of oxygen?

33. A team of biologists studying ABO blood groups tested 2000 people. They found 850 had antigen A, 780 had antigen B, and 370 had no antigen. What percentage of the people tested had antigen A? (Antigens are foreign substances involved with the production of antibodies.)

34. On a square centimeter of a leaf of a certain corn plant, there are 8928 stomata (small pores) on the upper surface and 15,872 on the under surface. What percentage of the total stomata is on the upper surface?

4-4 ADDITION AND SUBTRACTION OF EXPRESSIONS

Let's look at the expression $2x^2 - 7y + 6$. Each part connected by a plus or minus sign, together with its sign, is called a **term** of the expression.

Terms of $2x^2 - 7y + 6$ are
$+ 2x^2, \; - 7y, \; + 6.$

In the *first* term $2x^2$, we say that the factor 2 is the **coefficient**† and the factor x^2 is the **literal part**. The coefficient of the *second* term $-7y$ is -7. The literal part is y. The *third* term 6 is called a **constant term**. It has a fixed value. We say that 6 is the coefficient of this term.

EXAMPLE 1

a. The expression $t - t^2$ has two terms. Since $t = 1 \cdot t$, the coefficient of t is 1. The coefficient of $-t^2$ $(= -1 \cdot t^2)$ is -1. The literal parts are t and t^2.

b. $\frac{x}{10}$ has one term. Since $\frac{x}{10} = \frac{1}{10} \cdot x$, the coefficient is $\frac{1}{10}$.

Just as

$$4 \text{ apples} + 5 \text{ apples} = 9 \text{ apples},$$

we write $4x + 5x = 9x$. This is just a result of the distributive law, since

$$4x + 5x = (4 + 5)x = 9x.$$

We say $4x$ and $5x$ are **like** terms because they have the *same literal part*, x. We combine like terms, but we **do not** combine things such as

$$2x^2 + 4x. \qquad \left[\text{Not like terms, since } x^2 \neq x.\right]$$

---

†Some people call this the *numerical* coefficient.

EXAMPLE 2

a. $2ab$ and $-4ab$ are like terms because $ab = ab$.

b. $3xy^2$ and $2x^2y$ are *not* like terms because $xy^2 \neq x^2y$.

EXAMPLE 3

Combining like terms by using the distributive law.

a. $2x^2 + 7x^2 + x^2 = 2x^2 + 7x^2 + 1 \cdot x^2 = (2 + 7 + 1)x^2 = 10x^2$.

b. $9a - 14a = (9 - 14)a = -5a$.

c. $-4a^2b + a^2b = -4a^2b + 1 \cdot a^2b = (-4 + 1)a^2b = -3a^2b$.

d. $2xy + 3yx = 2xy + 3xy = 5xy$ (like terms because $xy = yx$).

To simplify

$$3x^2 - 2x + 4 + 12x^2 - 9x - 3,$$

we "shuffle" the terms until like terms are together.

$$\underbrace{3x^2 + 12x^2}_{\text{like terms}} \underbrace{- 2x - 9x}_{\text{like terms}} \underbrace{+ 4 - 3}_{\text{like terms}}$$

Then we combine:

$$15x^2 - 11x + 1.$$

EXAMPLE 4

Simplify.

$$\begin{aligned} & a^2 - 2ab + b^2 - 3a^2 + 4b^2 \\ &= a^2 - 3a^2 - 2ab + b^2 + 4b^2 && \text{[getting like terms together]} \\ &= -2a^2 - 2ab + 5b^2 && \text{[combining like terms].} \end{aligned}$$

At times an expression contains grouping symbols such as parentheses ( ), brackets [ ], or braces { }. These can be removed by the use of the distributive law.

EXAMPLE 5

Removing grouping symbols by using the distributive law.

a. $3a + 2(b - c) = 3a + 2b - 2c$. We multiplied *each* enclosed term by 2.

b. $-2(-3 + 4x)$.

We have to multiply *each* enclosed term by -2.

$$\begin{aligned} -2(-3 + 4x) &= (-2)(-3) + (-2)(4x) \\ &= 6 + (-8x) \\ &= 6 - 8x. \end{aligned}$$

c. $-8[x^2 - 2x + 1] = -8x^2 + 16x - 8$. We multiplied *each* enclosed term by -8.

$x + 3(y - 2) \neq x + 3y - 2$, *but* $x + 3(y - 2) = x + 3y - 6$. *Take care of* ***each*** *term inside the parentheses.*

Notice that

$$2 + (x - 2y) = 2 + (1)(x - 2y) = 2 + x - 2y,$$

[distributive law]

but

$$\begin{aligned} 2 - (x - 2y) &= 2 + (-1)(x - 2y) \\ &= 2 + (-1)(x) - (-1)(2y) \end{aligned}$$

[distributive law]

$$= 2 - x + 2y.$$

This shows that

> **RULE 1**
> If a plus sign is immediately in front of grouping symbols, then you can remove these symbols without changing the signs of the enclosed terms.

> **RULE 2**
> You can remove a minus sign immediately in front of grouping symbols by removing the grouping symbols **and** changing the sign of each of the enclosed terms.

EXAMPLE 6

Removing grouping symbols.

a. $5a + (3 - 4b + 3c) = 5a + 3 - 4b + 3c$ by Rule 1.

b. $-[-2 - 4a + 3b] = +2 + 4a - 3b = 2 + 4a - 3b$ by Rule 2.

c. $(5x - 2y) + 3 = 5x - 2y + 3$. Here we assume that a plus sign is in front of the parentheses.

EXAMPLE 7

Simplify.

a. $(3x - y) - (-4y - 2x + 3)$

$= 3x - y + 4y + 2x - 3$ [removing parentheses]

$= 3x + 2x - y + 4y - 3$ [getting like terms together]

$= 5x + 3y - 3$ [combining like terms].

b. $x^2 + 3[-6x + x^2] = x^2 - 18x + 3x^2$ [removing brackets]

$= x^2 + 3x^2 - 18x$ [getting like terms together]

$= 4x^2 - 18x$ [combining like terms].

c. $3(x^2 - 5) - 4[-a^2 + 2x^2] + 5\{x^2 + 3\}$

$= 3x^2 - 15 + 4a^2 - 8x^2 + 5x^2 + 15$ [removing grouping symbols]

$= 4a^2$ [combining like terms].

Note that we removed the brackets by multiplying each term within the brackets by -4.

EXAMPLE 8

Subtract $-2x + 6$ *from* 6.

$$6 - (-2x + 6) = 6 + 2x - 6 = 2x.$$

When grouping symbols appear within other grouping symbols, as in

$$-2[x - (5x - 3)],$$

it's best to remove them from the inside out. Here's what we mean. Since the parentheses are the innermost grouping symbols, we remove them first. Watch.

$$-2[x - (5x - 3)] = -2[x - 5x + 3] \quad \text{[removing parentheses]}$$
$$= -2[-4x + 3] \quad \text{[combining within brackets]}$$
$$= 8x - 6 \quad \text{[removing brackets].}$$

EXAMPLE 9

Simplify $2x - 3[-2(x - 1) - x]$.

Here we want to remove all symbols of grouping and combine like terms.

$$2x - 3[-2(x - 1) - x]$$
$$= 2x - 3[-2x + 2 - x] \quad \text{[removing parentheses]}$$
$$= 2x - 3[-3x + 2] \quad \text{[combining within brackets]}$$
$$= 2x + 9x - 6 \quad \text{[removing brackets]}$$
$$= 11x - 6 \quad \text{[combining].}$$

EXAMPLE 10

Simplify.

$$-2\{2 - 3[-(x + 1)]\} - x$$
$$= -2\{2 - 3[-x - 1]\} - x \quad \text{[removing parentheses]}$$
$$= -2\{2 + 3x + 3\} - x \quad \text{[removing brackets]}$$
$$= -2\{3x + 5\} - x \quad \text{[combining within braces]}$$
$$= -6x - 10 - x \quad \text{[removing braces]}$$
$$= -7x - 10 \quad \text{[combining].}$$

Completion Set 4-4

In Problems **1** *and* **2**, *fill in the blanks.*

1. The coefficient of $2x^2$ is ____; the coefficient of $-x$ is ____; the coefficient of x is ____.

2. Two terms are *like terms* if they have the same ________________ part.

3. Circle the pairs that are like terms.

a. $3x$, $5x$ b. $8y$, $9x$ c. $9x^2y^3$, $2x^3y^2$ d. $7a$, a

In Problems **4–6**, *fill in the blanks.*

4. $2x^2 - 5x - 1 + 7x^2 + 3x + 6 = 2x^2 + 7x^2 - 5x + 3x - 1 + 6$

$= \_\_\_\_ x^2 - \_\_\_\_ x + \_\_\_\_.$

5. $3x + 6y + 4x - 2y = \_\_\_\_ x + \_\_\_\_ y.$

6. $4ab + ba + 8 = \_\_\_\_ ab + 8.$

In Problems **7–10**, *fill in either "+" or "–".*

7. $3(x - 2y + 7) = 3x \_\_\_\_ 6y \_\_\_\_ 21.$

8. $-2(1 + 8x - 9y) = -2 \_\_\_\_ 16x \_\_\_\_ 18y.$

9. $9 + (4a + 3b - c) = 9 \_\_\_\_ 4a \_\_\_\_ 3b \_\_\_\_ c.$

10. $x - (-2y + z - 5) = x \_\_\_\_ 2y \_\_\_\_ z \_\_\_\_ 5.$

11. Fill in each blank with the word *parentheses*, *brackets*, or *braces*.

a. { } are called ____________________.

b. ( ) are called ____________________.

c. [ ] are called ____________________.

In Problems **12** *and* **13**, *fill in the blanks.*

12. $2[5x + 4 - x] = 2[\_\_\_\_ x + \_\_\_\_] = \_\_\_\_ x + \_\_\_\_.$

13. $3\{7 - 2(2x + 1)\} = 3\{7 - \_\_\_\_ x - \_\_\_\_\} = 3\{\_\_\_\_ - \_\_\_\_ x\}$

$= \_\_\_\_ - \_\_\_\_ x.$

Problem Set 4-4

In Problems **1–36**, *simplify*.

1. $4x^2 + 6x^2 + 2x^2$

2. $3t^3 + 2t^3 - t^3$

3. $8y - 12y - 3y$

4. $-5x^2 + 7x^2 - 2x^2$

5. $3x - 6 + 8x + 4$

6. $9x^2 - 3x - 4x^2 - 5x$

7. $4(9x + 3y + x)$

8. $3(a - 5b) - a$

9. $-2(7x - 8y + 2y)$

10. $5(a - 5b - a)$

11. $4y + (4y - 8)$

12. $2 - (6x + 7) - 2x^2$

13. $3x + 5y - (8y - x)$

14. $2(x^2 + 5) - 8x$

15. $(3a - 2b) + (4a - 6b + c)$

16. $(6x + 5) - (8x + 3)$

17. $2(x^2 - 3) + 3(x^2 + 7)$

18. $4(3 - a) - 2(a - 1)$

19. $5(xy + z) - (3xy + 7)$

20. $2[x + 3z] - 4(z + x) + 3\{x - 4z\}$

21. $6(x^2 - 2x) + (3x - 5) - 4(x^2 + 2)$

22. $2[(3x - 5) + 4]$

23. $2[-(5 - x) + x]$

24. $3[4x + (2 + 5x)]$

25. $5[3x + 2(5 - x)]$

26. $7[4 - (y + 8)]$

27. $-\{4a - (6 + 3a)\}$

28. $-2\{9(z^2 - 1)\}$

29. $5x^2 + 3[8(x^2 - 1)] + 2$

30. $[2a + 3(b - a)] + 7a$

31. $\{9a - (3b + a + c)\} - 4\{2b + 3c\}$

32. $4x^2 - [8x + 2(x + x^2)]$

33. $3\{4x - 2[5 - (x + 1)]\}$

34. $-\{2[3(2x + 5) + 6x]\}$

35. $9 - 2\{8x - 3[4(y - 2x) - 6(x - y)]\}$

36. $3\{5 - [xy - 2(xy + x)]\} - 4x$

37. Subtract $6x - 4$ from $4x - 8$.

38. Subtract $2(1 - x)$ from $9x^2 - 3x$.

4-5 MORE EQUATIONS

Let's see how we can solve equations by using what we've learned about combining like terms.

To solve an equation it's usually a good idea first to remove any symbols of grouping. Then clear of any fractions. Next, get the x's (or other unknown) on one side and the constants on the other. From then on it's smooth sailing.

EXAMPLE 1

Solve $9x - (2x - 34) = 4(1 - 2x)$.

$$9x - (2x - 34) = 4(1 - 2x)$$

$$9x - 2x + 34 = 4 - 8x \quad \text{[removing parentheses]}$$

$$7x + 34 = 4 - 8x \quad \text{[combining]}$$

$$7x + 34 + 8x = 4 - 8x + 8x \quad \text{[adding } 8x \text{ to both sides]}$$

$$15x + 34 = 4 \quad \text{[combining]}$$

$$15x = -30 \quad \text{[subtracting 34 from both sides]}$$

$$\boxed{x = -2} \quad \text{[dividing both sides by 15].}$$

Here is a summary of steps that you may follow for solving basic equations.

STEPS TO SOLVE EQUATIONS

1. Remove symbols of grouping.
2. Clear of any fractions.
3. Add or subtract to get the x's (or other unknown) on one side and the constants on the other.
4. Multiply or divide so that the value of x is obvious.

We shall use this four-step procedure in solving the equation in the next example.

EXAMPLE 2

Solve $3\left(\frac{7}{5}x - 1\right) = 2x$.

Step 1

$$3 \cdot \frac{7}{5}x - 3 = 2x \quad \text{[removing parentheses]}$$

$$\frac{21}{5}x - 3 = 2x$$

Step 2

$$5\left(\frac{21}{5}x - 3\right) = 5(2x) \qquad \text{[multiplying both sides by 5]}$$

$$21x - 15 = 10x \qquad \text{[removing parentheses]}$$

Step 3

$$21x - 15 - 10x = 10x - 10x \qquad \text{[subtracting } 10x \text{ from both sides]}$$

$$11x - 15 = 0 \qquad \text{[combining]}$$

$$11x = 15 \qquad \text{[adding 15 to both sides]}$$

Step 4

$$\boxed{x = \frac{15}{11}} \qquad \text{[dividing both sides by 11].}$$

EXAMPLE 3

Dr. Stein asked his faithful lab assistant Frank to help him to prepare 350 *milliliters (ml) of a chemical solution. It is to be made up of* 2 *parts alcohol and* 3 *parts acid. How much of each should Frank and Stein use?*

Let n = number of ml in each part.

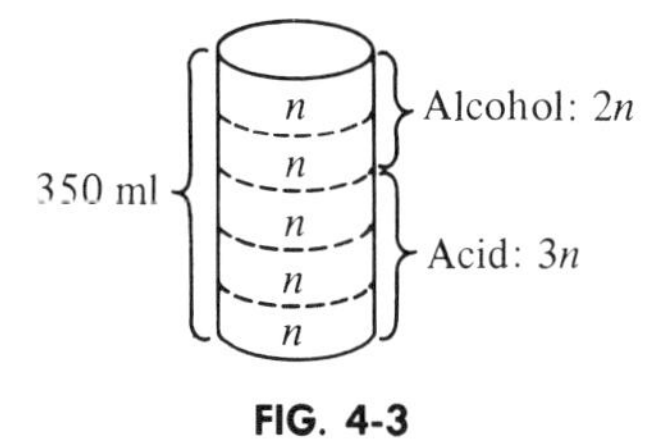

FIG. 4-3

From Fig. 4-3, we have

$$2n + 3n = 350$$

$$5n = 350$$

$$n = \frac{350}{5} = 70.$$

But $n = 70$ *is not* the answer to the original problem. *Each part* (n) has 70 ml. The amount of alcohol is $2n = 2(70) = 140$, and the amount of acid is $3n = 3(70) = 210$. Thus, Frank and Stein should use 140 ml of alcohol and 210 ml of acid.

Example 3 shows how helpful a picture can be in setting up a word problem.

Completion Set 4-5

In Problems **1–5**, *fill in the blanks.*

1. If $9x - 5x = 12$, then ____ $x = 12$ and so $x =$ ____.

2. If $6x = 4x + 8$, then ____ $x = 8$ and so $x =$ ____.

3. If $5(x - 2) = 4$, then $5x -$ ______ $= 4$ and $5x =$ ______, and so $x =$ ______.

4. If $3x + 2(3x - 1) = 7$, then $3x +$ ____$x - 2 = 7$. Therefore, ____$x - 2 = 7$ and ____$x = 9$, and so $x =$ ____.

5. If $(2x - 1) - (4x - 5) = 0$, then $-2x +$ ____ $= 0$. Thus, $-2x =$ ______ and so $x =$ ____.

Problem Set 4-5

In Problems **1–36**, *solve the equations.*

1. $6x + 4x = 20$

2. $5x - 10x = 15$

3. $9x - 12x = 0$

4. $9 = 3(2x + 4x)$

5. $5y - 2y + 3 = 12$

6. $2t + 6 = 6t + 2$

7. $6x = 4x$

8. $x = 3 - 2x$

9. $8x = 4x - 12$

10. $3x + 2x = 4x + 6$

11. $3x + 6 = 7x - 2$

12. $9x - 4 = 9 - 4x$

13. $2(y - 5) = y + 1$

14. $-3(y - 1) = 4y + 17$

15. $7(3 - 2z) = 3 - 5z$

16. $8 - 6z = 4(3z + 5)$

17. $(4x + 3) - (7 - x) = 7x$

18. $2(x - 1) - (3x + 7) = x$

19. $17x - 3(x - 5) = 2 + 4(3x + 2)$

20. $6 = -[3 - (x - 2)]$

21. $4[2 - 3(r + 6)] = 10r$

22. $2(3r + 2) = -[1 + (2 - r)]$

23. $t + \frac{t}{2} = 6$

24. $\frac{x}{2} + 1 = \frac{x}{3}$

25. $\frac{x}{3} = \frac{10 + x}{2}$

26. $\frac{z}{2} = \frac{z}{3}$

27. $3\left(\frac{x}{2} - 4\right) = 5x$

28. $\frac{2}{3}(x - 5) = 7$

29. $\frac{3}{4}(x - 1) = 7 + x$

30. $2\left(x - \frac{1}{5}\right) = 3x$

31. $\frac{2}{3}(x - 5) = 4x + \frac{1}{2}$

32. $\frac{x}{3} + 1 = 4\left(x - \frac{1}{2}\right)$

33. $2[3z + 4(z - 1)] = -(7 - z)$

34. $2\{4x - [8 + 2(5x - 4)]\} = 0$

35. $\frac{1}{2}\left(x + \frac{4}{3}\right) = 3 - (x + 1)$

36. $\frac{7}{2} - (x + 5) = \frac{x}{4}$

37. $F = \frac{9}{5}C + 32$ is a formula relating fahrenheit and celsius temperatures. Solve for C.

38. $V = E(x_b - x_a)$ is a formula that can be used in a certain situation involving electric intensity. Solve for x_b.

39. A chemist wants to prepare 280 grams of a compound by mixing 3 parts of chemical A with 5 parts of chemical B. How many grams of each chemical should he use?

40. A builder makes a certain type of concrete by mixing together 1 part cement, 3 parts sand, and 5 parts stone (by volume). If he wants 585 cubic feet of concrete, how many cubic feet of each ingredient does he need?

41. National Wigwam Co. makes two types of prefabricated houses: Early American and Little Big Horn. Last year they sold three times as many Early American models as they did Little Big Horn models. If a total of 2640 houses were sold last year, how many of each model were sold?

42. A student scored an 82 and an 88 on his first two math exams. What score does he have to get on his next exam so that the average of the three exams will be 90? (*Hint:* To find the average of three numbers, add the numbers and divide this sum by 3.)

43. The Dimwit Power Company is going to locate its new power plant along a road connecting the towns of Exton and Whyton, which are 10 miles apart. See Fig. 4-4.

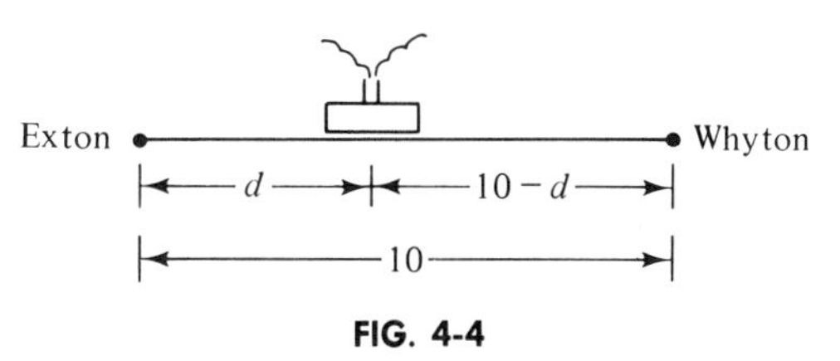

FIG. 4-4

For political reasons, Dimwit Power Company will buy coal from both towns. The price of coal per ton from Exton is \$26.65 plus \$0.45 per mile for delivery. The price per ton from Whyton is \$26.25 plus \$0.25 per mile for delivery. How far from Exton should the plant be located if the price of coal per ton delivered from Exton is to be equal to that from Whyton? (*Hint:* If d is the distance of the plant from Exton, then $10 - d$ is the distance from Whyton.)

44. An investment club bought a bond of an oil corporation for \$5000. The bond yields 8% per year. The club now wants to buy shares of stock in a utility company. The stock sells at \$20 per share and earns a dividend of \$0.50 per share per year. How many shares should the club buy so that its total investment in stocks and bonds yields 5% per year?

4-6 REVIEW

IMPORTANT TERMS AND SYMBOLS

equation *(p. 76)*
variable *(p. 76)*
unknown *(p. 76)*
solution of equation *(p. 77)*
root of equation *(p. 77)*
terms of expression *(p. 93)*
coefficient *(p. 93)*
literal part *(p. 93)*
constant term *(p. 93)*
like terms *(p. 93)*
grouping symbols *(p. 94)*
parentheses, ( ) *(p. 94)*
brackets, [ ] *(p. 94)*
braces, { } *(p. 94)*

REVIEW PROBLEMS

In Problems **1–10**, *simplify.*

1. $(3x + 2y - 5) + (8x - 4y + 2)$

2. $7x + 5(4x + 3)$

3. $6(a + 3b) - (8a - b - 4)$

4. $2(a + 4b) - 3(3b - 2a)$

5. $2[3(xy - 5) + 7(4 - xy)]$

6. $(4xy + 7) - 5[xy - (4 - 3xy)]$

7. $4x + [-5(2 - x) - 8]$

8. $\{1 - 2[x - (x - 1)]\} + 1$

9. $-3\{x^2 - [3(2x - 4) + 2x^2]\}$

10. $2\{x^2 + 3[x - (x^2 + 4)]\} + 7$

In Problems **11–24**, *solve.*

11. $4x + 1 = 3$

12. $9x - 7 = 11$

13. $5 = 8 - 2y$

14. $6y = 3y$

15. $8 - \frac{4x}{3} = 10$

16. $6x - \frac{1}{3} = 5$

17. $\frac{3}{4}z + 2 = \frac{1}{3}$

18. $\frac{1}{10} - \frac{2z}{5} = 4$

19. $9(3u + 2) = 3 - (u + 7)$

20. $4\left(u - \frac{5}{7}\right) = -3u$

21. $\frac{3}{2}(x - 8) = 2x + 4$

22. $3\{2x + 4[(7 - 2x) - 5x]\} = 0$

23. $5[3 - 2(3x - 4)] = 18 - 9x$

24. $7x - [8x + 4(x + 2)] = -(1 + x)$

25. The formula for the velocity v of a certain object moving along a straight line is $v = v_0 + at$. The initial velocity is v_0, the acceleration is a, and t is time. Solve for a.

26. $E = \frac{1}{2}mv^2$ is a formula for kinetic energy E. Here m is mass and v is velocity. Solve for m.

27. The formula $V = \frac{\pi}{4}d^2l$ gives the volume (in cubic millimeters) of fluid flowing per second from a capillary. Here d is the diameter (in millimeters) of the capillary and l is the linear velocity (in millimeters per second) of the fluid. If $V = .000015\pi$ and $d = .01$, find l.

28. In a certain heat experiment, 2 kilograms (kg) of water at 79°C is mixed with 4 kg of water at 40°C. The final temperature of the mixture is given by t, where

$$2000(1)(79 - t) = 4000(1)(t - 40).$$

Solve for t.

29. 96 is 80% of what number?

30. 28 is 35% of what number?

31. 9 is what percentage of 75?

32. 162 is what percentage of 900?

33. In a certain city, 40%, or 3300, of the registered voters went to the polls. How many registered voters were there?

34. A scientist proposed a new theory. Testing it with 65 experiments, he was successful in getting a desired result 52 times. What was the percentage of successful experiments?

35. A certain alloy is made up of 8 parts of metal A, 3 parts of metal B, and 1 part of metal C by weight. How much of each metal is needed to make 168 tons of the alloy?

36. In a poll of 360 people, four times as many favored a new product as compared to those who favored the best-selling brand. How many people favored the new product?

37. The Light My Fire Lamp Company will have fixed costs of $26,500 this year. If the manufacturing cost of each lamp is $11, how many lamps can be made this year for a total cost of $76,000?

38. A builder has a client who wants an L-shaped living and dining area in his new house. See Fig. 4-5. The area is to be a total of 385 square feet. What should the

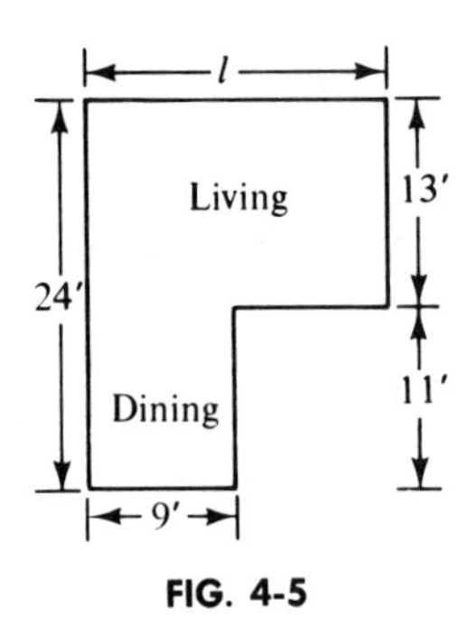

FIG. 4-5

"length" l of the living area be?

CHAPTER 5

Multiplication and Division of Expressions

5-1 MULTIPLICATION OF EXPRESSIONS

Let's look at the expressions

$$3x^2 \quad \text{and} \quad x^2y + 5.$$

Notice that $3x^2$ consists of *exactly one term.* For this reason, $3x^2$ is called a **monomial**. But $x^2y + 5$ has *more than one term.* It is called a **multinomial.**

EXAMPLE 1

Monomials	*Multinomials*
$4xy$	$7x - 3$
-3	$a + b - c$
$\dfrac{10}{x^3}$	$-2x^3 - 3x^2 + 4x - 5$

To multiply the monomials $3x^2$ and $4xy$, we have

$$
\begin{aligned}
(3x^2)(4xy) &= 3 \cdot x^2 \cdot 4 \cdot x \cdot y \\
&= 3 \cdot 4 \cdot x^2 \cdot x \cdot y && \text{[rearranging factors by commutative law]} \\
&= 12x^3y && \text{[law of exponents].}
\end{aligned}
$$

In short, to multiply monomials, just *multiply their coefficients and multiply their literal parts.*

EXAMPLE 2

Multiplying monomials.

a. $(2ab^2)(-4a^2b^2) = (2)(-4)(a \cdot a^2)(b^2 \cdot b^2) = -8a^3b^4$.

b. $-5x^2y(-3xy^2) = (-5)(-3)(x^2 \cdot x)(y \cdot y^2) = 15x^3y^3$.

c. $(2x)(3xy)(-2y^2) = (2)(3)(-2)(x \cdot x)(y \cdot y^2) = -12x^2y^3$. Another way to get the product is to start by multiplying the first two factors.

$$(2x)(3xy)(-2y^2) = (6x^2y)(-2y^2) = -12x^2y^3.$$

EXAMPLE 3

Find $(3a^3bc)(ab)^2$.

Here we multiply $3a^3bc$ *not* by ab, but by $(ab)^2$, which is a^2b^2. The squaring must be done first.

$$(3a^3bc)(ab)^2 = (3a^3bc)(a^2b^2) = 3a^5b^3c.$$

EXAMPLE 4

$$(-2x^2y)^3(-xy^2)^2 = (-8x^6y^3)(x^2y^4) = -8x^8y^7.$$

Let's now find $(2x)(x + 3)$. We note that the first factor, $2x$, is a monomial, but the second factor, $x + 3$, is a multinomial. We use the distributive law $a(b + c) = ab + ac$ and multiply the monomial by each term in the multinomial. Then we simplify.

$$\underbrace{(2x)}(x+3)=\underbrace{(2x)}(x)+\underbrace{(2x)}(3)=2x^2+6x.$$

$$\begin{array}{ccccccc} \uparrow & \uparrow\ \uparrow & & \uparrow\ \uparrow & & \uparrow\ \uparrow \\ a & (b+c)= & & a\ \ b & + & a\ \ c \end{array}$$

EXAMPLE 5

Multiplying monomials and multinomials.

a. $4ab(a^2 - 2b + 1) = (4ab)(a^2) - (4ab)(2b) + (4ab)(1)$

$= 4a^3b - 8ab^2 + 4ab.$

b. $-3x(x - 4) = (-3x)(x) - (-3x)(4)$

$= -3x^2 - (-12x) = -3x^2 + 12x.$

c. $(x^2y)^2(x - y + 2) = (x^4y^2)(x - y + 2)$

$= (x^4y^2)(x) - (x^4y^2)(y) + (x^4y^2)(2)$

$= x^5y^2 - x^4y^3 + 2x^4y^2.$

d. $5(3x^2 + 3) - 2x(x - 4) = 15x^2 + 15 - 2x^2 + 8x$

$= 13x^2 + 8x + 15.$

e. $2a[x(x + 1) - 7] = 2a[x^2 + x - 7] = 2ax^2 + 2ax - 14a.$

Let's find $(x + 2)(x + 3)$, a product of two multinomials. We use the distributive law $(a + b)c = ac + bc$, where $x + 2$ matches $a + b$ and $x + 3$ plays the role of c.

$$(x+2)\underbrace{(x+3)}=x\underbrace{(x+3)}+2\underbrace{(x+3)}=x^2+3x+2x+6=x^2+5x+6.$$

$$\begin{array}{ccccccc} \uparrow\ \uparrow & \uparrow & & \uparrow\ \uparrow & & \uparrow\ \uparrow \\ (a+b) & c & = & a\ \ c & + & b\ \ c \end{array}$$

That is, to find a product of two multinomials, *multiply each term in the first factor by the entire second factor* and then simplify.

EXAMPLE 6

Multiplying multinomials.

a. $(x^2 - 2)(x - 3) = x^2(x - 3) - 2(x - 3)$

$= x^3 - 3x^2 - 2x + 6.$

b. $(x + 1)(x^2 - x + 1) = x(x^2 - x + 1) + 1(x^2 - x + 1)$
$$= x^3 - x^2 + x + x^2 - x + 1$$
$$= x^3 + 1.$$

c. $2x(x + 1)(x - 1) = [2x(x + 1)](x - 1)$
$$= [2x^2 + 2x](x - 1)$$
$$= 2x^2(x - 1) + 2x(x - 1)$$
$$= 2x^3 - 2x^2 + 2x^2 - 2x$$
$$= 2x^3 - 2x.$$

d. $(x^2 + x - 1)(x^2 - 2x + 1)$
$$= x^2(x^2 - 2x + 1) + x(x^2 - 2x + 1) - 1(x^2 - 2x + 1)$$
$$= x^4 - 2x^3 + x^2 + x^3 - 2x^2 + x - x^2 + 2x - 1$$
$$= x^4 - x^3 - 2x^2 + 3x - 1.$$

Note that

$$x^2 - 2(x - 3) \neq (x^2 - 2)(x - 3),$$

and $$(x^2 - 2)x - 3 \neq (x^2 - 2)(x - 3).$$

For example,

$$x^2 - 2(x - 3) = x^2 - 2x + 6,$$

but $$(x^2 - 2)(x - 3) = x^3 - 3x^2 - 2x + 6,$$

as we showed in Example 6a.

Sometimes it may be convenient to do a multiplication problem in a vertical arrangement. For example, let's redo Example 6d. As shown below, we multiply the first term in the second row, x^2, by each term in the top row. This gives $x^4 - 2x^3 + x^2$. This is repeated with the second term and then with the third term in the second row. Then we add the results.

$$\begin{array}{rrrrr}
 & & x^2 & -2x & +1 \\
 & & x^2 & +x & -1 \\
\hline
x^4 & -2x^3 & +x^2 & & \\
 & x^3 & -2x^2 & +x & \\
 & & -x^2 & +2x & -1 \\
\hline
x^4 & -x^3 & -2x^2 & +3x & -1
\end{array}$$

(add)

Notice that to simplify the addition, we kept like terms in the same column.

We now turn to ways of handling "rate" problems. Actually, we could have put this in the last chapter. But perhaps it's better not to see too many types of word problems at one time.

EXAMPLE 7

*On the moon, a lunar rover traveled from point A to point B at the rate of 5 kilometers per hour (km/hr). It returned to A at the rate of 15 km/hr. The **total** traveling time was 2 hr. Find the distance from A to B.*

You may recall that

$$\boxed{\text{distance} = (\text{rate})(\text{time}).}$$

Two other forms of this are

$$\boxed{\text{time} = \frac{\text{distance}}{\text{rate}}, \qquad \text{rate} = \frac{\text{distance}}{\text{time}}.}$$

Now, let d be the distance from A to B (see Fig. 5-1). Then the time to go from A to

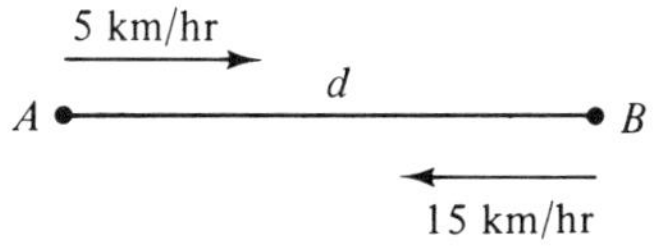

5 km/hr
d
A B
15 km/hr

FIG. 5-1

B at 5 km/hr is $\dfrac{\text{distance}}{\text{rate}}$ or $\dfrac{d}{5}$. From B to A the distance is also d, but the time at 15 km/hr is $\dfrac{\text{distance}}{\text{rate}} = \dfrac{d}{15}$. Thus

$$\begin{pmatrix}\text{time}\\ \text{from}\\ A \text{ to } B\end{pmatrix} + \begin{pmatrix}\text{time}\\ \text{from}\\ B \text{ to } A\end{pmatrix} = \text{total time}$$

$$\frac{d}{5} + \frac{d}{15} = 2$$

$$15\left[\frac{d}{5} + \frac{d}{15}\right] = 15 \cdot 2 \qquad \text{[multiplying both sides by 15, the L.C.D.]}$$

$$15 \cdot \frac{d}{5} + 15 \cdot \frac{d}{15} = 30$$

$$3d + d = 30$$

$$4d = 30$$

$$d = \frac{30}{4} = \frac{15}{2} = 7\frac{1}{2}.$$

Thus, the distance from A to B is $7\frac{1}{2}$ km. Now turn to the next page.

To show you that there are more ways than one to "skin a cat," here's another way to do the problem. It's fairly roundabout, because first we find the *time* it takes to go from A to B.

Let t = time to go from A to B at 5 km/hr. Because the total time traveled is 2 hr, then $2 - t$ is the time to go from B to A at 15 km/hr. See Fig. 5-2.

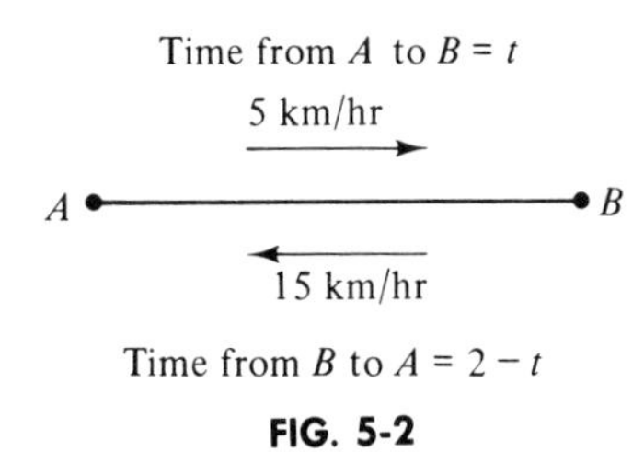

FIG. 5-2

Now, distance = (rate) (time) and

$$\begin{pmatrix}\text{distance}\\ \text{from}\\ A \text{ to } B\end{pmatrix} = \begin{pmatrix}\text{distance}\\ \text{from}\\ B \text{ to } A\end{pmatrix}.$$

Thus we have

$$\begin{aligned}(\text{rate})(\text{time}) &= (\text{rate})(\text{time})\\ 5t &= 15(2 - t)\\ 5t &= 30 - 15t\\ 20t &= 30\\ t &= \frac{30}{20} = \frac{3}{2}.\end{aligned}$$

Thus the distance from A to B = (rate)(time) = $(5)\left(\frac{3}{2}\right) = \frac{15}{2} = 7\frac{1}{2}$ km.

Completion Set 5-1

In Problems **1–4**, *fill the blanks.*

1. $(3xy)(4x)(2y^2) = (3 \cdot 4 \cdot 2)(x \cdot x)(y \cdot y^2) =$ ______________.

2. $(2x^2y)(3xy^3) = 2 \cdot$ ____ $\cdot x^2 \cdot$ ____ $\cdot y \cdot$ ____ = ______________.

3. The product $2x^2y(x - 2y + 1)$ is equal to $2x^2y$(____) $- 2x^2y$(_____) $+ 2x^2y$(____).

4. The product $(y + 2)(x - 4)$ is equal to $y(x - 4) +$ ________________.

In Problems **5–7**, *insert* T (= *true*) *or* F (= *false*).

5. $(2x)(3y)(4z) = (2x \cdot 3y)(2x \cdot 4z)$. ____

6. $a(ac)^2 = (a^2c)^2$. ____

7. $x(yx) = x^2y$. ____

Problem Set 5-1

In Problems **1–62**, *perform the indicated operations.*

1. $(3x)(5x^2)$
2. $(-4z)(10z)$
3. $(2x)(3xy)$
4. $-a(ac)$
5. $3ab(a^2b)$
6. $xy(xz)(xw)$
7. $2xy^2(-4x^3y^2)$
8. $-a^2b(-ac)$
9. $ab(a^2b)(bc^2)$
10. $x(xy^2)(y^2z)$
11. $2x^2yz^2\left(\frac{1}{2}xz^2\right)$
12. $(2xy^2)^3(2y)^2$
13. $(-3x)(4xy^2)(-2x^2y^3)$
14. $x(x^3y^2)^2(-2xy^2)^3$
15. $(10x^4)(3x^3)^2$
16. $(3y^3)^2(-y)^3$
17. $a(-bc)^2(cd)^2$
18. $(-xy)^3(-x)^2$
19. $x(x^2 - 4x + 7)$
20. $-2(x - 2y^2 + 7xy)$
21. $a^2b(-3 + ab - a)$
22. $2x(-x + 2y + 1)$
23. $-5xy(x^2 - y^2 + xy)$
24. $a(b - a)(ac)$
25. $(2x^2y)^2(x + 2y^2 - 3x^2)$
26. $x^2yz(xy - yz - xz)$
27. $-2xy(7 - 4x - 2y + x^2)$
28. $(ab)^2(a^2 - b^2c + ac^2)$
29. $(x + 2)(x + 5)$
30. $(x + 4)(x + 5)$
31. $(y - 2)(3y + 2)$
32. $(y - 2)(y + 2)$
33. $(3x - 1)(3x - 1)$
34. $(2x + 3)(3x + 2)$
35. $(4x + y)^2$
36. $(x^2 + 4)(x^4 + 4)$
37. $(3x^2)(2xy)(x + y)$
38. $(x + 3)(5x^2)$
39. $(t - 2)(t^2 + 2t + 4)$
40. $(x + 5)(x^2 - x - 1)$
41. $(x^2 - 2)(x^2 - 5x + 1)$
42. $(2 + x - y)(x - y)$
43. $(y^3 + 3y)(y^2 - y + 2)$
44. $(2ab + b^2)^2$

45. $(x + y + 1)(x + y - 1)$

46. $(2x - 1)(2x^3 - 3x + 1)$

47. $4x(2x - 1)(2x + 1)$

48. $-3x^2(x - 2)(x - 1)$

49. $(2ab + rt)^2$

50. $3x^2(x + 2)(x^2 - 2)$

51. $xy + y(y + x)$

52. $(x - 3)x - 4$

53. $(x^2 + 1)(2x) - (x^2 - 2)(2x)$

54. $x^2y - 2x - x^2y(1 + x)$

55. $x(x - 1) - 2(3 - x)$

56. $2xy(x^2y) - y^2(2x^3 - 2xy)$

57. $3x(x^2y - xy^2) - 3y(x^3 + 4x^2y)$

58. $2(x^3y - 2x^2) - 2x^2(x^2 - 2xy)$

59. $3(x - x^2) + (x + 1)(x - 1)$

60. $(x + 1)(x^2 - x + 1) - 1 - x^3$

61. $x^3 + y^3 - (x + y)(x^2 - xy + y^2)$

62. $(-x - 2)(x + 2) - (-1 + x)x$

63. Suppose that the lunar rover in Example 7 traveled from A to B at 6 km/hr and returned at 10 km/hr. If the total time was 3 hr, find the distance from A to B.

64. Suppose that the total time for the trip in Example 7 were 3 hr. Based on the rates given in that example, find the distance from A to B.

65. A traveling salesman drove 100 mi from Exton to Whyton in 2 hr. At first he averaged 55 mi/hr. But then he ran into bumpy road conditions for the rest of the trip. On that part he averaged 40 mi/hr. How long was he on the *bumpy* part of the road?

66. From two airports that are 300 mi apart, two airplanes leave at the same time and fly toward each other. One flies at 275 mi/hr and the other at 325 mi/hr. How long will it take for the planes to pass each other? *Hint*: When they pass, the sum of the distances traveled by the planes is 300 mi.

67. The water level in a certain reservoir is 6 ft deep, but the level is sinking at the rate of 4 in. a day. The water in another reservoir is 2 ft 9 in. deep and is rising $5\frac{3}{4}$ in. a day. When will the depths of the two reservoirs be the same? What will this depth be?

68. A pilot, flying against a headwind, traveled from A to B at 250 mi/hr. He flew back, with the wind, at 300 mi/hr. His trip from B to A took one hour less than the trip from A to B. Find the distance from A to B.

5-2 DIVISION

We now turn to division. To divide a monomial by a monomial, we use ordinary arithmetic and the laws of exponents. For example,

$$\begin{aligned}\frac{2x^3y^6}{8x^2y^8} &= \frac{2}{8} \cdot \frac{x^3}{x^2} \cdot \frac{y^6}{y^8} \\ &= \frac{1}{4} \cdot x \cdot \frac{1}{y^2} \\ &= \frac{x}{4y^2}.\end{aligned}$$

You may even think of this as cancelling common factors:

$$\frac{2x^3y^6}{8x^2y^8} = \frac{\cancel{2}\cancel{x^3}\cancel{y^6}}{\cancel{8}\cancel{x^2}\cancel{y^8}} = \frac{x}{4y^2}.$$

In fact, after some practice you may be able to do some steps in your head. Try this one.

$$\frac{-42x^3y^4}{7xy^5} = \frac{-6x^2}{y} = -\frac{6x^2}{y}.$$

EXAMPLE 1

Dividing monomials by monomials.

a. $$\frac{(3a^2b^3)(2a^2c^4)}{-2a^3b^7c^4d} = \frac{6a^4b^3c^4}{-2a^3b^7c^4d} \qquad \text{[simplifying numerator]}$$

$$= \frac{6}{-2}\cdot\frac{a^4}{a^3}\cdot\frac{b^3}{b^7}\cdot\frac{c^4}{c^4}\cdot\frac{1}{d}$$

$$= -3\cdot a\cdot\frac{1}{b^4}\cdot 1\cdot\frac{1}{d}$$

$$= -\frac{3a}{b^4d}.$$

b. $\left(\frac{x^2y^3}{2xy^2}\right)^3 = \left(\frac{xy}{2}\right)^3 = \frac{x^3y^3}{8}$. Here we simplified before cubing.

c. $$\frac{(2a^2x)^3(-2a^4y)}{(2xy^2)^2} = \frac{(8a^6x^3)(-2a^4y)}{4x^2y^4}$$

$$= \frac{-16a^{10}x^3y}{4x^2y^4}$$

$$= -\frac{4a^{10}x}{y^3}.$$

$$\frac{(ax)^2}{ay} \neq \frac{(\cancel{a}x)^2}{\cancel{a}y}, \quad \textit{but} \quad \frac{(ax)^2}{ay} = \frac{a^2x^2}{ay} = \frac{ax^2}{y}.$$

When the numerator of a fraction has more than one term, we can break up the fraction into simpler fractions as follows.

$$\boxed{\frac{a+b+c}{d} = \frac{a}{d} + \frac{b}{d} + \frac{c}{d}.}$$

That is, to divide a multinomial by a monomial, divide *each* term of the multinomial by the monomial. Thus,

$$\frac{8x^6 + x^4}{2x} = \frac{8x^6}{2x} + \frac{x^4}{2x}$$
$$= 4x^5 + \frac{x^3}{2}.$$

Note that $\frac{a + ax}{a} \neq \frac{\not{a} + ax}{\not{a}}$, *but* $\frac{a + ax}{a} = \frac{a}{a} + \frac{ax}{a} = 1 + x$. *You may cancel only common factors,* ***not*** *terms. Also,* $\frac{a}{a + ax} \neq \frac{a}{a} + \frac{a}{ax}$. *In general,*

$$\frac{a}{b + c} \neq \frac{a}{b} + \frac{a}{c}.$$

EXAMPLE 2

Dividing multinomials by monomials.

a. $\frac{4x^2 + 2x + 3}{2} = \frac{4x^2}{2} + \frac{2x}{2} + \frac{3}{2} = 2x^2 + x + \frac{3}{2}.$

b. $\frac{2(x + y)}{xy} = \frac{2x + 2y}{xy} = \frac{2x}{xy} + \frac{2y}{xy} = \frac{2}{y} + \frac{2}{x}.$

If some terms in the numerator have minus signs, we still use the same procedure as in Example 2. Thus,

$$\frac{a - b - c}{d} = \frac{a}{d} - \frac{b}{d} - \frac{c}{d}.$$

EXAMPLE 3

a. $$\frac{12a^2b^3 + 4ab - 6}{4ab} = \frac{12a^2b^3}{4ab} + \frac{4ab}{4ab} - \frac{6}{4ab}$$
$$= 3ab^2 + 1 - \frac{3}{2ab}.$$

b. $$\frac{5x^2y - x^3z}{-4xyz^2} = \frac{5x^2y}{-4xyz^2} - \frac{x^3z}{-4xyz^2}$$
$$= -\frac{5x}{4z^2} - \left(-\frac{x^2}{4yz}\right) = -\frac{5x}{4z^2} + \frac{x^2}{4yz}.$$

c.
$$\frac{3x - (2x^2y)^2 - 7x^2(x^2y^2)}{3x(2y^2)}$$
$$= \frac{3x - 4x^4y^2 - 7x^4y^2}{6xy^2}$$
$$= \frac{3x - 11x^4y^2}{6xy^2} = \frac{3x}{6xy^2} - \frac{11x^4y^2}{6xy^2} = \frac{1}{2y^2} - \frac{11x^3}{6}.$$

Some word problems are of a "mixture" type. Let's look at some.

EXAMPLE 4

A new insect spray, "Bug Off," is in the experimental stages. It contains the remarkable new "killer" ingredient K-57. A lab assistant has available two spray formulas: Formula A, of which 10% is K-57; and Formula B, of which 16% is K-57. So far, Formula A has proved too weak. On the other hand, Formula B seems too strong to be used near house pets. The lab assistant is told to mix Formula A with 400 milliliters (ml) of Formula B so that the result is 14% K-57. How many ml of Formula A should be used?

Let a be the number of ml of Formula A to be added to the 400 ml of Formula B. Then we end up with $a + 400$ ml, of which 14% must be K-57. That is, $.14(a + 400)$ is K-57. See Fig. 5-3. This K-57 comes from two sources: $.10a$ comes from Formula

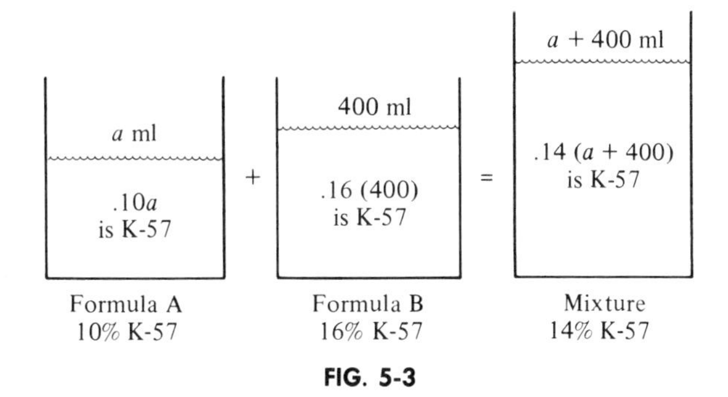

FIG. 5-3

A, and $.16(400)$ comes from Formula B. Thus,

$$.10a + .16(400) = .14(a + 400)$$
$$.10a + 64 = .14a + 56$$
$$8 = .04a$$
$$\boxed{a = \frac{8}{.04} = 200.}$$

Thus 200 ml of Formula A must be used.

EXAMPLE 5

Suppose that the lab assistant in Example 4 had needed exactly 500 ml of a 14% K-57 solution. How much of each formula would be used?

Let a be the number of ml of Formula A to be used. Then to get a total of 500 ml, there must be $500 - a$ ml of Formula B. See Fig. 5-4. The total amount of K-57 in

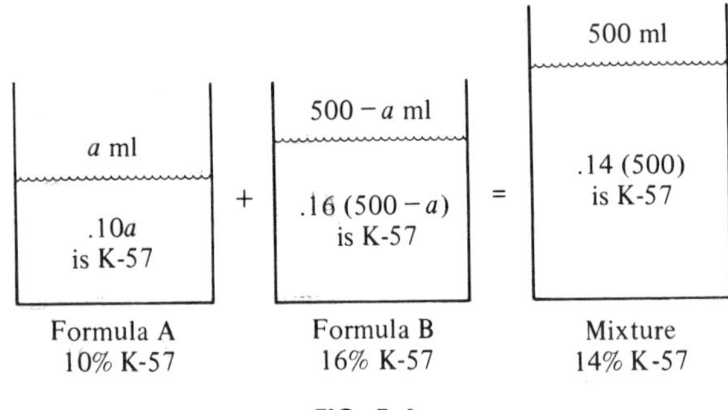

FIG. 5-4

the 500 ml of the 14% solution is .14(500). This K-57 comes from two sources: $.10a$ comes from Formula A, and $.16(500 - a)$ comes from Formula B. Thus,

$$.10a + .16(500 - a) = .14(500)$$
$$.10a + 80 - .16a = 70$$
$$-.06a = -10$$
$$\boxed{a = \frac{-10}{-.06} = 166\tfrac{2}{3}}$$

Thus, $500 - a = 500 - 166\frac{2}{3} = 333\frac{1}{3}$. The lab assistant should mix $166\frac{2}{3}$ ml of Formula A with $333\frac{1}{3}$ ml of Formula B.

Completion Set 5-2

In Problems **1–6**, *fill in the blanks.*

1. $\dfrac{5x^3}{25x^2} = \dfrac{5}{25} \cdot \dfrac{x^3}{x^2} = \_\_\_\_.$

2. $\dfrac{4x^4y^3z}{-2x^2y^3z^2} = \dfrac{4}{-2} \cdot \dfrac{x^4}{x^2} \cdot \dfrac{y^3}{y^3} \cdot \dfrac{z}{z^2} = \_\_\_\_\_\_\_\_.$

3. $\dfrac{-2x^3y}{-4x^4z} = \dfrac{-2}{-4} \cdot \dfrac{x^3}{x^4} \cdot \dfrac{y}{1} \cdot \dfrac{1}{z} =$ ________.

4. $\dfrac{9x^2 - 6x + 5}{3} =$ ____$x^2 -$ ____$x +$ ____.

5. $\dfrac{2x^2y - xy + y^2}{xy} =$ ____$x -$ ____ $+$ ____.

6. Suppose you mix solutions A and B to form 100 cubic centimeters (cm^3) of solution C.

If you use x cm^3 of solution A, then you will use ________ $-$ ____ cm^3 of solution B.

In Problems **7** *and* **8**, *insert* T (= *true*) *or* F (= *false*).

7. $\dfrac{x - y}{x} = 1 - \dfrac{y}{x}$. ____

8. $\dfrac{a + \not{x}}{\not{x}^2_{x}} = \dfrac{a + 1}{x}$. ____

Problem Set 5-2

In Problems **1–38**, *do the divisions.*

1. $\dfrac{2ab}{4a}$

2. $\dfrac{-3x^2y}{xy}$

3. $\dfrac{-14ab^2}{7a^2}$

4. $\dfrac{35x^3y^2}{-7xy}$

5. $\dfrac{6abc}{-6ab}$

6. $\dfrac{-ax^2y^5}{-x^3y^3}$

7. $\dfrac{-16x^2yz}{-32xy^2z^3}$

8. $\dfrac{-25ab^3c^2}{5ab^2}$

9. $\dfrac{15xy^6z^2}{10x^2y^3z}$

10. $\dfrac{2x^4yw^2}{4x^3yw^3}$

11. $\dfrac{-a^2b(abc^2)}{a^2b^4c}$

12. $\dfrac{(abc^2)^2(ab)}{2(ab)^2}$

13. $\dfrac{(3xy)^2}{y}$

14. $\dfrac{2xy^2}{(2xy)^2}$

15. $\left(\dfrac{8x^2y}{24xy^2}\right)^2$

16. $\left(\dfrac{3xy}{6y}\right)^2$

17. $\dfrac{4x^2 - 6x + 8}{2}$

18. $\dfrac{3x^2 - 8y + xy}{xy}$

19. $\dfrac{5 - x^2 + 2x}{x}$

20. $\dfrac{9x^2 - 15}{-3x}$

21. $\dfrac{10x^3 - 15x + 2}{5x}$

22. $\dfrac{3xy - 2y}{6}$ **23.** $\dfrac{4x + 2y}{-2x}$ **24.** $\dfrac{2x^2 - y^2x}{x^2}$

25. $\dfrac{xy + x^2}{xy}$ **26.** $\dfrac{x - xy}{x}$ **27.** $\dfrac{3x^2y - x^3y^2 + 1}{x^2y^2}$

28. $\dfrac{2 + 2y}{2y}$ **29.** $\dfrac{6x^2y - 2y^2 + 7x - 4}{-3x}$ **30.** $\dfrac{2x^2y^2 - 6xz^2 + 4x}{2xy^2z}$

31. $\dfrac{-20x^2y^2 + 5xy^2 - 2x}{2xy}$ **32.** $\dfrac{-6a^4b^2c^3 + 3a^5b^2c - 12a^6bc^2}{-6a^4b^2c^2}$

33. $\dfrac{2xy - (2xy^2)^2 - 3x^3y^3}{(xy)^2}$ **34.** $\dfrac{x(xy) + y(-x^2) + x^3y^3}{-x(xy)}$

35. $\dfrac{2x(x^3y^2)^2 + 3x^2(2y^2) - x}{-xy^4}$ **36.** $\dfrac{x(x - y) - y(2x)}{xy}$

37. $\dfrac{x(x^2 - 2x + 1)}{x^2}$ **38.** $\dfrac{(x - 2)^2}{x}$

39. The manager of the Doo Drop Inn, a local hamburger joint, finds that his "soy-burgers" aren't selling too well. Soyburgers are a blend of 70% hamburger meat and 30% soy protein. To boost sales he decides to increase the percentage of hamburger meat to 80%. He has 40 lb of raw soyburger meat. How much pure hamburger must be added to it so that the resulting blend is 80% hamburger?

40. A 6-gallon truck radiator is two-thirds full of water. How much of a 90% antifreeze solution (90% is antifreeze by volume) must be added to it to make a 10% antifreeze solution in the radiator?

41. A chemical manufacturer mixes a 20% acid solution (20% is acid by volume) with a 30% acid solution to get 700 gallons of a 24% acid solution. How many gallons of each does he use?

42. A chemical manufacturer wants to fill an order for 500 gallons of a 25% acid solution (25% is acid by volume). Solutions of 30% and 18% are available in stock. How many gallons of each must he mix to fill the order?

43. How many milliliters (ml) of water must be evaporated from 80 ml of a 12% salt solution (12% is salt by volume) so that what remains is a 20% salt solution?

5-3 DIVISION OF POLYNOMIALS

In this section we'll divide special types of expressions called *polynomials*. The expression $3x^2 - 8x + 9$ is a polynomial in x. In general, a **polynomial** in x is an expression in which *each term is either a constant or of the form* ax^n; here a is a fixed number and n is a positive integer. In $3x^2 - 8x + 9$, the term with the greatest power of x is $3x^2$. We say that the exponent 2 in this term is the **degree** of the polynomial. The coefficient 3 of this term is the **leading coefficient** of the polynomial.

$3x^2 - 8x + 9$ is a polynomial of degree 2
and has leading coefficient 3.

EXAMPLE 1

a. $y^6 + 7y^4 - 4y - 5$ is a polynomial in y.

Degree: 6. *Leading coefficient*: 1 (since $y^6 = 1 \cdot y^6$).

b. $3x^3 - 7x^8$ is a polynomial in x. The term with the greatest power of x is $-7x^8$, and so $3x^3 - 7x^8$ has

Degree: 8. *Leading coefficient*: -7.

c. $\dfrac{3}{x^2}$ is *not* a polynomial, because it is not a constant or of the form ax^n where n is a positive integer.

d. $6x$ is a polynomial in x.

Degree: 1. *Leading coefficient*: 6.

In Example 1, the polynomials in parts a and b are also multinomials. In c and d, both expressions are monomials, but only d is a polynomial.

To divide polynomials we use so-called "long division." Before we show you how, let's look at long division with numbers. In this way we can recall some words used in describing division.

Let's find $\frac{25}{6}$ (25 divided by 6):

$$\begin{array}{r} 4 \\ 6\overline{)25} \\ \underline{24} \\ 1 \end{array}$$

6 is the *divisor*;
25 is the *dividend*;
4 is the *quotient*;
1 is the *remainder*.

We usually write the answer as $4\frac{1}{6}$ or $4 + \frac{1}{6}$. Thus,

$$\frac{25}{6} = 4 + \frac{1}{6} = \text{quotient} + \frac{\text{remainder}}{\text{divisor}}.$$

You can check a division by making sure that

$$(\text{quotient})(\text{divisor}) + \text{remainder} = \text{dividend}.$$
$$(4)(6) + 1 = 25$$
$$25 = 25. \checkmark$$

Now we're ready for division of polynomials. For example,

$$\frac{6x^2 - 13x + 10}{3x - 2}.$$

We'll go through it step by step.

$$\begin{array}{r} 2x \qquad\qquad\quad \\ 3x - 2 \overline{\smash{)}\, 6x^2 - 13x + 10} \end{array}$$

divide $6x^2$ by $3x$

Divide $6x^2$ by $3x$ and get $2x$.

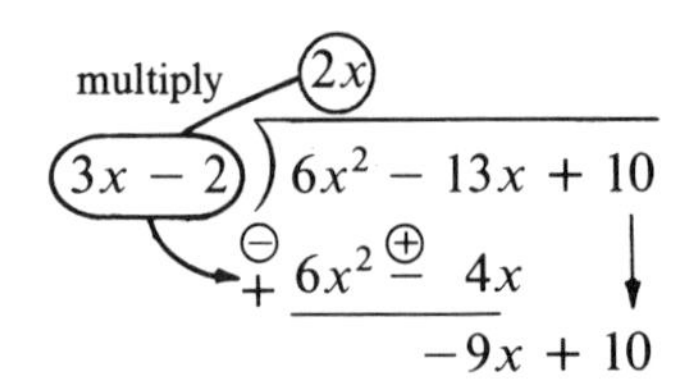

Multiply $2x$ by $3x - 2$ and get $6x^2 - 4x$. Then *subtract* this from $6x^2 - 13x$. So we change signs and proceed as in addition. The circles show the sign changes. We get $-9x$. Then bring down the $+10$.

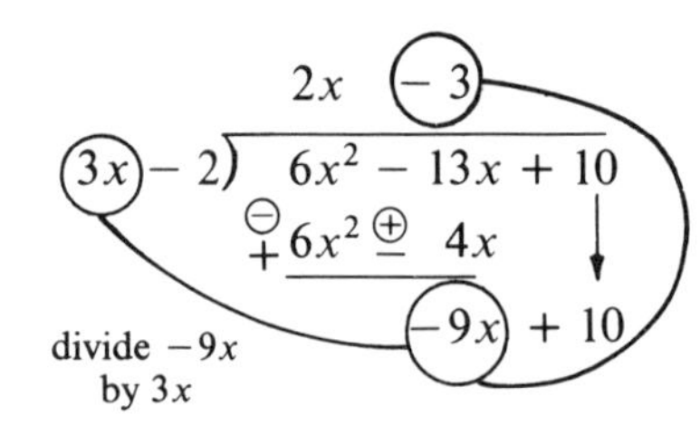

Repeat process again. Divide $-9x$ by $3x$ and get -3.

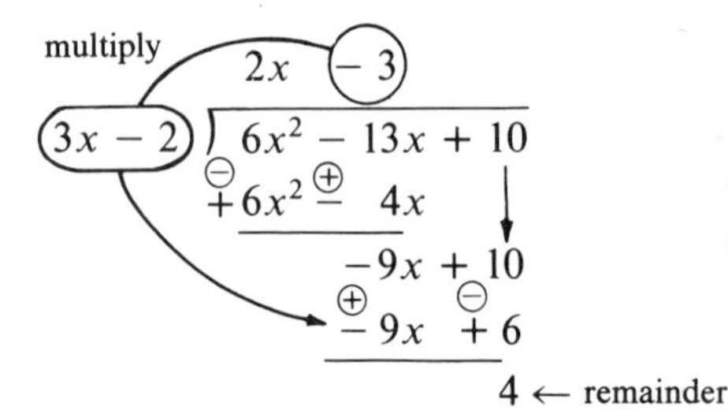

Multiply -3 by $3x - 2$ and get $-9x + 6$. Then *subtract* this from $-9x + 10$ and get 4, the remainder.

Thus,

$$\frac{6x^2 - 13x + 10}{3x - 2} = \text{quotient} + \frac{\text{remainder}}{\text{divisor}} = 2x - 3 + \frac{4}{3x - 2}.$$

Just as in arithmetic, we can check our division by making sure that

$$(\text{quotient})(\text{divisor}) + \text{remainder} = \text{dividend}.$$

$$\begin{aligned}(2x - 3)(3x - 2) + 4 &= 2x(3x - 2) - 3(3x - 2) + 4\\ &= 6x^2 - 13x + 10. \checkmark\end{aligned}$$

EXAMPLE 2

Divide $\dfrac{2 + 9x - 5x^2 - 2x^3}{x + 3}$.

When you write the divisor and dividend in long division, *make sure that the powers of x decrease from left to right and that the constant term is on the right*. So, write $2 + 9x - 5x^2 - 2x^3$ as $-2x^3 - 5x^2 + 9x + 2$.

$$\begin{array}{r}
-2x^2 + x + 6 \\
x + 3\,\overline{)\,-2x^3 - 5x^2 + 9x + 2} \\
\overset{\oplus}{-}2x^3 \overset{\oplus}{-} 6x^2 \qquad\qquad \\
\hline
x^2 + 9x \qquad \\
\overset{\ominus}{+}\; x^2 \overset{\ominus}{+} 3x \qquad \\
\hline
6x + 2 \\
\overset{\ominus}{+}\, 6x \overset{\ominus}{+} 18 \\
\hline
-16
\end{array}$$

Thus,

$$\frac{2 + 9x - 5x^2 - 2x^3}{x + 3} = -2x^2 + x + 6 - \frac{16}{x + 3}.$$

EXAMPLE 3

Divide $2y^4 - 5y^3 - y - 4$ *by* $2y - 1$.

In the dividend $2y^4 - 5y^3 - y - 4$, the powers of y decrease from left to right and the constant term is on the right. But there is a term "missing": the y^2-term. When powers of y or the constant term are missing, we fill in with zeros. Thus we write the dividend as $2y^4 - 5y^3 + 0y^2 - y - 4$.

$$
\begin{array}{r l}
 & \quad y^3 - 2y^2 - y - 1 \\
2y - 1 \big) & \quad 2y^4 - 5y^3 + 0y^2 - y - 4 \\
 & \overset{\ominus}{+} 2y^4 \overset{\oplus}{-} y^3 \\
\hline
 & \qquad -4y^3 + 0y^2 \\
 & \qquad \overset{\oplus}{-} 4y^3 \overset{\ominus}{+} 2y^2 \\
\hline
 & \qquad\qquad -2y^2 - y \\
 & \qquad\qquad \overset{\oplus}{-} 2y^2 \overset{\ominus}{+} y \\
\hline
 & \qquad\qquad\qquad -2y - 4 \\
 & \qquad\qquad\qquad \overset{\oplus}{-} 2y \overset{\ominus}{+} 1 \\
\hline
 & \qquad\qquad\qquad\qquad -5
\end{array}
$$

Thus,

$$\frac{2y^4 - 5y^3 - y - 4}{2y - 1} = y^3 - 2y^2 - y - 1 - \frac{5}{2y - 1}.$$

Completion Set 5-3

1. Circle the expressions that are polynomials.

a. $3x + 4$ b. $9y$ c. $\frac{1}{x}$ d. $-5z^3$

In Problems **2–9**, *fill in the blanks.*

2. The degree of the polynomial $7x^3 - 4x^2 + 5$ is ____.

3. The leading coefficient of the polynomial $5x - 3x^2$ is ______.

4. When dividing $1 - x + x^2$ by $x + 1$, we write the dividend as ____________________.

5. When dividing $x^2 + 1$ by $x - 5$, we write the dividend as ____________________.

6. When dividing $1 - x^2$ by $x + 2$, we write the dividend as ____________________.

7. In the division $x - 3\overline{)3x^2 - 5x + 4}$, the first term of the quotient is ______.

8. In the division $2x + 5\overline{)-8x^2 + x + 1}$, the first term of the quotient is _______.

9. To check a long division we see if

dividend = (____________________)(divisor) + remainder.

Problem Set 5-3

In Problems **1–10**, *do the divisions and check your answers.*

1. $\dfrac{2x^2 + 3x - 4}{x - 2}$ **2.** $\dfrac{9x^2 - 6x - 6}{3x - 1}$

3. $\dfrac{x + 3}{x + 2}$ **4.** $\dfrac{x}{x + 1}$

5. $\dfrac{4x^2 - 7x - 5}{4x + 1}$ **6.** $\dfrac{5x^2 + 26x + 8}{x + 5}$

7. $\dfrac{4x^2 - 5}{2x + 3}$ **8.** $\dfrac{3 + 6x - 10x^2}{5x - 3}$

9. $\dfrac{x^3 + 2x^2 - 5x + 2}{x - 1}$ **10.** $\dfrac{3x^3 + 2x^2 + 3x + 1}{3x + 2}$

In Problems **11–18**, *do the divisions.*

11. $\dfrac{1 + 2x - x^2 - 2x^3 - 3x^4}{3x + 5}$ **12.** $\dfrac{8x^4 - 4x^2 + x}{2x + 1}$

13. $\dfrac{8x^3 - 2x^2 + 4x - 3}{4x - 1}$ **14.** $\dfrac{3x^3 + x^2 - 6x + 1}{3x + 1}$

15. $\dfrac{x^4 - 2x^2 + 1}{x - 1}$ **16.** $\dfrac{x^4 - 8x^2 + 16}{x + 2}$

17. $\dfrac{81 - x^4}{x + 3}$ **18.** $\dfrac{x^3}{x - 4}$

5-4 SYNTHETIC DIVISION[†]

There is an easy way to find the quotient and remainder when a polynomial is divided by $x - a$. The method is called **synthetic division**. To show you how it works, we'll use it to divide $3x^3 - 5x^2 - 3x - 4$ by $x - 2$.

A three-line arrangement is used. On the first line the coefficients in the dividend $3x^3 - 5x^2 - 3x - 4$ are written. Off to the right we set off the 2 that appears in the divisor $x - 2$ (that is, the a in $x - a$).

coefficients of dividend

$$
\begin{array}{llllll}
\text{line 1} \rightarrow 3 & -5 & -3 & -4 & \underline{|2} \leftarrow \text{the } a \text{ in } x - a \\
\text{line 2} \rightarrow \underline{} & & & & \\
\text{line 3} \rightarrow & & & &
\end{array}
$$

[†]This section can be omitted if Section 10-2 is not covered.

The third line begins with the leading coefficient, 3, of the dividend.

$$\begin{array}{rrrr|l} 3 & -5 & -3 & -4 & \underline{2} \\ \downarrow & & & & \\ \hline 3 & & & & \end{array}$$

Next, we multiply the 3 on the third line by the 2 (= a) and put this result, 6, in the second line under the coefficient -5. Then we *add* the -5 and 6 and write their sum, 1, on the third line.

$$\begin{array}{rrrr|l} 3 & -5 & -3 & -4 & \underline{2} \\ \downarrow 3\cdot 2 & 6 & & & \\ \hline 3 & 1 & & & \end{array}$$

In the same way, we multiply the 1 by 2 (= a) and put the result, 2, under the -3. Again, add and put the sum on the third line.

$$\begin{array}{rrrr|l} 3 & -5 & -3 & -4 & \underline{2} \\ \downarrow & 6 & 2 & & \\ \hline 3 & 1 & -1 & & \end{array}$$

Now, multiply -1 by 2 and put the result, -2, under the -4 and add.

$$\begin{array}{rrrr|l} 3 & -5 & -3 & -4 & \underline{2} \\ \downarrow & 6 & 2 & -2 & \\ \hline 3 & 1 & -1 & -6 & \end{array}$$

The last number on the third line, -6, is the constant remainder. The other numbers on the third line are the coefficients of the quotient. The first number, 3, is the coefficient of the leading term of the quotient. The degree of the quotient is one less than the degree of the dividend. Thus the quotient has degree $3 - 1 = 2$ and must be $3x^2 + x - 1$. Here's a summary.

$$\textit{dividend} = 3x^3 - 5x^2 - 3x - 4$$

$$\begin{array}{rrrr|l} 3 & -5 & -3 & -4 & \underline{2} \\ \downarrow & 6 & 2 & -2 & \\ \hline 3 & 1 & -1 & -6 & \end{array}$$

$$\textit{quotient} = 3x^2 + x - 1 \qquad \textit{remainder} = -6$$

Keep two things in mind. *First*, the dividend must be arranged in order of decreasing powers of x and the constant term must be on the right. *Second*, if any powers of x or the constant term are "missing" in the dividend, zeros must be inserted in the proper places on the first line.

EXAMPLE 1

Use synthetic division to find the quotient and remainder when $x^4 - 3x^3 - 2x + 1$ is divided by $x + 1$.

Since the x^2-term is missing, we think of the dividend as $x^4 - 3x^3 + 0x^2 - 2x + 1$. Also, $x - a = x + 1 = x - (-1)$, and so $a = -1$. Since the dividend is of degree 4, the quotient will have degree 3.

$$\begin{array}{rrrrr|l} 1 & -3 & 0 & -2 & 1 & \underline{-1} \\ \downarrow & -1 & 4 & -4 & 6 & \\ \hline 1 & -4 & 4 & -6 & 7 & \end{array}$$

$$\text{quotient} = x^3 \quad -4x^2 + 4x \quad -6 \qquad \text{remainder} = 7$$

Thus the quotient is $x^3 - 4x^2 + 4x - 6$ and the remainder is 7.

EXAMPLE 2

Use synthetic division to find $\dfrac{2x + 2x^2 + 3x^3 - 4x^4 + x^5}{x - 3}$.

The dividend, in order of decreasing exponents, is $x^5 - 4x^4 + 3x^3 + 2x^2 + 2x + 0$. Since $x - a = x - 3$, clearly $a = 3$. The quotient will have degree 4.

$$\begin{array}{rrrrrr|l} 1 & -4 & 3 & 2 & 2 & 0 & \underline{3} \\ \downarrow & 3 & -3 & 0 & 6 & 24 & \\ \hline 1 & -1 & 0 & 2 & 8 & 24 & \end{array}$$

The quotient is $x^4 - x^3 + 0x^2 + 2x + 8$ or $x^4 - x^3 + 2x + 8$. The remainder is 24. Thus,

$$\frac{2x + 2x^2 + 3x^3 - 4x^4 + x^5}{x - 3} = x^4 - x^3 + 2x + 8 + \frac{24}{x - 3}.$$

EXAMPLE 3

Use synthetic division to find $\dfrac{4t^4 + 8t^3 - 3t - 6}{t + 2}$.

The quotient will have degree 3.

$$\begin{array}{rrrrr|l} 4 & 8 & 0 & -3 & -6 & \underline{-2} \\ \downarrow & -8 & 0 & 0 & 6 & \\ \hline 4 & 0 & 0 & -3 & 0 & \end{array}$$

The quotient is $4t^3 + 0t^2 + 0t - 3$ or $4t^3 - 3$. The remainder is 0. Now,

$$\text{dividend} = (\text{quotient})(\text{divisor}) + \text{remainder}.$$

Thus,

$$\begin{aligned} 4t^4 + 8t^3 - 3t - 6 &= (4t^3 - 3)(t + 2) + 0 \\ &= (4t^3 - 3)(t + 2). \end{aligned}$$

This means that $t + 2$ is a factor of $4t^4 + 8t^3 - 3t - 6$. In fact, *whenever a remainder is* 0, *then the divisor is a factor of the dividend.*

Problem Set 5-4

In Problems **1–12**, *find the quotient and remainder when*

1. $x^3 - 3x^2 + 3x - 4$ is divided by $x - 2$.

2. $2x^3 + 4x^2 - 6x - 2$ is divided by $x + 3$.

3. $3x^4 - 3x^3 + 2x^2 - 8x + 1$ is divided by $x + 1$.

4. $2x^3 - 8x^2 - 6x + 3$ is divided by $x - 5$.

5. $2t^4 + 5t^3 + 6t - 2$ is divided by $t + 3$.

6. $x^4 - 2x^3 + 6x - 4$ is divided by $x - 2$.

7. $12 - 20x + 3x^2 - 6x^3 + x^4$ is divided by $x - 6$.

8. $4y^5 + 15y^4 - 6y^3 - 10y^2 - 8y$ is divided by $y + 4$.

9. $x^3 - 5x^2 + 7x$ is divided by $x - 3$.

10. $x^3 - 8$ is divided by $x - 2$.

11. $x^3 - 1$ is divided by $x - 1$.

12. $x^5 + 1$ is divided by $x + 1$.

13. By using synthetic division, show that $x - 3$ is a factor of $x^4 - x^3 + 2x^2 - 72$. *Hint*: See Example 3.

14. As in Problem 13, show that $x - \frac{1}{4}$ is a factor of $4x^3 + 7x^2 + 2x - 1$.

5-5 REVIEW

IMPORTANT TERMS

monomial *(p. 107)*

polynomial *(p. 120)*

degree of a polynomial *(p. 121)*

multinomial *(p. 107)*

leading coefficient *(p. 121)*

synthetic division *(p. 125)*

REVIEW PROBLEMS

In Problems **1–38**, *do the indicated operations.*

1. $(2x^2yz)(xy^3z^6)$

2. $3xy(-2xy^2)$

3. $8ab^2(3a^2b)^2$

4. $-xy^3(-3xz)$

5. $(2x^2y)^2(xy)^3(xy^2)$

6. $(xy)^2(xz)^2(yz)^2$

7. $x(x^2 - 2x + 4)$

8. $x^2y(xy - xz + yz)$

9. $a^2b(-2a^2b + 2ab - 3)$

10. $(2xy)^2(x - y + 2xy)$

11. $(x + 3)(x - 4)$

12. $(y - 4)(y^2 - 3y + 5)$

13. $(x + 4)(x - 4)$

14. $(3 - x)(3 + x)$

15. $(x - 3)(x - 2)$

16. $(x - 3)^2$

17. $(x + 2y)^2$

18. $(x^2 + 1)(x^2 - 2)$

19. $(x^2 + 3)(x - 4)$

20. $(x - 3y)^2$

21. $(3x - 1)(2x^3 - 3x^2 + 5)$

22. $(y - 4)(y^2 - 3y + 5)$

23. $(x + y + 2)(2x - 3y + 1)$

24. $(1 + x + y)(1 - x - y)$

25. $3x(x - 4) - 2(x^2 - 9)$

26. $2x^2(x^2 - xy) + 2(x^4 - x^3y)$

27. $\dfrac{ax^2y^5}{x^3y^3}$

28. $\dfrac{2xy^2}{4y^3}$

29. $\dfrac{(2ab)(a^2b)}{10a^2}$

30. $\dfrac{(ab)^2(2ab^2)}{2ab^3}$

31. $\dfrac{(-3x^2y)(2xy)}{(4x^2y)^2}$

32. $\dfrac{(xy^2)^2(-3x)^3}{9x(xy)}$

33. $\dfrac{x^2 - 5x + 7}{x}$

34. $\dfrac{-3x^3 - 5x^2 + 6}{30x}$

35. $\dfrac{x^2y - 5xy^3 + 7xy}{xy^2}$

36. $\dfrac{6x^2y - 3y^2 + 2x - 2}{-4x}$

37. $\dfrac{2x^2y + (2xy^2w)^2 - 4x^3y^3w}{-2xy}$

38. $\dfrac{3xy^2 - xy^2 + 3}{xy^2}$

In Problems **39–44**, *do the long divisions.*

39. $\dfrac{6x^3 + 3x^2 - 5x - 1}{2x - 1}$

40. $\dfrac{3x^3 - 5x^2 + x + 1}{3x + 1}$

41. $\dfrac{x^4 + x^3 + 8x - 30}{x + 3}$

42. $\dfrac{5 - x + 4x^2 - 3x^3 - 4x^4}{4x + 3}$

43. $\dfrac{3x + 3x^3 - 2x^4}{2x + 1}$

44. $\dfrac{9 - x^3}{x - 2}$

In Problems **45–48**, *use synthetic division to find the quotient and remainder when*

45. $x^2 + 5x + 6$ is divided by $x + 2$.

46. $x^4 + 2x^2 + 1$ is divided by $x - 1$.

47. $2x^3 - 5x^2 + 3x$ is divided by $x - 2$.

48. $x^4 - 2x^2 + 1$ is divided by $x + 1$.

49. A man can row 7 mi/hr in still water. The current of a stream is 2 mi/hr. How far upstream can he row if he is to be back at his starting point in 2 hr? *Hint*: Upstream he goes 5 mi/hr (which is rate − current), and downstream he goes 9 mi/hr (which is rate + current).

50. How many gallons of antifreeze that is 70% alcohol (by volume) must be added to 10 gallons of a 35% solution to get a 50% solution?

51. A company manufactures Unplug drain cleaner. The cleaner consists of a chemical compound and metal shavings. The chemical compound not only loosens grease, but when dissolved in water it gives off heat, which speeds up the reaction, and it reacts with the metal to generate hydrogen, which also loosens dirt and grease. The company markets two forms of the cleaner: Industrial Strength Unplug, of which 9% is metal shavings (by weight); and Household Strength Unplug, of which 6% is metal shavings. The Holly Inns Corp., a motel chain, has placed an order with the company to supply them with 12,000 kilograms (kg) of new Motel Strength Unplug, which is 8% metal shavings. To fill the order, the company will mix the industrial and household forms of Unplug. How many kg of each should go into the mixture?

CHAPTER 6

Special Products and Factoring

6-1 SPECIAL PRODUCTS

In mathematics certain products occur so often that we find it worthwhile to memorize their patterns. But just "knowing" these so-called *special products* is not enough. You have to be so familiar with them that you can recognize them in any form.

All of the special products come from the distributive law. We'll show this in the first special product. You can check the others for yourself.

To begin, let's look at the square of $a + b$. We call $a + b$ a **binomial** because it has exactly two terms.

$$(a + b)^2 = (a + b)(a + b)$$

$$= a(a + b) + b(a + b) = a^2 + ab + ba + b^2$$

$$(a + b)^2 = a^2 + 2ab + b^2.$$

Since a is the first term of $a + b$ and b is the second term, then

$$(a+b)^2 = \begin{pmatrix}\text{square of}\\ \text{first term}\end{pmatrix} + \begin{bmatrix}\text{twice the}\\ \text{product of}\\ \text{the terms}\end{bmatrix} + \begin{pmatrix}\text{square of}\\ \text{second term}\end{pmatrix}.$$

Similarly, $(a - b)^2 = a^2 - 2ab + b^2$. These are our first two special products.

SQUARE OF BINOMIAL

$$(a + b)^2 = a^2 + 2ab + b^2;$$

$$(a - b)^2 = a^2 - 2ab + b^2.$$

Note that the square of a binomial has exactly *three* terms. Such expressions are called **trinomials**.

$(a + b)^2 \neq a^2 + b^2$ *and* $(a - b)^2 \neq a^2 - b^2$. *For example,*

$$(6 + 3)^2 \neq 6^2 + 3^2 \text{ because } 81 \neq 45.$$

EXAMPLE 1

Square of a binomial: $(a + b)^2 = a^2 + 2ab + b^2$.

a. $(x + 3)^2$. Here x plays the role of a, and 3 plays the role of b.

$$(x + 3)^2 = x^2 + 2(x)(3) + 3^2 = x^2 + 6x + 9.$$

b. $(2x + 4)^2$. Here $2x$ plays the role of a.

$$(2x + 4)^2 = (2x)^2 + 2(2x)(4) + 4^2 = 4x^2 + 16x + 16.$$

c. $(x^2 + 4y)^2$. Here x^2 is a and $4y$ is b.

$$\begin{aligned}(x^2 + 4y)^2 &= (x^2)^2 + 2(x^2)(4y) + (4y)^2\\ &= x^4 + 8x^2y + 16y^2.\end{aligned}$$

d. $-3y(2a + b)^2$. First we square the binomial.

$$\begin{aligned}-3y(2a + b)^2 &= (-3y)\left[(2a)^2 + 2(2a)(b) + b^2\right]\\ &= (-3y)[4a^2 + 4ab + b^2]\\ &= -12a^2y - 12aby - 3b^2y.\end{aligned}$$

EXAMPLE 2

Square of a binomial: $(a - b)^2 = a^2 - 2ab + b^2$.

a. $(x - 4)^2 = x^2 - 2(x)(4) + 4^2 = x^2 - 8x + 16$.

b. $(3 - 2x)^2$. Here 3 plays the role of a, and $2x$ plays the role of b.

$$\begin{aligned}(3 - 2x)^2 &= 3^2 - 2(3)(2x) + (2x)^2 \\ &= 9 - 12x + 4x^2.\end{aligned}$$

c. $(x^3 - 2y^2)^2$. Here x^3 is a and $2y^2$ is b.

$$\begin{aligned}(x^3 - 2y^2)^2 &= (x^3)^2 - 2(x^3)(2y^2) + (2y^2)^2 \\ &= x^6 - 4x^3y^2 + 4y^4.\end{aligned}$$

d. $(-m + n)^2 = (n - m)^2$

$= n^2 - 2mn + m^2$.

Our next special product involves the product of the sum and difference of two terms.

PRODUCT OF THE SUM AND DIFFERENCE

$$(a + b)(a - b) = a^2 - b^2.$$

That is, the product of the sum and difference of two terms is equal to the square of the first term, minus the square of the second term.

EXAMPLE 3

Product of the sum and difference: $(a + b)(a - b) = a^2 - b^2$.

a. $(x + 5)(x - 5) = x^2 - 5^2 = x^2 - 25$.

b. $(7 + y)(7 - y) = 7^2 - y^2 = 49 - y^2$.

c. $(4x - 3)(4x + 3)$. Here $4x$ plays the role of a, and 3 plays the role of b.

$$(4x - 3)(4x + 3) = (4x)^2 - 3^2 = 16x^2 - 9.$$

d. $(x^2 - 1)(x^2 + 1) = (x^2)^2 - 1^2 = x^4 - 1$.

e. $(2x^2 + 3y)(3y - 2x^2)$. This will have the special form if we rewrite the first factor.

$$(3y + 2x^2)(3y - 2x^2) = (3y)^2 - (2x^2)^2$$
$$= 9y^2 - 4x^4.$$

f. $(x - \sqrt{5})(x + \sqrt{5}) = x^2 - (\sqrt{5})^2 = x^2 - 5.$

Another special product involving binomials is

$$(x + a)(x + b) = x^2 + (a + b)x + ab.$$

EXAMPLE 4

The product of the binomials $(x + a)(x + b)$.

a. $(x + 3)(x + 4) = x^2 + (3 + 4)x + (3 \cdot 4)$
$= x^2 + 7x + 12.$

b. $(x + 5)(x + 1) = x^2 + (5 + 1)x + (5 \cdot 1)$
$= x^2 + 6x + 5.$

c. $(x - 2)(x + 6)$. Here -2 is a since $x - 2 = x + (-2)$.

$$(x - 2)(x + 6) = x^2 + (-2 + 6)x + (-2)(6)$$
$$= x^2 + 4x - 12.$$

d. $(y + 1)(y - 5)$. Here -5 is b.

$$(y + 1)(y - 5) = y^2 + (1 - 5)y + (1)(-5)$$
$$= y^2 - 4y - 5.$$

e. $(x - 3)(x - 2)$. Here -3 is a and -2 is b.

$$(x - 3)(x - 2) = x^2 + (-3 - 2)x + (-3)(-2)$$
$$= x^2 - 5x + 6.$$

The products in Example 4 can be found another way. We show you how by redoing Example 4c: $(x - 2)(x + 6)$.

Step 1. Multiply the first terms in the binomials to get the first term of the result.

$$(x - 2)(x + 6) = \underline{x^2 \qquad\qquad}$$

Step 2. The middle term of the result is the product of the inner terms of the binomials, *plus* the product of the outer terms.

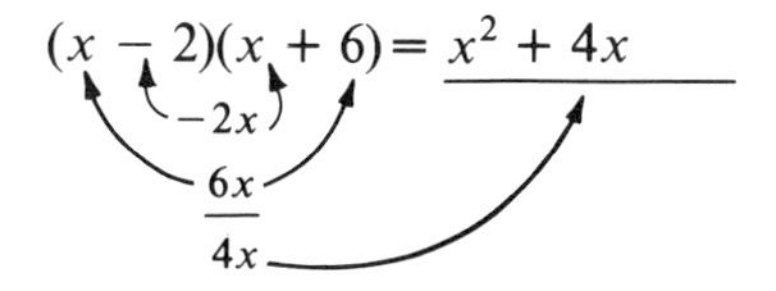

Step 3. Multiply the last terms in the binomials to get the last term of the result.

$$(x - 2)(x + 6) = \underline{x^2 + 4x - 12.}$$

The 3-step method above can be used for multiplying many types of binomials.

EXAMPLE 5

Find $(4x + 1)(5x - 3)$ *by the 3-step method.*

Step 1. $(4x + 1)(5x - 3) = \underline{20x^2 \qquad\qquad}$

Step 2. $(4x + 1)(5x - 3) = \underline{20x^2 - 7x \qquad}$

$$\begin{array}{r} 5x \\ -12x \\ \hline -7x \end{array}$$

Step 3. $(4x + 1)(5x - 3) = \underline{20x^2 - 7x - 3.}$

In Example 5 we spread our work over three lines so that you could follow it. But you really should do it in one line.

After some practice you should be able to find special products in your head. Memorize the formulas and the 3-step method until you know them "cold."

EXAMPLE 6

Problems involving special products.

a. $(x^2 + 1)(x - 1)(x + 1) = (x^2 + 1)[(x - 1)(x + 1)]$
$= (x^2 + 1)(x^2 - 1) = x^4 - 1.$

b. $(x + 2)(x - 2) + (x + 3)^2 = [x^2 - 4] + [x^2 + 6x + 9]$
$= 2x^2 + 6x + 5.$

c. $(x - 3)^2 - (4 - x)^2 = x^2 - 6x + 9 - (16 - 8x + x^2)$
$= x^2 - 6x + 9 - 16 + 8x - x^2$
$= 2x - 7.$

d. $(x - 1)(x + 3)^2 = (x - 1)(x^2 + 6x + 9)$
$= x(x^2 + 6x + 9) - 1(x^2 + 6x + 9)$
$= x^3 + 6x^2 + 9x - x^2 - 6x - 9$
$= x^3 + 5x^2 + 3x - 9.$

In Example 6c, don't let the minus sign in front of $(4 - x)^2$ throw you off. You should keep the parentheses until after squaring this binomial.

$$(x - 3)^2 - (4 - x)^2 \neq x^2 - 6x + 9 - 16 - 8x + x^2.$$

Completion Set 6-1

*In Problems **1–3**, insert "+" or "−".*

1. $(x + 1)^2 = x^2$ ____ $2x$ ____ $1.$

2. $(x - 3)^2 = x^2$ ____ $6x$ ____ $9.$

3. $(x - 8)(x + 8) = x^2$ ____ $64.$

*In Problems **4–13**, fill in the blanks.*

4. $(x + 5)^2 = x^2 + (2)($____$)x + ($____$)^2 = x^2 +$ ______ $x +$ ______.

5. $(2x - 3)^2 =$ _______ − ______ $x + 9.$

6. $(x + 5)(x - 5) = x^2 - ($____$)^2 = x^2 -$ ______.

7. $(x + 3)(x - 3) = (\_\_\_)^2 - (\_\_\_)^2 = \_\_\_\_ - \_\_\_.$

8. $(x + 3)(x + 2) = x^2 + 5x + \_\_\_.$

9. $(x + 1)(x + 5) = x^2 + \_\_\_\ x + 5.$

10. $(x + 7)(x + 9) = x^2 + \_\_\_\_\ x + \_\_\_\_.$

11. $(x + 5)(x - 2) = x^2 + \_\_\_\ x - \_\_\_\_.$

12. $2x(x - 3)^2 = 2x(x^2 - 6x + \_\_\_) = \_\_\_\_ - \_\_\_\_\_\_ + \_\_\_\_.$

13. An expression having exactly two terms is called a ____________; one having exactly three terms is a ____________.

Problem Set 6-1

In Problems **1–62**, *find the products by inspection. Do as many as you can in your head.*

1. $(x + 4)^2$
2. $(x + 6)^2$
3. $(y + 10)^2$
4. $(z + 5)^2$
5. $(4x + 1)^2$
6. $(2x + 10)^2$
7. $(x + \frac{1}{2})^2$
8. $(2 + x)^2$
9. $(x - 6)^2$
10. $(x - 2)^2$
11. $(3x - 2)^2$
12. $(3 - x)^2$
13. $(2x - y)^2$
14. $(2x - 3)^2$
15. $(3x + 3)^2$
16. $(2 - t)^2$
17. $(2 - 4y)^2$
18. $(3y - 4x)^2$
19. $(x + 3)(x - 3)$
20. $(x + 2)(x - 2)$
21. $(x - 9)(x + 9)$
22. $(7 - x)(7 + x)$
23. $(1 + x)(1 - x)$
24. $(4 + 2t)(4 - 2t)$
25. $(3x^2 + 5)(3x^2 - 5)$
26. $(2 - 3x)(2 + 3x)$
27. $(2y + 3x)(2y - 3x)$
28. $(y - 8)(y + 8)$
29. $(6 - x)(x + 6)$
30. $(x + yz)(x - yz)$
31. $(x + 8)(x + 3)$
32. $(x - 6)(x - 1)$
33. $(x + 4)(x + 1)$
34. $(x + 3)(x + 6)$
35. $(x - 2)(x + 1)$
36. $(x + 5)(x + 4)$
37. $(t + 7)(t - 5)$
38. $(x - 2)(x + 14)$
39. $(x - 2)(x - 3)$
40. $(x - 6)(x - 2)$
41. $(x - 4)(x - 5)$
42. $(x - 1)(x - 2)$
43. $(y - 2)(y + 3)$
44. $(t - 4)(t + 3)$
45. $(x + 3)(2x - 4)$

46. $(2x + 1)(2x + 2)$ **47.** $(4x - 2)(2x - 1)$ **48.** $(x - 3)(4x - 3)$

49. $(2x + 1)(3x - 3)$ **50.** $(2x + 1)(x - 2)$ **51.** $(5x - 4)(2x - 1)$

52. $(x + \sqrt{13})(x - \sqrt{13})$ **53.** $(2 - 5t)(1 + 7t)$ **54.** $(4 - t)(4 + 3t)$

55. $(xy^2 + a)^2$ **56.** $(x^3 - 1)(x^3 - 1)$ **57.** $(x^2 - y^2)(x^2 + y^2)$

58. $(x^2 - 2)^2$ **59.** $(x - \sqrt{7})(x + \sqrt{7})$ **60.** $(ab - c)(ab + c)$

61. $(-2x + 1)^2$ **62.** $(t - 2s)(t + 2s)$

*In Problems **63–80**, perform the operations.*

63. $4(x + 3)^2$ **64.** $x(x - 2)^2$ **65.** $2x(y - 3)(y + 3)$

66. $-2a(a + 4)^2$ **67.** $(a^2b - 2m^2n)^2$ **68.** $(x - 1)(x + 1)(x)$

69. $(x + 2)(x - 2)(x^2 + 4)$ **70.** $(x + 1)(x + 1)(x + 2)$

71. $(a - b)^2 - (b - a)^2$ **72.** $(x + y)^2 - (y - x)^2$

73. $(2x - 3)^2 + (x + 1)(x + 2)$ **74.** $(t + 3)(t - 3) - (t + 3)^2$

75. $(x + 2)(x - 2)^2$ **76.** $a(a - b)^2 - b(b - a)^2$

77. $[4x(x - 3)]^2$ **78.** $[2x^2(x - y)]^2$

79. $3x(x + 2) - (\sqrt{3}x - 1)(\sqrt{3}x + 1)$ **80.** $(5x^2 + 1)(\sqrt{5}x + 1)(\sqrt{5}x - 1)$

81. Use the formula

$$(a + b)^3 = a^3 + 3a^2b + 3ab^2 + b^3$$

to find $(x + 2)^3$.

82. Use the formula

$$(a - b)^3 = a^3 - 3a^2b + 3ab^2 - b^3$$

to find $(x - 1)^3$.

6-2 FACTORING

In this section we'll write expressions as products of factors. This is called *factoring*. It will come in handy when we solve equations in the next chapter.

Let's factor $3x + 3y$. Here 3 is a factor of *each* of the terms $3x$ and $3y$. This *common factor* 3 can be "yanked out" of $3x + 3y$ by the distributive law.

$$3x + 3y = 3(x + y).$$

Thus $3x + 3y$ is factored. We say that a *common factor was removed* from each term.

In the expression $4x - 6y + 2$, we see that 2 is a factor not only of $4x$ (since $4x = 2 \cdot 2x$), but also $6y$ (since $6y = 2 \cdot 3y$) and 2 (since $2 = 2 \cdot 1$). Thus we can remove the common factor 2 from each term.

$$4x - 6y + 2 = 2(2x - 3y + 1).$$

Common factors can be numbers, letters, or both.

EXAMPLE 1

Removing a common factor.

a. $5x - 30 = 5(x - 6)$.

b. $8x + 12y = 4(2x + 3y)$.

c. $6x + 2 = 2(3x + 1)$.

d. $ax - ab = a(x - b)$.

e. $xy - 3xz + x = x(y - 3z + 1)$.

*Be careful when removing common factors. You can remove only factors that are common to **all** terms. In $6x + 6y + z$, the 6 is a factor of only two terms.*

$$6x + 6y + z \neq 6(x + y + z).$$

Let's factor $4bx^2 + 16bx^6$. Both 4 and b are factors of both terms.

$$4bx^2 + 16bx^6 = 4b(x^2 + 4x^6).$$

But this is not completely factored. We can factor some more, since x^2 is common to x^2 and $4x^6 (= 4 \cdot x^2 \cdot x^4)$. Thus,

$$4bx^2 + 16bx^6 = 4b(x^2 + 4x^6) = 4bx^2(1 + 4x^4).$$

We've now factored as much as possible. *Always factor completely.*

EXAMPLE 2

Factoring completely.

a. $7ax + 14bx = 7x(a + 2b)$.

b. $x^4 + x^3 + x^2 = x^2(x^2 + x + 1)$. *Note*: If only an x were factored out, this would give $x(x^3 + x^2 + x)$, which is not completely factored.

c. $3a^2b^5 - 4a^3b^3 = a^2b^3(3b^2 - 4a)$.

d. $9x^2y + 3xy^2 - 6xy = 3xy(3x + y - 2)$.

EXAMPLE 3

Completely factor $(x - 1)^2(x + 2) + (x - 1)(x + 2)^2$.

Here the factors $(x - 1)$ and $(x + 2)$ are common to both terms.

$$\begin{aligned}(x - 1)^2(x + 2) + (x - 1)(x + 2)^2 &= (x - 1)(x + 2)[(x - 1) + (x + 2)] \\ &= (x - 1)(x + 2)(2x + 1).\end{aligned}$$

From the last section we know that $(a + b)(a - b) = a^2 - b^2$. Thus the *difference of two squares* will factor into a sum and difference.

$$\boxed{a^2 - b^2 = (a + b)(a - b).}$$

EXAMPLE 4

Factoring the difference of two squares.

a. $x^2 - 4 = x^2 - 2^2 = (x + 2)(x - 2)$.

b. $y^2 - \frac{1}{4} = y^2 - (\frac{1}{2})^2 = (y + \frac{1}{2})(y - \frac{1}{2})$.

c. $4x^2 - 1 = (2x)^2 - 1^2 = (2x + 1)(2x - 1)$.

d. $9x^2 - 16y^2 = (3x + 4y)(3x - 4y)$.

e. $z^2 - 5 = z^2 - (\sqrt{5})^2 = (z + \sqrt{5})(z - \sqrt{5})$. Remember, $5 = (\sqrt{5})^2$.

f. $x^4 - 1 = (x^2)^2 - 1^2 = (x^2 + 1)(x^2 - 1)$. Although $x^2 + 1$ doesn't factor, $x^2 - 1$ does, and so we have

$$x^4 - 1 = (x^2 + 1)(x + 1)(x - 1).$$

Sometimes we can factor trinomials that have leading coefficient 1 such as $x^2 - 4x - 12$, into a product of two binomials of the form $(x + a)$ and $(x + b)$. Suppose

$$x^2 - 4x - 12 = (x + a)(x + b).$$

Now we must determine the values of a and b that do the job.

By the 3-step rule of the last section,

$$(x + a)(x + b) = x^2 + ax + bx + ab = x^2 + (a + b)x + ab.$$

Thus, $$x^2 - 4x - 12 = x^2 + (a + b)x + ab.$$

Matching the expression on the right side with $x^2 - 4x - 12$, we must have

$$a + b = -4 \quad \text{and} \quad ab = -12.$$

There are several choices for a and b such that their product is -12:

-1 and 12	1 and -12
-2 and 6	2 and -6
-3 and 4	3 and -4

But since the sum of a and b must be -4, we choose $a = 2$ and $b = -6$. Thus,

$$x^2 - 4x - 12 = (x + 2)(x - 6).$$

You can always check yourself: multiply the right-hand side and see if you get the left-hand side.

EXAMPLE 5

Completely factor $x^2 - 5x + 4$.

Here we need to find two numbers whose product is $+4$ and whose sum is -5. The numbers -4 and -1 will do the job.

$$x^2 - 5x + 4 = (x - 4)(x - 1).$$

EXAMPLE 6

Factoring trinomials.

a. $x^2 + 9x + 20 = (x + 5)(x + 4)$.

b. $y^2 + 6y + 9 = (y + 3)(y + 3) = (y + 3)^2$. Thus $y^2 + 6y + 9$ is the square of a binomial.

c. $z^2 + 6z - 16 = (z - 2)(z + 8)$.

d. $z^2 - 6z - 16 = (z + 2)(z - 8)$. Don't confuse this with part c.

Let's tackle a trinomial like $6x^2 + 5x - 4$. If it factors into a product of two binomials, then we must have something like

$$6x^2 + 5x - 4 = (\underline{\quad} + \underline{\quad})(\underline{\quad} + \underline{\quad}).$$

(arrows: the product of the second terms is -4; the product of the first terms is $6x^2$)

The product of the first terms of the binomials must be $6x^2$, and the product of the last terms must be -4. There are many combinations of binomials that will do this. One is $(x + 4)(6x - 1)$. But here the sum of the products of the inner terms and outer terms, $23x$, is not equal to the middle term of $6x^2 + 5x - 4$, which is $5x$. By trial and error we try different combinations until we hit upon one that works.

$(x + 4)(6x - 1)$	$(2x + 2)(3x - 2)$
$(x - 4)(6x + 1)$	$(2x - 2)(3x + 2)$
$(x + 1)(6x - 4)$	$(2x + 4)(3x - 1)$
$(x - 1)(6x + 4)$	$(2x - 4)(3x + 1)$
$(x + 2)(6x - 2)$	$(2x + 1)(3x - 4)$
$(x - 2)(6x + 2)$	$(2x - 1)(3x + 4)$

The combination that works is $(2x - 1)$ and $(3x + 4)$.

$$6x^2 + 5x - 4 = (2x - 1)(3x + 4). \text{ Check it.}$$

EXAMPLE 7

Factoring trinomials completely.

a. $2x^2 + 5x + 2 = (2x + 1)(x + 2)$.

b. $8y^2 - 6y - 9 = (2y - 3)(4y + 3)$.

c. $9y^2 - 12y + 4 = (3y - 2)(3y - 2) = (3y - 2)^2$, a square of a binomial.

d. $x^4 - 3x^2 + 2 = (x^2 - 1)(x^2 - 2) = (x + 1)(x - 1)(x + \sqrt{2})(x - \sqrt{2})$.

When factoring, first remove any common factors, as the next example shows.

EXAMPLE 8

Factoring completely.

a. $4x^3 - 16x = 4x(x^2 - 4) = 4x(x + 2)(x - 2)$.

b. $8ay^2 + 8ay + 2a = 2a(4y^2 + 4y + 1)$

$$= 2a(2y + 1)(2y + 1) = 2a(2y + 1)^2.$$

c. $7a^5 + 7a^3 = 7a^3(a^2 + 1)$.

d. $6x^5 - 4x^3 - 2x = 2x(3x^4 - 2x^2 - 1)$

$= 2x(3x^2 + 1)(x^2 - 1)$

$= 2x(3x^2 + 1)(x + 1)(x - 1).$

Completion Set 6-2

In Problems **1–16**, *fill in the blanks.*

1. $5x + 25 = 5(x + 5)$ is an example of removing a common ______________.

2. $6x - 9y + 3z = 3($______ $-$ ______ $+$ ____$)$.

3. $x^6 + x^4 - x^3 = x^3($______ $+$ ____ $-$ ____$)$.

4. $20x - 8y =$ ____$(5x - 2y)$.

5. $2x^4 + 2x^2 =$ ________ $(x^2 + 1)$.

6. $x^2 + 4x + 3 = (x + 3)(x +$ ____$)$.

7. Fill in signs: $x^2 - 7x + 12 = (x$ ____ $3)(x$ ____ $4)$.

8. Fill in signs: $x^2 - 6x - 7 = (x$ ____ $1)(x$ ____ $7)$.

9. Fill in signs: $x^2 + 2x - 24 = (x$ ____ $4)(x$ ____ $6)$.

10. $2x^2 - 5x - 3 = (2x +$ ____$)(x -$ ____$)$.

11. $6x^2 + 17x + 12 = ($____ $x + 3)($____ $x + 4)$.

12. $x^2 - 4 = (x +$ ____$)(x -$ ____$)$.

13. $9x^2 - 25 = ($____ $x + 5)($____ $x - 5)$.

14. $2x^4 - 18x^2 =$ ________ $(x^2 - 9) =$ ________ $(x +$ ____$)(x -$ ____$)$.

15. To factor $2x^3 + 10x^2 + 12x$, you should first remove the factor ______.

16. If $x^4 - y^4$ is written as $(x^2 + y^2)(x^2 - y^2)$, is it completely factored? ______

Problem Set 6-2

In Problems **1–76**, *factor completely.*

1. $8x + 8$
2. $9x - 9$
3. $14y - 8$
4. $9 - 21x$
5. $10x - 5y + 25$
6. $12x^2 + 24y - 4$
7. $5cx + 9x$
8. $16mx + 4m$
9. $4y - 16y^2$
10. $8a - 4ab$
11. $6xy + 3xz$
12. $4xyz - 5yz$
13. $2x^3 - x^2$
14. $5x^8 - 4x^7$
15. $2x^3y^3 + x^5y^5$
16. $m^2y - y^2m$
17. $4m^2x^3 - 8mx^4$
18. $25a^5x^9 - 15a^4x^{10}$
19. $9a^4y^3 + 3a^2y^5 - 6a^3y^4z$
20. $by^5 - 2b^3y^4 - 8b^2y^2$
21. $x^2 - 1$
22. $x^2 - 49$
23. $x^2 + 4x + 3$
24. $x^2 + 5x + 6$
25. $x^2 + 7x + 10$
26. $x^2 - 5x + 6$
27. $x^2 - 9x + 20$
28. $x^2 + 3x - 10$
29. $y^2 + 2y - 24$
30. $y^2 - 10y + 9$
31. $y^2 - 3$
32. $4 - y^2$
33. $x^2 + 12x + 36$
34. $x^2 - 3x - 28$
35. $x^2 - 4x - 32$
36. $x^2 - 8x + 12$
37. $y^2 - 10y + 25$
38. $y^2 + 8y + 16$
39. $25x^2 - 16$
40. $4x^2 - 49y^2$
41. $y^2 - \frac{4}{9}$
42. $x^2y^2 - \frac{1}{4}$
43. $3x^2 + 7x + 2$
44. $5x^2 - 12x + 4$
45. $2y^2 - 7y + 3$
46. $7y^2 + 9y + 2$
47. $16x^2 + 8x + 1$
48. $4x^2 - 4x + 1$
49. $9 - 4x^2y^2$
50. $a^2b^2 - c^2d^2$
51. $4y^2 + 7y - 2$
52. $8y^2 + 2y - 3$
53. $6x^2 - 11x - 10$
54. $5x^2 + 14x - 3$
55. $12x^2 + x - 6$
56. $10x^2 - 19x + 6$
57. $2x^2 + 4x - 6$
58. $a^2x^2 + a^2x - 20a^2$
59. $3x^3 + 18x^2 + 27x$
60. $3x^4 - 15x^3 + 18x^2$
61. $16s^2t^3 - 4s^2t$
62. $a^2b^2 - a^4b^4$
63. $4y^2 - 6y - 18$
64. $30y^2 + 55y + 15$
65. $(x + 3)^3(x - 1) + (x + 3)^2(x - 1)^2$
66. $(x + 5)^2(x + 1)^3 + (x + 5)^3(x + 1)^2$
67. $(x + 4)(2x + 1) + (x + 4)$
68. $(x - 3)(2x + 3) - (2x + 3)(x + 5)$
69. $x^4 - 16$
70. $81x^4 - y^4$
71. $y^8 - 1$
72. $t^4 - 4$
73. $x^4 + x^2 - 2$
74. $x^4 - 5x^2 + 4$
75. $x^5 - 2x^3 + x$
76. $4x^3 - 6x^2 - 4x$

77. Use the formula

$$a^3 + b^3 = (a + b)(a^2 - ab + b^2)$$

to factor $x^3 + 8$.

78. Use the formula

$$a^3 - b^3 = (a - b)(a^2 + ab + b^2)$$

to factor $8x^3 - 27$.

79. The total area of a certain closed cylinder of radius r and altitude h is given by $2\pi r^2 + 2\pi rh$. Completely factor this expression.

6-3 BINOMIAL THEOREM†

Let's look at some powers of the binomial $a + b$. From Sec. 6-1 we know that

$$(a + b)^2 = a^2 + 2ab + b^2. \qquad (6\text{-}1)$$

Now we'll find $(a + b)^3$.

$$\begin{aligned}(a + b)^3 &= (a + b)(a + b)^2 \\ &= (a + b)(a^2 + 2ab + b^2).\end{aligned}$$

By the distributive law,

$$\begin{aligned}(a + b)(a^2 + 2ab + b^2) &= a(a^2 + 2ab + b^2) + b(a^2 + 2ab + b^2) \\ &= a^3 + 2a^2b + ab^2 + a^2b + 2ab^2 + b^3 \\ &= a^3 + 3a^2b + 3ab^2 + b^3.\end{aligned}$$

Thus,

$$(a + b)^3 = a^3 + 3a^2b + 3ab^2 + b^3. \qquad (6\text{-}2)$$

There are similarities in the *expansions* of $(a + b)^2$ and $(a + b)^3$ given in Eqs. (6-1) and (6-2). In both of them, the number of terms is one more than the power to which $a + b$ is raised:

$(a + b)^2$ has three terms;

$(a + b)^3$ has four terms.

---

†This section is not needed for the rest of the book.

The first and last terms of $(a + b)^2$ are the *squares* of a and b; the first and last terms of $(a + b)^3$ are the *cubes* of a and b. In both expansions the powers of a *decrease* from left to right, whereas from the second term on, the powers of b *increase*. Also, for the terms involving both a and b, the sum of the exponents of a and b equals the power to which $a + b$ is raised. For example, the second term in $(a + b)^2$ is $2ab = 2a^1b^1$ and $1 + 1 = 2$. Similarly, the second term in $(a + b)^3$ is $3a^2b = 3a^2b^1$ and $2 + 1 = 3$.

We could determine the expansions of $(a + b)^4$, $(a + b)^5$, etc., by successive multiplication. However, there is a formula, called the *binomial theorem*, that gives the expansion of $(a + b)^n$ where n is any positive integer. The patterns that we observed in $(a + b)^2$ and $(a + b)^3$ carry over to $(a + b)^n$.

BINOMIAL THEOREM

If n is a positive integer, then

$$(a + b)^n = a^n + \frac{n}{1}a^{n-1}b + \frac{n(n-1)}{1\cdot 2}a^{n-2}b^2 + \frac{n(n-1)(n-2)}{1\cdot 2\cdot 3}a^{n-3}b^3 + \ldots + \frac{n(n-1)(n-2)\cdot\ldots\cdot 2}{1\cdot 2\cdot\ldots\cdot(n-1)}ab^{n-1} + b^n.$$

Notice that in the next-to-last term above, the numerator of the coefficient is $n(n - 1)(n - 2) \cdot \ldots \cdot 2$, which is the product of the integers from n to 2, and the denominator is $1 \cdot 2 \cdot \ldots \cdot (n - 1)$, which is the product of the integers from 1 to $n - 1$.

In the formula, the first and last terms of the expansion of $(a + b)^n$ are a^n and b^n, and there are $n + 1$ terms. Also, the powers of a *decrease* from left to right (from n to 1). From the second term on, the powers of b *increase* from 1 to n. For the terms involving both a and b, the sum of the exponents of a and b is n.

EXAMPLE 1

Using the binomial theorem for $(a + b)^n$.

a. $(x + 3)^4$.

Here $n = 4$, x plays the role of a, and 3 plays the role of b. The expansion will have five terms.

$$(a + b)^4 = a^4 + \frac{4}{1}a^3b + \frac{4\cdot 3}{1\cdot 2}a^2b^2 + \frac{4\cdot 3\cdot 2}{1\cdot 2\cdot 3}ab^3 + b^4.$$

$$\begin{aligned}(x + 3)^4 &= x^4 + \frac{4}{1}x^3(3) + \frac{4\cdot 3}{1\cdot 2}x^2(3)^2 + \frac{4\cdot 3\cdot 2}{1\cdot 2\cdot 3}x(3)^3 + 3^4\\ &= x^4 + 4x^3(3) + 6x^2(9) + 4x(27) + 81\\ &= x^4 + 12x^3 + 54x^2 + 108x + 81.\end{aligned}$$

b. $(y-2)^5 = [y+(-2)]^5$.

Here $n = 5$, y plays the role of a, and -2 plays the role of b. The expansion will have six terms.

$$(a+b)^5 = a^5 + \frac{5}{1}a^4b + \frac{5\cdot 4}{1\cdot 2}a^3b^2 + \frac{5\cdot 4\cdot 3}{1\cdot 2\cdot 3}a^2b^3 + \frac{5\cdot 4\cdot 3\cdot 2}{1\cdot 2\cdot 3\cdot 4}ab^4 + b^5.$$

$$(y-2)^5 = y^5 + \frac{5}{1}y^4(-2) + \frac{5\cdot 4}{1\cdot 2}y^3(-2)^2 + \frac{5\cdot 4\cdot 3}{1\cdot 2\cdot 3}y^2(-2)^3 + \frac{5\cdot 4\cdot 3\cdot 2}{1\cdot 2\cdot 3\cdot 4}y(-2)^4 + (-2)^5$$

$$= y^5 + 5y^4(-2) + 10y^3(4) + 10y^2(-8) + 5y(16) + (-32)$$

$$= y^5 - 10y^4 + 40y^3 - 80y^2 + 80y - 32.$$

c. $(3x-1)^4 = [3x+(-1)]^4$.

Here $n = 4$, $3x$ plays the role of a, and -1 plays the role of b.

$$(3x-1)^4 = (3x)^4 + \frac{4}{1}(3x)^3(-1) + \frac{4\cdot 3}{1\cdot 2}(3x)^2(-1)^2 + \frac{4\cdot 3\cdot 2}{1\cdot 2\cdot 3}(3x)(-1)^3 + (-1)^4$$

$$= 3^4x^4 + 4(3)^3x^3(-1) + 6(3)^2x^2(1) + 4(3)x(-1) + 1$$

$$= 81x^4 - 108x^3 + 54x^2 - 12x + 1.$$

EXAMPLE 2

Using the binomial theorem, find the first three terms in the expansion of $(x^2+4)^8$.

Here $n = 8$, x^2 plays the role of a, and 4 plays the role of b.

first term: $(x^2)^8 = x^{16}$.

second term: $\frac{8}{1}(x^2)^7(4) = 32x^{14}$.

third term: $\frac{8\cdot 7}{1\cdot 2}(x^2)^6(4)^2 = 28x^{12}(16) = 448x^{12}$.

Completion Set 6-3

Fill in the blanks.

1. In the expansion of $(x+y)^6$, the first term is ______ and the last term is ______.

2. The number of terms in the expansion of $(4x - 1)^8$ is ________.

3. The first term in the expansion of $(x^2 + 1)^3$ is ______.

Problem Set 6-3

In Problems **1–8**, *use the binomial theorem to expand each expression.*

1. $(x + 4)^3$
2. $(x - 3)^3$
3. $(y - 2)^4$
4. $(y + 3)^5$
5. $(3x + 1)^5$
6. $(x + h)^6$
7. $(2z - y)^4$
8. $(z^2 - 2)^3$

In Problems **9–16**, *find the first three terms in the binomial expansion of each expression.*

9. $(x + 1)^{100}$
10. $(x - 1)^{45}$
11. $(2x - 3)^7$
12. $(x + 2)^8$
13. $(y^2 - 5)^{10}$
14. $(y^3 - 6)^{12}$
15. $(3z^2 + 1)^5$
16. $(2z^2 + 2)^5$

6-4 REVIEW

IMPORTANT TERMS

binomial *(p. 131)*
square of a binomial *(p. 132)*
trinomial *(p. 132)*
factoring *(p. 138)*
common factor *(p. 138)*
binomial theorem *(p. 146)*

REVIEW PROBLEMS

In Problems **1–20**, *find the special products.*

1. $(x + 6)^2$
2. $(x - 7)^2$
3. $(x - 5)^2$
4. $(x + 3y)^2$
5. $(2x + 4y)^2$
6. $(3x - 6)^2$
7. $(x - 8)(x + 8)$
8. $(9 - 2x)(9 + 2x)$
9. $(3x + 2)(3x - 2)$
10. $(1 - 3x)(3x + 1)$
11. $(2x - 4y)(2x + 4y)$
12. $(a - 2b)(a + 2b)$
13. $(x - 6)(x + 4)$
14. $(x + 3)(x - 2)$
15. $(x - 6)(x - 7)$

16. $(x + 5)(x + 8)$ **17.** $(2x - 3)(2x - 4)$ **18.** $(3x - 2)(2x + 3)$

19. $(y^2 + 4)(y^2 - 4)$ **20.** $(y^3 + 1)(y^3 - 1)$

In Problems **21–26**, *perform the indicated operations.*

21. $2x^2(x - 3)(x + 4)$ **22.** $-3x(x + 3)^2$

23. $(y - \sqrt{2})(y + \sqrt{2}) - (y - 4)^2$ **24.** $5y(1 + 2y)^2$

25. $(4x + 3)(4x - 3)(x + 2)$ **26.** $3(3x - 1)^2 + 5x(x + 2)^2$

In Problems **27–44**, *factor completely.*

27. $6x^3y^4 + 4xy^6$ **28.** $10abc^8 - 15ab^2c$ **29.** $x^2 - 11x + 30$

30. $x^2 + 6x + 8$ **31.** $16 - y^2$ **32.** $y^2 - 7$

33. $x^3 - x^2 - 56x$ **34.** $2x^2 + 4x - 96$ **35.** $3x^2 + 10x - 8$

36. $2x^2 + 9x - 35$ **37.** $8x^2 - 50$ **38.** $x^2 - \frac{1}{16}$

39. $15y^2 + 2y - 8$ **40.** $8y^2 + 6y + 1$ **41.** $x^4 - 2x^2 - 8$

42. $x^4 + 10x^2 + 25$ **43.** $x^3(x - 6)^2 + x^4(x - 6)$

44. $(x + 4)^3(x + 6)^4 + (x + 4)^4(x + 6)^3$

In Problems **45–48**, *use the binomial theorem to expand each expression.*

45. $(x - 4)^4$ **46.** $(x + y)^5$

47. $(2x + 1)^5$ **48.** $(x^3 + 2y)^3$

CHAPTER 7

Quadratic Equations

7-1 SOLUTION BY FACTORING

In Chapter 4 we solved equations where the highest power of x was 1. They are called **first-degree equations** or **linear equations** and have the form $ax + b = 0$.† We solved them by performing operations until x was by itself.

Now we want to solve equations of the form $ax^2 + bx + c = 0$.† These are called **second-degree equations** or **quadratic equations**. Here the highest power of x is 2.

QUADRATIC EQUATION
$ax^2 + bx + c = 0$

EXAMPLE 1

Quadratic equations: $ax^2 + bx + c = 0$.

a. $3x^2 - 5x + 2 = 0$ $\quad a = 3, b = -5, c = 2.$

b. $3y^2 - 5y + 2 = 0$ $\quad$ Here the variable is y.

---

†We assume $a \neq 0$.

c. $x^2 - 4 = 0$ — $a = 1, b = 0, c = -4.$

d. $x^2 - x = 0$ — $a = 1, b = -1, c = 0.$

e. $2x^2 = x - 4$ — Rewrite this as $2x^2 - x + 4 = 0.$ Thus $a = 2, b = -1, c = 4.$

Factoring is useful in solving many quadratic equations. We'll use it to solve

$$x^2 + 5x + 6 = 0.$$

Factoring the left side gives

$$(x + 3)(x + 2) = 0.$$

But it is a fact that

> If the product of two or more factors is *zero*, then at least one of the factors must be zero.

Applying this to the two factors of our equation gives

$$\text{either } x + 3 = 0 \quad \textit{or} \quad x + 2 = 0.$$

Thus,

$$\text{if } x + 3 = 0, \text{ then } x = -3,$$
$$\text{and} \quad \text{if } x + 2 = 0, \text{ then } x = -2.$$

Thus there are *two* solutions:

$$\boxed{x = -3, -2.}$$

Let's check both of these values in the original equation.

$x = -3$	$x = -2$
$x^2 + 5x + 6 = 0$	$x^2 + 5x + 6 = 0$
$(-3)^2 + 5(-3) + 6 = 0$	$(-2)^2 + 5(-2) + 6 = 0$
$9 - 15 + 6 = 0$	$4 - 10 + 6 = 0$
$0 = 0 \checkmark$	$0 = 0 \checkmark$

Both check. Thus we see that *a quadratic equation can have two solutions.*

EXAMPLE 2

Solve $2x^2 - 4x = 0$ *by factoring.*

We factor the left side and set each factor equal to 0.

$$2x^2 - 4x = 0$$
$$2x(x - 2) = 0.$$

There are three factors on the left: 2, x, and $x - 2$. Thus,

$$2 = 0 \quad \text{or} \quad x = 0 \quad \text{or} \quad x - 2 = 0.$$

Certainly $2 \neq 0$.

$x = 0$ is one solution.

If $x - 2 = 0$, then $x = 2$.

$$\boxed{x = 0, 2.}$$

EXAMPLE 3

Solve $0 = 6t^2 + t - 2$ *by factoring.*

$$0 = 6t^2 + t - 2$$
$$0 = (2t - 1)(3t + 2).$$

$2t - 1 = 0$	$3t + 2 = 0$
$2t = 1$	$3t = -2$
$t = \frac{1}{2}$	$t = -\frac{2}{3}$.

$$\boxed{t = \frac{1}{2}, -\frac{2}{3}.}$$

EXAMPLE 4

Solve $\frac{x^2}{2} - x = x - 2$ *by factoring.*

First, to clear of fractions we multiply both sides by 2:

$$2\left(\frac{x^2}{2} - x\right) = 2(x - 2)$$
$$x^2 - 2x = 2x - 4.$$

Now we'll write this equation so that one side is 0.

$$x^2 - 4x + 4 = 0$$

$$(x - 2)(x - 2) = 0.$$

$$x - 2 = 0 \quad \Big| \quad x - 2 = 0$$

$$x = 2 \quad \Big| \quad x = 2.$$

$$\boxed{x = 2.}$$

This quadratic equation has only one solution. Since more than one factor gave rise to the same solution, 2, we say that 2 is a **repeated root**.

EXAMPLE 5

Solve $(3x - 4)(x + 1) = -2$.

The left side is factored, but the right side is -2. We want one side to be 0. Thus we multiply first and combine terms.

$$(3x - 4)(x + 1) = -2$$

$$3x^2 - x - 4 = -2$$

$$3x^2 - x - 2 = 0$$

$$(3x + 2)(x - 1) = 0.$$

$$3x + 2 = 0 \quad \Big| \quad x - 1 = 0$$

$$3x = -2 \quad \Big| \quad x = 1.$$

$$x = -\frac{2}{3} \quad \Big|$$

$$\boxed{x = -\frac{2}{3}, 1.}$$

In the equation of Example 5, $(3x - 4)(x + 1) = -2$, *you should* ***not*** *set each factor equal to* -2. *Thus* $3x - 4 \neq -2$ *and* $x + 1 \neq -2$.

Sometimes factoring can be used to solve equations that are not quadratic, as Example 6 demonstrates.

EXAMPLE 6

Solve $x^4 - 4x^2 = 0$.

This is called a fourth-degree equation. Can you tell why?

$$x^4 - 4x^2 = 0$$
$$x^2(x^2 - 4) = 0$$
$$x^2(x + 2)(x - 2) = 0$$
$$x \cdot x(x + 2)(x - 2) = 0.$$

Setting all *four* factors equal to 0, we have

$x = 0$	$x = 0$	$x + 2 = 0$	$x - 2 = 0$
		$x = -2$	$x = 2.$

$$\boxed{x = 0, -2, 2.}$$

Here 0 is a repeated root. Sometimes the pair 2 and -2 is written ± 2 which is read "plus or minus 2." Thus we could write the answer as $x = 0, \pm 2$.

EXAMPLE 7

Solve $x^2 = 3$.

$$x^2 = 3$$
$$x^2 - 3 = 0$$
$$(x + \sqrt{3})(x - \sqrt{3}) = 0.$$

$x + \sqrt{3} = 0$	$x - \sqrt{3} = 0$
$x = -\sqrt{3}$	$x = \sqrt{3}.$

$$\boxed{x = \pm\sqrt{3}.}$$

A more general form of the equation $x^2 = 3$ is $u^2 = k$. In the same manner as above, we can show that

$$\boxed{\begin{array}{l}\text{if } u^2 = k, \text{ then} \\ \quad u = \pm\sqrt{k}.\end{array}}$$

For example, the solution of $y^2 = 4$ is $y = \pm\sqrt{4} = \pm 2$.

EXAMPLE 8

Solve $(x + 2)^2 = 7$.

This equation has the special form $u^2 = k$, where u is $x + 2$ and k is 7. From Example 7 we know that $u = \pm \sqrt{k}$. Thus,

$$(x + 2)^2 = 7$$

$$x + 2 = \pm\sqrt{7}$$

$$\boxed{x = -2 \pm \sqrt{7}\,.}$$

Here we have two solutions: one when we use the "+" sign, $-2 + \sqrt{7}$, and another when we use the "−" sign, $-2 - \sqrt{7}$.

Completion Set 7-1

In Problems **1–10**, *fill in the blanks.*

1. $2x^2 - 3x = 7$ can be written as $2x^2 - 3x -$ ____ $= 0$.

2. $4x^2 - 5x + 4 = x^2 - 2x - 6$ can be written as $3x^2 -$ ____ $x +$ ____ $= 0$.

3. To solve $(x - 2)(x + 3) = 0$, we set ____________ $= 0$ and ____________ $= 0$.

Thus, $x =$ ____ or $x =$ _____.

4. The solutions of $(x + 7)(x - 6) = 0$ are $x =$ _____ and $x =$ ____.

5. The solutions of $(x + 1)(x - 2)(x + 3)(x - 4) = 0$ are $x = -1$, $x =$ ____,

$x =$ _____, and $x =$ ____.

6. To solve $x^2 - 100x = 0$ we factor the left side and get (____)(____ − _______) = 0. Setting the first factor equal to 0 gives $x =$ ____. Doing the same to the second factor gives $x =$ _______.

7. If $x^2 = 36$, then $x = \pm\sqrt{(\qquad)} = \pm$ ____.

8. If $4y^2 = 100$, then $y^2 = \dfrac{100}{(\quad)} = $ _____. Thus, $y = \pm\sqrt{(\qquad)} = \pm$ _____.

9. If $x = 7 \pm 3$, then $x = 7 +$ ____ $= 10$ or $x = 7 -$ ____ $=$ ____.

10. If $(x + 5)^2 = 9$, then $x +$ ____ $= \pm$ ____. Thus, $x =$ _____ $\pm$ ____.

Problem Set 7-1

In Problems **1–50**, *solve by the methods of this section.*

1. $x^2 + 3x + 2 = 0$
2. $x^2 - 5x + 6 = 0$
3. $x^2 + 9x + 14 = 0$
4. $x^2 + 8x + 15 = 0$
5. $t^2 - 7t + 12 = 0$
6. $t^2 - 4t + 4 = 0$
7. $z^2 + 2z - 3 = 0$
8. $z^2 + z - 12 = 0$
9. $x^2 - 12x + 36 = 0$
10. $x^2 - 1 = 0$
11. $x^2 - 8x = 0$
12. $t^2 = 16$
13. $2x^2 + 10x = 0$
14. $0 = 3x - x^2$
15. $0 = 3t^2 - 6t$
16. $x^2 = 32$
17. $4 - x^2 = 0$
18. $10x^2 - x - 3 = 0$
19. $x^2 = 25$
20. $x^2 = 8$
21. $x^2 = 0$
22. $9x^2 = 36$
23. $\dfrac{x^2}{3} = 4$
24. $\dfrac{x^2}{7} = 1$
25. $9z^2 = 81$
26. $3x^2 - 12x + 12 = 0$
27. $6x^2 + 7x - 3 = 0$
28. $3 = z^2$
29. $\dfrac{2}{3}t^2 = 6$
30. $7t^2 = 7$
31. $2x^2 - 14 = 0$
32. $x^2 + 2 = 18$
33. $2x^2 + 7x = 4$
34. $x^2 = 2x + 3$
35. $-x^2 + 3x + 10 = 0$
36. $2x^2 + 3x - 2 = 0$
37. $4x^2 + 4x = -1$
38. $9x^2 - 1 = 0$
39. $6(x^2 + 2x) + 6 = 0$
40. $4x^2 - 12x + 9 = 0$
41. $t(t + 4) = 5$
42. $2y^2 = 4y$
43. $3x^2 + 5(2 - x) = 2x^2 + 4$
44. $2(x^2 - 5) - 3(x^2 - 7) = 3$
45. $4 - x^2 = (x + 1)^2 + 3$
46. $(2x - 5)(x + 5) = -22$
47. $\dfrac{x^2}{2} - x - 4 = 0$
48. $x^2 + \dfrac{7}{2}x - 2 = 0$
49. $6y^2 + \dfrac{5}{2}y + \dfrac{1}{4} = 0$
50. $\dfrac{x^2}{2} + \dfrac{10}{3}x + 2 = 0$

In the manner of Example 8, *solve Problems* **51–56.**

51. $(x - 3)^2 = 16$
52. $(x + 5)^2 = 6$
53. $(x + 4)^2 = 8$
54. $(w - 6)^2 = 1$
55. $\left(y + \dfrac{1}{2}\right)^2 = 1$
56. $(x - 4)^2 = \dfrac{9}{4}$

In Problems **57–68**, *solve.*

57. $x(x - 1)(x + 2) = 0$

58. $x^2(x - 4) = 0$

59. $(x - 2)^2(x + 1)^2 = 0$

60. $x(x - 1)(x + 1) = 0$

61. $7x^2(x - 2)^2(x + 3)(x - 4) = 0$

62. $x(x^2 - 1)(x^2 - 4) = 0$

63. $x(x^2 - 1)(x^2 - 1) = 0$

64. $x^3 - x = 0$

65. $x^3 - 64x = 0$

66. $x^3 - 4x^2 - 5x = 0$

67. $3y^3 + 18y^2 + 24y = 0$

68. $3x^4 + 11x^3 - 4x^2 = 0$

To solve $x^4 - 5x^2 + 4 = 0$, *we can factor the left side.*

$$(x^2 - 1)(x^2 - 4) = 0$$

$$(x - 1)(x + 1)(x - 2)(x + 2) = 0. \qquad \textit{Thus,} \quad \boxed{x = \pm 1, \pm 2.}$$

Now solve Problems **69–74** *by factoring.*

69. $x^4 - 10x^2 + 9 = 0$

70. $x^4 - 29x^2 + 100 = 0$

71. $x^2(x^4 - 2x^2 + 1) = 0$

72. $x^5 - 6x^3 + 8x = 0$

73. $x^4 - 13x^2 + 36 = 0$

74. $x^5 - 4x^3 + 4x = 0$

7-2 COMPLETING THE SQUARE†

In the last section you solved quadratic equations by factoring. This is a good method, but sooner or later you'll run into a quadratic equation that you can't factor. For example, it's hard to factor the left side of $3x^2 - 6x - 6 = 0$.

You may still solve this equation, but other methods are used. One way is the **method of completing the square**. It involves rewriting the equation until one side factors as a square of a binomial.

The "rewriting" procedure may seem strange to you right now, but the reasons for each step will later be obvious to you.

Method of completing the square for $3x^2 - 6x - 6 = 0$.

We first add 6 to both sides so that only terms involving x remain on the left side.

$$3x^2 - 6x = 6.$$

†This section can be omitted if Sections 7-6 and/or 16-2 are not covered.

Next, we get the coefficient of the x^2-term to be 1. Thus we divide both sides by 3.

$$\frac{3x^2 - 6x}{3} = \frac{6}{3}$$

$$\frac{3x^2}{3} - \frac{6x}{3} = 2$$

$$x^2 - 2x = 2.$$

Take half of the coefficient of the x-term and square the result.

$$\text{Half of } -2\text{:} \quad \frac{-2}{2} = -1. \qquad \text{Squaring:} \quad (-1)^2 = 1.$$

Add this square, 1, to both sides.

$$x^2 - 2x + 1 = 2 + 1$$

$$x^2 - 2x + 1 = 3.$$

Look at the left side of the equation. It's the square of the binomial $x - 1$.

$$(x - 1)^2 = 3.$$

This was the reason for doing all the operations. The equation now has the form $u^2 = k$. Thus,

$$x - 1 = \pm\sqrt{3}$$

$$\boxed{x = 1 \pm \sqrt{3}\,.}$$

EXAMPLE 1

Solve $4x^2 + 12x + 3 = 0$ *by completing the square.*

$$4x^2 + 12x + 3 = 0$$

$$4x^2 + 12x = -3 \qquad \text{[getting constant on right]}$$

$$\frac{4x^2 + 12x}{4} = \frac{-3}{4} \qquad \text{[dividing both sides by 4]}$$

$$x^2 + 3x = -\frac{3}{4}.$$

$$\text{Half of 3: } \frac{3}{2}. \qquad \text{Squaring: } \left(\frac{3}{2}\right)^2 = \frac{9}{4}.$$

$$x^2 + 3x + \frac{9}{4} = -\frac{3}{4} + \frac{9}{4} \qquad \left[\text{adding } \frac{9}{4} \text{ to both sides}\right]$$

$$\left(x + \frac{3}{2}\right)^2 = \frac{6}{4} \qquad \text{[factoring left side]}$$

$$x + \frac{3}{2} = \pm\sqrt{\frac{6}{4}}$$

$$\boxed{x = -\frac{3}{2} \pm \frac{\sqrt{6}}{2}.} \qquad \left[\text{since } \sqrt{\frac{6}{4}} = \frac{\sqrt{6}}{\sqrt{4}} = \frac{\sqrt{6}}{2}\right]$$

Completion Set 7-2

In Problems **1–5**, *fill in the blanks.*

1. One way to solve a quadratic equation is by the method of completing the ____________.

2. To complete the square in $4x^2 - 6x - 7 = 0$, first get the ____________ term on the right side.

3. To complete the square in $3x^2 + 6x = 5$, one of the things to do is divide both sides by the number ____.

4. If $x^2 + 10x + 25 = 2$, then $(x + \_\_\_\_)^2 = 2$ or $x + \_\_\_\_ = \pm\sqrt{2}$. Thus, $x =$ _____ $\pm$ _____.

5. Suppose $x^2 + 8x = 1$. To complete the square, we take half of ____, which is ____, square it and get _____. We add _____ to both sides of the equation and get

$$x^2 + 8x + \_\_\_\_\_ = 1 + \_\_\_\_\_.$$

Thus,

$$(x + \_\_\_\_)^2 = \_\_\_\_\_$$

$$x + \_\_\_\_ = \pm \_\_\_\_\_$$

$$x = \_\_\_\_\_ \pm \_\_\_\_\_.$$

In Problems **6–8**, *fill in the blanks to complete the square.*

6. $x^2 + 6x + \_\_\_\_ = 3 + \_\_\_\_.$

7. $x^2 - 10x + \_\_\_\_\_\_ = 7 + \_\_\_\_\_\_.$

8. $y^2 + y + \_\_\_\_ = 3 + \_\_\_\_.$

Problem Set 7-2

In Problems **1–14**, *solve by completing the square.*

1. $x^2 + 6x - 1 = 0$

2. $2x^2 + 4x - 8 = 0$

3. $5x^2 - 20x + 5 = 0$

4. $x^2 - 12x + 8 = 0$

5. $y^2 - 3y - 1 = 0$

6. $x^2 + 5x + 5 = 0$

7. $x^2 + x - 4 = 0$

8. $4w^2 - 8w - 3 = 0$

9. $2x^2 - 14x + 1 = 0$

10. $9x^2 + 9x - 2 = 0$

11. $x^2 + \frac{1}{2}x - 1 = 0$

12. $x^2 + 2x - \frac{1}{4} = 0$

13. $7x + 3(x^2 - 5) = x - 3$

14. $2x(4x - 1) = 4 + 2x$

7-3 THE QUADRATIC FORMULA

Another way to solve quadratic equations is to substitute numbers into a special formula called the **quadratic formula.**†

> **QUADRATIC FORMULA**
> If $ax^2 + bx + c = 0$, then
> $$x = \frac{-b \pm \sqrt{b^2 - 4ac}}{2a}.$$

EXAMPLE 1

Solve $2x^2 + 5x - 3 = 0$ *by the quadratic formula.*

Here $a = 2$, $b = 5$, and $c = -3$.

$$x = \frac{-b \pm \sqrt{b^2 - 4ac}}{2a} = \frac{-5 \pm \sqrt{(5)^2 - 4(2)(-3)}}{2(2)}$$

$$= \frac{-5 \pm \sqrt{25 + 24}}{4} = \frac{-5 \pm \sqrt{49}}{4} = \frac{-5 \pm 7}{4}.$$

†The way we obtain the quadratic formula is explained in Section 7-6.

Thus, solutions are

$$x = \frac{-5+7}{4} = \frac{2}{4} = \frac{1}{2} \quad \text{and} \quad x = \frac{-5-7}{4} = \frac{-12}{4} = -3.$$

$$\boxed{x = \frac{1}{2}, -3}$$

EXAMPLE 2

Solve $4y^2 + 9 = 12y$.

First we put the equation in the form $ay^2 + by + c = 0$.

$$4y^2 - 12y + 9 = 0.$$

Here $a = 4$, $b = -12$, and $c = 9$. By the quadratic formula,

$$y = \frac{-b \pm \sqrt{b^2 - 4ac}}{2a} = \frac{-(-12) \pm \sqrt{(-12)^2 - 4(4)(9)}}{2(4)}$$

$$= \frac{12 \pm \sqrt{144 - 144}}{8} = \frac{12 \pm 0}{8}.$$

Notice that in $y = \frac{12 \pm 0}{8}$ we get the same value whether we use the "+" sign or the "−" sign. Thus,

$$y = \frac{12}{8} = \boxed{\frac{3}{2}}$$

is the only solution. We consider $\frac{3}{2}$ to be a repeated root.

EXAMPLE 3

Solve $2x(x - 3) = x^2 - 1$.

$2x^2 - 6x = x^2 - 1$ [removing parentheses]

$x^2 - 6x = -1$ [subtracting x^2 from both sides]

$x^2 - 6x + 1 = 0$ [adding 1 to both sides].

This equation is quadratic. Factoring does not work here, so we shall solve it by the quadratic formula.

$$a = 1, \qquad b = -6, \qquad c = 1$$

$$x = \frac{-b \pm \sqrt{b^2 - 4ac}}{2a} = \frac{-(-6) \pm \sqrt{(-6)^2 - 4(1)(1)}}{2(1)}$$

$$= \frac{6 \pm \sqrt{36 - 4}}{2} = \frac{6 \pm \sqrt{32}}{2}.$$

But $\sqrt{32} = \sqrt{16 \cdot 2} = \sqrt{16} \cdot \sqrt{2} = 4\sqrt{2}$. Thus,

$$x = \frac{6 \pm 4\sqrt{2}}{2} = \frac{6}{2} \pm \frac{4\sqrt{2}}{2} = \boxed{3 \pm 2\sqrt{2}.}$$

Don't foul up the quadratic formula. Make sure you use it correctly. ***Don't write***

$$x = -b \pm \frac{\sqrt{b^2 - 4ac}}{2a}.$$

The expression $b^2 - 4ac$ in the quadratic formula is called the **discriminant** of the quadratic equation $ax^2 + bx + c = 0$. If you look back at Examples 1 and 3 in this section, you will see that the discriminants 49 and 32 are positive. We have two different solutions to each equation. But in Example 2, the discriminant is 0 and the equation has only one solution (repeated root). In general,

a positive discriminant gives two different solutions, but a zero discriminant gives only one solution.

Negative discriminants will be considered in the next section.

Completion Set 7-3

In Problems **1–9**, *fill in the blanks.*

1. One way to solve $x^2 + 2x - 4 = 0$ is by the ________________ formula.

2. If $3x^2 - 4x + 7 = 0$, then $a = 3$, $b =$ ______, and $c =$ ______.

3. If $x^2 + 7x = 3$, then $a = 1$, $b =$ ______, and $c =$ ______.

4. If $x^2 + 3x = 4x - 2$, then $a = 1$, $b =$ ______, and $c =$ ____.

5. If $3x - 5x^2 = 0$, then $b = 3$, $a =$ ______, and $c =$ ____.

6. If $x^2 - 2x + 1 = 0$, then $x = \dfrac{-(\quad) \pm \sqrt{(\quad)^2 - 4(\quad)(\quad)}}{2(1)} =$

$\dfrac{(\quad) \pm \sqrt{(\quad)}}{2}$. Thus, $x =$ ____.

7. Insert $b^2 - 4ac$ or $\sqrt{b^2 - 4ac}$. The discriminant of $ax^2 + bx + c = 0$ is

____________________.

8. The discriminant of $x^2 + 3x + 1 = 0$ is $(\_\_\_\_)^2 - 4(\_\_\_\_)(\_\_\_\_) =$ ____.

9. Insert the word *positive* or *zero*. If the discriminant of $ax^2 + bx + c = 0$ is

__________, then the equation has only one solution.

Problem Set 7-3

In Problems **1–12**, *solve by the quadratic formula.*

1. $x^2 + 3x + 1 = 0$

2. $x^2 - 4x + 2 = 0$

3. $x^2 - 6x + 9 = 0$

4. $x^2 + x - 3 = 0$

5. $2y^2 + 3y - 4 = 0$

6. $3x^2 + 6x - 2 = 0$

7. $4x^2 = -20x - 25$

8. $9z(z - 1) = 3z - 4$

9. $5x(x + 2) + 6 = 3$

10. $(4x - 1)(2x + 3) = 18x - 4$

11. $2 - 2x - 3x^2 = 0$

12. $1 + 8x - 4x^2 = 0$

In Problems **13–44**, *solve by either factoring or the quadratic formula. The choice is yours for each problem.*

13. $x^2 - 36 = 0$

14. $x^2 + 4x - 1 = 0$

15. $x^2 + 6x - 7 = 0$

16. $x^2 - x - 20 = 0$

17. $2z^2 + 3z = 0$

18. $4z^2 - 25 = 0$

19. $x^2 + 4x + 2 = 0$

20. $x^2 - 5x + 3 = 0$

21. $y^2 + 16y + 64 = 0$

22. $3y^2 - y - 4 = 0$

23. $3x^2 - 4x - 2 = 0$

24. $3 + x - 4x^2 = 0$

25. $5x^2 - 21x + 4 = 0$

26. $2x^2 - 6x + 2 = 0$

27. $4 + 4y - y^2 = 0$

28. $4y^2 + 12y + 9 = 0$

29. $1 - 6y + 9y^2 = 0$

30. $7 - y - y^2 = 0$

31. $x^2 + 9 = 7x$

32. $x^2 - 16 = 8 - 2x$

33. $x^2 + 3x = 12 - 2x - x^2$

34. $3x(2x - 5) = -4x - 3$

35. $y(y + 4) = 5$

36. $(y + 1)^2 = 2y^2$

37. $z^2 = 2(z - 1)(z + 2)$

38. $z^4 - 6z^2 + 7 = z^2(z^2 + 1)$

39. $(2x + 1)^2 = 8x$

40. $2x(x + 1) = (x - 1)(x + 1)$

41. $\frac{2x^2 - 5x}{3} = x - 1$

42. $\frac{x^2}{3} + 2x = x + 1$

43. $\frac{x^2}{3} = \frac{11}{6}x + 1$

44. $5x^2 - \frac{7}{2}x = \frac{x + 2}{2}$

7-4 COMPLEX NUMBERS

When you use the quadratic formula, sometimes the square root of a negative number pops up. For example, solving

$$x^2 + x + 1 = 0$$

gives

$$x = \frac{-1 \pm \sqrt{1^2 - 4(1)(1)}}{2(1)} = \frac{-1 \pm \sqrt{1 - 4}}{2}$$
$$= \frac{-1 \pm \sqrt{-3}}{2}.$$

$\sqrt{-3}$ is a number whose square is -3. This number cannot be real, since the square of any real number cannot be negative. By working only with $\sqrt{-1}$ we can handle all square roots of negative numbers.

We use the symbol i to stand for $\sqrt{-1}$. It is called the **imaginary unit**. It has the property that $i^2 = -1$. We can handle square roots of other negative numbers by using the following:

> If a is a positive number,
>
> $\sqrt{-a} = i\sqrt{a}$.

EXAMPLE 1

Square roots of negative numbers.

a. $\sqrt{-4} = i\sqrt{4} = i(2) = 2i$.

b. $\sqrt{-16} = i\sqrt{16} = 4i$.

c. $\sqrt{-8} = i\sqrt{8} = i(2\sqrt{2}) = 2i\sqrt{2}$.

d. $-\sqrt{-3} = -(i\sqrt{3}) = -i\sqrt{3}$.

Any number of the form bi, where b is a real number, is called a **pure imaginary number**. Thus, $2i$ and $-3i$ are pure imaginary. From Example 1 you can see that square roots of negative numbers are also pure imaginary. Since $0 = 0i$, then 0 is not only a real number but also pure imaginary.

Combining a real number and a pure imaginary number by addition gives a so-called *complex number*, such as $4 + 2i$.

> A **complex number** is one of the form $a + bi$, where a and b are real numbers and i is the imaginary unit.

EXAMPLE 2

Complex numbers: $a + bi$.

a. $2 + 3i$. Here $a = 2$ and $b = 3$.

b. $-3 - 5i$. Here $a = -3$ and $b = -5$, since $-3 - 5i = -3 + (-5)i$.

c. 6. Here $a = 6$ and $b = 0$, since $6 = 6 + 0i$. Every real number is also a complex number.

d. $7i$. Here $a = 0$ and $b = 7$, since $7i = 0 + 7i$.

From Example 2(c), you can see that every real number is also a complex number. Complex numbers that are not real are called **imaginary numbers**. Thus $2 + 3i$ is an imaginary number. Also, $7i$ is imaginary as well as pure imaginary.

Some quadratic equations have imaginary numbers as solutions.

EXAMPLE 3

Solve $x^2 + 2x + 6 = 0$.

$$x = \frac{-2 \pm \sqrt{2^2 - 4(1)(6)}}{2(1)}$$

$$= \frac{-2 \pm \sqrt{-20}}{2} = \frac{-2 \pm i\sqrt{20}}{2}$$

$$= \frac{-2 \pm 2i\sqrt{5}}{2} = -\frac{2}{2} \pm \frac{2i\sqrt{5}}{2}$$

$$\boxed{= -1 \pm i\sqrt{5}\,.}$$

Thus, the solutions are $x = -1 + i\sqrt{5}$ and $x = -1 - i\sqrt{5}$, which are imaginary numbers.

In Example 3, notice that the discriminant is negative, -20. Whenever this happens, the solutions of a quadratic equation will be two different imaginary numbers. Here is a summary of the possibilities of the discriminant and the corresponding type of solutions of a quadratic equation.

$b^2 - 4ac$	*Type of Solutions*
positive	two different real solutions
zero	one real solution (repeated root)
negative	two different imaginary solutions

The discriminant is useful because it tells you something about the solutions of a quadratic equation without actually having to find the solutions.

EXAMPLE 4

Describe the solutions and solve.

a. $2x^2 - 4x + 5 = 0$.

First find the discriminant.

$$b^2 - 4ac = (-4)^2 - 4(2)(5) = 16 - 40 = -24.$$

Since the discriminant is negative, there are two different imaginary solutions.

$$x = \frac{-(-4) \pm \sqrt{-24}}{2(2)} = \frac{4 \pm 2i\sqrt{6}}{4}$$

$$= \frac{4}{4} \pm \frac{2i\sqrt{6}}{4} = \boxed{1 \pm \frac{i\sqrt{6}}{2}\,.}$$

b. $x^2 - 6x + 6 = 0$.

The discriminant is

$$b^2 - 4ac = (-6)^2 - 4(1)(6) = 36 - 24 = 12.$$

A positive discriminant means two different real solutions.

$$x = \frac{-(-6) \pm \sqrt{12}}{2(1)} = \frac{6 \pm 2\sqrt{3}}{2} = \boxed{3 \pm \sqrt{3}\,.}$$

c. $x^2 - 10x + 25 = 0$.

The discriminant is

$$b^2 - 4ac = (-10)^2 - 4(1)(25) = 100 - 100 = 0.$$

A discriminant of 0 means one real solution.

$$x = \frac{-(-10) \pm \sqrt{0}}{2(1)} = \frac{10}{2} = \boxed{5.}$$

Whenever a discriminant is 0 we get a repeated root.

EXAMPLE 5

Solve $x^2 + 16 = 0$.

We certainly don't need to use the quadratic formula on so easy a problem.

$$x^2 + 16 = 0$$

$$x^2 = -16 \qquad [\text{form } u^2 = k^\dagger]$$

$$x = \pm\sqrt{-16} = \boxed{\pm 4i.} \qquad [\text{since } u = \pm\sqrt{k}\,]$$

Completion Set 7-4

In Problems **1–6**, *fill in the blanks.*

1. The symbol that stands for $\sqrt{-1}$ is ____.

---

†A discussion of the form $u^2 = k$ is found on page 154.

2. Insert *real* or *imaginary*. The number $\sqrt{-10}$ is ______________________.

3. Using i, we can write $\sqrt{-49}$ as (____)$\sqrt{49}$ = ______.

4. A number of the form $a + bi$ is called a __________________ number.

5. Insert *positive*, *negative*, or *zero*. If the discriminant of a quadratic equation is __________________, the roots are imaginary.

6. i has the property that $i^2 =$ ____.

Problem Set 7-4

In Problems **1–12**, *write each number in terms of* i.

1. $\sqrt{-81}$ **2.** $\sqrt{-36}$ **3.** $\sqrt{-\frac{1}{16}}$

4. $\sqrt{-.04}$ **5.** $\sqrt{-12}$ **6.** $\sqrt{-50}$

7. $\sqrt{-32}$ **8.** $\sqrt{-28}$ **9.** $\sqrt{-2}$

10. $-\sqrt{-9}$ **11.** $-\sqrt{-25}$ **12.** $\sqrt{-27}$

In Problems **13–28**, *solve the equation.*

13. $x^2 - 4x + 5 = 0$ **14.** $x^2 - 2x + 2 = 0$ **15.** $x^2 + 2x + 3 = 0$

16. $2x^2 - 4x + 5 = 0$ **17.** $x^2 - 2x + 4 = 0$ **18.** $t^2 - 3t + 1 = 0$

19. $x^2 + 4 = 0$ **20.** $x^2 + 8 = 0$ **21.** $6r^2 + 8r + 3 = 0$

22. $x^2 = x - 1$ **23.** $-3r^2 + 5r = 4$ **24.** $x(x + 5) = 5(x - 5)$

25. $1 + x^2 = 0$ **26.** $(x + 1)^2 = -4$ **27.** $x(4x - 3) = x^2 - 2$

28. $\frac{x^2}{5} + 2 = x$

In Problems **29–36**, *first find the discriminant. Next, use just the discriminant to determine the type of solutions. Then solve.*

29. $3x^2 - 5x - 2 = 0$ **30.** $x^2 + 4x + 5 = 0$ **31.** $9t^2 + 12t + 4 = 0$

32. $4t^2 + 12t + 9 = 0$ **33.** $3x^2 - 4x + 3 = 0$ **34.** $4x^2 - 17x + 15 = 0$

35. $(3x - 1)(x + 2) = 2x$ **36.** $x(x - 4) - 3(2 - x) + 7 = 0$

7-5 APPLICATIONS OF QUADRATIC EQUATIONS

Here are some word problems that lead to quadratic equations.

EXAMPLE 1

A rectangular observation deck overlooking a scenic valley is to be built. See Fig. 7-1(a). *It is to have dimensions* 6 *meters* (*m*) *by* 12 *m. For commercial reasons it is decided to add on a rectangular shelter which would take up* 40 m^2 *of the deck. The remainder of the deck will be used as a walkway. It will have uniform width. How wide should this walkway be?*

A diagram of the deck is shown in Fig. 7-1(b). Let w = width (in m) of the walkway.

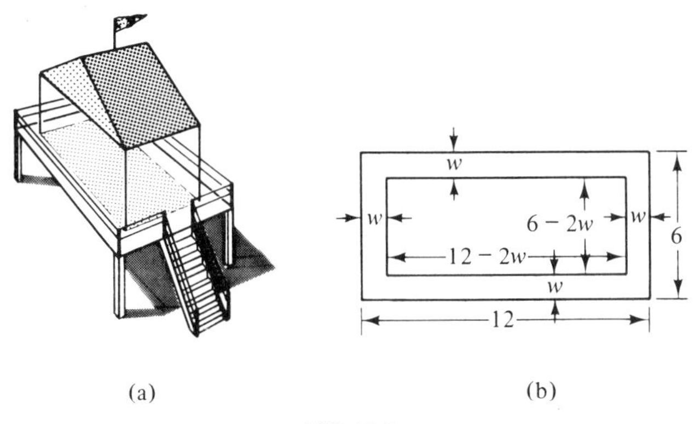

FIG. 7-1

Then the part of the deck for the shelter has dimensions $12 - 2w$ by $6 - 2w$. Since this area must be 40 m^2, where area = (length)(width), we have

$$(12 - 2w)(6 - 2w) = 40$$

$$72 - 36w + 4w^2 = 40 \quad \text{[multiplying]}$$

$$4w^2 - 36w + 32 = 0$$

$$w^2 - 9w + 8 = 0 \quad \text{[dividing both sides by 4]}$$

$$(w - 8)(w - 1) = 0$$

$$w = 8, 1.$$

Although $w = 8$ is a solution to the equation, it is not a solution to our problem, because one of the dimensions of the deck itself is only 6 m. Thus the only possible solution is 1 m.

EXAMPLE 2

A subscription TV cable service estimates that 1000 *subscribers will pay* \$3.00 *each to watch an off-Broadway show in their own homes. They also believe that the number of subscribers will increase by* 50 *for each* \$.10 *decrease in the charge. If the firm must receive* \$3125 *to meet its costs and to earn a profit, how much should they charge each subscriber?*

Method 1.

For each \$.10 decrease in the charge, the number of subscribers increases by 50. If there are n decreases, the charge decreases by $.10n$ and the number of subscribers increases by $50n$. Thus there will be $1000 + 50n$ subscribers, each of which pays $3.00 - .10n$. Now,

$$\text{(number of subscribers)(charge per subscriber)} = \text{total income.}$$

Consequently,

$$(1000 + 50n)(3.00 - .10n) = 3125$$

$$3000 + 50n - 5n^2 = 3125$$

$$5n^2 - 50n + 125 = 0 \qquad \text{[multiplying both sides by } -1 \text{ and simplifying]}$$

$$n^2 - 10n + 25 = 0 \qquad \text{[dividing both sides by 5]}$$

$$(n - 5)^2 = 0$$

$$n = 5.$$

Thus, there are five \$.10 decreases (= \$.50), and the charge should be 3.00 − .50 or \$2.50.

Method 2.

Another way of handling the problem is as follows. Suppose that c is the charge to each subscriber. Then the total decrease from the \$3 charge is $3 - c$. Thus the number of \$.10 decreases is $\frac{3 - c}{.10}$. Since each \$.10 decrease gives 50 more subscribers, the total increase in subscribers will be $50\left(\frac{3 - c}{.10}\right)$. Thus the total number of subscribers will be $1000 + 50\left(\frac{3 - c}{.10}\right)$. Now,

$$\text{total income} = \text{(charge per subscriber)(number of subscribers).}$$

Thus,

$$3125 = c\left[1000 + 50\left(\frac{3 - c}{.10}\right)\right]$$

$$= c[1000 + 500(3 - c)]$$

$$= c[1000 + 1500 - 500c]$$

$$= c[2500 - 500c]$$

$$3125 = 2500c - 500c^2.$$

Consequently,

$$500c^2 - 2500c + 3125 = 0.$$

Dividing both sides by 125, we obtain

$$4c^2 - 20c + 25 = 0.$$

By the quadratic formula,

$$\begin{aligned} c &= \frac{-(-20) \pm \sqrt{(-20)^2 - 4(4)(25)}}{2(4)} \\ &= \frac{20 \pm \sqrt{0}}{8} \\ &= \frac{20}{8} = 2.50 \end{aligned}$$

Thus the charge should be \$2.50.

Discussion.

Both methods give the same result. However, the nice thing about Method 2 is that the unknown is the charge you're looking for. Unfortunately, the equation that you get is very messy. In Method 1, the unknown is not the charge but the equation is much simpler.

EXAMPLE 3

The board of directors of Fuddy Duddy Corporation agrees to redeem some of its stock in two years. At that time \$1,123,600 *will be needed. Suppose that they presently set aside* \$1,000,000. *At what annual rate of interest,*† *compounded annually, will the money have to be invested so that its future value will be enough to redeem the stock?*

Let r be the annual rate of interest (in decimal form). At the end of the first year, the accumulated amount will be \$1,000,000 plus the interest on this, which is $1{,}000{,}000r$, for a total of

$$1{,}000{,}000 + 1{,}000{,}000r$$

or, after factoring,

$$1{,}000{,}000(1 + r).$$

At the end of the second year, the accumulated amount will be $1{,}000{,}000(1 + r)$ plus

†Interest is explained in Chapter 1, Section 5.

the interest on this, which is $[1{,}000{,}000(1 + r)]r$, for a total of

$$1{,}000{,}000(1 + r) + 1{,}000{,}000(1 + r)r.$$

This must equal \$1,123,600.

$$1{,}000{,}000(1 + r) + 1{,}000{,}000(1 + r)r = 1{,}123{,}600.$$

Factoring out $1{,}000{,}000(1 + r)$ on the left side, we have

$$1{,}000{,}000(1 + r)[1 + r] = 1{,}123{,}600$$
$$1{,}000{,}000(1 + r)^2 = 1{,}123{,}600$$
$$(1 + r)^2 = \frac{1{,}123{,}600}{1{,}000{,}000}$$
$$(1 + r)^2 = \frac{11{,}236}{10{,}000}$$
$$(1 + r)^2 = \frac{2809}{2500}.$$

Thus,

$$1 + r = \pm\sqrt{\frac{2809}{2500}} \qquad [\text{if } u^2 = k, \text{ then } u = \pm\sqrt{k}\,]$$
$$= \pm\frac{\sqrt{2809}}{\sqrt{2500}} = \pm\frac{53}{50}.$$

Either $r = -1 + \frac{53}{50} = \frac{3}{50} = .06$, or $r = -1 - \frac{53}{50} = -2.06$. We throw out -2.06, since we don't want r to be negative. Thus $r = .06 = 6$ percent is the rate that we want.

Problem Set 7-5

1. A ball is thrown up from the ground at a speed of 56 ft/sec. Its height s (in feet) from the ground after t sec is given by

$$s = 56t - 16t^2.$$

When the ball hits the ground, s will be 0. Find how many seconds it takes to hit the ground.

2. A ball is dropped from a cliff 144 ft above the ground. The distance s (in feet) it falls in t seconds is given by

$$s = 16t^2.$$

How many seconds does it take for the ball to hit the ground?

3. Given the formula of motion,

$$s = v_0t + \frac{1}{2}at^2,$$

find t (in seconds) to one decimal place if $s = 15$ meters (m), $v_0 = 18$ m/sec, and $a = -9.8$ m/sec^2. Here s, v_0, and a are displacement, initial velocity, and acceleration, respectively.

4. A certain projectile is fired. Its height s (in ft) above ground level after t sec is given by

$$s = 152t - 16t^2.$$

The projectile will reach a maximum height and fall down. At what times will the projectile be 280 ft above ground level?

5. An economics instructor told his class that the demand equation for a certain product is $p = 400 - x^2$ and its supply equation is $p = 20x + 100$. If the $400 - x^2$ is set equal to the $20x + 100$, then the *positive* solution to the resulting equation gives the *equilibrium quantity*. The instructor asked his class to find this quantity. What answer should the class give?

6. A lumber company owns a forest that is of rectangular shape, 1 mi by 2 mi. If the company cuts a uniform strip of trees along the outer edges of this forest, how wide should the strip be if $\frac{3}{4}$ sq mi of forest is to remain?

7. A rectangular plot, 4 meters (m) by 8 m, is to be used for a garden. It is decided to put a pavement inside the entire border so that 12 m^2 of the plot is left for flowers. How wide should the pavement be?

8. A company parking lot is 120 ft long and 80 ft wide. Because of an increase in personnel, it is decided to double the area of the lot by adding strips of equal width to one end and one side. Find the width of one such strip.

9. A real estate firm owns the Shantytown Garden Apartments, consisting of 70 apartments. At \$125 per month each apartment can be rented. However, for each \$5 per month increase there will be two vacancies with no possibility of filling them. If the firm wants to receive \$8990 per month from rents, what rent should be charged for each apartment?

10. Imperial Educational Services (I.E.S.) wants to offer a workshop in pollution control to key personnel at Acme Corporation. I.E.S. will offer the course to thirty persons at a charge of \$50 each. Moreover, I.E.S. will agree to reduce the charge for *everybody* by \$1.00 for each person over the thirty who attends, up to a total group size of fifty. It has been determined that the greatest revenue that I.E.S. can receive is \$1600. What group size will give this revenue?

11. An open box is to be made from a square piece of tin by cutting out a 3-in. square from each corner and folding up the sides. See Fig. 7-2. The box is to contain 75

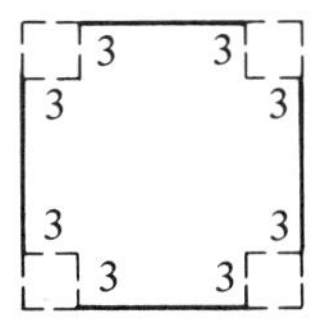

FIG. 7-2

cubic inches. Find the dimensions of the square piece of tin that must be used.

12. For security reasons a company will enclose a rectangular area of 11,200 ft^2 in the rear of its plant. One side will be bounded by the building and the other three sides by fencing. See Fig. 7-3. If 300 ft of fencing will be used, what will be the

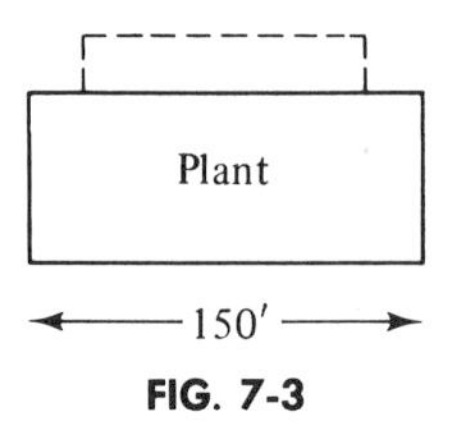

FIG. 7-3

dimensions of the rectangular area?

13. The Dandy Candy Company makes the ever-popular Dandy Bar. That's the bar with the slogan "Although I live in Zanzibar, I'd go to New York for a Dandy Bar." The rectangular-shaped bar is 10 centimeters (cm) long, 5 cm wide, and 2 cm thick. See Fig. 7-4. Because of increasing costs, the company has decided to cut the volume of

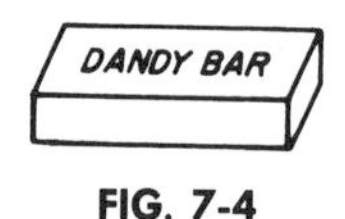

FIG. 7-4

the bar by a drastic 28 percent. The thickness will be the same, but the length and width will be reduced by equal amounts. What will be the length and width of the new bar?

14. The Dandy Candy Company, besides making the Dandy Bar, also makes a washer-shaped candy (a candy with a hole in it), called Life Rings. See Fig. 7-5. Because of

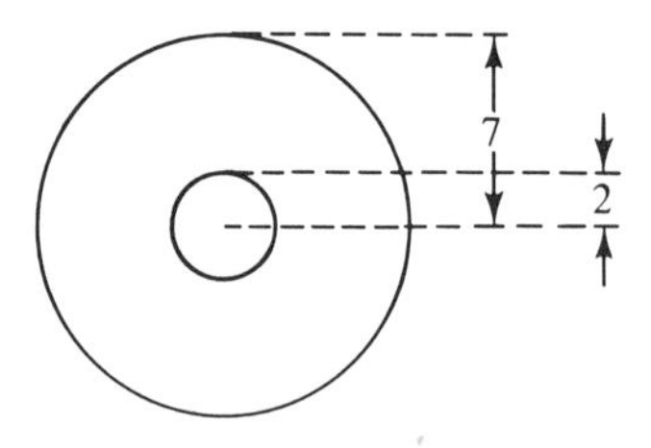

FIG. 7-5

increasing costs, the company will cut the volume of candy in each piece by 20 percent. To do this they will keep the same thickness and outer radius, but will make the inner radius larger. At present the thickness is 2 millimeters (mm), the inner radius is 2 mm, and the outer radius is 7 mm. Find the inner radius of the new-style candy. *Hint:* The volume V of a solid disc is $\pi r^2 h$, where r is the radius and h is the height.

15. An oil dealer decides to tear down two cylindrical storage tanks and replace them by one new tank. The old tanks are each 16 ft high. One has a radius of 15 ft and the other a radius of 20 ft. The new tank will also be 16 ft high. Find its radius if it is to hold the same volume of oil as the old tanks combined. *Hint:* The volume V of a cylindrical tank is $V = \pi r^2 h$, where r is the radius and h is the height.

16. A man deposits \$50 in a bank and in two years he has \$56.18. If the bank compounds interest annually, what annual rate of interest does it pay?

17. In two years a company will initiate an expansion program. It has decided to invest \$1,000,000 now so that in two years the total value of the investment will be \$1,102,500, the amount required for the expansion. What is the annual rate of interest, compounded annually, that the company must receive to achieve its purpose?

18. Researchers at a university need a 10,000-m^2 rectangular plot on which to grow three hybrids of corn. The plot is to be enclosed by fencing. Fencing will also be used to separate the different hybrids. See Fig. 7-6. If 600 m of fence are to be used for the

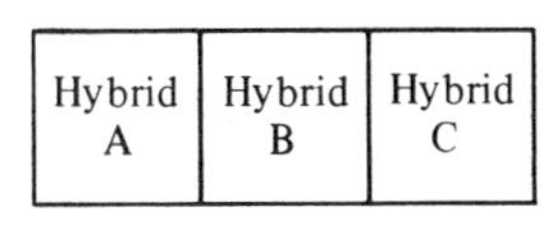

FIG. 7-6

project, what are the dimensions of this plot?

7-6 SUPPLEMENT ON QUADRATIC FORMULA†

In Section 7-3 we gave you the quadratic formula. Here is how the formula comes about. We'll use the method of completing the square.

Suppose that $ax^2 + bx + c = 0$ is a quadratic equation. Then

$$ax^2 + bx = -c$$

$$\frac{ax^2 + bx}{a} = \frac{-c}{a}$$

$$x^2 + \frac{b}{a}x = -\frac{c}{a}.$$

Half of $\frac{b}{a}$: $\quad \frac{\frac{b}{a}}{2} = \frac{b}{2a}$. $\quad$ Squaring: $\left(\frac{b}{2a}\right)^2 = \frac{b^2}{4a^2}$.

$$x^2 + \frac{b}{a}x + \frac{b^2}{4a^2} = \frac{b^2}{4a^2} - \frac{c}{a}. \tag{7-1}$$

†This topic is not needed in the rest of the book.

The left side is $\left(x + \frac{b}{2a}\right)^2$ and the right side is

$$\frac{b^2}{4a^2} - \frac{c}{a} = \frac{b^2}{4a^2} - \frac{4ac}{4a^2} = \frac{b^2 - 4ac}{4a^2}.$$

Thus, Eq. (7–1) becomes

$$\left(x + \frac{b}{2a}\right)^2 = \frac{b^2 - 4ac}{4a^2}$$

$$x + \frac{b}{2a} = \pm\sqrt{\frac{b^2 - 4ac}{4a^2}}$$

$$x = -\frac{b}{2a} \pm \frac{\sqrt{b^2 - 4ac}}{2a}$$

$$\text{or} \qquad x = \frac{-b \pm \sqrt{b^2 - 4ac}}{2a},$$

which is the quadratic formula.

7-7 REVIEW

IMPORTANT TERMS AND SYMBOLS

quadratic equation *(p. 150)*
second-degree equation *(p. 150)*
repeated root *(p. 153)*
completing the square *(p. 157)*
quadratic formula *(p. 160)*
discriminant *(p. 162)*
complex numbers *(p. 165)*
imaginary unit *(p. 164)*
$\sqrt{-1}$, i *(p. 164)*
imaginary numbers *(p. 165)*
pure imaginary numbers *(p. 165)*

REVIEW PROBLEMS

In Problems **1–10**, *solve by factoring.*

1. $x^2 - 10x + 25 = 0$ **2.** $x^2 - 2x - 8 = 0$ **3.** $x^2 - 2x - 24 = 0$

4. $x^2 + 6x + 9 = 0$ **5.** $12x^2 - 20x + 3 = 0$ **6.** $4x^2 - 5x - 6 = 0$

7. $x^2 - 12 = 0$ **8.** $x^2 - 28 = 0$ **9.** $2x^3 - x^2 = 0$

10. $x^4 - 9x^2 = 0$

In Problems **11–14**, *solve by completing the square.*†

11. $x^2 - 10x + 1 = 0$ **12.** $x^2 + 8x - 5 = 0$ **13.** $4x^2 + 12x - 2 = 0$

14. $2x^2 - 2x - 3 = 0$

In Problems **15–20**, *solve by the quadratic formula.*

15. $x^2 - 6x + 7 = 0$ **16.** $x^2 + 3x - 5 = 0$ **17.** $4x^2 + 4x + 1 = 0$

18. $3x^2 - 2x + 6 = 0$ **19.** $2x - 5 - 2x^2 = 0$ **20.** $25 - 20x + 4x^2 = 0$

In Problems **21–32**, *solve.*

21. $16x^2 - 9 = 0$ **22.** $25x^2 - 3 = 1$ **23.** $y^2 + 2y - 24 = 0$

24. $y^2 + 4y - 21 = 0$ **25.** $(z + 4)^2 = 36$ **26.** $4z(z + 2) = -8 - 4z$

27. $3(t - 1) = 2t^2$ **28.** $100t^2 = 100t$ **29.** $(x + 1)(x + 2) = 4$

30. $x^2 + 5 = -(1 - x)$ **31.** $4x^2 + 10x = -\frac{25}{4}$ **32.** $\frac{3}{4}(x^2 - 2) = x$

33. The formula $S = 2\pi r^2 + 2\pi rh$ gives the total surface area S of a cylinder of radius r and height h. Find r so that a cylinder of height 2 in. will have an area of 48π in.^{2}.

34. An open box is to be made from a square piece of tin by cutting out a 6-in. square from each corner and turning up the sides. The box will contain 150 in.^{3}. Find the *area* of the original square.

35. Imperial Educational Services (I.E.S.) is offering a workshop in data processing to key personnel at Zeta Corporation. The price per person is $50, and currently fifty persons will attend. Suppose I.E.S. offers to reduce the charge for *everybody* by $.50 for each person over the fifty who attends. How many people, over the fifty, can attend so that the cost to Zeta Corporation is the same as that for fifty people?

36. A square plot, 12 yd by 12 yd, is to be used for a garden. It is decided to put a pavement of uniform width inside the plot bordering three of the sides so that 80 yd^2 of the plot is left for flowers. How wide should the pavement be?

---

†These problems refer to Section 7-2.

CHAPTER 8

FRACTIONS

8-1 REDUCTION, MULTIPLICATION, DIVISION

With the fraction $-\frac{2}{3}$, which can be thought of as $-\frac{+2}{+3}$, we can associate three signs:

the sign of the numerator $(+)$,
the sign of the denominator $(+)$, and
the sign in front of the fraction $(-)$.

From Chapter 2 (p. 40), you know that

$$-\frac{2}{3} = \frac{-2}{3}.$$

Notice that $\frac{-2}{3}$ can be obtained from $-\frac{2}{3}$ by changing *two* signs of $-\frac{2}{3}$: the sign in front of the fraction and the sign of the numerator. You also know that

$$-\frac{2}{3} = \frac{2}{-3}.$$

Again, by changing *two* signs of $-\frac{2}{3}$, namely, the sign of the fraction and the sign of the denominator, we can obtain $\frac{2}{-3}$.

In general, **by changing any two signs of a fraction, we get a fraction equal** (equivalent) **to the original one**. Thus, $-\frac{-6}{7}$ can be written in the following ways:

$$+\frac{+6}{+7}, \qquad -\frac{+6}{-7}, \qquad +\frac{-6}{-7}.$$

EXAMPLE 1

Changing signs of a fraction.

a. $\dfrac{-x}{y} = -\dfrac{x}{y}$.

b. $\dfrac{-x}{-y} = \dfrac{x}{y}$.

c. $-\dfrac{x(-y)}{z} = -\dfrac{-(xy)}{z} = \dfrac{xy}{z}$.

d. $\dfrac{w(-x)}{(-y)(-z)} = \dfrac{-(wx)}{yz} = -\dfrac{wx}{yz}$.

e. $\dfrac{-a-b}{c} = \dfrac{-(a+b)}{c} = -\dfrac{a+b}{c}$.

f. $\dfrac{y}{(-1)ab} = \dfrac{y}{-ab} = -\dfrac{y}{ab}$.

If the numerator and denominator of a fraction have a *common factor*, then we can *reduce* the fraction. This is really "cancellation."

$$\boxed{\frac{ab}{ac} = \frac{b}{c}} \quad \text{or} \quad \boxed{\frac{\cancel{a}b}{\cancel{a}c} = \frac{b}{c}.}$$

To reduce a fraction, first factor completely both the numerator and denominator. Then cancel any common factors.

EXAMPLE 2

Reducing fractions.

a. $\dfrac{2x+6}{6x} = \dfrac{\cancel{2}(x+3)}{\underset{3}{\cancel{6}}x} = \dfrac{x+3}{3x}$.

b. $\dfrac{3x-9}{x-3} = \dfrac{3\cancel{(x-3)}}{\cancel{x-3}} = 3$.

c. $\dfrac{4x^2}{8x^2-4x^3} = \dfrac{\cancel{4x^2}}{\cancel{4x^2}(2-x)} = \dfrac{1}{2-x}$.

EXAMPLE 3

Reducing fractions.

a. $$\frac{x^2-1}{x^2+2x+1} = \frac{(x-1)\cancel{(x+1)}}{\underset{x+1}{\cancel{(x+1)^2}}} = \frac{x-1}{x+1}.$$

b. $$\frac{2x^2-2x-12}{4x^2-8x-12} = \frac{\cancel{2}(x^2-x-6)}{\underset{2}{\cancel{4}}(x^2-2x-3)} = \frac{(x+2)\cancel{(x-3)}}{2(x+1)\cancel{(x-3)}} = \frac{x+2}{2(x+1)}.$$

c. $$\frac{(y-2)^3(y-4)^2}{(y-2)^2(y-4)^4} = \frac{y-2}{(y-4)^2}.$$

*Don't go hog wild with cancellation. Remember, you can cancel only **factors**, and those factors must be common to the **entire** numerator and **entire** denominator.* ***Don't do things like***

$$\frac{4+3\cancel{x}}{\cancel{x}} \qquad [x \text{ is } \textbf{not} \text{ a factor of the numerator}]$$

or

$$\frac{\cancel{x+5}}{x+5\cancel{(x+5)}} \qquad [x+5 \text{ is } \textbf{not} \text{ a factor of the denominator}].$$

The expressions $2-1$ and $1-2$ look almost alike, but they do have different values. $2-1=1$, but $1-2=-1$. One value is the opposite of the other: $2-1=-(1-2)$. In general,

$$\boxed{a-b=-(b-a)=(-1)(b-a).}$$

This fact can sometimes be used to reduce fractions in a "sneaky" way.

EXAMPLE 4

"Sneaky" reductions.

a. $$\frac{3-x}{x-3} = \frac{-(x-3)}{x-3} = \frac{(-1)\cancel{(x-3)}}{\cancel{x-3}} = -1.$$

Another way to reduce the fraction is by rewriting the denominator:

$$\frac{3-x}{x-3} = \frac{\cancel{3-x}}{(-1)\cancel{(3-x)}} = \frac{1}{-1} = -1.$$

b. $\frac{x^2 + 3x - 4}{1 - x^2} = \frac{(x - 1)(x + 4)}{(1 - x)(1 + x)} = \frac{\cancel{(x - 1)}(x + 4)}{(-1)\cancel{(x - 1)}(1 + x)} = -\frac{x + 4}{1 + x}.$

$$\frac{x - 1}{x + 1} \neq \frac{-(x + 1)}{x + 1}, \qquad \frac{1 - x}{x + 1} \neq \frac{-(x + 1)}{x + 1},$$

$$\text{and} \quad \frac{1 + x}{x + 1} \neq \frac{-(x + 1)}{x + 1}.$$

In arithmetic we multiply two fractions by using the next rule:

MULTIPLICATION OF FRACTIONS

$$\frac{a}{b} \cdot \frac{c}{d} = \frac{ac}{bd} \begin{array}{l} \leftarrow \textit{product of numerators} \\ \leftarrow \textit{product of denominators} \end{array}$$

This rule works not only when a, b, c, and d are numbers, but also when they are expressions.

EXAMPLE 5

Multiplication.

a. $\frac{x}{x - 1} \cdot \frac{x + 1}{x - 2} = \frac{x(x + 1)}{(x - 1)(x - 2)}.$

b. $\frac{3}{-x} \cdot \frac{x^2 + 4}{x} = \frac{3(x^2 + 4)}{-x^2} = -\frac{3(x^2 + 4)}{x^2}.$

c. $z^3 \cdot \frac{z + 2}{z + 1} \cdot \frac{z - 3}{4} = \frac{z^3}{1} \cdot \frac{z + 2}{z + 1} \cdot \frac{z - 3}{4} = \frac{z^3(z + 2)(z - 3)}{4(z + 1)}.$

To multiply $\frac{x}{x + 1}$ by $\frac{x + 1}{x - 5}$, we have

$$\frac{x}{x + 1} \cdot \frac{x + 1}{x - 5} = \frac{x\cancel{(x + 1)}}{\cancel{(x + 1)}(x - 5)} = \frac{x}{x - 5}.$$

Here we multiplied first and then simplified the result. But we can do the cancellation right away.

$$\frac{x}{\cancel{x + 1}} \cdot \frac{\cancel{x + 1}}{x - 5} = \frac{x}{x - 5}.$$

EXAMPLE 6

Multiplication.

a. $\cancel{x+2} \cdot \dfrac{x-4}{\cancel{x+2}} = x - 4.$

b. $$\frac{3x-3}{x} \cdot \frac{5}{x^2-1} = \frac{3(x-1)}{x} \cdot \frac{5}{(x-1)(x+1)} \qquad \text{[factoring]}$$

$$= \frac{3\cancel{(x-1)}}{x} \cdot \frac{5}{\cancel{(x-1)}(x+1)} = \frac{15}{x(x+1)}.$$

c. $$\frac{x^3}{4x^2-9} \cdot \frac{2x+3}{x^2+x} = \frac{\overset{x^2}{\cancel{x^3}}}{(2x-3)\cancel{(2x+3)}} \cdot \frac{\cancel{2x+3}}{\cancel{x}(x+1)}$$

$$= \frac{x^2}{(2x-3)(x+1)}.$$

EXAMPLE 7

Multiplication.

a. $$\frac{x^2+5x+4}{x^2+5x+6} \cdot \frac{x^2+3x+2}{x^2+2x-8} = \frac{(x+1)\cancel{(x+4)}}{(x+3)\cancel{(x+2)}} \cdot \frac{(x+1)\cancel{(x+2)}}{\cancel{(x+4)}(x-2)}$$

$$= \frac{(x+1)^2}{(x+3)(x-2)}.$$

b. $$\frac{6}{x^2-3x+2} \cdot \frac{2-x}{x+3} = \frac{6}{(x-1)\cancel{(x-2)}} \cdot \frac{(-1)\cancel{(x-2)}}{x+3}$$

$$= \frac{-6}{(x-1)(x+3)} = -\frac{6}{(x-1)(x+3)}.$$

After factoring $x^2 - 3x + 2$, we wrote $2 - x$ as $(-1)(x - 2)$ so that we could cancel.

To divide two fractions, we "invert" the divisor (which is a fraction) and then multiply as follows:

DIVISION OF FRACTIONS

$$\frac{\frac{a}{b}}{\frac{c}{d}} = \frac{a}{b} \cdot \frac{d}{c} = \frac{ad}{bc}.$$

Notice the switch from division to multiplication. We multiply $\frac{a}{b}$ by the *reciprocal* of the divisor $\frac{c}{d}$. Sometimes we use the division symbol "÷" to denote the original division:

$$\frac{\frac{a}{b}}{\frac{c}{d}} = \frac{a}{b} \div \frac{c}{d}.$$

Thus,

$$\frac{\frac{a}{b}}{\frac{c}{d}} = \frac{a}{b} \div \frac{c}{d} = \frac{a}{b} \cdot \frac{d}{c} = \frac{ad}{bc}.$$

EXAMPLE 8

Division.

a. $$\frac{\frac{4y}{3z}}{\frac{z}{9}} = \frac{4y}{3z} \div \frac{z}{9} = \frac{4y}{3z} \cdot \frac{9}{z} = \frac{4y}{\cancel{3}z} \cdot \frac{\overset{3}{\cancel{9}}}{z} = \frac{12y}{z^2}.$$

b. $$\frac{\frac{x}{x+2}}{\frac{x+3}{x-5}} = \frac{x}{x+2} \cdot \frac{x-5}{x+3} = \frac{x(x-5)}{(x+2)(x+3)}.$$

c. $$\frac{\frac{4x}{x^2-1}}{\frac{2x^2+8x}{x-1}} = \frac{4x}{x^2-1} \cdot \frac{x-1}{2x^2+8x} = \frac{\overset{2}{\cancel{4x}}}{\cancel{(x-1)}(x+1)} \cdot \frac{\cancel{x-1}}{\cancel{2x}(x+4)}$$

$$= \frac{2}{(x+1)(x+4)}.$$

EXAMPLE 9

Division.

a. $$\frac{\frac{x}{x-3}}{2x} = \frac{\frac{x}{x-3}}{\frac{2x}{1}} = \frac{\cancel{x}}{x-3} \cdot \frac{1}{2\cancel{x}} = \frac{1}{2(x-3)}.$$

b. $$\frac{4x^2-4x+1}{\frac{x}{2x-1}} = \frac{\frac{4x^2-4x+1}{1}}{\frac{x}{2x-1}} = \frac{4x^2-4x+1}{1} \cdot \frac{2x-1}{x}$$

$$= \frac{(2x-1)^2}{1} \cdot \frac{2x-1}{x} = \frac{(2x-1)^3}{x}.$$

The properties in this section are helpful in solving certain equations, as the next example shows.

EXAMPLE 10

Solve.

a. $-\dfrac{x+1}{-3} = 2.$

$$-\frac{x+1}{-3} = 2$$

$$\frac{x+1}{3} = 2$$

$$x + 1 = 6 \qquad \text{[multiplying both sides by 3]}$$

$$\boxed{x = 5.}$$

b. $\dfrac{x}{\frac{1}{4}} = \dfrac{3}{5}.$

$$\frac{x}{\frac{1}{4}} = \frac{3}{5}$$

$$4x = \frac{3}{5} \qquad \left[\text{since } \frac{x}{\frac{1}{4}} = \frac{x}{1} \cdot \frac{4}{1} = 4x\right]$$

$$x = \frac{\frac{3}{5}}{4} \qquad \text{[dividing both sides by 4]}$$

$$x = \frac{3}{5} \cdot \frac{1}{4} = \boxed{\frac{3}{20}.}$$

Completion Set 8-1

In Problems **1–6**, *fill in the blanks with* T (= *True*) *or* F (= *False*).

1. $\dfrac{-a}{b} = -\dfrac{a}{-b}$. ____

2. $\dfrac{a-b}{c} = -\dfrac{b+a}{c}$. ____

3. $\dfrac{1}{a} \cdot \dfrac{1}{b} = \dfrac{1}{ab}$. ____

4. $\dfrac{\frac{1}{a}}{b} = \dfrac{1}{a} \cdot \dfrac{1}{b} = \dfrac{1}{ab}$. ____

5. $\dfrac{-x}{1-x} = \dfrac{x}{x-1} \cdot \_\_\_$

6. $\dfrac{x}{\frac{a}{b}} = \dfrac{ax}{b} \cdot \_\_\_$

In Problems **7–13**, *fill in the blanks.*

7. $\dfrac{3x^2 - 12x}{3x^3} = \dfrac{(\qquad)(x-4)}{3x^3} = \dfrac{x-4}{(\qquad)}.$

8. $\dfrac{x^2 - 4}{x^2 + 5x + 6} = \dfrac{(x-2)(x+2)}{(x+2)(\qquad)} = \dfrac{(\qquad)}{x+3}.$

9. $\dfrac{10 - x}{x - 10} = \dfrac{(-1)(\quad - \quad)}{x - 10} = \_\_\_.$

10. $\dfrac{x}{x+4} \cdot \dfrac{x}{x+2} = \dfrac{\qquad}{(x+4)(x+2)}.$

11. $\dfrac{x-1}{x+2} \cdot \dfrac{x+3}{x-1} = \dfrac{\qquad}{x+2}.$

12. $\dfrac{\frac{7}{x}}{\frac{2}{x+1}} = \dfrac{7}{x} \cdot \dfrac{\qquad}{2}.$

13. $\dfrac{\frac{x-1}{x+2}}{\frac{x+1}{x-2}} = \dfrac{x-1}{x+2} \cdot \left(\qquad\right).$

Problem Set 8-1

In Problems **1–6**, *write each fraction so that minus signs do not appear in the numerator and in the denominator. For example, write* $\dfrac{ab}{-c}$ *as* $-\dfrac{ab}{c}$.

1. $\dfrac{-x}{-y}$

2. $-\dfrac{-wx}{-z}$

3. $-\dfrac{a}{b(-c)}$

4. $-\dfrac{ab}{(-c)(-d)}$

5. $\dfrac{x}{-y-z}$

6. $\dfrac{-x(y+z)}{(-a)(b)(-c)}$

In Problems **7–20**, *reduce.*

7. $\dfrac{5x - 15}{5x - 25}$ **8.** $\dfrac{2x - 4}{4x - 8}$ **9.** $\dfrac{12x^2 + 6x}{3x^2 - 9x}$

10. $\dfrac{6x + 2x^2}{8x - 8x^2}$ **11.** $\dfrac{y^4 + y^2}{y^6 - 4y^5}$ **12.** $\dfrac{x^4 - x^2}{x^2 + x}$

13. $\dfrac{x + 1}{x^2 + 7x + 6}$ **14.** $\dfrac{z^2 - 9}{z^2 - 6z + 9}$ **15.** $\dfrac{x^2 - 3x - 10}{x^2 - 4x - 5}$

16. $\dfrac{2x^2 + 11x + 12}{2x^2 + 8x}$ **17.** $\dfrac{3z^2 + z - 2}{6z^2 - z - 2}$ **18.** $\dfrac{a^2 - b^2}{b^2 - a^2}$

19. $\dfrac{25 - x^2}{x^2 - 2x - 15}$ **20.** $\dfrac{x^2 + 3x - 4}{2 - x - x^2}$

In Problems **21–36**, *do the multiplications and simplify your answers.*

21. $\dfrac{2x}{x - 1} \cdot \dfrac{3x}{x + 2}$ **22.** $\dfrac{x - 3}{x^2} \cdot \dfrac{x - 3}{x^4}$ **23.** $\dfrac{-x}{x + 2} \cdot \dfrac{x + 2}{x^2}$

24. $\dfrac{3}{x + 1} \cdot \dfrac{(x + 1)^2}{-(x + 3)}$ **25.** $\dfrac{5x - 10}{5(x + 3)} \cdot \dfrac{x}{2x - 4}$ **26.** $\dfrac{4}{6x^2} \cdot \dfrac{3x + 9}{x^2 + 3x}$

27. $(x^2 - 9) \cdot \dfrac{x - 3}{4x + 12}$ **28.** $\dfrac{5}{(x + 1)^2} \cdot (x^2 + 2x + 1)$

29. $\dfrac{(x - 1)^2}{x + 3} \cdot \dfrac{x^3}{x - 1} \cdot \dfrac{(x + 3)^2}{x - 1}$ **30.** $\dfrac{2x - 1}{x^2 + 4x} \cdot \dfrac{x}{4x^2 - 1} \cdot \dfrac{2x + 1}{6}$

31. $\dfrac{x^2 + 6x + 8}{x^2 - 5x + 6} \cdot \dfrac{x - 2}{x^2 + 5x + 4}$ **32.** $\dfrac{2x^2 + 3x + 1}{x^2 - 4x - 5} \cdot \dfrac{x^2 - 25}{4x^2 - 1}$

33. $\dfrac{8x^2 + 32}{x^2 + 2x} \cdot \dfrac{x^2 + 4x + 4}{8x^2 - 32}$ **34.** $\dfrac{x^3 + 8x^2 + 15x}{x^2 + 4x + 3} \cdot \dfrac{x - 1}{x^2 + 5x}$

35. $\dfrac{x^2 - 8x + 7}{(x + 2)^2} \cdot \dfrac{x^2 + x - 2}{49 - x^2}$ **36.** $\dfrac{8 - 4x}{6x^3} \cdot \dfrac{4 - x}{x^2 - 6x + 8}$

In Problems **37–54**, *do the divisions and simplify your answers.*

37. $\dfrac{\frac{x^2}{6}}{\frac{x}{3}}$ **38.** $\dfrac{\frac{4x^3}{9x}}{\frac{x}{18}}$ **39.** $\dfrac{\frac{2m}{n^3}}{\frac{4m}{n^2}}$

40. $\dfrac{\frac{c + d}{c}}{\frac{c - d}{2c}}$ **41.** $\dfrac{\frac{4x}{3}}{2x}$ **42.** $\dfrac{-9x^3}{\frac{x}{3}}$

43. $\dfrac{x - 5}{\frac{x^2 - 7x + 10}{x - 2}}$ **44.** $\dfrac{\frac{x^2 + 6x + 9}{x}}{x + 3}$ **45.** $\dfrac{\frac{2x - 4}{-6x}}{\frac{x - 2}{3x^2}}$

46. $\dfrac{\dfrac{10x^3}{x^2-1}}{\dfrac{5x}{x+1}}$ 47. $\dfrac{\dfrac{x^2-4}{x^2+2x-3}}{\dfrac{x^2-x-6}{x^2-9}}$ 48. $\dfrac{\dfrac{x^2+7x+10}{x^2-2x-8}}{\dfrac{x^2+6x+5}{x^2-3x-4}}$

49. $\dfrac{\dfrac{2x^2+5x+3}{6x^2+4x}}{\dfrac{x^2-1}{3x^2-x-2}}$ 50. $\dfrac{\dfrac{2x^2+5x-3}{4x^2-1}}{\dfrac{x^2+4x+3}{6x^2+x-1}}$ 51. $\dfrac{\dfrac{(x+2)^2}{3x-2}}{\dfrac{9x+18}{4-9x^2}}$

52. $\dfrac{\dfrac{3-2x-x^2}{-20x^3}}{\dfrac{x^2-1}{x^4+x^3}}$ 53. $\dfrac{\frac{3}{5}(x^2+4x+4)}{\frac{9}{8}(x^2-4)}$ 54. $\dfrac{\frac{8}{3}(x^2+x-20)}{\dfrac{3x+15}{4}}$

In Problems **55–66**, *solve.*

55. $-\dfrac{-x}{5}=2$ 56. $-\dfrac{4x}{-5}=8$ 57. $-\dfrac{y-4}{-2}=y$

58. $\dfrac{-y}{3}-\dfrac{y}{-4}=1$ 59. $\dfrac{-(x+1)}{2}-\dfrac{-7x}{-4}=4$ 60. $\dfrac{x}{-2}-\dfrac{-x}{3}=\dfrac{x}{-4}$

61. $2x=\dfrac{3}{7}$ 62. $5x=\dfrac{6}{5}$ 63. $\dfrac{y}{\frac{2}{3}}=5$

64. $\dfrac{y}{\frac{1}{3}}=9$ 65. $\dfrac{x-1}{\frac{4}{5}}=16$ 66. $\dfrac{x}{\frac{3}{4}}=\dfrac{\frac{3}{2}}{6}$

67. In a scale drawing, the scale is $\frac{1}{4}$ in. = 2 ft. The scale length (in inches) of a 56-ft object is given by l, where

$$\frac{l}{\frac{1}{4}}=\frac{56}{2}.$$

Find l.

8-2 ADDITION AND SUBTRACTION OF FRACTIONS

To add fractions having a *common denominator*, we use the next rule:

$$\frac{a}{c}+\frac{b}{c}=\frac{a+b}{c}.$$

That is, the sum is a fraction whose denominator is the common denominator and whose numerator is the sum of the numerators of the fractions.

There are similar rules for handling sums and differences involving any number of fractions having a common denominator. For example,

$$\boxed{\frac{a}{d} - \frac{b}{d} + \frac{c}{d} = \frac{a - b + c}{d}.}$$

EXAMPLE 1

a. $$\frac{x^2 - 5}{x - 2} + \frac{2x - 3}{x - 2} = \frac{(x^2 - 5) + (2x - 3)}{x - 2}$$
$$= \frac{x^2 + 2x - 8}{x - 2}$$
$$= \frac{\cancel{(x - 2)}(x + 4)}{\cancel{x - 2}} \qquad \text{[simplifying]}$$
$$= x + 4.$$

b. $$\frac{x^2 - 5x + 4}{x^2 + 2x - 3} - \frac{x^2 + 2x}{x^2 + 5x + 6}$$
$$= \frac{\cancel{(x - 1)}(x - 4)}{\cancel{(x - 1)}(x + 3)} - \frac{x\cancel{(x + 2)}}{\cancel{(x + 2)}(x + 3)} \qquad \text{[factoring and cancelling]}$$
$$= \frac{x - 4}{x + 3} - \frac{x}{x + 3} \qquad \text{[denominators are now the same]}$$
$$= \frac{x - 4 - x}{x + 3} = \frac{-4}{x + 3} = -\frac{4}{x + 3}.$$

c. $$\frac{x^2 + x - 5}{x - 7} - \frac{x^2 - 2}{x - 7} + \frac{8 - 4x}{x^2 - 9x + 14}$$
$$= \frac{x^2 + x - 5}{x - 7} - \frac{x^2 - 2}{x - 7} + \frac{-4\cancel{(x - 2)}}{\cancel{(x - 2)}(x - 7)} \qquad \text{[denominators are now equal]}$$
$$= \frac{(x^2 + x - 5) - (x^2 - 2) + (-4)}{x - 7}$$
$$= \frac{x^2 + x - 5 - x^2 + 2 - 4}{x - 7}$$
$$= \frac{x - 7}{x - 7} = 1.$$

$$\frac{a}{b} + \frac{a}{c} \neq \frac{a}{b + c} \quad \textit{and} \quad \frac{a}{b} + \frac{c}{d} \neq \frac{a + c}{b + d}.$$

To add or subtract fractions with different denominators, we first rewrite the fractions as equivalent fractions that have the same denominator. Then we add or subtract as we did before.

The key to rewriting fractions is the rule

$$\boxed{\frac{a}{b} = \frac{ac}{bc}.}$$

That is, *multiplying both numerator and denominator of a fraction by the same number gives an equivalent fraction.* But the number you multiply by can't be 0, because that would give a zero denominator. Remember: Division by 0 is **not** allowed.

Let's consider adding fractions with different denominators. To add

$$\frac{2}{x(x-3)} + \frac{3x}{x(x+4)},$$

we can begin by rewriting the first fraction as

$$\frac{2(x+4)}{x(x-3)(x+4)} \qquad [\text{multiplying numerator and denominator by } x+4]$$

and rewriting the second one as

$$\frac{3x(x-3)}{x(x+4)(x-3)} \qquad [\text{multiplying numerator and denominator by } x-3].$$

Since these fractions have the same denominator, we can combine them.

$$\begin{aligned}\frac{2}{x(x-3)} + \frac{3x}{x(x+4)} &= \frac{2(x+4)}{x(x-3)(x+4)} + \frac{3x(x-3)}{x(x+4)(x-3)}\\ &= \frac{2(x+4)+3x(x-3)}{x(x-3)(x+4)}\\ &= \frac{2x+8+3x^2-9x}{x(x-3)(x+4)}\\ &= \frac{3x^2-7x+8}{x(x-3)(x+4)}.\end{aligned}$$

We could have rewritten the original fractions with other common denominators. But we chose to rewrite them as fractions with the denominator

$x(x-3)(x+4)$, which is called the **least common denominator** (L.C.D.) of the fractions $\dfrac{2}{x(x-3)}$ and $\dfrac{3x}{x(x+4)}$.

In general,

To find the L.C.D. *of two or more fractions, multiply all the different factors in the denominators of the fractions, each raised to the highest power to which that factor occurs in any one denominator.*

EXAMPLE 2

Find the L.C.D. *of the fractions*

$$\frac{2x}{(x+1)(x-2)}, \quad \frac{x-3}{x^2(x+1)}, \quad \textit{and} \quad \frac{x^2+x}{x(x-2)^2}.$$

There are three different factors in the denominators:

$$x+1, \quad x-2, \quad \text{and} \quad x.$$

The factor $x+1$ occurs at most one time in any denominator. The factor $x-2$ occurs at most two times (in the third fraction). The factor x occurs at most two times (in the second fraction). The L.C.D. is the product of these factors, each raised to the highest power to which it occurs in any denominator. Thus, the L.C.D. is

$$(x+1)(x-2)^2x^2.$$

EXAMPLE 3

Find the L.C.D. *of the fractions*

$$\frac{2}{x^2}, \quad \frac{x}{x+1}, \quad \frac{3x}{x^3(x-1)}, \quad \textit{and} \quad \frac{5}{\underbrace{x^2+2x+1}_{(x+1)^2}}.$$

The factor x occurs at most three times, the factor $x+1$ occurs at most twice, and the factor $x-1$ occurs at most once. Thus, the L.C.D. is

$$x^3(x+1)^2(x-1).$$

To add or subtract fractions with different denominators, here's what to do:

For each fraction, multiply both its numerator and denominator by a quantity that makes its denominator equal to the L.C.D. *of the fractions. Then combine the fractions and simplify if possible.*

EXAMPLE 4

Find $\dfrac{3}{x+2} + \dfrac{x-1}{x-6}$.

The L.C.D. is $(x + 2)(x - 6)$. To get the denominator of the first fraction equal to the L.C.D., we multiply the numerator and denominator by $x - 6$. In the second fraction we multiply numerator and denominator by $x + 2$.

$$\frac{3}{x+2} + \frac{x-1}{x-6}$$

$$= \frac{3(x-6)}{(x+2)(x-6)} + \frac{(x-1)(x+2)}{(x-6)(x+2)} \qquad \text{[putting the L.C.D. in each fraction]}$$

$$= \frac{3(x-6) + (x-1)(x+2)}{(x+2)(x-6)} \qquad \text{[combining fractions with common denominators]}$$

$$= \frac{3x - 18 + x^2 + x - 2}{(x+2)(x-6)}$$

$$= \frac{x^2 + 4x - 20}{(x+2)(x-6)}.$$

EXAMPLE 5

Find $3x - 4 + \dfrac{2}{x-1}$.

Since $3x - 4 = \dfrac{3x-4}{1}$, the L.C.D. of $\dfrac{3x-4}{1}$ and $\dfrac{2}{x-1}$ is just $x - 1$.

$$3x - 4 + \frac{2}{x-1} = \frac{(3x-4)(x-1)}{x-1} + \frac{2}{x-1}$$

$$= \frac{(3x-4)(x-1) + 2}{x-1}$$

$$= \frac{3x^2 - 7x + 4 + 2}{x-1}$$

$$= \frac{3x^2 - 7x + 6}{x-1}.$$

EXAMPLE 6

Find $\dfrac{2}{x} - \dfrac{3}{xy} + \dfrac{4}{xz^2}$.

The L.C.D. is xyz^2.

$$\frac{2}{x} - \frac{3}{xy} + \frac{4}{xz^2} = \frac{2(yz^2)}{xyz^2} - \frac{3(z^2)}{xyz^2} + \frac{4(y)}{xz^2y}$$

$$= \frac{2yz^2 - 3z^2 + 4y}{xyz^2}.$$

EXAMPLE 7

Find $\dfrac{6x - 17}{x^2 - 5x + 6} - \dfrac{1}{x - 3} + 3$.

The first denominator factors into $(x - 3)(x - 2)$. Thus, the denominators of the three terms are $(x - 3)(x - 2)$, $x - 3$, and 1. The L.C.D. is $(x - 3)(x - 2)$.

$$\frac{6x - 17}{\underbrace{x^2 - 5x + 6}_{(x-3)(x-2)}} - \frac{1}{x - 3} + 3$$

$$= \frac{6x - 17}{(x - 3)(x - 2)} - \frac{x - 2}{(x - 3)(x - 2)} + \frac{3(x - 3)(x - 2)}{(x - 3)(x - 2)}$$

$$= \frac{6x - 17 - (x - 2) + 3\overbrace{(x - 3)(x - 2)}^{x^2 - 5x + 6}}{(x - 3)(x - 2)}$$

$$= \frac{6x - 17 - x + 2 + 3x^2 - 15x + 18}{(x - 3)(x - 2)}$$

$$= \frac{3x^2 - 10x + 3}{(x - 3)(x - 2)} = \frac{(3x - 1)\cancel{(x - 3)}}{\cancel{(x - 3)}(x - 2)}$$

$$= \frac{3x - 1}{x - 2}.$$

EXAMPLE 8

Find $\dfrac{3}{x - 2} + \dfrac{2}{2 - x}$.

You might be tempted to say that the L.C.D. is $(x - 2)(2 - x)$. However, we can rewrite the second fraction so that its denominator is $(x-2)$.

$$\frac{3}{x-2}+\frac{2}{2-x}=\frac{3}{x-2}+\frac{2}{-(x-2)}$$
$$=\frac{3}{x-2}-\frac{2}{x-2}$$
$$=\frac{3-2}{x-2}=\frac{1}{x-2}.$$

EXAMPLE 9

Find $\dfrac{x-2}{x^2+6x+9}-\dfrac{x+2}{2(x^2-9)}$.

Since $x^2+6x+9=(x+3)^2$ and $2(x^2-9)=2(x+3)(x-3)$, the L.C.D. is $2(x+3)^2(x-3)$.

$$\frac{x-2}{(x+3)^2}-\frac{x+2}{2(x+3)(x-3)}$$
$$=\frac{(x-2)(2)(x-3)}{(x+3)^2(2)(x-3)}-\frac{(x+2)(x+3)}{2(x+3)(x-3)(x+3)}$$
$$=\frac{(x-2)(2)(x-3)-(x+2)(x+3)}{2(x+3)^2(x-3)}$$
$$=\frac{2(x^2-5x+6)-[x^2+5x+6]}{2(x+3)^2(x-3)}$$
$$=\frac{2x^2-10x+12-x^2-5x-6}{2(x+3)^2(x-3)}$$
$$=\frac{x^2-15x+6}{2(x+3)^2(x-3)}.$$

Completion Set 8-2

In Problems **1–5**, *fill in the blanks.*

1. $\dfrac{x+1}{x-2}+\dfrac{3}{x-2}=\dfrac{x+(\quad)}{x-2}$.

2. $\dfrac{x}{x-1}-\dfrac{1}{x-1}=\dfrac{(\qquad\qquad)}{x-1}=$ ____.

3. The L.C.D. of the fractions

$$\frac{2x}{x-1}, \quad \frac{5}{x^2}, \quad \text{and} \quad \frac{9}{x(x-1)}$$

is $(x - 1)($______$)$.

4. The L.C.D. of the fractions

$$\frac{2}{x+1}, \quad \frac{x}{x-1}, \quad \text{and} \quad \frac{x-3}{x^2-1}$$

is (____________)(____________).

5. To add $\frac{3}{x-1} + \frac{4}{x+2}$, we first note that the L.C.D. of the fractions is (____________) · (____________). Then we multiply the numerator and denominator of the first fraction by ____________, and multiply the numerator and denominator of the second fraction by ____________. Then we combine.

In Problems **6–10**, *insert* T (= *True*) *or* F (= *False*).

6. $\frac{1}{x} + \frac{1}{y} = \frac{xy}{x+y}$. ____

7. $\frac{1}{x} - \frac{x-y}{x^2} = \frac{x-x+y}{x^2}$. ____

8. $\frac{1}{a} + \frac{1}{b} = \frac{2}{ab}$. ____

9. $\frac{1}{x} - \frac{1}{y} = \frac{y-x}{xy}$. ____

10. $1 + \frac{1}{x} = \frac{x}{x} + \frac{1}{x} = \frac{x+1}{x}$. ____

Problem Set 8-2

In Problems **1–8**, *find the* L.C.D.

1. $\frac{6}{(x-4)^2}, \frac{7}{(x-4)^5}$

2. $\frac{x}{x+1}, \frac{2}{x-3}$

3. $\frac{4}{x^2y}, \frac{5}{xy^3}$

4. $\frac{x+1}{x^2y^3z}, \frac{y-1}{xyz^4}$

5. $\frac{3x}{x^2+6x+9}, \frac{x^2}{x^2-9}$

6. $\frac{2}{x^2+3x-4}, \frac{1}{x-1}$

7. $\frac{1}{2x+2}, \frac{x}{x^2+x}, \frac{2}{x+1}$

8. $\frac{x}{4x+2}, \frac{4}{4x^2-1}, \frac{x}{3}$

In Problems **9–51**, *perform the indicated operations and simplify.*

9. $\dfrac{x+1}{x-3} + \dfrac{4}{x-3}$

10. $\dfrac{x^2}{x-2} + \dfrac{x-6}{x-2}$

11. $\dfrac{3x}{x+1} + \dfrac{4}{x+1} - \dfrac{x+2}{x+1}$

12. $\dfrac{2x}{x^2-1} - \dfrac{2}{x^2-1}$

13. $\dfrac{3x^2+6x}{x^2+x-2} - \dfrac{x^2+2x+1}{x^2-1}$

14. $\dfrac{3x+4}{x+2} + \dfrac{x^2-9}{x^2+5x+6}$

15. $\dfrac{2}{x} + \dfrac{3}{y}$

16. $3 + \dfrac{x}{y}$

17. $\dfrac{x-4}{6} - \dfrac{x-2}{9}$

18. $\dfrac{x-2}{3} + 1$

19. $\dfrac{3}{2x} - \dfrac{2}{xy}$

20. $\dfrac{4}{x} + \dfrac{2}{y} - \dfrac{x+1}{xy}$

21. $\dfrac{x}{2} + \dfrac{2}{x}$

22. $\dfrac{1}{x} - \dfrac{2}{3x} + \dfrac{4}{3}$

23. $\dfrac{5}{x-2} + \dfrac{3}{x-3}$

24. $\dfrac{y}{y-2} - \dfrac{3}{y}$

25. $\dfrac{5y}{x^2} - \dfrac{2}{xy} + \dfrac{3}{y}$

26. $\dfrac{x}{a^2} + \dfrac{y}{ab}$

27. $\dfrac{x+3}{x-1} + 4$

28. $\dfrac{a}{b} + \dfrac{c}{d}$

29. $\dfrac{x}{x-y} + \dfrac{y}{x+y}$

30. $\dfrac{2}{x+1} - \dfrac{3}{x-1}$

31. $\dfrac{x+3}{x-3} - \dfrac{x-3}{2(x+3)}$

32. $\dfrac{4}{2x-1} + \dfrac{x}{x+2}$

33. $\dfrac{6x+12}{x^2+5x+4} + \dfrac{x}{x+4}$

34. $\dfrac{5}{x^2+3x-4} + \dfrac{1}{x+4}$

35. $\dfrac{1}{x^2-1} - \dfrac{1}{x-1} + \dfrac{1}{x+1}$

36. $\dfrac{x}{x+1} - \dfrac{2x}{x^2+3x+2}$

37. $\dfrac{x-1}{x^2+6x+9} + \dfrac{2}{x^2-9}$

38. $x^2 + 2 - \dfrac{x^4}{x^2-2}$

39. $\dfrac{x+1}{x^2+7x+10} - \dfrac{2x}{x^2+6x+5}$

40. $\dfrac{y}{3y^2-5y-2} - \dfrac{2}{3y^2-7y+2}$

41. $\dfrac{2x-6}{x^2-5x+6} + \dfrac{x^2+8x+16}{x^2+6x+8}$

42. $\dfrac{2}{x^3(x-3)} + \dfrac{3}{x(x-3)^2}$

43. $2x + 3 + \dfrac{2}{x-1}$

44. $\dfrac{3}{x-1} - \dfrac{4}{1-x}$

45. $\dfrac{x-2}{x^2+x} + \dfrac{3}{x^3+2x^2} - \dfrac{2x-3}{x^2+3x+2}$

46. $\dfrac{2}{x^2-5x+6} - \dfrac{1}{x^2-3x+2} + \dfrac{4}{x^2-4x+3}$

47. $\dfrac{y}{2y^2 + 7y + 3} - \dfrac{2}{4y^2 + 4y + 1}$

48. $\dfrac{3}{x + 3} + \dfrac{1}{x - 3} - \dfrac{4}{x + 2}$

49. $\dfrac{y}{x^2 + 2xy + y^2} + \dfrac{3x}{x^2 - y^2} - \dfrac{2}{x + y}$

50. $1 - \dfrac{2y^2}{x^2 - y^2} + \dfrac{2xy}{x^2 + y^2}$

51. $\dfrac{1}{x + y} - \dfrac{1}{y - x} + \dfrac{3}{x^2 - y^2}$

8-3 COMBINED OPERATIONS

We are now going to simplify some fractions whose numerator or denominator involves operations with fractions. There are two methods that are used:

Method 1. Perform the indicated operations on the numerator and on the denominator. Then divide the numerator by the denominator.

Method 2. Multiply the numerator and denominator of the given fraction by the L.C.D. of the fractions that appear in the numerator or denominator. This gives an equivalent fraction by the rule

$$\frac{a}{b} = \frac{ac}{bc} \quad \text{on p. 189.}$$

In Example 1 both methods will be shown. Method 2 is used in Examples 2 and 3.

EXAMPLE 1

Simplify $\dfrac{2 - \dfrac{3}{x}}{4x}$.

Method 1.

This is a fraction with numerator $2 - \dfrac{3}{x}$ and denominator $4x$. To simplify, we'll combine $2 - \dfrac{3}{x}$ into one fraction and then divide that result by $4x$.

$$\frac{2 - \dfrac{3}{x}}{4x} = \frac{\dfrac{2x}{x} - \dfrac{3}{x}}{4x} = \frac{\dfrac{2x - 3}{x}}{4x}$$

$$= \frac{\dfrac{2x - 3}{x}}{\dfrac{4x}{1}} = \frac{2x - 3}{x} \cdot \frac{1}{4x} = \frac{2x - 3}{4x^2}.$$

Method 2.

We multiply the numerator and denominator by the L.C.D. of the fractions that appear in the numerator or denominator. The only fraction is $\frac{1}{x}$, and so the L.C.D. is x. *Note the use of the distributive law below.*

$$\frac{2-\frac{3}{x}}{4x}=\frac{x\left(2-\frac{3}{x}\right)}{x(4x)}=\frac{x(2)-x\left(\frac{3}{x}\right)}{4x^2}$$

$$=\frac{2x-3}{4x^2}.$$

EXAMPLE 2

Simplify $\dfrac{\frac{x}{x-1}}{\frac{1}{x-1}+\frac{1}{x+1}}$.

We multiply the numerator and denominator by the L.C.D. of the fractions which appear. The L.C.D. is $(x-1)(x+1)$.

$$\frac{\frac{x}{x-1}}{\frac{1}{x-1}+\frac{1}{x+1}}=\frac{\cancel{(x-1)}(x+1)\left[\frac{x}{\cancel{x-1}}\right]}{(x-1)(x+1)\left[\frac{1}{x-1}+\frac{1}{x+1}\right]}$$

$$=\frac{(x+1)(x)}{\cancel{(x-1)}(x+1)\left(\frac{1}{\cancel{x-1}}\right)+(x-1)\cancel{(x+1)}\left(\frac{1}{\cancel{x+1}}\right)}$$

$$=\frac{(x+1)(x)}{(x+1)+(x-1)}$$

$$=\frac{(x+1)\cancel{(x)}}{2\cancel{x}}=\frac{x+1}{2}.$$

EXAMPLE 3

Simplify $\dfrac{3-\frac{1}{2x}}{6x+\frac{11x}{x-2}}$.

The L.C.D. of the fractions is $2x(x-2)$.

$$\frac{3-\dfrac{1}{2x}}{6x+\dfrac{11x}{x-2}} = \frac{2x(x-2)\left[3-\dfrac{1}{2x}\right]}{2x(x-2)\left[6x+\dfrac{11x}{x-2}\right]}$$

$$= \frac{2x(x-2)(3) - \cancel{2x}(x-2)\left(\dfrac{1}{\cancel{2x}}\right)}{2x(x-2)(6x) + 2x\cancel{(x-2)}\left(\dfrac{11x}{\cancel{x-2}}\right)}$$

$$= \frac{6x(x-2)-(x-2)}{12x^2(x-2)+22x^2} = \frac{6x(x-2)-(x-2)(1)}{12x^3-24x^2+22x^2}$$

$$= \frac{(x-2)[(6x)-(1)]}{12x^3-2x^2} \qquad \text{[factoring } x-2 \text{ from numerator]}$$

$$= \frac{(x-2)\cancel{(6x-1)}}{2x^2\cancel{(6x-1)}} = \frac{x-2}{2x^2}.$$

Completion Set 8-3

In Problems **1–3**, *fill in the blanks.*

1. To simplify $\dfrac{\dfrac{3}{x}+\dfrac{x}{2}}{\dfrac{x}{x-1}}$, we can first add ____ to ____ and then divide the result by ____________.

2. $\dfrac{\dfrac{x}{3}}{\dfrac{1}{x}+\dfrac{1}{x}} = \dfrac{\dfrac{x}{3}}{\dfrac{(\quad)}{x}} = \dfrac{x}{3}\cdot\dfrac{x}{(\quad)} = (\qquad).$

3. To simplify $\dfrac{3-\dfrac{1}{x-3}}{\dfrac{1}{x+2}}$, we could first multiply the numerator and denominator by ____________________.

Problem Set 8-3

In Problems **1–14**, *simplify.*

1. $\dfrac{\dfrac{1}{x}+\dfrac{3}{x}}{4}$

2. $\dfrac{\dfrac{x}{x-1}-\dfrac{1}{x-1}}{x-1}$

3. $\dfrac{4-\dfrac{6}{x}}{2}$

4. $\dfrac{x-1}{1-\dfrac{1}{x}}$

5. $\dfrac{7}{3x-\dfrac{1}{2}}$

6. $\dfrac{x-\dfrac{1}{x}}{x+1}$

7. $\dfrac{3-\dfrac{1}{y}}{2+\dfrac{1}{x}}$

8. $\dfrac{\dfrac{a}{b}+2}{\dfrac{b}{a}-2}$

9. $\dfrac{\dfrac{1}{x}+\dfrac{x}{2x-3}}{\dfrac{x-1}{x}}$

10. $\dfrac{\dfrac{x^2-9}{4}}{\dfrac{1}{x}-\dfrac{1}{3}}$

11. $\dfrac{x+1+\dfrac{1}{x+3}}{\dfrac{x+2}{3}}$

12. $\dfrac{\dfrac{x}{x^2+4x+3}}{\dfrac{1}{x^2-1}+1}$

13. $\dfrac{3x+\dfrac{x}{x-3}}{3x-\dfrac{x}{x-3}}$

14. $\dfrac{\dfrac{1}{x^4}+\dfrac{1}{x^2}+1}{\dfrac{1}{x^3}+\dfrac{1}{x}+x}$

8-4 FRACTIONAL EQUATIONS

Sometimes you'll have to solve an equation having the unknown in a denominator. We call these **fractional equations.**

Fractional equations can be tricky little devils. Some have no solutions. In others, you may think you have the solution when in fact you really don't. In the following examples you'll see how to handle them.

EXAMPLE 1

Solve $\dfrac{6}{x-3} = 5.$

We clear the equation of fractions by multiplying both sides by $x - 3$.

$$\frac{6}{x-3} = 5$$

$$(x-3)\cdot\frac{6}{x-3} = (x-3)\cdot 5$$

$$\cancel{(x-3)}\cdot\frac{6}{\cancel{x-3}} = 5x - 15$$

$$6 = 5x - 15$$

$$21 = 5x$$

$$\frac{21}{5} = x.$$

We're not done yet! Recall that we're not allowed to multiply both sides of an equation by 0. Thus we must check that when we multiplied both sides by $x - 3$, the value $x = \frac{21}{5}$ does not make $x - 3$ zero.

$$\text{If } x = \frac{21}{5}, \quad \text{then} \quad x - 3 = \frac{21}{5} - 3 \neq 0.$$

Now we can say the solution is $\boxed{x = \frac{21}{5}}$.

EXAMPLE 2

$$\textit{Solve} \quad \frac{1}{x+2} + \frac{1}{x-2} = \frac{4}{x^2 - 4}. \tag{8-1}$$

We clear the equations of fractions by multiplying both sides by their L.C.D. Since $x^2 - 4 = (x + 2)(x - 2)$, clearly the L.C.D. is just $(x + 2)(x - 2)$.

$$\frac{1}{x+2} + \frac{1}{x-2} = \frac{4}{(x+2)(x-2)}$$

$$(x+2)(x-2)\left[\frac{1}{x+2} + \frac{1}{x-2}\right] = (x+2)(x-2)\left[\frac{4}{(x+2)(x-2)}\right]$$

$$\frac{\cancel{(x+2)}(x-2)}{\cancel{x+2}} + \frac{(x+2)\cancel{(x-2)}}{\cancel{x-2}} = \frac{\cancel{(x+2)}\cancel{(x-2)}\cdot 4}{\cancel{(x+2)}\cancel{(x-2)}} \qquad \text{[distributive law]}$$

$$(x - 2) + (x + 2) = 4 \tag{8-2}$$

$$2x = 4$$

$$x = 2.$$

Now we check that we didn't multiply both sides by 0.

$$\text{If } x = 2, \quad \text{then} \quad (x + 2)(x - 2) = (2 + 2)(2 - 2) = 0.$$

Thus, we multiplied by 0. This means we *cannot guarantee* that $x = 2$ is a solution. Looking at the original equation, we see that when $x = 2$ there is division by 0, so we conclude that **there is no solution.**† Although $x = 2$ is a solution of Eq. (8-2), it is not a solution of the given equation and is sometimes called an *extraneous solution* of Eq. (8-1). In general, multiplying both sides of an equation by an expression involving a variable may lead to extraneous solutions.

---

†The solution set is a set with no elements in it, { }. It is called the *empty set* or *null set*. We can write the empty set with the symbol ∅.

EXAMPLE 3

Solve $\dfrac{y+1}{y+3} + \dfrac{y+5}{y-2} = \dfrac{7(2y+1)}{y^2+y-6}$.

Since $y^2 + y - 6 = (y+3)(y-2)$, we see that the L.C.D. is $(y+3)(y-2)$. Multiplying both sides by the L.C.D., we get

$$(y+3)(y-2)\left[\frac{y+1}{y+3} + \frac{y+5}{y-2}\right] = (y+3)(y-2)\cdot\frac{7(2y+1)}{(y+3)(y-2)}$$

$$\frac{\cancel{(y+3)}(y-2)(y+1)}{\cancel{y+3}} + \frac{(y+3)\cancel{(y-2)}(y+5)}{\cancel{y-2}}$$

$$= \frac{\cancel{(y+3)}\cancel{(y-2)}\cdot 7(2y+1)}{\cancel{(y+3)}\cancel{(y-2)}}$$

$$(y-2)(y+1) + (y+3)(y+5) = 7(2y+1)$$

$$y^2 - y - 2 + y^2 + 8y + 15 = 14y + 7$$

$$2y^2 + 7y + 13 = 14y + 7$$

$$2y^2 - 7y + 6 = 0 \qquad \text{[quadratic]}$$

$$(2y-3)(y-2) = 0.$$

$2y - 3 = 0$	$y - 2 = 0$
$2y = 3$	$y = 2.$
$y = \dfrac{3}{2}$	

Now we check to make sure that we didn't multiply both sides by 0.

If $y = \dfrac{3}{2}$, then $(y+3)(y-2) = (\dfrac{3}{2}+3)(\dfrac{3}{2}-2) \neq 0.$

If $y = 2$, then $(y+3)(y-2) = (2+3)(2-2) = 0.$

We *don't accept* $y = 2$, since this would give division by 0 in the original equation. The only solution is

$$\boxed{y = \frac{3}{2}.}$$

EXAMPLE 4

Solve $\dfrac{2}{x+1} - \dfrac{3}{x-1} = \dfrac{1}{x-4}$.

Multiplying both sides by the L.C.D., $(x + 1)(x - 1)(x - 4)$, and simplifying, we get

$$(x+1)(x-1)(x-4)\left[\frac{2}{x+1} - \frac{3}{x-1}\right] = (x+1)(x-1)(x-4)\cdot\frac{1}{x-4}$$

$$2(x-1)(x-4) - 3(x+1)(x-4) = 1(x+1)(x-1)$$

$$2(x^2 - 5x + 4) - 3(x^2 - 3x - 4) = x^2 - 1$$

$$2x^2 - 10x + 8 - 3x^2 + 9x + 12 = x^2 - 1$$

$$-x^2 - x + 20 = x^2 - 1$$

$$0 = 2x^2 + x - 21$$

$$0 = (2x + 7)(x - 3).$$

$$2x + 7 = 0 \quad\Big|\quad x - 3 = 0$$

$$x = -\frac{7}{2} \quad\Big|\quad x = 3.$$

Checking to make sure that we didn't multiply by 0, we have:

If $x = -\frac{7}{2}$, then

$$(x+1)(x-1)(x-4) = \left(-\frac{7}{2}+1\right)\left(-\frac{7}{2}-1\right)\left(-\frac{7}{2}-4\right) \neq 0.$$

If $x = 3$, then

$$(x+1)(x-1)(x-4) = (3+1)(3-1)(3-4) \neq 0.$$

Thus the solution is

$$\boxed{x = -\frac{7}{2}, 3.}$$

EXAMPLE 5

A construction firm has a government contract to build a swimming pool for the use of certain public officials. According to the contract, it must be completed within the next 21 *days. The foreman of the job knows that his regular crew would take* 45 *days to build it. To meet the deadline, the foreman decides to use a second crew, who can build the pool by themselves in* 30 *days. How long will it take both crews to construct the pool if they work together?*

In one day the regular crew does $\frac{1}{45}$ of the construction work and the second crew does $\frac{1}{30}$. When both crews work, in one day a total of $\frac{1}{45} + \frac{1}{30}$ of the construction work is done. Let n = the number of days it takes when both crews work. Then in one day, $\frac{1}{n}$ of the work is done. Thus, $\frac{1}{45} + \frac{1}{30}$ would be $\frac{1}{n}$ of the work.

$$\frac{1}{45} + \frac{1}{30} = \frac{1}{n}$$

$$90n\left[\frac{1}{45} + \frac{1}{30}\right] = 90n\left(\frac{1}{n}\right) \qquad \text{[multiplying both sides by } 90n\text{, the L.C.D.]}$$

$$2n + 3n = 90$$

$$5n = 90$$

$$n = 18.$$

Since we did not multiply by 0, it will take 18 days to build the pool.

EXAMPLE 6

The rate of the current in a stream is 3 miles per hour. A man rowed upstream for 3 miles and then returned to his starting point. The round trip took a total of one hour and twenty minutes. How fast could the man row in still water?

Let r be the rate (in miles per hour) at which the man can row in still water. Since the rate of the current is 3 mi/hr, the man's rate upstream was $r - 3$, and downstream it was $r + 3$. Since time $= \dfrac{\text{distance}}{\text{rate}}$ and distance = 3 for each rate, we have

$$\left(\begin{matrix}\text{time}\\ \text{upstream}\end{matrix}\right) + \left(\begin{matrix}\text{time}\\ \text{downstream}\end{matrix}\right) = \text{total time}$$

$$\frac{3}{r-3} + \frac{3}{r+3} = \frac{4}{3}. \qquad \left[1 \text{ hr } 20 \text{ min} = \frac{4}{3} \text{ hr}\right]$$

Multiplying both sides by $3(r - 3)(r + 3)$ and simplifying, we obtain

$$3(3)(r + 3) + 3(3)(r - 3) = 4(r - 3)(r + 3)$$

$$9r + 27 + 9r - 27 = 4[r^2 - 9]$$

$$18r = 4r^2 - 36$$

$$0 = 4r^2 - 18r - 36$$

$$0 = 2r^2 - 9r - 18 \qquad \text{[dividing both sides by 2]}$$

$$0 = (2r + 3)(r - 6).$$

$2r + 3 = 0$	$r - 6 = 0$
$r = -\frac{3}{2}$	$r = 6.$

The values $-\frac{3}{2}$ and 6 do not make $3(r-3)(r+3)$ equal 0. Thus they are solutions of the original equation. But r is a rate, a positive number, and so we choose $r = 6$ mi/hr.

Completion Set 8-4

In Problems **1–3**, *fill in the blanks.*

1. To solve $\frac{x+1}{x} + \frac{1}{x+1} = 3$, we can multiply both sides by the L.C.D., which is $x($______________$)$.

2. To solve $1 - \frac{1}{x} - \frac{2}{x^2} = 0$, we can multiply both sides by the L.C.D., which is ______. We then obtain $x^2 - x - 2 = 0$. Factoring, we get $(x -$ ____$)(x +$ ____$) = 0$. Thus, $x =$ ____ or $x =$ ______. For these values of x we did not multiply by 0, and so our solution is $x =$ ____, ______.

3. To solve $1 = \frac{x}{x^2}$, we can multiply both sides by x^2. We then get $x^2 = x$ or $x^2 - x = 0$. Thus $x(x-1) = 0$ and either $x = 0$ or $x = 1$. But the only solution to the original equation is $x =$ ____.

In Problems **4** *and* **5**, *fill in* T (= *True*) *or* F (= *False*).

4. If $\frac{1}{x} = 3$, then $x = -3$. ____

5. Suppose $\frac{x-1}{x-1} = 0$. Multiplying both sides by $x - 1$, we get $x - 1 = 0$, and so the solution of the given equation is $x = 1$. ____

Problem Set 8-4

In Problems **1–34**, *solve the fractional equations.*

1. $\frac{3}{x} = 12$

2. $\frac{1}{x} + \frac{1}{5} = \frac{4}{5}$

3. $\frac{x}{3x-4} = 3$

4. $\frac{4}{x-1} = 2$

5. $\frac{10}{3r} - \frac{9r+2}{6r} = 3$

6. $\frac{4x}{7-x} = 1$

7. $\frac{x}{3} = \frac{6}{x} - 1$

8. $\frac{2x-3}{4x-5} = 6$

9. $1 + \frac{2}{x} = \frac{2(x+1)}{x}$

10. $\frac{1}{2x} - \frac{2x-3}{4x} = \frac{3x}{4}$

11. $\frac{3}{4y} - \frac{5}{6y} = \frac{1}{6}$

12. $\frac{1}{x} - \frac{2}{3x} = 3x$

13. $\frac{1}{x^2} + \frac{6}{x} + 8 = 0$

14. $\frac{1}{x^2} + \frac{1}{x} - 12 = 0$

15. $\frac{1}{x-1} = \frac{2}{x-2}$

16. $\frac{x}{x-1} = \frac{4}{x}$

17. $\frac{3}{x-4} + \frac{x-3}{x} = 2$

18. $\frac{4-x}{x} + \frac{8}{4+x} = 1$

19. $\frac{2}{x-2} = \frac{x+1}{x+4}$

20. $\frac{4}{x-3} = \frac{3}{x-4}$

21. $\frac{x^2}{x-1} + 1 = \frac{1}{x-1}$

22. $\frac{9}{x-3} = \frac{3x}{x-3}$

23. $\frac{3x-2}{2x+3} = \frac{3x-1}{2x+1}$

24. $\frac{y-3}{y+3} = \frac{y-3}{y+2}$

25. $\frac{y-6}{y} - \frac{6}{y} = \frac{y+6}{y-6}$

26. $\frac{x+1}{x} - \frac{6}{x+5} = \frac{3}{x}$

27. $\frac{2}{x-1} - \frac{6}{2x+1} = 5$

28. $x + \frac{2x}{x-2} = \frac{4}{x-2}$

29. $\frac{3}{x} - \frac{4}{x+2} = \frac{5}{3}$

30. $\frac{x}{x+3} - \frac{x}{x-3} = \frac{3x-4}{x^2-9}$

31. $\frac{3x+4}{x+2} - \frac{3x-5}{x-4} = \frac{12}{x^2-2x-8}$

32. $\frac{4}{x-4} - \frac{3}{x-3} = \frac{1}{x-5}$

33. $\frac{3}{x+1} + \frac{4}{x} = \frac{12}{x+2}$

34. $\frac{2}{x^2-1} - \frac{1}{x(x-1)} = \frac{2}{x^2}$

35. An object is 120 in. from a wall. In order to focus the image of the object on the wall, a converging lens with a focal length of 24 in. is used. The lens is placed between the object and the wall at a distance of p in. from the object, where

$$\frac{1}{p} + \frac{1}{120-p} = \frac{1}{24}.$$

Find p to one decimal place. *Note*: In your computation, assume $\sqrt{2880} = 53.67$.

36. An important equation for lenses is

$$\frac{1}{p} + \frac{1}{q} = \frac{1}{f},$$

where f is focal length, p is object distance, and q is image distance. Suppose that for a converging lens the focal length is 12 centimeters (cm) and the object distance is 24 cm. Find the image distance.

37. The total capacitance C_T of an electric circuit containing two capacitors C_1 and C_2 in

series is given by

$$\frac{1}{C_T} = \frac{1}{C_1} + \frac{1}{C_2}.$$

Find C_2 if $C_1 = 2$ (microfarads) and $C_T = \frac{2}{3}$ (microfarads).

P
|

38. Solve the equation

$$\frac{p_1 v_1}{t_1} = \frac{p_2 v_2}{t_2}$$

for t_1. This equation relates to gases.

? |

39. Entering into a storage tank are three pipes: A, B, and C. Pipe A can fill the tank in 2 hr, pipe B in 3 hr, and pipe C in 4 hr. How long will it take to fill the tank if all three pipes are used?

40. Water is flowing into a tank by means of pipes A and B. Pipe A can fill the tank in 2 hr, and pipe B can fill it in 5 hr. However, water is also flowing out of the tank into another tank by a pipe C by which the original tank can be completely emptied in 4 hr. How long would it take to fill the original tank if it were initially empty and pipes A, B, and C were all opened?

41. A boat traveled 36 mi upstream on a river where the rate of the current was 3 mi/hr. It then returned. The round trip took five hours. Find the speed of the boat in still water.

42. A man rowed downstream for 10 mi and then rowed upstream for the same period of time. However, he only covered 5 mi going back. If the rate of the stream was $1\frac{1}{4}$ mi/hr, find how fast the man can row in still water.

8-5 REVIEW

IMPORTANT TERMS

L.C.D. *(p. 190)*
extraneous solution *(p. 200)*
fractional equation *(p. 199)*

REVIEW PROBLEMS

In Problems **1–26**, *perform the indicated operations and simplify your answers.*

1. $\frac{2}{x} + \frac{4}{x-6}$

2. $\frac{x-7}{x+4} - \frac{6}{2x+8}$

3. $\frac{x^2-64}{x^3} \cdot \frac{x^2}{2x+16}$

4. $\frac{x-2}{4} \cdot \frac{8x+4}{x^2+2x-8}$

5. $\dfrac{x+2}{\dfrac{2x+4}{3}}$

6. $\dfrac{\dfrac{x^3}{x-1}}{x^5}$

7. $\dfrac{3}{x-2} - \dfrac{x+2}{x-3}$

8. $\dfrac{2}{3x} + \dfrac{3}{x} - \dfrac{4}{5x}$

9. $\dfrac{9x}{x^2+2x+1} \cdot \dfrac{(x+1)^3}{-3}$

10. $\dfrac{-(x+3)}{x^2+x} \cdot \dfrac{x}{-(x^2-9)}$

11. $2 + \dfrac{x}{x-1} - \dfrac{x-1}{x^2-1}$

12. $\dfrac{2}{x-y} + \dfrac{2}{y-x}$

13. $\dfrac{x^2+5x+6}{x^2-2x-8} \cdot \dfrac{x^2-16}{x^2+7x+12}$

14. $\dfrac{4x^2+4x+1}{x^2-2x-3} \cdot \dfrac{x^2+2x+1}{2x^2+3x+1}$

15. $\dfrac{\dfrac{8-4x}{2x}}{\dfrac{x^2-4x+4}{x^2-2x}}$

16. $\dfrac{\dfrac{4x^2-9}{(x+1)^3}}{\dfrac{4x+6}{(x+1)^2}}$

17. $\dfrac{x+2}{x^2+4x+4} + \dfrac{x-3}{x+2}$

18. $\dfrac{6}{y+3} - \dfrac{2y}{y-3} + \dfrac{3}{y^2-9}$

19. $\dfrac{\dfrac{2x+2}{x^3-x}}{\dfrac{x-1}{x^2}}$

20. $\dfrac{\dfrac{3x^2-12x-15}{x^2+5x+4}}{\dfrac{30-6x}{x+4}}$

21. $\dfrac{x+1}{x^2+x-12} \cdot \dfrac{9-x^2}{x^2+3x+2}$

22. $\dfrac{4-x^2}{25-x^2} \cdot \dfrac{x-5}{x-2}$

23. $\dfrac{x+2}{\dfrac{x}{x+1} + \dfrac{4}{x}}$

24. $\dfrac{\dfrac{1}{x+2} - \dfrac{1}{x-2}}{\dfrac{2}{x-2}}$

25. $\dfrac{1 - \dfrac{7}{x^2-9}}{\dfrac{x-4}{3-x}}$

26. $\dfrac{\dfrac{2}{x+3} + 1}{1 - \dfrac{2}{x+4}}$

13

In Problems **27–34**, *solve the equations.*

27. $\dfrac{2}{x+5} = \dfrac{4}{x-5}$

28. $\dfrac{2x}{x-3} - \dfrac{x+1}{x+2} = 1$

29. $\dfrac{x}{x-1} - \dfrac{9}{x+3} = 0$

30. $\dfrac{1}{x^2} - \dfrac{9}{x} + 8 = 0$

31. $\frac{x+1}{x} + \frac{2x}{x-2} = \frac{5x+1}{x}$

32. $\frac{6x+7}{2x+1} - \frac{6x+1}{2x} = 1$

33. $\frac{x+2}{x-5} = \frac{7}{x-5}$

34. $\frac{x}{x-1} - \frac{2}{x} + \frac{x-2}{x^2-x} = 0$

35. In studies of photosynthesis, the formula

$$P = \frac{bE}{1 + aE}$$

occurs. Solve for a.

36. A chemical company can fill a tank car with an industrial solvent with their regular pump in 20 minutes. Another pump, one that the company keeps in reserve, can fill the tank car in 30 minutes. How many minutes would it take to fill the tank car if both pumps were used together?

37. A trucker on a 130-mile run decreased his speed by 10 miles per hour on the last 80-mile stretch because of a rainstorm. What was his original speed if his entire run took 3 hours?

CHAPTER 9

Graphs and Functions

9-1 RECTANGULAR COORDINATE SYSTEM

On TV and in newspapers you've probably seen how information can be given by means of graphs. In this chapter you'll see how graphs are used in algebra.

Our graphs will be drawn on a so-called **rectangular** (or *Cartesian*) **coordinate plane**. This is a plane (think of a sheet of paper) on which two number lines are placed, one horizontal and one vertical. See Fig. 9-1. The origins of the

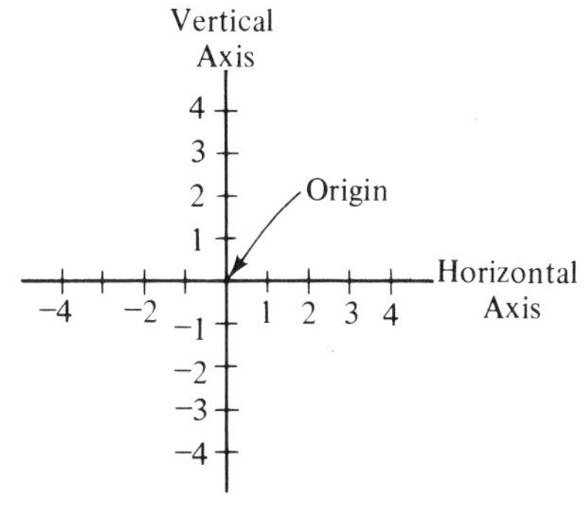

FIG. 9-1

number lines meet at a point called the **origin** of the coordinate plane.

The number lines in Fig. 9-1 are called **coordinate axes**. The *horizontal axis* has its positive numbers to the right of the origin, while the *vertical axis* has them above the origin. The unit distance on the vertical axis does not have to be the same as that on the horizontal axis.

Every point in the plane can be labeled to indicate its position. For example, let's label point P in Fig. 9-2(a).

First, from P we draw a perpendicular to the horizontal axis. It hits this axis at 4. Next, we draw a perpendicular from P to the vertical axis. It hits this axis at 2.

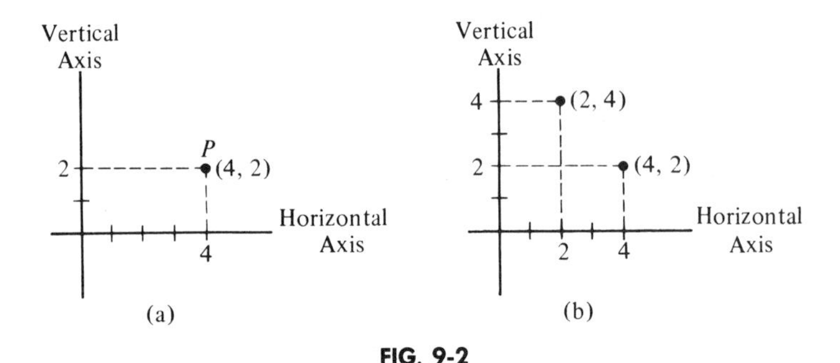

FIG. 9-2

Thus P determines two numbers, 4 and 2. We say that the **rectangular coordinates** of P are given by the **ordered pair** (4, 2). The word "ordered" is important because

$$(4, 2) \neq (2, 4) \quad \left[\text{see Fig. 9-2(b)}\right].$$

The first number of the ordered pair (4, 2) is called the **first coordinate** or **abscissa** of P. The second number is the **second coordinate** or **ordinate**.

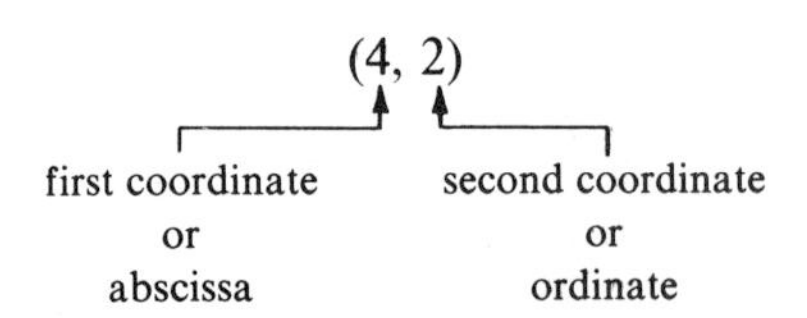

In Fig. 9-3 the coordinates of several points appear. Notice a few things.

- The origin has coordinates (0, 0).
- Any point on the horizontal axis has its *second* coordinate equal to 0.
- Any point on the vertical axis has its *first* coordinate equal to 0.

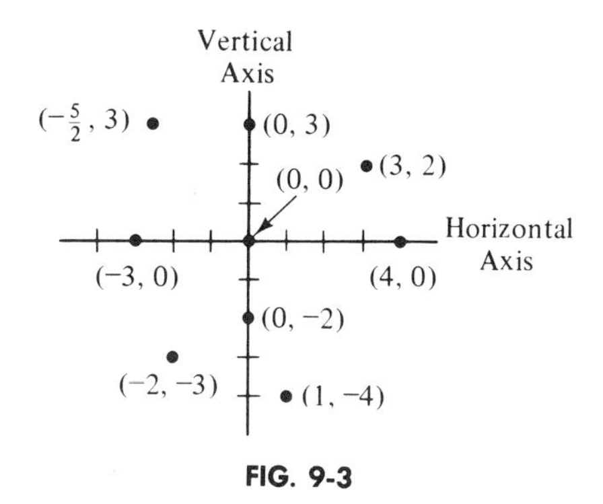

FIG. 9-3

The coordinate axes divide the plane into four parts called **quadrants**. They are numbered as in Fig. 9-4. Any point (a_1, b_1) in Quadrant I has both its

Quadrant II •(a_2, b_2) $a_2 < 0,\ b_2 > 0$	Quadrant I •(a_1, b_1) $a_1 > 0,\ b_1 > 0$
Quadrant III •(a_3, b_3) $a_3 < 0,\ b_3 < 0$	Quadrant IV •(a_4, b_4) $a_4 > 0,\ b_4 < 0$

FIG. 9-4

coordinates positive. Any point (a_2, b_2) in Quadrant II has its first coordinate negative and its second coordinate positive. Look at Fig. 9-4 and see what must be true for points in Quadrants III and IV.

EXAMPLE 1

a. $(4, -2)$ lies in Quadrant IV, since $4 > 0$ and $-2 < 0$.

b. $(1, 8)$ lies in Quadrant I, since $1 > 0$ and $8 > 0$.

c. $(-3, -1)$ lies in Quadrant III.

d. $\left(-\frac{1}{2}, \frac{5}{8}\right)$ lies in Quadrant II.

e. The points on the axes do not lie in any quadrant.

Completion Set 9-1

In Problems **1–4**, *fill in the blanks.*

1. The origin of a rectangular coordinate system has coordinates (____, ____).

2. The coordinate axes divide a rectangular coordinate plane into four regions called

____________________.

3. Another name for the first coordinate is ____________________.

4. Another name for the second coordinate is ____________________.

In Problems **5–8**, *insert the word "first" or "second."*

5. A point on the horizontal axis has its ______________ coordinate equal to 0.

6. A point on the vertical axis has its ______________ coordinate equal to 0.

7. A point in the second quadrant has a negative ______________ coordinate.

8. A point in the fourth quadrant has a positive ______________ coordinate.

Problem Set 9-1

In Problems **1–6**, *give the coordinates of the indicated points. See* Fig. 9-5.

1. P **2.** Q **3.** R **4.** S **5.** T **6.** U

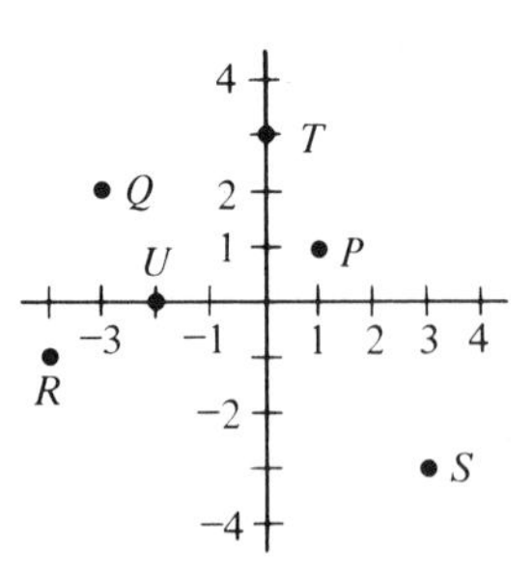

FIG. 9-5

Construct a rectangular coordinate plane. Then locate and label (that is, plot) each of the points in Problems **7–18**.

7. $(0, 0)$ **8.** $(2, 3)$ **9.** $(1, 4)$ **10.** $(0, -2)$

11. $(-2, 0)$ **12.** $(-1, -2)$ **13.** $(-1, 2)$ **14.** $(2, -3)$

15. $(-4, -3)$ **16.** $(-1, 1)$ **17.** $\left(0, -\frac{1}{2}\right)$ **18.** $\left(\frac{3}{2}, -\frac{1}{2}\right)$

In Problems **19–30**, *give the quadrant in which the point lies.*

19. $(3, 2)$ **20.** $(-3, 2)$ **21.** $(2, -3)$ **22.** $(-4, -8)$

23. $(-6, 1)$ **24.** $(4, 1)$ **25.** $(-1, -2)$ **26.** $(5, -5)$

27. $\left(-\frac{3}{5}, \frac{2}{3}\right)$ **28.** $\left(\frac{4}{7}, -\frac{2}{5}\right)$ **29.** $(\sqrt{2}, \sqrt{3})$ **30.** $(-.1, -.2)$

9-2 GRAPHS OF EQUATIONS

Up to now we have worked with equations having one variable. Now we'll look at some in two variables.

For example, $y = x^2$ is an equation in the variables x and y. Here a solution is a pair of numbers: a value of x *and* a value of y that make the equation true. For instance, $x = 2$ and $y = 4$ is a solution:

$$y = x^2$$
$$4 = 2^2$$
$$4 = 4. \checkmark$$

In Table 9-1 are more solutions. We got them by first choosing values of x and then substituting them into $y = x^2$ to get corresponding values of y. For example,

$$\text{if } x = -\frac{1}{2}, \quad \text{then} \quad y = x^2 = \left(-\frac{1}{2}\right)^2 = \frac{1}{4},$$

and so a solution is

$$x = -\frac{1}{2}, \quad y = \frac{1}{4}.$$

By no means are all solutions listed. That's impossible to do because there are infinitely many.

Notice in Table 9-1 that we associated ordered pairs with the solutions. In each ordered pair the first number is the x-value of a solution, while the second is the y-value.

TABLE 9-1

x	$y(= x^2)$	
0	0	$\leftrightarrow$ (0, 0)
$\frac{1}{2}$	$\frac{1}{4}$	$\leftrightarrow$ $(\frac{1}{2}, \frac{1}{4})$
$-\frac{1}{2}$	$\frac{1}{4}$	$\leftrightarrow$ $(-\frac{1}{2}, \frac{1}{4})$
1	1	$\leftrightarrow$ (1, 1)
-1	1	$\leftrightarrow$ $(-1, 1)$
2	4	$\leftrightarrow$ (2, 4)
-2	4	$\leftrightarrow$ $(-2, 4)$
3	9	$\leftrightarrow$ (3, 9)
-3	9	$\leftrightarrow$ $(-3, 9)$

Now, from the last section we know that ordered pairs can be thought of as points in a plane. In Fig. 9-6(a) we've plotted the ordered pairs in Table 9-1. *Each point represents a solution of* $y = x^2$.

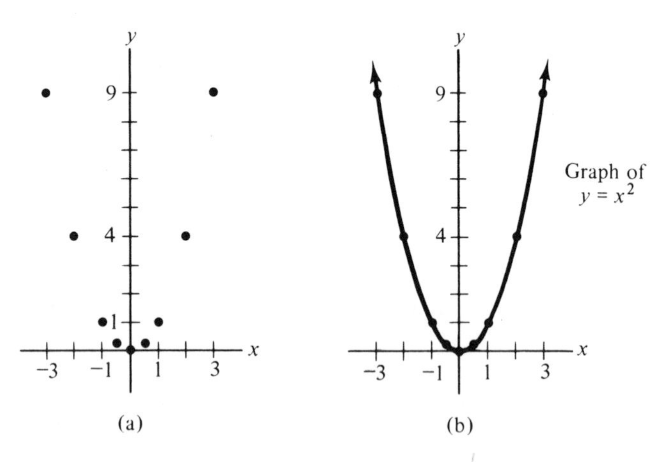

FIG. 9-6

The horizontal axis is labeled the ***x*-axis** (or more simply, x) and the vertical axis the ***y*-axis**. This is natural because the first coordinate of each point is an

x-value and the second coordinate is a y-value. Sometimes we say the first coordinate is the *x-coordinate*, and the second coordinate is the *y-coordinate*. When we look geometrically at solutions of an equation in x and y, we usually choose the horizontal axis as the x-axis.

If we were able to plot all the solutions of $y = x^2$, we would get a picture something like Fig. 9-6(b). It is called the *graph* of $y = x^2$. **The graph of an equation is the geometric representation of its solutions.** To get Fig. 9-6(b), all we did was connect the points in Fig. 9-6(a) by a smooth curve. It should be clear that the graph extends upward indefinitely, and every point on the graph gives a solution to the equation.

Of course, the more points that we originally plot, the better is our graph. Always plot enough points so that the general shape of the graph is clear. When in doubt, plot more points.

EXAMPLE 1

Graph the equation $y = 2x - 1$.

We choose some values for x and find the corresponding y-values.

$$\text{If } x = 0, \quad \text{then} \quad y = 2x - 1 = 2(0) - 1 = -1.$$

Thus $(0, -1)$ lies on the graph.

$$\text{If } x = 1, \quad \text{then} \quad y = 2x - 1 = 2(1) - 1 = 1.$$

Thus, $(1, 1)$ lies on the graph. Continuing in this way, we get the table in Fig. 9-7. Plotting those points and connecting them, we get a straight line.

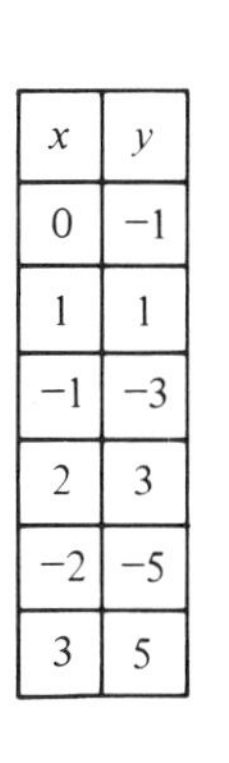

x	y
0	−1
1	1
−1	−3
2	3
−2	−5
3	5

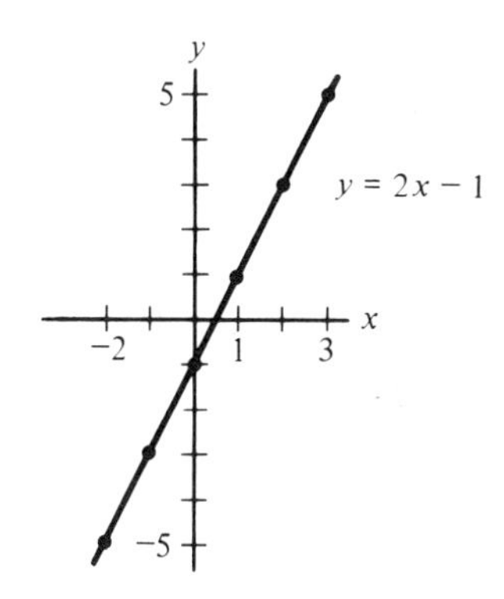

FIG. 9-7

EXAMPLE 2

Graph $y = 2x^3$.

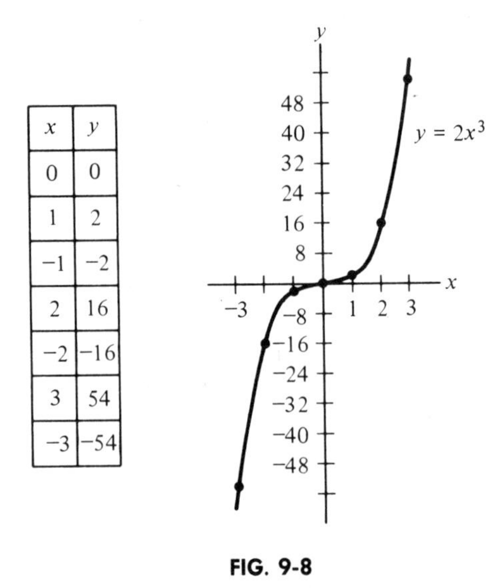

x	y
0	0
1	2
−1	−2
2	16
−2	−16
3	54
−3	−54

FIG. 9-8

In Fig. 9-8, notice our choice of the unit distance on the y-axis. It makes the graphing easier to handle.

EXAMPLE 3

Graph $x = \frac{1}{2}y^2$.

We'll show two ways to do this problem: a messy way and a better way.

Messy Way.

If we choose a value of x, say $x = 1$, then to find the values of y we have to solve the quadratic equation $1 = \frac{1}{2}y^2$. Although nothing is wrong with this, it can be a little messy. Solving gives $y = \pm\sqrt{2}$. Thus the points $(1, \sqrt{2})$ and $(1, -\sqrt{2})$ lie on the graph.

Better Way.

Choose values of y and then find the corresponding values of x. For example,

$$\text{if } y = 1, \quad \text{then} \quad x = \frac{1}{2}y^2 = \frac{1}{2}(1)^2 = \frac{1}{2}.$$

Thus, $\left(\frac{1}{2}, 1\right)$ lies on the graph. With this method we do not have to solve a quadratic equation. See Fig. 9-9.

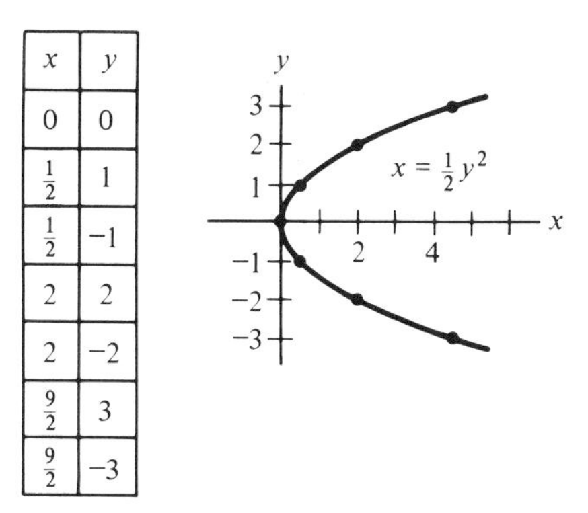

x	y
0	0
$\frac{1}{2}$	1
$\frac{1}{2}$	−1
2	2
2	−2
$\frac{9}{2}$	3
$\frac{9}{2}$	−3

FIG. 9-9

Moral.

Sometimes it is better to choose values for one variable rather than the other.

EXAMPLE 4

Graph $y = 2$.

We can think of $y = 2$ as an equation in the variables x and y. Now you may say to yourself, "But I don't see any x!" Well, you can if you write the equation in the form $y = 2 + 0x$. No matter what x is, any solution must have its y-value equal to 2. For example,

$$\text{if } x = 4, \quad \text{then} \quad y = 2 + 0x = 2 + 0(4) = 2.$$

Thus (4, 2) lies on the graph. In fact, every point on the graph will be of the form $(x, 2)$, where x is a real number. See Fig. 9-10. *The graph of* $y = 2$ *is a horizontal line.*

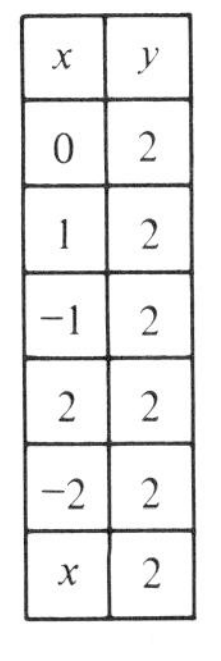

x	y
0	2
1	2
−1	2
2	2
−2	2
x	2

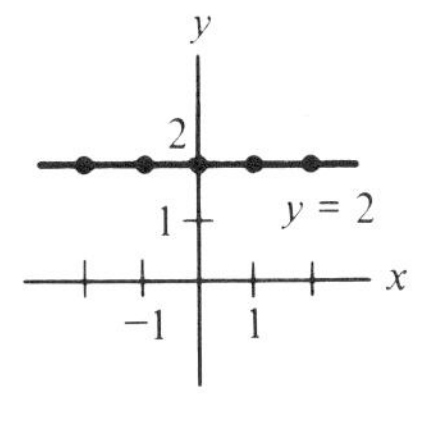

FIG. 9-10

EXAMPLE 5

Graph $x^2 + y^2 = 4$.

Let's solve this equation for y.

$$x^2 + y^2 = 4$$

$$y^2 = 4 - x^2 \qquad [\text{form } u^2 = k]$$

$$y = \pm\sqrt{4 - x^2} \qquad [u = \pm\sqrt{k}\,].$$

If $x = 0$, then $y = \pm\sqrt{4 - x^2} = \pm\sqrt{4 - (0)^2} = \pm 2$. Thus, both $(0, 2)$ and $(0, -2)$ lie on the graph.

If $x = 1$, then $y = \pm\sqrt{4 - (1)^2} = \pm\sqrt{3} \approx \pm 1.73$. (The symbol "$\approx$" means "approximately equals," and the value 1.73 comes from the table in the back of the book.)

Similarly, if $x = -1$, then $y = \pm\sqrt{4 - (-1)^2} = \pm\sqrt{3} \approx \pm 1.73$.

Other values of x and y appear in Fig. 9-11. Notice that we do not choose any

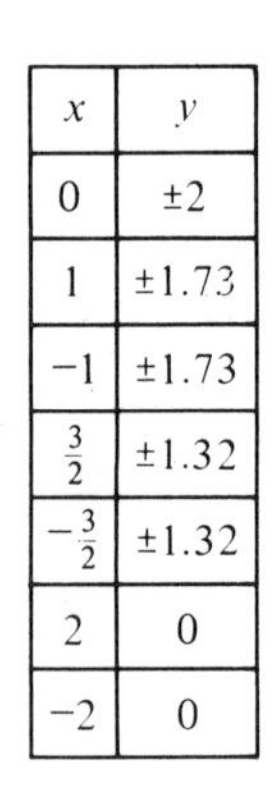

x	y
0	±2
1	±1.73
−1	±1.73
$\frac{3}{2}$	±1.32
$-\frac{3}{2}$	±1.32
2	0
−2	0

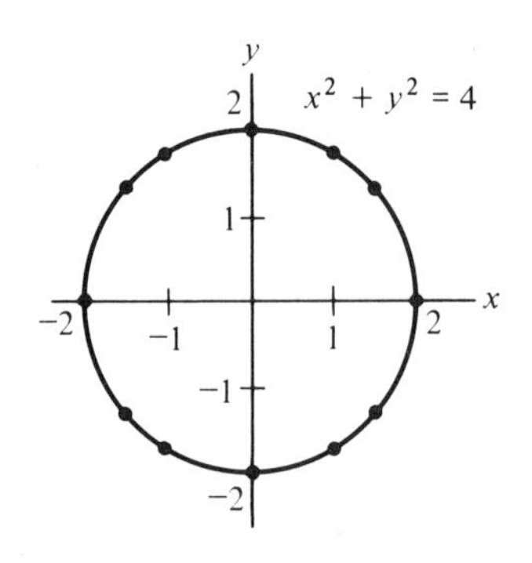

FIG. 9-11

values of x greater than 2 or less than -2. This is because we don't want $4 - x^2$ to be negative, for then y (which is $\pm\sqrt{4 - x^2}$) would be imaginary. We don't plot points with imaginary coordinates in a rectangular coordinate plane.

From Fig. 9-11, we see that *the graph is a circle with center at* $(0, 0)$ *and radius* 2.

Problem Set 9-2

In Problems **1–28**, *graph the equations.*

1. $y = x$ **2.** $y = -x$ **3.** $y = 3x - 2$

4. $y = 2x + 3$ **5.** $x = y + 1$ **6.** $x + y + 1 = 0$

7. $y = -1$

8. $x = 2$

9. $y = 2 - \frac{x}{2}$

10. $y = \frac{x}{4} + 1$

11. $y = 2x^2$

12. $y = \frac{1}{2}x^2$

13. $x = y^2$

14. $x = y^2 - 2$

15. $x = 3$

16. $y = 3$

17. $y = x^3$

18. $y = \frac{x^3}{3}$

19. $y = 0$

20. $x = 0$

21. $y = -x^3$

22. $y = 2 - x^3$

23. $y = x^2 - 4x + 3$

24. $y = x^2 - 2x - 3$

25. $y = 2x - x^2$

26. $y = 4x - x^2$

27. $x^2 + y^2 = 9$

28. $y^2 = 25 - x^2$

29. Graph $s = t^2 - 4t$. Choose t for the horizontal axis.

30. Graph $p + q = 1$. Choose p for the horizontal axis.

9-3 FUNCTIONS

Look at the equation

$$y = x + 2.$$

If we replace x by different numbers, we get corresponding values of y. For example,

$$\text{if } x = 0, \quad \text{then} \quad y = 0 + 2 = 2;$$
$$\text{if } x = 1, \quad \text{then} \quad y = 1 + 2 = 3.$$

Notice that for *each* value of x that "goes into" the equation, only *one* value of y "comes out."

Think of the equation $y = x + 2$ as defining a rule: add 2 to x. This rule assigns to each *input number* x exactly one *output number* y.

$$\underset{\text{input number}}{x} \longrightarrow \underset{\text{output number}}{y} \quad (= x + 2)$$

This rule is an example of a *function*.

> A **function** is a rule that assigns to each input number exactly one output number. The set of all input numbers to which the rule applies is called the **domain** of the function. The set of all output numbers is called the **range**.

For the function defined by $y = x + 2$, the input x can be any real number. Thus the domain of this function is all real numbers. To the input number 0 is assigned the output number 2:

$$x \longrightarrow y \quad \text{where} \quad y = x + 2$$
$$0 \longrightarrow 2 \quad \text{where} \quad 2 = 0 + 2.$$

Thus 2 is in the range.

A variable that represents input numbers for a function is called an **independent variable**. One that represents output numbers is a **dependent variable**. Its value *depends* on the value of the independent variable. We say that the dependent variable is a *function of* the independent variable. Thus, in the equation $y = x + 2$ the independent variable is x, the dependent variable is y, and y is a function of x.

Not all equations give y as a function of x, as Example 1 shows.

EXAMPLE 1

Let $y^2 = x$.

a. Suppose x is an input number, say $x = 9$. Then $y^2 = 9$ and so $y = \pm 3$. Thus with the input number 9 there are assigned not one but *two* output numbers, $+3$ and -3. Consequently, y is **not** a function of x.

b. Now, suppose y is an input number.

$$y \longrightarrow x(\text{where } x = y^2)$$

It determines exactly one output number, x. For example, if $y = 3$, then $x = y^2 = 3^2 = 9$. Thus x is a function of y. Since y can be any real number, the domain is all real numbers. The independent variable is y, and the dependent variable is x.

For some equations, either variable may be a function of the other, as Example 2 demonstrates.

EXAMPLE 2

$x + y - 1 = 0$.

Suppose x is an input number. If we solve the equation for y, then

$$y = 1 - x.$$

With each input number x, we can associate exactly one output number y (where $y = 1 - x$). Thus y is a function of x.

Now let's choose y as an input number. Solving $x + y - 1 = 0$ for x, we have

$$x = 1 - y.$$

With each input number y we can associate exactly one output number x (where $x = 1 - y$). Thus, x is a function of y.

Usually, letters such as f, g, h, F, G, etc. are used to name functions. Suppose we let f be the name of our original function defined by $y = x + 2$. Then the notation

$f(x)$, which is read "f of x," means the output number in the range of f that corresponds to the input number x in the domain.	input ↓ $f(x)$ ↑ output

Thus $f(x)$ is the same as y. But since $y = x + 2$, we may write

$$f(x) = x + 2.$$

To find $f(3)$, which is the output corresponding to the input 3, replace each x in $f(x) = x + 2$ by 3:

$$f(3) = 3 + 2 = 5.$$

Also,

$$f(8) = 8 + 2 = 10,$$
$$f(-4) = -4 + 2 = -2,$$
$$f(0) = 0 + 2 = 2.$$

Sometimes output numbers, such as $f(3)$, $f(8)$, etc., are called *functional values*. They are in the range of f.

*$f(x)$ **does not** mean "f times x."*

Functions may also be defined by "functional notation." For example, $g(x) = x^3 + x^2$ defines the function g which assigns to an input number x the output number $x^3 + x^2$.

$$g: x \longrightarrow x^3 + x^2$$

Some functional values for $g(x) = x^3 + x^2$ are

$$g(0) = 0^3 + 0^2 = 0,$$
$$g(2) = 2^3 + 2^2 = 12,$$
$$g(-1) = (-1)^3 + (-1)^2 = -1 + 1 = 0,$$
$$g(t) = t^3 + t^2,$$
$$g(x + 1) = (x + 1)^3 + (x + 1)^2.$$

Note that $g(x + 1)$ was found by replacing each x in $x^3 + x^2$ by $x + 1$. That is, g adds the cube and the square of an input number.

Since the equation $g(x) = x^3 + x^2$ defines a function, we shall feel free to call $g(x)$ a function itself. For the same reason, we speak of the *function* $y = x + 2$.

Let's be specific about the domain (input numbers) of a function. We shall consider input numbers to be all real numbers *except* those for which the functional values would involve either division by 0 or imaginary numbers.

For example, suppose h is the function given by

$$h(x) = \frac{1}{x - 1}.$$

The denominator of $\frac{1}{x - 1}$ is 0 if x is 1, so 1 cannot be an input number. Thus the

domain of h is all real numbers except 1.

EXAMPLE 3

Domains of functions.

a. $f(x) = \dfrac{x}{x^2 - x - 2}.$

We can't have division by zero, so we'll first find any values of x that make the denominator 0. These will not be input numbers.

$$x^2 - x - 2 = 0$$
$$(x - 2)(x + 1) = 0. \qquad \text{Thus,} \quad x = 2, -1.$$

Thus the

domain of f is all real numbers *except* 2 and -1.

b. $g(x) = \sqrt{x - 2}$.

We don't want the functional values to involve imaginary numbers. To keep away from square roots of negative numbers, we must make sure that $x - 2$ is 0 or positive.

If $x = 2$, then $x - 2 = 0$.

If x is greater than 2, then $x - 2$ is positive.

But if x is less than 2, then $x - 2$ is negative, and so $\sqrt{x - 2}$ is imaginary.

Thus the

domain of g is all real numbers x such that $x \geqslant 2$.

EXAMPLE 4

Domains and functional values.

a. $f(x) = 2x + 3$.
Here we don't have to rule out any values of x, so the domain of f is all real numbers.

Let's find some functional values.

$$f(x) = 2x + 3$$

Find $f(4)$: $f(4) = 2(4) + 3 = 11.$

Find $f(t)$: $f(t) = 2(t) + 3 = 2t + 3.$

Find $f(x + 1)$: $f(x + 1) = 2(x + 1) + 3 = 2x + 5.$

b. $g(x) = 3x^2 - x + 5$.

The domain of g is all real numbers.

$$g(x) = 3x^2 - x + 5$$

Find $g(z)$: $g(z) = 3(z)^2 - z + 5 = 3z^2 - z + 5.$

Find $g(r^2)$: $g(r^2) = 3(r^2)^2 - r^2 + 5 = 3r^4 - r^2 + 5.$

Find $g(x + h)$: $g(x + h) = 3(x + h)^2 - (x + h) + 5$

$$= 3(x^2 + 2hx + h^2) - x - h + 5$$

$$= 3x^2 + 6hx + 3h^2 - x - h + 5.$$

Don't get confused by notation. In Example 4b, *we found $g(x + h)$ by replacing each x in $g(x) = 3x^2 - x + 5$ by $x + h$. Don't write the function and then add h.*

$$g(x + h) \neq 3x^2 - x + 5 + h.$$

Also, don't use the distributive law on $g(x + h)$. It is not a multiplication.

$$g(x + h) \neq g(x) + g(h).$$

EXAMPLE 5

$h(x) = 2.$

The domain of h is all real numbers. All functional values will equal 2. For example,

$$h(10) = 2, \quad h(-387) = 2, \quad h(5.3) = 2, \quad h(x + 3) = 2.$$

We call h a *constant function*. In fact,

> *Any function of the form $h(x) = c$, where c is a fixed number, is called a* ***constant function***.

Sometimes functions are given in unusual ways. Let's look at one.

EXAMPLE 6

$$F(s) = \begin{cases} s^2 + 1, & \text{if } s > 0 \\ -3, & \text{if } s = 0 \\ 2 - s, & \text{if } s < 0. \end{cases}$$

Here s represents input numbers. We see that s can be any real number.

Find $F(2)$: Since $2 > 0$, we substitute $s = 2$ in $F(s) = s^2 + 1$.

$$F(2) = 2^2 + 1 = 5.$$

Find $F(0)$: Since $s = 0$, we have $F(0) = -3$.

Find $F(-1)$: Since $-1 < 0$, we substitute $s = -1$ in $F(s) = 2 - s$.

$$F(-1) = 2 - (-1) = 3.$$

Completion Set 9-3

In Problems **1–6**, *fill in the blanks.*

1. A function is a rule that assigns to each input number exactly ______ output number(s).

2. If $f(x) = 3x - 5$, then $f(2) = 3(\_\_\_\_) - 5 = \_\_\_\_$.

3. If $g(r) = r^2 + r + 1$, then $g(5) = (\_\_\_\_)^2 + \_\_\_\_ + 1 = \_\_\_\_\_$.

4. If $h(u) = 2u$, then $h(t + 1) = 2(\_\_\_\_\_\_\_\_\_)$.

5. If $f(x) = \dfrac{3}{x - 5}$, then the domain of f is all real numbers except ____.

6. If $f(x) = 3$, then $f(x)$ is called a ________________ function and all functional values are equal to ____.

In Problems **7–10**, *insert* T (= *True*) *or* F (= *False*) *in the blanks.*

7. 12 is in the domain of $f(x) = 5x + 3$. ____

8. 6 is in the domain of $g(x) = \sqrt{25 - x^2}$. ____

9. 0 is in the domain of $h(z) = \dfrac{z}{z^2 - 9}$. ____

10. -3 is in the domain of $F(t) = \dfrac{t}{t^2 - 9}$. ____

Problem Set 9-3

In Problems **1–22**, *give the domain of each of the functions.*

1. $f(x) = \dfrac{3}{x}$

2. $g(x) = \dfrac{4}{x^2}$

3. $g(x) = \dfrac{x}{3}$

4. $f(x) = \dfrac{x+2}{5}$

5. $h(x) = \sqrt{x}$

6. $f(r) = 7r - 2$

7. $H(z) = 10$

8. $h(t) = (2t + 1)^2$

9. $F(t) = 3t^2 + 5$

10. $G(s) = \dfrac{s+1}{s}$

11. $H(x) = \dfrac{x}{x+2}$

12. $f(y) = \dfrac{8-y}{4-7y}$

13. $f(x) = \dfrac{3x-1}{2x+5}$

14. $f(x) = \sqrt{x+3}$

15. $g(x) = \sqrt{x-5}$

16. $h(x) = \dfrac{x^3-1}{x^2-4x+4}$

17. $G(y) = \dfrac{4}{y^2-y}$

18. $H(z) = \dfrac{-3}{z^2+2z-8}$

19. $f(x) = \dfrac{x+1}{x^2+6x+5}$

20. $f(x) = \dfrac{4}{x} + \dfrac{x}{x-3}$

21. $h(s) = \dfrac{4-s^2}{2s^2-7s-4}$

22. $G(r) = \dfrac{2}{r^2+1}$

In Problems **23–40**, *find the functional values of the given functions.*

23. $f(x) = 5x;\quad f(0),\quad f(3),\quad f(-4),\quad f\left(\frac{12}{5}\right)$

24. $g(x) = 2x - 5;\quad g(-1),\quad g(4),\quad g\left(-\frac{1}{2}\right),\quad g(100)$

25. $h(t) = 4 - 3t;\quad h(1),\quad h(3),\quad h\left(-\frac{2}{3}\right),\quad h\left(\frac{1}{2}\right)$

26. $H(s) = s^2 - 3;\quad H(4),\quad H(-8),\quad H(\sqrt{2}),\quad H\left(\frac{2}{3}\right)$

27. $f(x) = 7x;\quad f(4),\quad f(s),\quad f(t+1),\quad f(x+3)$

28. $G(x) = 2 - x^2;\quad G(-8),\quad G(8),\quad G(u),\quad G(u^2)$

29. $g(u) = u^2 + u;\quad g(3),\quad g(-2),\quad g(2v),\quad g(x^2)$

30. $h(v) = \dfrac{1}{\sqrt{v}};\quad h(1),\quad h(16),\quad h\left(\frac{1}{4}\right),\quad h(1-x)$

31. $f(x) = 12;\quad f(2),\quad f(-1.5),\quad f(t+8),\quad f(\sqrt{17})$

32. $H(x) = (x+4)^2;\quad H(0),\quad H(2),\quad H(-4),\quad H(t-4)$

33. $f(x) = x^2 + 2x + 1;\quad f(1),\quad f(-1),\quad f(w),\quad f(x+h)$

34. $f(x) = \dfrac{x^2}{x+3};\quad f(0),\quad f(1),\quad f(z),\quad f(t^3)$

35. $g(x) = \dfrac{x-5}{x^2+4};\quad g(3),\quad g(5),\quad g(3x),\quad g(x+h)$

36. $g(x) = |x-3|;\quad g(10),\quad g(3),\quad g(-3),\quad g(-5)$

37. $F(p) = \sqrt{25 - p^2}$; $F(5), F(3), F(4), F(p^4)$

38. $F(t) = \begin{cases} 1, \text{ if } t > 0 \\ 0, \text{ if } t = 0; \\ -1, \text{ if } t < 0 \end{cases}$ $F(10),\ F(-\sqrt{3}),\ F(0),\ F(-\frac{18}{5})$

39. $G(x) = \begin{cases} x, & \text{if } x \geqslant 3 \\ 2 - x, & \text{if } x < 3 \end{cases};$ $G(8),\ G(3),\ G(-1),\ G(1)$

40. $G(x) = \dfrac{1}{x-2} + \dfrac{1}{x+3};$ $G(-2),\ G(3),\ G(0),\ G(-4)$

In Problems **41–44**, *for the given function find (a)* $f(x + h)$, *and (b)* $\dfrac{f(x+h) - f(x)}{h}$, *and simplify your answers.*

41. $f(x) = 3x - 4$

42. $f(x) = 2x + 1$

43. $f(x) = 2$

44. $f(x) = x^2$

In Problems **45–48** *is y a function of x? Is x a function of y?*

45. $y - 3x - 4 = 0$

46. $x^2 + y = 0$

47. $y = 7x^2$

48. $x^2 + y^2 = 1$

49. In a lab experiment, different weights were attached to a spring, causing it to stretch. Table 9-2 gives the recorded data.

TABLE 9-2

WEIGHT (lb) w	AMOUNT OF STRETCH (in.) s
1	$\frac{1}{2}$
4	2
6	3

Since for each value of w there is exactly one value of s, we can think of s as a function of w. Here w's are inputs and s's are outputs. Let's say $s = F(w)$. Find $F(4)$, $F(1)$, and $F(6)$.

50. The distance s (in ft) that an object falls as a function of elapsed time t (in sec) is given by $s = f(t) = 16t^2$.

a. Find $f(0)$, $f(1)$, and $f(2)$.

b. From a practical standpoint, what would you define the domain of f to be?

51. Suppose that consumers will buy x units of a certain product at a price of p dollars per unit. Then the function f given by $p = f(x) = 80 - 2x$ is called a *demand function* for the product. We say price per unit p is a function of quantity demanded x. Find $f(30)$. What does this number represent? Also find $f(25)$, $f(10)$, and $f(1)$.

52. The formula for the area A of a circle of radius r is $A = \pi r^2$. Is the area a function of the radius?

53. In the study of circular motion, the formula

$$a = \frac{v^2}{20}$$

occurs. Is a a function of v?

54. The formula for the circumference C of a circle of diameter d is $C = \pi d$. Is C a function of d? Is d a function of C?

55. Suppose that a ball is thrown up in the air and the equation $s = 48t - 16t^2$ gives the height s (in ft) of the ball after t seconds. Find the heights when $t = 1$ and when $t = 2$. Is t a function of s? Is s a function of t?

9-4 GRAPHS OF FUNCTIONS

Often it is helpful to see what a function "looks like." This is done by graphing.

To graph the function $f(x) = 3x - 4$, we let $y = f(x)$ and graph this equation.

$$y = f(x) = 3x - 4$$

or

$$y = 3x - 4.$$

We choose various values of the independent variable x (input numbers) and find corresponding values of y (output numbers). See Table 9-3. Plotting the points and connecting them, we get the graph of f. See Fig. 9-12. We always

TABLE 9-3

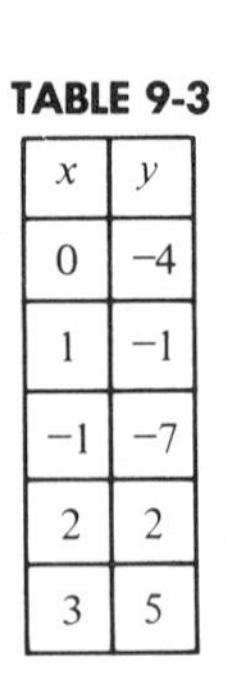

x	y
0	−4
1	−1
−1	−7
2	2
3	5

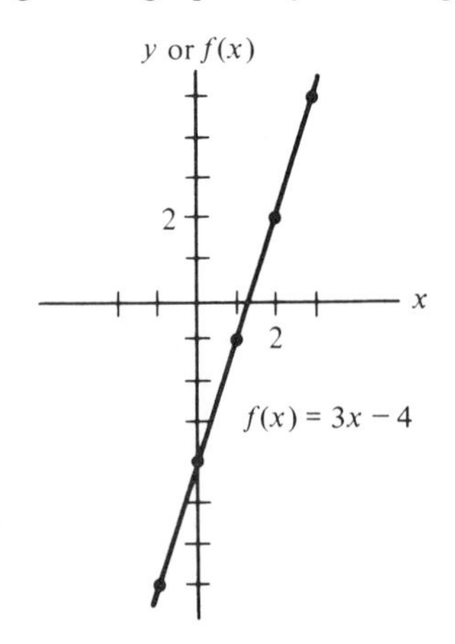

FIG. 9-12

label the horizontal axis with the independent variable. The vertical axis can be labeled as either the y-axis or the $f(x)$-axis and can be called the *function-value axis*.

EXAMPLE 1

Graph $g(x) = |x|$.

Recall that $|x|$ is the absolute value of x and was discussed on p. 33. See Fig. 9-13.

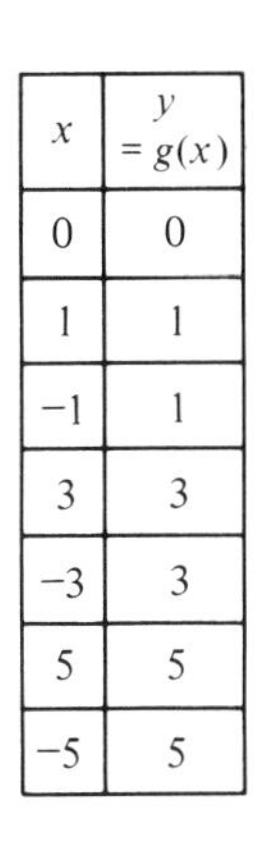

x	$y = g(x)$
0	0
1	1
−1	1
3	3
−3	3
5	5
−5	5

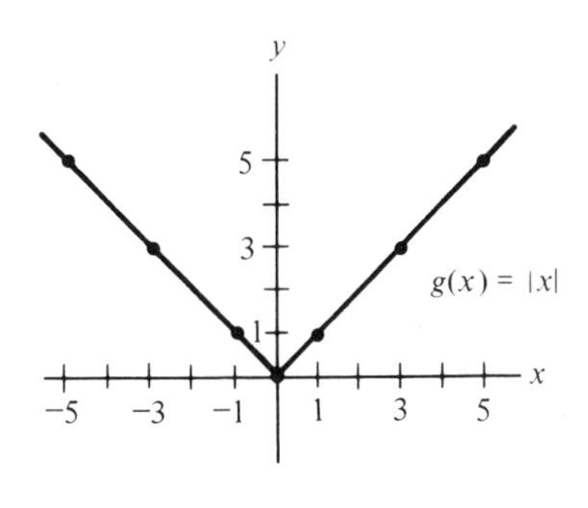

FIG. 9-13

In Fig. 9-14 is the graph of some function $y = f(x)$. Corresponding to the input number x on the horizontal axis is the output number $f(x)$ on the vertical

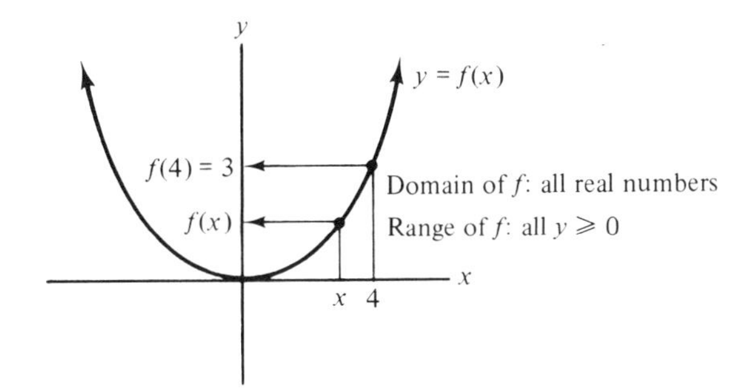

FIG. 9-14

axis. For example, $f(4) = 3$. Since there is an output number for any value of x, the domain of f is all real numbers. Notice that the y-coordinates of all points on the graph are nonnegative. Here the range of f is all $y \geqslant 0$. This shows that we may find the domain and range of a function by looking at its graph. Thus, from Fig. 9-13 we see that the domain of $g(x) = |x|$ is all real numbers, and its range is all nonnegative numbers ($y \geqslant 0$).

EXAMPLE 2

Graph $G(u) = \sqrt{u}$.

Here u is the independent variable, so the horizontal axis is labeled as the u-axis.

The function-value axis is labeled $G(u)$. See Fig. 9-15. We cannot choose negative

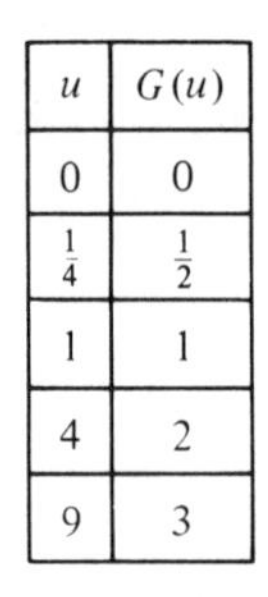

u	$G(u)$
0	0
$\frac{1}{4}$	$\frac{1}{2}$
1	1
4	2
9	3

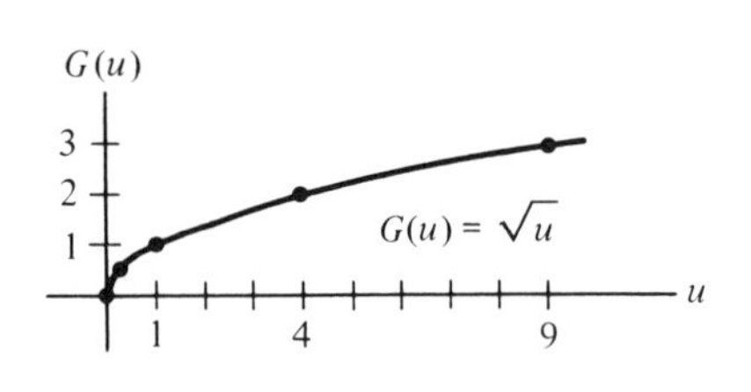

FIG. 9-15

values for u, since we don't want to get involved with imaginary numbers. Thus, the domain of G is all nonnegative numbers. From the graph, the range of G is clearly all nonnegative numbers.

EXAMPLE 3

Graph $f(x) = \dfrac{1}{x}$.

The domain of f is all numbers except $x = 0$, so the graph will have *no* point corresponding to $x = 0$. See Fig. 9-16. From the graph we see that the range of f is

x	$f(x)$
$\frac{1}{4}$	4
$-\frac{1}{4}$	-4
$\frac{1}{2}$	2
$-\frac{1}{2}$	-2
1	1
-1	-1
3	$\frac{1}{3}$
-3	$-\frac{1}{3}$
5	$\frac{1}{5}$
-5	$-\frac{1}{5}$

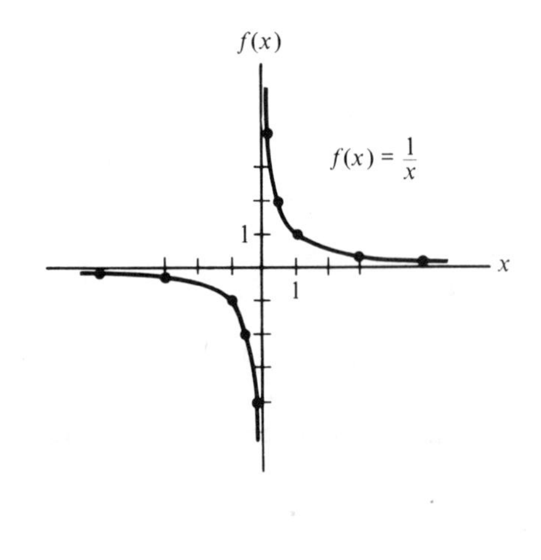

FIG. 9-16

all real numbers except 0.

There is an easy way to tell whether a graph represents a function. Look at the leftmost diagram in Fig. 9-17(a). It is the graph of some equation in x and y.

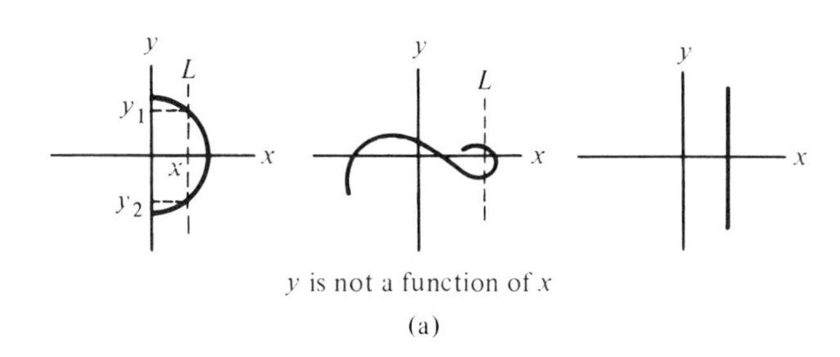

y is not a function of x

(a)

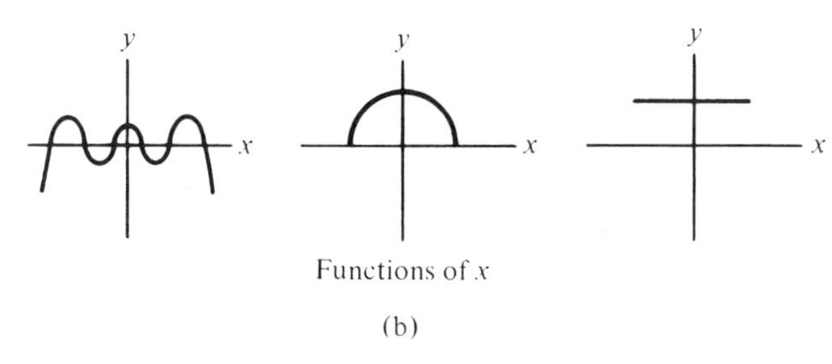

Functions of x

(b)

FIG. 9-17

Notice that if x is an input number, then there corresponds two output numbers, namely y_1 and y_2. Thus the equation *does not* give y as a function of x.

In general, if you can draw a *vertical* line L that meets the graph of an equation in at least two points, then you *do not* have a function of x. When you can't draw such a line, then you do have a function of x. Thus the graphs in Fig. 9-17(a) do not represent functions of x, but those in Fig. 9-17(b) do.

Problem Set 9-4

In Problems **1–14**, *graph each function and give the domain and range.*

1. $f(x) = 2x + 2$

2. $g(x) = 4 - x$

3. $g(x) = |x - 2|$

4. $f(x) = |x| - 2$

5. $G(t) = 2\sqrt{t}$

6. $F(t) = \sqrt{4t}$

7. $f(x) = 4$

8. $f(x) = 0$

9. $h(x) = 4 - x^2$

10. $h(x) = (x + 1)^2$

11. $F(z) = \dfrac{3}{z}$

12. $F(w) = -\dfrac{4}{w}$

13. $f(x) = \dfrac{2}{x - 4}$

14. $f(x) = \dfrac{1}{x + 2}$

15. In Fig. 9-18 is the graph of $y = f(x)$. (a) Find $f(0)$, $f(2)$, $f(4)$, and $f(-2)$. (b) What is the domain of f? (c) What is the range of f?

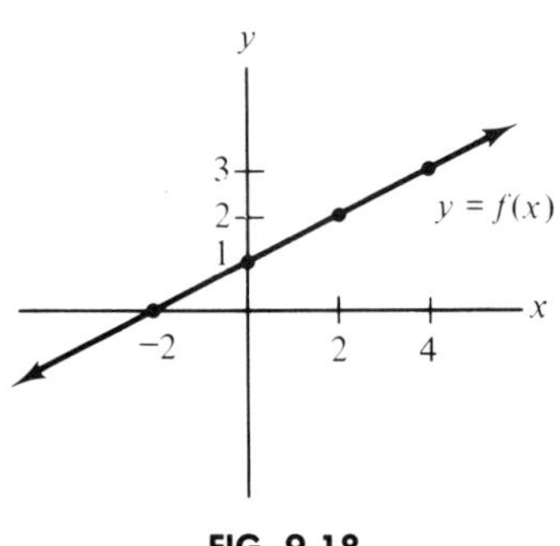

FIG. 9-18

16. In Fig. 9-19 is the graph of $y = f(x)$. (a) Find $f(0)$ and $f(2)$. (b) What is the domain of f? (c) What is the range of f?

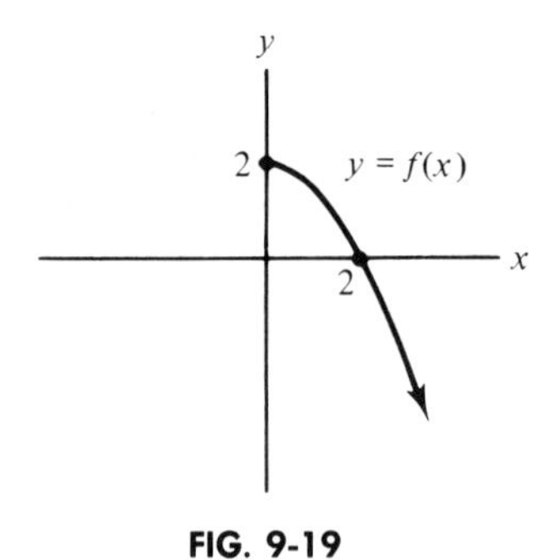

FIG. 9-19

17. In Fig. 9-20 is the graph of $y = f(x)$. (a) Find $f(0)$, $f(1)$, and $f(-1)$. (b) What is the domain of f? (c) What is the range of f?

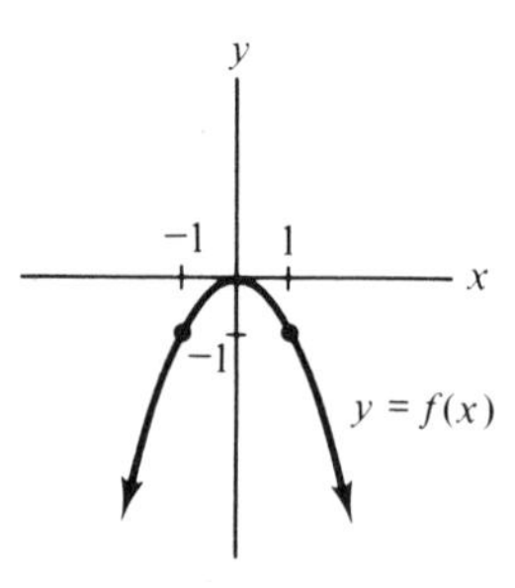

FIG. 9-20

18. The electric potential V (in volts) as a function of its distance r (in meters) from an object having a certain electric charge is given by

$$V = \frac{100}{r}.$$

Here r must be positive. Sketch the graph of this function.

19. Table 9-4 is called a *demand schedule*. It gives a correspondence between the quantity q of a certain product that consumers will buy (demand) at the price p per unit.

(a) Suppose we think of q as an input number and p as an output number. With each input number there corresponds exactly one output number. Thus, price per unit p is a function of the quantity demanded q. Call this function f. Find $f(5)$, $f(10)$, $f(20)$, and $f(25)$.

TABLE 9-4 DEMAND SCHEDULE

QUANTITY DEMANDED q	PRICE PER UNIT p
5	20
10	10
20	5
25	4

(b) Draw a rectangular coordinate plane and label the horizontal axis q and the vertical axis p. Plot each quantity-price pair [for example, plot (5, 20), (10, 10), etc.]. Connect the points by a smooth curve. In this way we can approximate points in between the given data. This curve is called a *demand curve*.

20. The dividends paid per share (in cents) by a corporation during the last ten years are as follows:

year	1	2	3	4	5	6	7	8	9	10
dividend	5	5	5	6	$11\frac{1}{2}$	15	20	22	$27\frac{1}{4}$	30

These data give rise to a function f—namely, the one defined by thinking of the year y as input and the corresponding dividend d as output. Thus, $d = f(y)$. For example, $f(1) = 5$, $f(2) = 5$, $f(8) = 22$, etc. Plot the year-dividend pairs (1, 5), (2, 5), (8, 22), etc., and connect the points by a smooth curve.

21. In Fig. 9-21, which graphs represent functions of x?

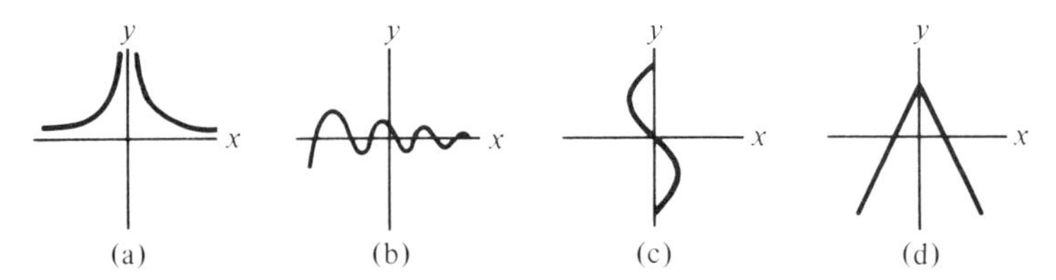

FIG. 9-21

9-5 NEW FUNCTIONS FROM OLD ONES

There are some simple ways of taking two functions and getting a third function from them. For example, suppose that f and g are the functions

$$f(x) = x^2 \quad \text{and} \quad g(x) = 3x.$$

We can get a third function h by *adding* f and g:

$$\begin{aligned} h(x) &= f(x) + g(x) \\ &= x^2 + 3x. \end{aligned}$$

Thus,

$$h(2) = f(2) + g(2) = 2^2 + 3(2) = 10$$

and

$$h(-\tfrac{1}{2}) = (-\tfrac{1}{2})^2 + 3(-\tfrac{1}{2}) = \tfrac{1}{4} - \tfrac{3}{2} = -\tfrac{5}{4}.$$

Besides adding $f(x)$ and $g(x)$, we can also subtract, multiply, or divide them.

EXAMPLE 1

Let $f(x) = 3x^2 + 4$ and $g(x) = x^2 - x + 5$. Find $H(x) = f(x) - g(x)$ and use the result to find $H(0)$ and $H(3)$.

$$\begin{aligned} H(x) &= f(x) - g(x) \\ &= (3x^2 + 4) - (x^2 - x + 5) \\ H(x) &= 2x^2 + x - 1. \qquad \text{[simplifying]} \\ H(0) &= 2(0)^2 + 0 - 1 = -1. \\ H(3) &= 2(3)^2 + 3 - 1 = 20. \end{aligned}$$

EXAMPLE 2

Let $f(x) = x + 1$ and $g(x) = x + 2$. Find $F(x) = f(x) \cdot g(x)$, $F(5)$, and $F(-3)$.

$$\begin{aligned} F(x) &= f(x) \cdot g(x) \\ &= (x + 1)(x + 2) \\ F(x) &= x^2 + 3x + 2. \\ F(5) &= 5^2 + 3(5) + 2 = 42. \\ F(-3) &= (-3)^2 + 3(-3) + 2 = 2. \end{aligned}$$

EXAMPLE 3

Let $f(x) = x$ and $g(x) = x - 5$. Find $G(x) = \dfrac{f(x)}{g(x)}$, $G(0)$, and $G(-1)$.

$$G(x) = \frac{f(x)}{g(x)}$$

$$G(x) = \frac{x}{x - 5}.$$

$$G(0) = \frac{0}{0 - 5} = \frac{0}{-5} = 0.$$

$$G(-1) = \frac{-1}{-1 - 5} = \frac{-1}{-6} = \frac{1}{6}.$$

Notice that the domain of $G(x) = \dfrac{x}{x - 5}$ is all real numbers *except* 5.

Here's another way to combine functions. Suppose that $f(x) = x^2$ and $g(x) = 3x$. Now, $g(2) = 3(2) = 6$. Thus g *sends* 2 into 6.

$$2 \xrightarrow{g} 6$$

Let's see what f does to 6.

$$f(6) = 6^2 = 36,$$

so f sends 6 into 36.

$$6 \xrightarrow{f} 36$$

By applying g and then f, we get a new function h which sends 2 into 36:

$$2 \xrightarrow{g} 6 \xrightarrow{f} 36.$$

$$h$$

Thus $h(2) = 36$. Let's replace the 6 and the 36 in the diagram above. Since

$$6 = g(2) \quad \text{and} \quad 36 = f(6) = f[g(2)],$$

our diagram looks like

$$2 \xrightarrow{g} g(2) \xrightarrow{f} f[g(2)].$$

$$h$$

Consequently,

$$h(2) = f[g(2)].$$

To be more general, let's replace the 2 by the variable x. Then we have

$$h(x) = f[g(x)].$$

h is called the **composition function** *of f with g.*

We can get a nice form for $h(x)$. Since $f(x) = x^2$ and $g(x) = 3x$, then by the formula in the box we have

$$h(x) = f[g(x)] = f[3x] = (3x)^2 = 9x^2,$$

or simply

$$h(x) = 9x^2.$$

Thus, $h(2) = 9(2)^2 = 36$, as we saw before. The formula in the box works for other functions f and g, as the next examples show.

EXAMPLE 4

Let $f(x) = 2x + 3$ *and* $g(x) = x - 4$. *Find* $h(x) = f[g(x)]$, $h(4)$, *and* $h\left(-\frac{1}{2}\right)$.

$$\begin{aligned} h(x) &= f[g(x)] \\ &= f[x - 4] = 2(x - 4) + 3 \\ h(x) &= 2x - 5. \\ h(4) &= 2(4) - 5 = 3. \\ h\left(-\tfrac{1}{2}\right) &= 2\left(-\tfrac{1}{2}\right) - 5 = -6. \end{aligned}$$

EXAMPLE 5

Let $f(x) = 2x^2$ *and* $g(x) = 3x + 1$. *Find* $f[g(x)]$ *and* $g[f(x)]$.

$$f[g(x)] = f[3x + 1] = 2(3x + 1)^2.$$

$$g[f(x)] = g[2x^2] = 3(2x^2) + 1 = 6x^2 + 1.$$

Generally, $f[g(x)] \neq g[f(x)]$. *In Example* 5, *notice* $f[g(x)] = 2(3x + 1)^2$ *while* $g[f(x)] = 6x^2 + 1$.

Sometimes we can think of a function as a composition. For example, let's consider the function $f(x) = (2x - 1)^3$, which is the cube of $2x - 1$. Suppose we let $g(x) = x^3$ and $h(x) = 2x - 1$. Then

$$g[h(x)] = g[2x - 1] = (2x - 1)^3 = f(x).$$

Thus $f(x) = g[h(x)]$, which is a composition of two functions.

Completion Set 9-5

In Problems **1–6**, $f(x) = x + 1$ *and* $g(x) = 2x$. *Fill in the blanks.*

1. If $h(x) = f(x) + g(x)$, then $h(x) = (x + 1) + (2x) = 3x + 1$ and $h(4) = 3(\_\_\_\_) + 1 = \_\_\_\_\_$.

2. If $H(x) = f(x) - g(x)$, then $H(x) = (x + 1) - (2x) = -x + \_\_\_\_$ and $H(2) = \_\_\_\_\_$.

3. If $F(x) = f(x) \cdot g(x)$, then $F(x) = (x + 1) \cdot (\_\_\_\_\_)$ and $F(0) = \_\_\_\_$.

4. If $G(x) = \dfrac{f(x)}{g(x)}$, then $G(x) = \dfrac{(\qquad\qquad)}{2x}$ and $G(1) = \_\_\_\_$.

5. $f[g(x)] = f[2x] = \_\_\_\_\_ + 1$.

6. $g[f(x)] = g[x + 1] = \_\_\_\_(x + 1)$.

Problem Set 9-5

In Problems **1–5**,

$$h(x) = f(x) + g(x), \qquad H(x) = f(x) - g(x),$$

$$F(x) = f(x) \cdot g(x), \qquad G(x) = \frac{f(x)}{g(x)},$$

$$k(x) = f[g(x)], \qquad K(x) = g[f(x)].$$

1. If $f(x) = x + 1$ and $g(x) = x + 4$, find

a. $h(x)$ b. $h(0)$ c. $H(x)$ d. $H(5)$
e. $F(x)$ f. $F(-2)$ g. $G(x)$ h. $G(-1)$
i. $k(x)$ j. $k(3)$ k. $K(x)$ l. $K(5)$

2. If $f(x) = 8x$ and $g(x) = 8 + x$, find

a. $h(x)$ b. $h(-2)$ c. $H(x)$ d. $H(4)$
e. $F(x)$ f. $F(-1)$ g. $G(x)$ h. $G(2)$
i. $k(x)$ j. $k(1)$ k. $K(x)$ l. $K(2)$

3. If $f(x) = x^2$ and $g(x) = x^2 + x$, find

a. $h(x)$ b. $h(\frac{1}{2})$ c. $H(x)$ d. $H(-\frac{1}{2})$
e. $F(x)$ f. $F(-1)$ g. $G(x)$ h. $G(-\frac{1}{2})$
i. $k(x)$ j. $k(2)$ k. $K(x)$ l. $K(-3)$

4. If $f(x) = x^2 - 1$ and $g(x) = 4$, find

a. $h(x)$ b. $h(\frac{1}{2})$ c. $H(x)$ d. $H(-5)$
e. $F(x)$ f. $F(4)$ g. $G(x)$ h. $G(1)$
i. $k(x)$ j. $k(100)$ k. $K(x)$ l. $K(18)$

5. If $f(x) = -2x$ and $g(x) = x^3$, find

a. $h(x)$ b. $h(t)$ c. $H(x)$ d. $H(-2)$
e. $F(x)$ f. $F(-\frac{1}{2})$ g. $G(x)$ h. $G(r + 1)$
i. $k(x)$ j. $k(-2)$ k. $K(x)$ l. $K(2)$

In Problems **6–11**, *find* $f[g(x)]$ *and* $g[f(x)]$.

6. $f(x) = x + 5, \quad g(x) = 8 - 2x$

7. $f(x) = \sqrt{x}, \quad g(x) = x - 1$

8. $f(x) = \dfrac{1}{x}, \quad g(x) = 3$

9. $f(x) = \dfrac{1}{x}, \quad g(x) = x^2 - 1$

10. $f(x) = x^2 - 1, \quad g(x) = x^2 + 1$

11. $f(x) = 3x + 4, \quad g(x) = 4 - 3x$

9-6 SUPPLEMENT ON FUNCTIONS†

You saw that a function is a special kind of input-output operation. However, we can think of functions in another way—as a set of ordered pairs.

---

†This topic is not needed in the rest of the book.

For example, the set

$$\{(2, 7), (3, 8), (4, 9)\}$$

can be thought of as a function in the following way. The ordered pair (2, 7) indicates that with the input number 2 we associate the output number 7.

$$(2, 7): \quad 2 \longrightarrow 7$$

Likewise,

$$(3, 8): \quad 3 \longrightarrow 8$$

$$(4, 9): \quad 4 \longrightarrow 9.$$

The domain of this function is the set of all first coordinates of the ordered pairs.

$$\text{Domain is } \{2, 3, 4\}.$$

Not any set of ordered pairs gives a function. For example,

$$\{(1, 3), (1, 4), (2, 5)\}$$

is *not* a function, because with the number 1 there are associated *two* different numbers, 3 and 4.

A function *cannot* be a set that contains two ordered pairs with the same first coordinates and different second coordinates. In fact, a function is just the opposite.

> *A function can be thought of as a set of ordered pairs that* **does not** *contain two ordered pairs with the same first coordinates and different second coordinates.*

9-7 REVIEW

IMPORTANT TERMS AND SYMBOLS

rectangular coordinate plane *(p. 209)*
coordinate axes *(p. 210)*
origin *(p. 210)*
ordered pair *(p. 210)*
first coordinate *(p. 210)*
second coordinate *(p. 210)*
abscissa *(p. 210)*
range *(p. 219)*
functional value *(p. 221)*
independent variable *(p. 220)*
dependent variable *(p. 220)*
constant function *(p. 224)*
function-value axis *(p. 228)*
$f(x) + g(x)$ *(p. 234)*

ordinate *(p. 210)*

quadrant *(p. 211)*

graph *(p. 215)*

function *(p. 219)*

$f(x)$ *(p. 221)*

domain *(p. 219)*

$f(x) - g(x)$ *(p. 234)*

$f(x) \cdot g(x)$ *(p. 234)*

$\dfrac{f(x)}{g(x)}$ *(p. 235)*

composition of functions *(p. 236)*

$f[g(x)]$ *(p. 236)*

REVIEW PROBLEMS

In Problems **1–6**, *give the domains of the functions.*

1. $f(x) = \dfrac{x}{x^2 - 3x + 2}$

2. $g(x) = x^2 + 3x$

3. $F(t) = 7t + 4t^2$

4. $G(x) = 18$

5. $h(x) = \dfrac{\sqrt{x}}{x - 1}$

6. $H(s) = \dfrac{\sqrt{s - 5}}{4}$

In Problems **7–12**, *find the functional values for the given functions.*

7. $f(x) = 3x^2 - 4x + 7$; $f(0)$, $f(-3)$, $f(5)$, $f(t)$

8. $g(x) = 4$; $g(4)$, $g(\frac{1}{100})$, $g(-156)$, $g(x + 4)$

9. $G(x) = \sqrt{x - 1}$; $G(1)$, $G(10)$, $G(t + 1)$, $G(x^2)$

10. $F(x) = \dfrac{x - 3}{x + 4}$; $F(-1)$, $F(0)$, $F(5)$, $F(x + 3)$

11. $h(u) = \dfrac{\sqrt{u + 4}}{u}$; $h(5)$, $h(-4)$, $h(x)$, $h(u - 4)$

12. $H(s) = \dfrac{(s - 4)^2}{3}$; $H(-2)$, $H(7)$, $H(\frac{1}{2})$, $H(x^2)$

In Problems **13** *and* **14**, *for the given function find (a)* $f(x + h)$, *and (b)* $\dfrac{f(x + h) - f(x)}{h}$, *and simplify your answers.*

13. $f(x) = 3 - 7x$

14. $f(x) = x^2 + 4$

In Problems **15–22**, *graph the given equation or function.*

15. $y = 5x - 4$

16. $x = 4$

17. $f(x) = 5$

18. $g(x) = 6 - x^2$

19. $x = y^2$

20. $x^2 + y^2 = 1$

21. $F(t) = \dfrac{6}{t}$

22. $G(u) = \sqrt{u + 4}$

In Problems **23** *and* **24**, $h(x) = f(x) + g(x)$, $H(x) = f(x) - g(x)$, $F(x) = f(x) \cdot g(x)$, $G(x) = \dfrac{f(x)}{g(x)}$, $k(x) = f[g(x)]$, *and* $K(x) = g[f(x)]$.

23. If $f(x) = 3x - 1$ and $g(x) = 2x + 3$, find

a. $h(x)$	b. $h(4)$	c. $H(x)$	d. $H(-2)$
e. $F(x)$	f. $F(1)$	g. $G(x)$	h. $G(2)$
i. $k(x)$	j. $k(5)$	k. $K(x)$	l. $K(-3)$

24. If $f(x) = x^2$ and $g(x) = 2x + 1$, find

a. $h(x)$	b. $h(1)$	c. $H(x)$	d. $H(-3)$
e. $F(x)$	f. $F(2)$	g. $G(x)$	h. $G(-2)$
i. $k(x)$	j. $k(0)$	k. $K(x)$	l. $K(-4)$

In Problems **25–28**, *find* $f[g(x)]$ *and* $g[f(x)]$.

25. $f(x) = \dfrac{1}{x}$, $g(x) = x - 1$

26. $f(x) = \dfrac{x + 1}{4}$, $g(x) = \sqrt{x}$

27. $f(x) = x + 2$, $g(x) = x^3$

28. $f(x) = 2$, $g(x) = 3$

29. Suppose that manufacturers will supply x units of a product at a price of p dollars per unit. Then the function f given by $p = f(x) = .5x + 10$ is called a *supply function* for the product. We say that price per unit p is a function of the quantity supplied x. Find $f(10)$. What does this number represent? Also, find $f(20)$, $f(100)$, and $f(500)$.

30. Table 9-5 is called a *supply schedule*. It gives a correspondence between the quantity q of a certain product that manufacturers will supply at the price p per unit.

(a) Suppose we think of q as an input number and p as an output number. With each input number there corresponds exactly one output number. Thus price per unit p is a function of quantity supplied q. Call this function f. Find $f(30)$, $f(100)$, $f(150)$, $f(190)$, and $f(210)$.

TABLE 9-5 SUPPLY SCHEDULE

QUANTITY SUPPLIED q	PRICE PER UNIT p
30	10
100	20
150	30
190	40
210	50

(b) Draw a rectangular coordinate plane and label the horizontal axis q and the vertical axis p. Plot each quantity-price pair [for example, plot (30, 10), (100, 20), etc.]. Connect these points by a smooth curve. In this way we can approximate the points in between the given data. This curve is called a *supply curve*.

31. In Fig. 9-22 is the graph of $y = f(x)$. (a) Find $f(0)$, $f(1)$, $f(10)$, $f(-30)$. (b) What is

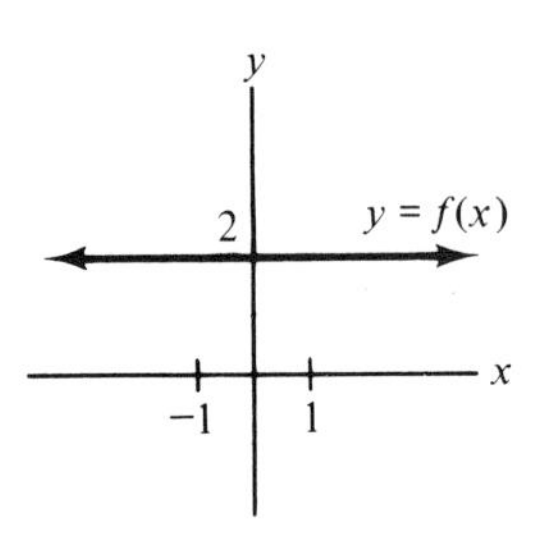

FIG. 9-22

the domain of f? (c) What is the range of f?

32. In Fig. 9-23, which graphs represent functions of x?

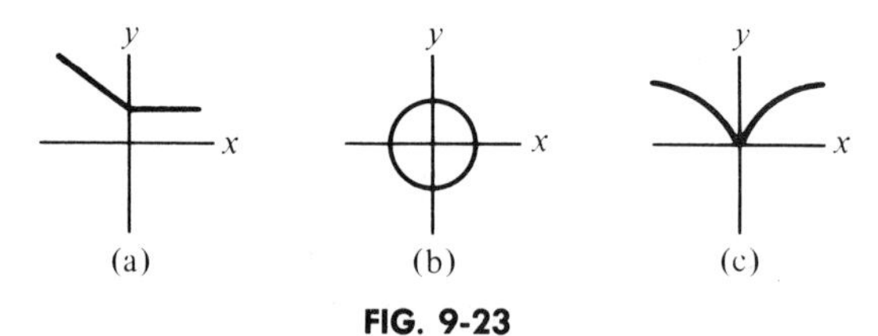

FIG. 9-23

CHAPTER 10

Zeros of Polynomial Functions†

10-1 ZEROS

In this chapter we'll be working only with *polynomial functions*. These are functions defined by polynomials. Here are two examples along with their special names, which depend on the degree of the polynomial.

$$f(x) = 2x - 6 \qquad \textit{linear} \text{ function (degree 1)}$$
$$g(x) = 3x^2 - x - 5 \qquad \textit{quadratic} \text{ function (degree 2).}$$

Any value of x for which a polynomial function $f(x)$ is equal to 0 is called a **zero** of $f(x)$. For example, 3 is a zero of $f(x) = 2x - 6$ since $f(3) = 2(3) - 6 = 0$.

To find zeros of $f(x)$ we set $f(x) = 0$ and solve for x.

EXAMPLE 1

Find all zeros of the following polynomial functions.

a. $f(x) = 7 - 4x$.

---

†This chapter may be omitted.

Setting $f(x) = 0$ gives

$$\begin{aligned} 7 - 4x &= 0 \\ -4x &= -7 \\ x &= \frac{7}{4}. \end{aligned}$$

The only zero is $\frac{7}{4}$. This means $f(\frac{7}{4}) = 0$.

b. $f(x) = x^4 - 2x^3 - 3x^2$.

Setting $f(x) = 0$ and factoring $f(x)$ gives

$$\begin{aligned} x^4 - 2x^3 - 3x^2 &= 0 \\ x^2(x^2 - 2x - 3) &= 0 \\ x^2(x - 3)(x + 1) &= 0. \end{aligned} \tag{10-1}$$

The roots of this equation are $x = 0$, $x = 3$, and $x = -1$, where 0 is a repeated root. Thus the zeros of $f(x)$ are 0, 3, and -1. This means $f(0) = 0$, $f(3) = 0$, and $f(-1) = 0$. Here we say that 0 is a *repeated zero*.

In Example 1(b), note from Eq. (10-1) that $x - 3$ is a factor of $f(x)$. Also, we found that 3 is a zero of $f(x)$. This illustrates the following rule:

If $x - r$ is a factor of $f(x)$, then r is a zero of $f(x)$.

Also, $x + 1$ is a factor of $f(x)$. Writing $x + 1$ in the form $x - r$, we have $x + 1 = x - (-1)$. From the rule, -1 is a zero. Finally, the factor x (which is $x - 0$) gives 0 (which is r) as a zero.

The "reverse" of the statement in the above box is also true.

If r is a zero of $f(x)$, then $x - r$ is a factor of $f(x)$.

Thus in Example 1(b), since 3 is a zero of $f(x)$, then $x - 3$ must be a factor of $f(x)$.

From what we have said, it should be clear that there is another way to find zeros of a function $f(x)$. First factor $f(x)$. Then find the values of x that make any of the factors 0. Each of these values is a zero of $f(x)$.

EXAMPLE 2

Find all zeros of $f(x) = 2x^3 - 2x^2 - 2x$.

We factor $f(x)$.

$$\begin{aligned} f(x) &= 2x^3 - 2x^2 - 2x \\ &= 2x(x^2 - x - 1). \end{aligned}$$

The factor x is zero when $x = 0$. Thus 0 is a zero of $f(x)$. Any other zero of $f(x)$ must be a zero of the factor $x^2 - x - 1$. Let's use the quadratic formula on $x^2 - x - 1 = 0$.

$$x^2 - x - 1 = 0$$

$$x = \frac{-(-1) \pm \sqrt{(-1)^2 - 4(1)(-1)}}{2(1)} = \frac{1 \pm \sqrt{5}}{2}.$$

Thus the zeros of $f(x)$ are 0, $\frac{1 + \sqrt{5}}{2}$, and $\frac{1 - \sqrt{5}}{2}$.

In this chapter the only zeros we'll consider are those that are real numbers. They are called *real zeros*.

EXAMPLE 3

Find all real zeros of $g(x) = x^4 - 1$.

$$\begin{aligned} g(x) = x^4 - 1 &= (x^2 - 1)(x^2 + 1) \\ &= (x - 1)(x + 1)(x^2 + 1). \end{aligned}$$

The factors $x - 1$ and $x + 1$ give rise to the zeros 1 and -1. Any other zero of $g(x)$ is a zero of the factor $x^2 + 1$. But $x^2 + 1$ is never 0 if x is a real number.† Thus the only real zeros of $g(x)$ are 1 and -1.

There is a connection between the real zeros and the graph of a function $f(x)$. The real zeros are the first coordinates of the points where the graph

†Setting $x^2 + 1 = 0$ gives $x = \pm i$.

touches the x-axis. To illustrate, Fig. 10-1 shows graphically that the real zeros

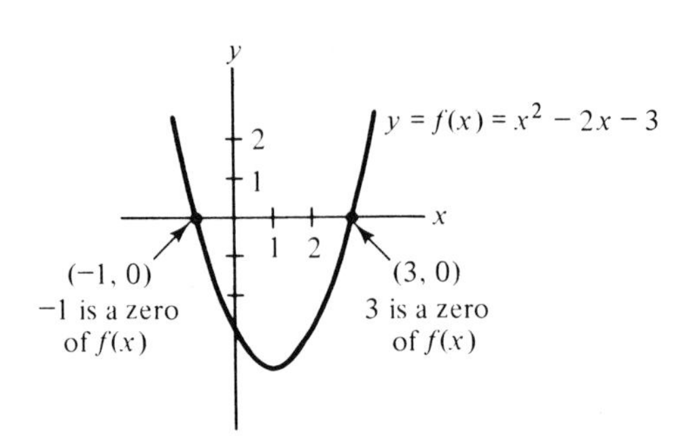

FIG. 10-1

of $f(x) = x^2 - 2x - 3 = (x - 3)(x + 1)$ are 3 and -1.

Completion Set 10-1

In Problems **1–6**, *fill in the blanks. Assume that $f(x)$ is a polynomial function.*

1. If r is a number and $f(r) = 0$, then r is called a ____________ of $f(x)$.

2. If $x - 3$ is a factor of $f(x)$, then ____ is a zero of $f(x)$.

3. If $x + 4$ is a factor of $f(x)$, then ______ is a zero of $f(x)$.

4. If x is a factor of $f(x)$, then ____ is a zero of $f(x)$.

5. If -2 is a zero of $f(x)$, then ______________ is a factor of $f(x)$.

6. The zeros of $f(x) = x^2(x - 5)(x + 7)^2$ are ____, ____, and ____.

Problem Set 10-1

In Problems **1–20**, *find all real zeros of the function.*

1. $f(x) = 3x - 7$

2. $f(x) = 4x - 4$

3. $f(x) = 8 - \dfrac{x}{3}$

4. $f(x) = \dfrac{2}{3}x + 3$

5. $g(x) = 8(x - 2)$

6. $f(x) = x^2 - x$

7. $f(x) = x^2 - 36$

8. $h(x) = x^2 + 2x$

9. $f(x) = (x^3 + x)(x + 2)$

10. $f(x) = (x + 1)^2(x - 2)^3$

11. $g(y) = 2y^4 - 32$

12. $f(x) = 3x^2 - 3x - 60$

13. $f(x) = 2x^2 + 3x - 5$

14. $f(z) = z(z + 2) + 1$

15. $f(x) = x^3 + 2x^2 - 3x$

16. $f(x) = x^4 + 2x^3 + x^2$

17. $f(x) = x^3 - 3x^2 + x$

18. $f(x) = (x^2 - 1)(x^2 - 3)$

19. $f(x) = x^2 + 8x^3 - 4x^4$

20. $f(x) = (x^2 + 1)(x^2 - 1)(x^2 + 2)$

21. In a calculus class a student had to find zeros of the following two functions:

a. $f'(x) = 24x^3 - 24x^2$
b. $f''(x) = 72x^2 - 48x.$

Here $f'(x)$ is read "f prime of x" and $f''(x)$ is read "f double prime of x." Find the zeros of these functions.

10-2 RATIONAL ZEROS

From the last section you should know that saying that $x - r$ is a factor of $f(x)$ and saying that r is a zero of $f(x)$ really mean the same thing. With these facts you may simplify the work of finding zeros.

Suppose that we want to determine whether $\frac{1}{4}$ is a zero of

$$f(x) = 4x^3 + 7x^2 + 2x - 1.$$

We could compute $f(\frac{1}{4})$ and see if $f(\frac{1}{4}) = 0$, but rather than do that, let's see if $x - \frac{1}{4}$ is a factor of $f(x)$. We'll use synthetic division to divide $f(x)$ by $x - \frac{1}{4}$. If the remainder is 0, then the divisor $x - \frac{1}{4}$ is a factor of the dividend $f(x)$.†

$$\begin{array}{cccc|c} 4 & 7 & 2 & -1 & \frac{1}{4} \\ \downarrow & 1 & 2 & 1 & \\ \hline 4 & 8 & 4 & 0 & \end{array}$$

The remainder is 0. This means that $x - \frac{1}{4}$ is a factor of $f(x)$, and so $\frac{1}{4}$ is a zero of $f(x)$.

---

†We talked about this in Section 4 of Chapter 5.

Our choice of trying $\frac{1}{4}$ was not by chance. There is a rule that gives information about *possible* rational† zeros of certain functions.

RATIONAL ZERO RULE

Suppose that $f(x)$ is a polynomial function whose coefficients are *integers*:

$$f(x) = a_n x^n + a_{n-1}x^{n-1} + \ldots + a_1 x + a_0.$$

If $\dfrac{p}{q}$ is a rational zero of $f(x)$, then p is a factor of a_0, and q is a factor of a_n.

The rational zero rule says that the only rational numbers that can be zeros of $f(x)$ are those whose numerators are factors of the constant term a_0 and whose denominators are factors of the leading coefficient a_n.

For the function above, $f(x) = 4x^3 + 7x^2 + 2x - 1$, we saw that $\frac{1}{4}$ is a rational zero. Notice that the numerator 1 is a factor of the constant term -1, and the denominator 4 is a factor of the leading coefficient 4.

EXAMPLE 1

Apply the rational zero rule to $f(x) = 2x^3 - 5x^2 - 2x + 8$.

The coefficients of $f(x)$ are integers, so the rule applies. The numerator p of a rational zero must be a factor of the constant term 8.

Possible values of p: $\pm 1, \pm 2, \pm 4, \pm 8$.

The denominator q must be a factor of the leading coefficient 2.

Possible values of q: $\pm 1, \pm 2$.

We shall display the possible values of p and q as follows:

p	± 1	± 2	± 4	± 8
q	± 1	± 2		

Thus the only possible rational zeros of $f(x)$ are

$$\pm\frac{1}{1}, \pm\frac{1}{2}, \pm\frac{2}{1}, \pm\frac{2}{2}, \pm\frac{4}{1}, \pm\frac{4}{2}, \pm\frac{8}{1}, \pm\frac{8}{2},$$

†Remember that a rational number is one that can be written as a quotient of two integers, such that the denominator is not 0. See page 28. In the rule we'll assume that the rational numbers are in lowest terms.

or (leaving out duplications)

$$1, \frac{1}{2}, 2, 4, 8, -1, -\frac{1}{2}, -2, -4, -8.$$

The rational zero rule does not tell us that $f(x)$ has rational zeros. It just gives those rational numbers that are ***candidates*** *for zeros. Sometimes none of these turns out to be a zero.*

EXAMPLE 2

Apply the rational zero rule to

a. $f(x) = 6x^4 - x^3 - 5.$

All coefficients are integers, so the rule applies.

Possible numerators p are factors of -5, and possible denominators q are factors of 6.

p	± 1	± 5		
q	± 1	± 2	± 3	± 6

Possible rational zeros:

$$\pm 1, \pm \frac{1}{2}, \pm \frac{1}{3}, \pm \frac{1}{6}, \pm 5, \pm \frac{5}{2}, \pm \frac{5}{3}, \pm \frac{5}{6}.$$

b. $f(x) = x^4 - 5x^3 + 7x - 4.$

The constant term is -4 and the leading coefficient is 1.

p	± 1	± 2	± 4
q	± 1		

Possible rational zeros: $\pm 1, \pm 2, \pm 4.$

Example 2(b) shows that when the leading coefficient of $f(x)$ is 1 (or -1) and all coefficients are integers, the only possible rational zeros must be integers. This is because the denominator of any possible rational zero must be 1 (or -1).

Once the possible rational zeros of $f(x)$ are found, using synthetic division we may test each one to see if it really is a zero.

The nice thing about the rational zero rule is that it gives you a starting point to find zeros of $f(x)$ when the factors of $f(x)$ are not obvious.

EXAMPLE 3

Find all real zeros of $f(x) = 2x^3 - x^2 - 6x + 3$.

Possible numerators p and denominators q of rational zeros are given by

p	± 1	± 3
q	± 1	± 2

Possible rational zeros: ± 1, ± 3, $\pm \frac{1}{2}$, $\pm \frac{3}{2}$. Let's see if 1 is a zero.

$$\begin{array}{rrrr|l} 2 & -1 & -6 & 3 & 1 \\ \downarrow & 2 & 1 & -5 & \\ \hline 2 & 1 & -5 & -2 & \end{array} \quad \leftarrow \text{Remainder is not } 0$$

No luck! Let's try 3.

$$\begin{array}{rrrr|l} 2 & -1 & -6 & 3 & 3 \\ \downarrow & 6 & 15 & 27 & \\ \hline 2 & 5 & 9 & 30 & \end{array} \quad \leftarrow \text{Remainder is not } 0$$

Again no luck! We'll try $\frac{1}{2}$.

$$\begin{array}{rrrr|l} 2 & -1 & -6 & 3 & \frac{1}{2} \\ \downarrow & 1 & 0 & -3 & \\ \hline 2 & 0 & -6 & 0 & \end{array}$$

Here the remainder is 0. Thus $x - \frac{1}{2}$ is a factor of $f(x)$, so $\frac{1}{2}$ is a zero of $f(x)$. Although we could now test the other possible rational zeros of $f(x)$, we'll use another approach. From our synthetic division we see that the result of dividing $f(x)$ by $x - \frac{1}{2}$ is $2x^2 - 6$:

$$\frac{f(x)}{x - \frac{1}{2}} = 2x^2 - 6.$$

Consequently,

$$f(x) = \left(x - \tfrac{1}{2}\right)(2x^2 - 6).$$

Any other zeros of $f(x)$ must be zeros of the function $2x^2 - 6$. This function has a degree lower than that of $f(x)$ and is called a **depressed function** of f. Since $2x^2 - 6$ is quadratic, we can easily find its zeros.

$$2x^2 - 6 = 0, \qquad 2x^2 = 6, \qquad x^2 = 3, \qquad x = \pm\sqrt{3}\,.$$

Thus the zeros of $f(x)$ are $\frac{1}{2}$, $\pm\sqrt{3}$. If we had tested the other possible rational zeros, they would not have worked. Thus, the only way the irrational zeros $\pm\sqrt{3}$ could turn up was by solving the equation $2x^2 - 6 = 0$.

As shown in Example 3, after finding a zero of a function, we'll examine the resulting depressed function for other zeros.

EXAMPLE 4

Find all real zeros of $f(x) = x^4 + 5x^3 + 12x^2 + 13x + 5$.

Possible rational zeros: ± 1, ± 5.

By trial and error we find that 1 and 5 are not zeros. Let's try -1 and -5.

$$\begin{array}{ccccc|l} 1 & 5 & 12 & 13 & 5 & -1 \\ \downarrow & -1 & -4 & -8 & -5 & \\ \hline 1 & 4 & 8 & 5 & 0 & \end{array}$$

Thus, -1 is a zero of $f(x)$ and

$$f(x) = (x + 1)(x^3 + 4x^2 + 8x + 5).$$

Any other zeros of $f(x)$ must be zeros of the depressed function

$$g(x) = x^3 + 4x^2 + 8x + 5.$$

Possible rational zeros of $g(x)$ are ± 1, ± 5. But we ruled out 1 and 5 before. If we try -5, we find that it is not a zero. We'll try -1 *again*.

$$\begin{array}{cccc|l} 1 & 4 & 8 & 5 & -1 \\ \downarrow & -1 & -3 & -5 & \\ \hline 1 & 3 & 5 & 0 & \end{array}$$

Thus -1 is a zero and

$$\begin{aligned} f(x) &= (x + 1)g(x) \\ &= (x + 1)(x + 1)(x^2 + 3x + 5) \\ &= (x + 1)^2(x^2 + 3x + 5). \end{aligned}$$

Notice that -1 is a repeated zero of $f(x)$. Any more zeros of $f(x)$ will be zeros of the depressed function $h(x) = x^2 + 3x + 5$. Using the quadratic formula on $x^2 + 3x + 5 = 0$, we get the negative discriminant

$$3^2 - 4(1)(5) = -11,$$

and so $h(x)$ has no real zeros. Thus the only real zero of $f(x)$ is -1.

EXAMPLE 5

Find all real zeros of $f(x) = 3x^4 + x^3 + x^2 + x - 2$.

The possible rational zeros are ± 1, ± 2, $\pm \frac{1}{3}$, $\pm \frac{2}{3}$. We find that 1, 2, and $\frac{1}{3}$ fail, but $\frac{2}{3}$ works.

3	1	1	1	−2	$\lfloor \frac{2}{3}$
↓	2	2	2	2	
3	3	3	3	0	

Thus $\frac{2}{3}$ is a zero and

$$f(x) = \left(x - \tfrac{2}{3}\right)(3x^3 + 3x^2 + 3x + 3).$$

We can factor out a 3 from the second factor. This will make our work easier.

$$f(x) = \left(x - \tfrac{2}{3}\right)(3)(x^3 + x^2 + x + 1).$$

Possible rational zeros of $x^3 + x^2 + x + 1$ are ± 1. From before we know that 1 isn't a zero, so we try -1.

1	1	1	1	$\lfloor -1$
↓	−1	0	−1	
1	0	1	0	

Thus -1 is a zero and

$$f(x) = 3\left(x - \tfrac{2}{3}\right)(x + 1)(x^2 + 1).$$

Since $x^2 + 1$ has no real zeros, the only real zeros of $f(x)$ are $\frac{2}{3}$ and -1.

Sometimes you can go through the list of possible rational zeros of $f(x)$ and find that none works. This means that any zero would have to be irrational or imaginary. A graph of $f(x)$ would indicate the irrational zeros.

Also, if the constant term of $f(x)$ is 0, then you can always factor out some power of x and work with the other factor. In this case 0 is always a zero of $f(x)$. For example, to find zeros of $f(x) = 3x^4 - 10x^2 - 4x$ we first write

$$f(x) = x(3x^3 - 10x - 4).$$

Clearly 0 is a zero, and any others will be zeros of $3x^3 - 10x - 4$.

Completion Set 10-2

In Problems **1** *and* **2**, *fill in the blanks.*

1. Any rational zero of $f(x) = 8x^3 - 4x^2 + x + 3$ will have a numerator that is a factor of ____ and a denominator that is a factor of ____. Thus, the possibilities for the numerator are $\pm$ ______ and $\pm$ ______, and the possibilities for the denominator are $\pm$ ______, $\pm$ ______, $\pm$ ______, and $\pm$ ______.

2. If $f(x) = x^4 - 26x - 3$, the denominator of any rational zero must be $\pm$ ______. This means that any rational zero must also be what kind of number? ____________. The possible rational zeros of $f(x)$ are $\pm$ ______ and $\pm$ ______.

Problem Set 10-2

In Problems **1–8**, *determine whether the given number is a zero of the given function. Use synthetic division.*

1. 2; $f(x) = x^3 - 6x^2 + 12x - 8$

2. $-\frac{1}{2}$; $f(x) = 4x^3 + 4x^2 - 7x + 2$

3. -3; $f(x) = x^3 - 5x - 24$

4. -4; $f(x) = x^3 + 2x^2 - 3x + 20$

5. $-\frac{1}{3}$; $f(x) = 3x^3 - 2x^2 - 7x - 2$

6. 3; $f(x) = 2x^3 - x^2 - 45$

7. $\frac{1}{4}$; $f(x) = -8x^4 - 2x^3 - 15x^2 - 4x + 2$

8. $\frac{3}{2}$; $f(x) = 2x^4 - x^3 + x^2 - 8x + 3$

In Problems **9–14**, *apply the rational zero rule and make a list of all possible rational zeros.*

9. $f(x) = 4x^3 - 8x + 3$

10. $f(x) = x^3 - 6x^2 + 8x + 10$

11. $f(x) = x^4 - 3x^3 + 2x - 7$

12. $f(x) = 5x^4 - x^2 + 10$

13. $f(x) = 9x^3 + 6x + 1$

14. $f(x) = 8x^3 - 6x^2 + 12x - 9$

In Problems **15–32**, *find all real zeros of* $f(x)$.

15. $f(x) = 4x^3 - 4x^2 + x - 1$

16. $f(x) = 2x^3 - x^2 - 45$

17. $f(x) = 2x^3 + 3x^2 - 1$

18. $f(x) = x^3 - 3x^2 + 4x - 4$

19. $f(x) = x^3 + x^2 - 8x - 12$

20. $f(x) = x^3 + 2x^2 - 3x + 20$

21. $f(x) = 3x^3 + x^2 - 5x + 2$

22. $f(x) = x^3 - 4x^2 + x + 6$

23. $f(x) = x^3 + x^2 - 5x - 5$

24. $f(x) = 2x^3 - 3x^2 + 4x + 3$

25. $f(x) = 2x^3 - 3x^2 - 5x + 3$

26. $f(x) = 3x^3 - 10x - 4$

27. $f(x) = x^4 + x^3 - 7x^2 - 13x - 6$

28. $f(x) = x^3 - 6x - 4$

29. $f(x) = 8x^4 + 2x^3 + 15x^2 + 4x - 2$

30. $f(x) = x^4 - 4x^3 + 5x^2 - 4x + 4$

31. $f(x) = 2x^4 - x^3 - 9x^2 + 4x + 4$

32. $f(x) = 5x^4 + 2x^3 - 10x^2 - 4x$

33. In a calculus class a student had to find the zeros of the following two functions:

a. $f'(x) = x^3 - 3x + 2$

b. $f''(x) = 3x^2 - 3$.

Here $f'(x)$ is read "f prime of x" and $f''(x)$ is read "f double prime of x." Find the zeros of these functions.

10-3 REVIEW

IMPORTANT TERMS

polynomial function *(p. 243)*

zero of function *(p. 243)*

real zero *(p. 245)*

rational zero rule *(p. 248)*

depressed function *(p. 250)*

REVIEW PROBLEMS

In Problems **1–8**, *find all real zeros of* $f(x)$.

1. $f(x) = x^3 + 5x^2 - 8x - 12$

2. $f(x) = x^3 - 6x^2 + 12x - 8$

3. $f(x) = 2x^4 - 9x^3 - 3x^2 - 9x - 5$

4. $f(x) = x^4 + 4x^3 + 5x^2 + 14x + 24$

5. $f(x) = 3x^3 + 2x^2 - 21x - 14$

6. $f(x) = 5x^3 - 2x^2 + 10x - 4$

7. $f(x) = 16x^3 - 28x^2 + 16x - 3$

8. $f(x) = x^5 - 6x^3 + 5x$

CHAPTER 11

STRAIGHT LINES

11-1 SLOPE OF A LINE AND POINT-SLOPE FORM

Some of the graphs you drew in Chapter 9 were straight lines. One feature of a straight line is its "steepness." For example, in Fig. 11-1 line L_1 rises faster as it

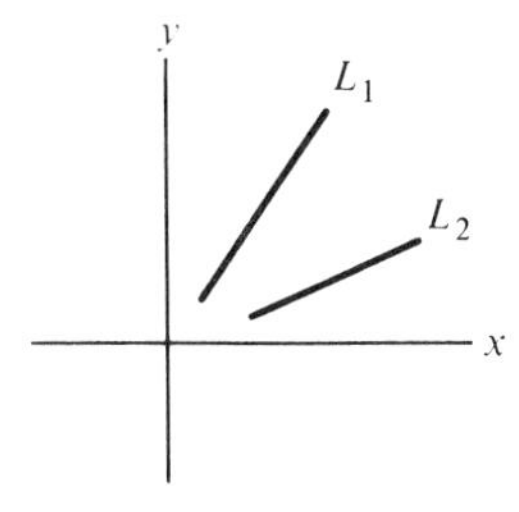

FIG. 11-1

goes from left to right than line L_2. We say that L_1 is steeper.

There's a way to measure steepness of a line. We choose two points on the line and see how the y-values change as the x-values change. Look at Fig. 11-2.

As x changes from 2 to 4 (an increase of two units), notice that y changes from 1

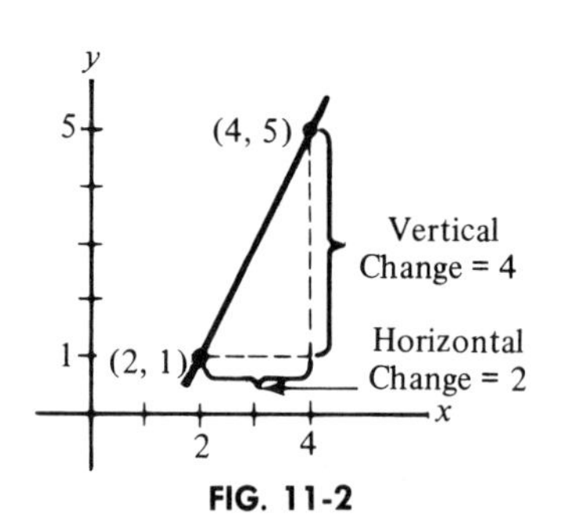

FIG. 11-2

to 5 (an increase of four units). Let's find how y changes for each one-unit increase in x. To do this, we shall divide the change in y (vertical change) by the change in x (horizontal change):

$$\frac{\text{change in } y}{\text{change in } x} = \frac{5-1}{4-2} = \frac{4}{2} = \frac{2}{1} = 2.$$

This means that for each one-unit increase in x, there is a two-unit increase in y. We say that the *slope* of the line is 2.†

SLOPE OF A LINE

Suppose that (x_1, y_1) and (x_2, y_2) are two different points on a line. Then the **slope** m of the line is the number

$$m = \frac{y_2 - y_1}{x_2 - x_1} \left(= \frac{\text{change in } y}{\text{change in } x}\right).$$

EXAMPLE 1

Find the slope of the line passing through (3, 4) *and* (5, 7).

The line is shown in Fig. 11-3. In the slope formula, let's choose (3, 4) as (x_1, y_1) and

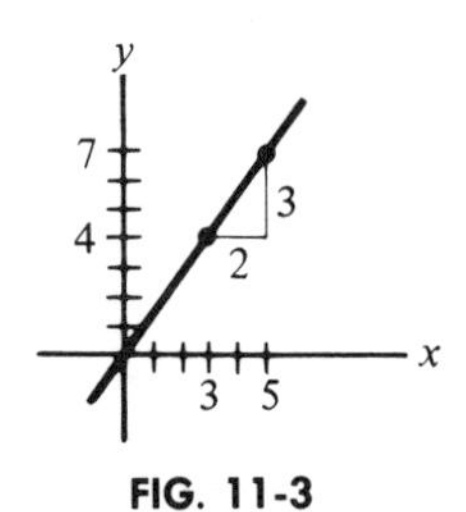

FIG. 11-3

†Choosing two other different points on this line would also give a slope of 2.

(5, 7) as (x_2, y_2). Then the slope is

$$m = \frac{y_2 - y_1}{x_2 - x_1} = \frac{7 - 4}{5 - 3} = \frac{3}{2}.$$

This means that if x increases by one unit, then y increases by $\frac{3}{2}$ units. In other words, if x increases by two units, then y increases by three units. It doesn't matter which point is picked as (x_1, y_1). If we choose (5, 7) as (x_1, y_1) and (3, 4) as (x_2, y_2), then

$$m = \frac{y_2 - y_1}{x_2 - x_1} = \frac{4 - 7}{3 - 5} = \frac{-3}{-2} = \frac{3}{2} \text{ as before}.$$

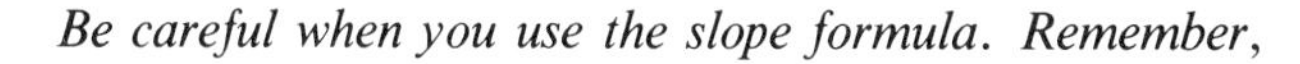

Be careful when you use the slope formula. Remember,

$$m = \frac{y_2 - y_1}{x_2 - x_1}. \qquad \textbf{Don't write } m = \frac{y_2 - y_1}{x_1 - x_2}.$$

Always align the subscripts correctly.

EXAMPLE 2

Find the slope of the line through (4, 5) *and* (6, 1).

Let $(4, 5) = (x_1, y_1)$ and $(6, 1) = (x_2, y_2)$. Then

$$m = \frac{y_2 - y_1}{x_2 - x_1} = \frac{1 - 5}{6 - 4} = \frac{-4}{2} = -2.$$

Here the slope is negative, -2. This means that if x increases one unit, then y *decreases* two units. Thus the line must *fall* from left to right (see Fig. 11-4).

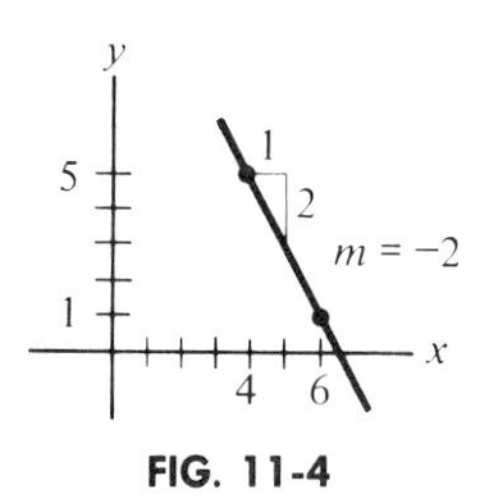

FIG. 11-4

From Examples 1 and 2 we see that the slope tells us about the "rising" or "falling" nature of a line.

Positive slope: line *rises* from left to right. (Fig. 11-3)
Negative slope: line *falls* from left to right. (Fig. 11-4)

EXAMPLE 3

a. The slope of the *horizontal* line through (3, 4) and (5, 4) is [see Fig. 11-5(a)]

$$m = \frac{y_2 - y_1}{x_2 - x_1} = \frac{4 - 4}{5 - 3} = \frac{0}{2} = 0.$$

This means that the change in y is 0 for any change in x. In fact, the slope of *every* horizontal line is 0.

Zero slope: horizontal line

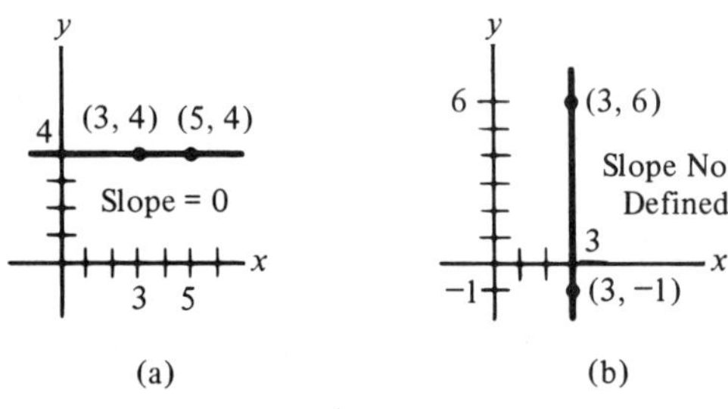

FIG. 11-5

b. If we use the slope formula on the *vertical* line through (3, − 1) and (3, 6) [see Fig. 11-5(b)], we get

$$m = \frac{y_2 - y_1}{x_2 - x_1} = \frac{6 - (-1)}{3 - 3} = \frac{7}{0}, \text{ which is not defined}.$$

Here there is no change in x at all. In fact, the slope of *every* vertical line is not defined.

No slope: vertical line

Don't confuse a line having slope 0 *with a line having no slope. Having no slope* ***does not*** *mean having a slope of* 0.

In Fig. 11-6 are lines with different slopes. Notice that **the closer the slope is**

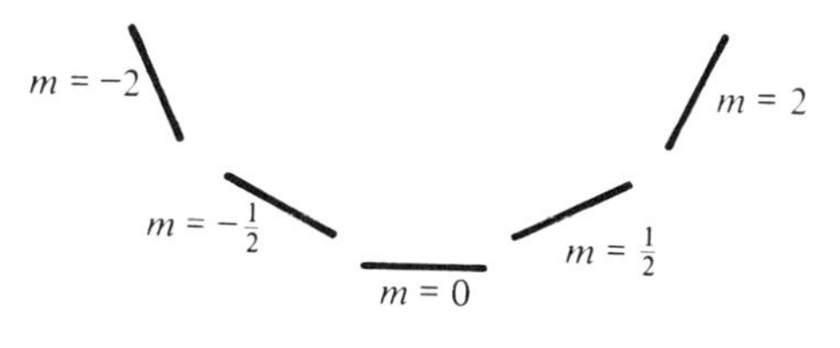

FIG. 11-6

to 0, the more horizontal is the line.

Now that you know what slope is, let's do something with it. Assume that you are given the slope of a line and the coordinates of one point on it. Then you can find an equation whose graph is that line. Here's how.

Suppose that the given line has a slope of 3 and the point (1, 2) lies on it. See Fig. 11-7. Let (x, y) be *any* other point on the line. Then by applying the

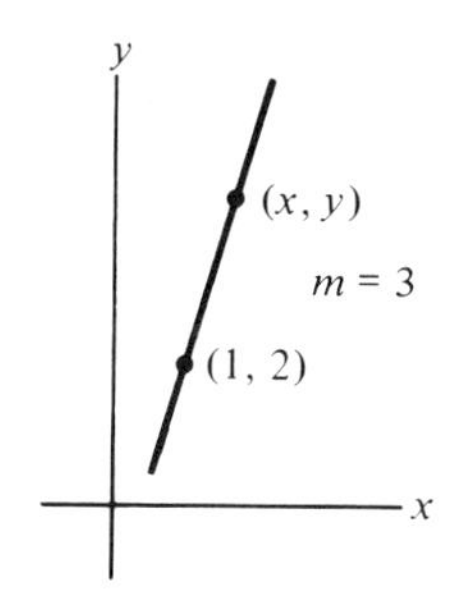

FIG. 11-7

slope formula to the points (1, 2) and (x, y), we have

$$\frac{y-2}{x-1} = 3.$$

Multiplying both sides by $x - 1$ gives

$$y - 2 = 3(x - 1). \tag{11-1}$$

This equation is called a *point-slope form* of an equation of the line.

We can write Eq. (11-1) as

$$\begin{aligned} y - 2 &= 3x - 3 \\ y &= 3x - 1. \end{aligned}$$

The coordinates of every point on the line make the equation $y = 3x - 1$ true. We say that these coordinates *satisfy* the equation. Also, every point whose coordinates satisfy $y = 3x - 1$ is on the line. For example, if $x = 2$, then $y = 3x - 1 = 3(2) - 1 = 5$. Thus (2, 5) is on the line.

Let's get more general. In the above discussion, replace 3 by m and (1, 2) by (x_1, y_1). Thus, in the manner of Eq. (11-1), we have:

$$y - y_1 = m(x - x_1)$$

is the **point-slope form** of an equation of the line passing through (x_1, y_1) and having slope m.

EXAMPLE 4

Find an equation of the line with slope -3 *which passes through* $(2, -1)$.

Using a point-slope form, we set $m = -3$ and $(x_1, y_1) = (2, -1)$.

$$y - y_1 = m(x - x_1)$$
$$y - (-1) = -3(x - 2)$$
$$y + 1 = -3x + 6.$$

We can write our answer as

$$y = -3x + 5. \qquad (11\text{-}2)$$

If we want to sketch the line, we need one more point besides $(2, -1)$. If $x = 0$, then from Eq. (11-2) we get $y = -3(0) + 5 = 5$. See Fig. 11-8.

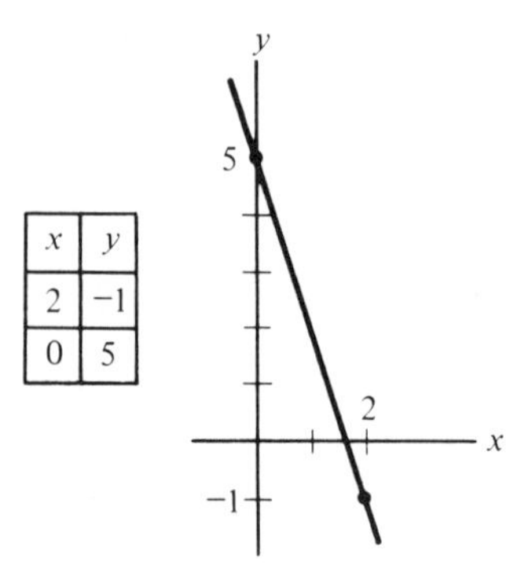

x	y
2	-1
0	5

FIG. 11-8

You can also find an equation of a line if you know only two points on it. First find the slope. Then use a point-slope form with either point as (x_1, y_1).

EXAMPLE 5

Find an equation of the line through (4, −2) *and* (−3, 8).

We first find the slope. Let $(x_1, y_1) = (4, -2)$ and $(x_2, y_2) = (-3, 8)$.

$$m = \frac{y_2 - y_1}{x_2 - x_1} = \frac{8 - (-2)}{-3 - 4} = -\frac{10}{7}.$$

Using a point-slope form with (4, −2) as (x_1, y_1) and $m = -\frac{10}{7}$, we obtain

$$y - y_1 = m(x - x_1)$$
$$y - (-2) = -\frac{10}{7}(x - 4)$$
$$y + 2 = -\frac{10}{7}x + \frac{40}{7}.$$

Simplifying gives

$$y = -\frac{10}{7}x + \frac{26}{7},$$

which is an equation of the line. If we had chosen (−3, 8) as (x_1, y_1), we would get

$$y - 8 = -\frac{10}{7}(x + 3).$$

This also simplifies to $y = -\frac{10}{7}x + \frac{26}{7}$.

Completion Set 11-1

In Problems **1–8**, *fill in the blanks.*

1. Does the line in Fig. 11-9 have a positive slope or a negative slope? ______________

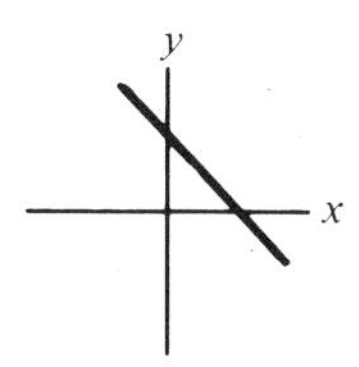

FIG. 11-9

2. The slope of every ______________ line is 0, but the slope of every ______________ line is not defined.

3. The slope of the line passing through (6, − 4) and (7, 9) is $\dfrac{9 - (\quad)}{7 - (\quad)}$ = ______.

4. If the points (3, 4) and (8, 6) lie on a line, then the slope of the line equals ____.

5. Insert *parallel* or *perpendicular*. A line whose slope is 0 is ______________ to the x-axis.

6. Insert *rises* or *falls*. If a line has a positive slope, then the line __________ from left to right.

7. The equation $y - y_1 = m(x - x_1)$ is called a point-__________ form of an equation of a line.

8. A point-slope form of the line through (1, −2) having slope 4 is $y - (-2) =$ ______________.

Problem Set 11-1

In Problems **1–8**, *find the slope of the line passing through the given points.*

1. (6, 3), (8, 6)

2. (−3, 4), (0, 1)

3. (−2, 2), (3, − 2)

4. (2, − 4), (3, − 4)

5. (−2, 4), (−2, 8)

6. (0, 0), (−3, 4)

7. (3, 6), (−1, 6)

8. (0, 6), (1, 1)

In Problems **9–16**, *find an equation of the line that has the given properties.*

9. Passes through (1, 1) and has slope 1.

10. Passes through (2, 4) and has slope 3.

11. Passes through (−2, 5) and has slope $-\frac{1}{4}$.

12. Passes through (3, 5) and has slope 0.

13. Passes through (3, −6) and has slope $\frac{1}{3}$.

14. Has slope $\frac{3}{5}$ and passes through (0, 2).

15. Has slope -5 and passes through origin.

16. Has slope $\frac{1}{2}$ and passes through $(-\frac{1}{2}, -\frac{3}{2})$.

In Problems **17–24**, *find an equation of the line passing through the given points.*

17. (2, 4), (8, 7)

18. (2, -5), (3, 4)

19. (0, 0), (5, -5)

20. (-2, 5), (3, 5)

21. (3, -1), (2, -9)

22. (2, 3), (0, 0)

23. (2, -7), (3, -7)

24. (-1, -1), (-3, -3)

25. A straight line passes through (1, 2) and (-3, 8). Find the point on it that has a first coordinate of 5.

26. A straight line has slope -3 and passes through (4, -1). Find the point on it that has a second coordinate of -2.

11-2 MORE EQUATIONS OF STRAIGHT LINES

In the last section we looked at a point-slope form of an equation of a line. We'll look at other forms in this section.

In Fig. 11-10 the line crosses the y-axis at $(0, b)$. The number b is called the

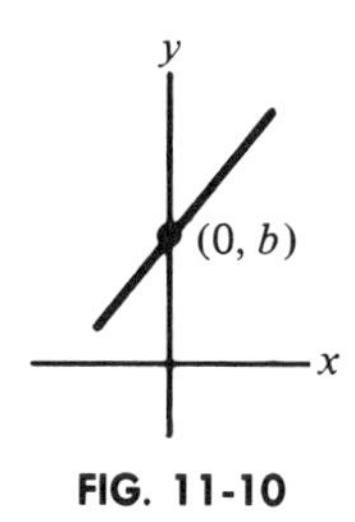

FIG. 11-10

***y*-intercept** of the line.

Suppose that you know the slope m and y-intercept b of a line. Then $(0, b)$ is on the line and, using a point-slope form, we get

$$y - y_1 = m(x - x_1)$$
$$y - b = m(x - 0).$$

If we solve for y, we'll get an important form of an equation of the line.

$$y - b = mx$$
$$y = mx + b.$$

We say that

> $$y = mx + b$$
>
> is the **slope-intercept** form of an equation of the line with slope m and y-intercept b.

EXAMPLE 1

Find an equation of the line with slope 3 and y-intercept 4.

$$y = mx + b$$
$$y = 3x + 4.$$

EXAMPLE 2

Discuss and draw the graph of $y = -\frac{2}{3}x + 7$.

The form of $y = -\frac{2}{3}x + 7$ is $y = mx + b$ with $m = -\frac{2}{3}$ and $b = 7$. Thus the graph is a straight line with slope $-\frac{2}{3}$ and y-intercept 7. To draw the graph. all we really need is two points on it. See Fig. 11-11. Note that we used the y-intercept to

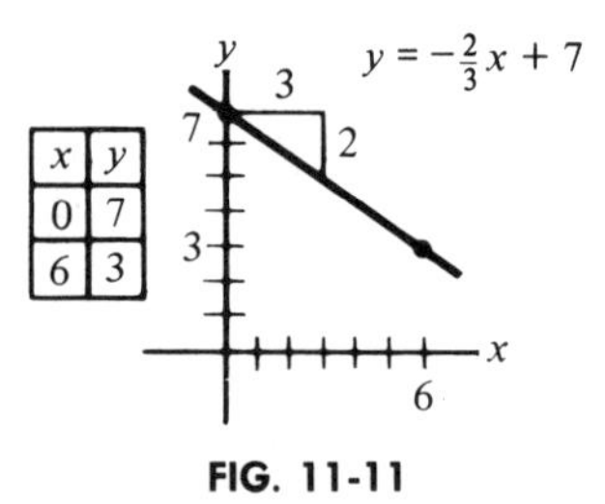

FIG. 11-11

get one of them. Observe that since the slope is $-\frac{2}{3}$, then as x increases by three units, y *decreases* by two units.

Here's an important note. The function $f(x) = mx + b$ is called a **linear function** (it has degree 1). If $y = f(x)$, then $y = mx + b$ and so

the graph of a linear function is a straight line.

EXAMPLE 3

Graph $f(x) = 2x - 1$.

Since f is a linear function, its graph *is a straight line*. Thus we need to plot only two points to draw the line. See Fig. 11-12.

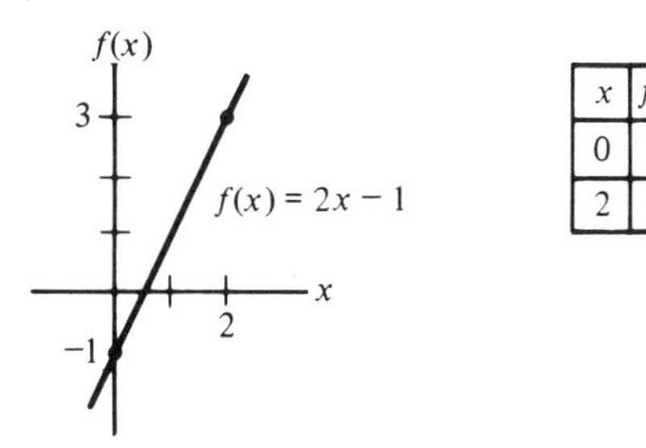

x	$f(x)$
0	−1
2	3

FIG. 11-12

Let's turn to horizontal and vertical lines. In Fig. 11-13(a), look at the

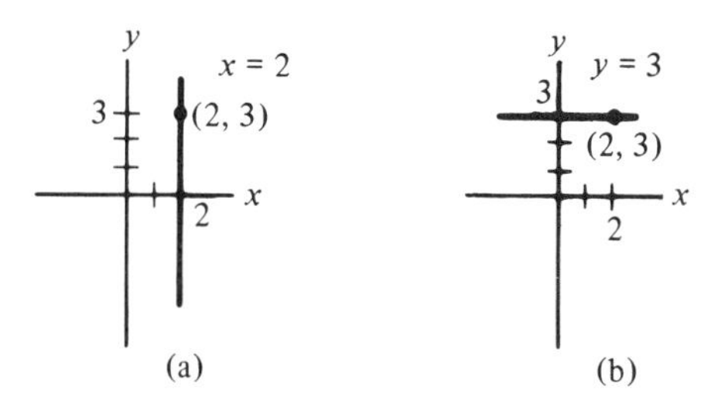

FIG. 11-13

vertical line through (2, 3). This line consists of all points (x, y) where $x = 2$. Thus an equation of the line is $x = 2$. More generally,

> An equation of the *vertical* line through (a, b) is
>
> $$x = a.$$

Similarly, in Fig. 11-13(b), an equation for the horizontal line through (2, 3) is $y = 3$. In general,

An equation of the *horizontal* line through (a, b) is $y = b$.

It turns out that every straight line is the graph of an equation that can be put in the form

$$Ax + By + C = 0,$$

where A and B are not both zero. This is called a **general linear equation in x and y.** We say that x and y are *linearly related*. For example, we can get a general linear equation for the line $y = 7x - 2$:

$$y = 7x - 2$$
$$-7x + y + 2 = 0$$
$$(-7)x + (1)y + (2) = 0 \qquad [A = -7, B = 1, C = 2].$$

On the other hand, the graph of every general linear equation is a straight line. For example,

$$3x + 4y - 5 = 0$$

can be solved for y:

$$3x + 4y - 5 = 0$$
$$4y = -3x + 5$$
$$y = -\frac{3}{4}x + \frac{5}{4}.$$

This is the slope-intercept form of a line ($m = -\frac{3}{4}$, y-intercept $\frac{5}{4}$).

EXAMPLE 4

Find the slope of the line $16x + 8y - 3 = 0$.

We solve for y to get the slope-intercept form.

$$16x + 8y - 3 = 0$$
$$8y = -16x + 3$$
$$y = -2x + \frac{3}{8}.$$

The slope is -2.

EXAMPLE 5

Sketch the graph of $2x - 3y + 6 = 0$.

Method 1.

Since this is a general linear equation, its graph is a straight line. All we have to do is get two points on the graph. If $x = 0$, then $y = 2$. If $y = 0$, then $x = -3$. The graph is given in Fig. 11-14(a).

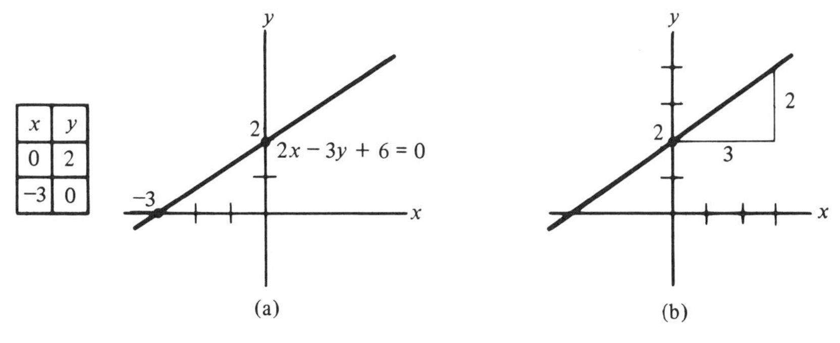

FIG. 11-14

Method 2.

Here we use the slope and y-intercept to sketch the line. First we solve for y to get the slope-intercept form.

$$\begin{aligned} 2x - 3y + 6 &= 0 \\ -3y &= -2x - 6 \\ y &= \frac{2}{3}x + 2. \end{aligned}$$

Hence the line intersects the y-axis at 2 and has slope $\frac{2}{3}$. Thus as x increases by three units, y increases by two units. Using this information, we sketch the graph as in Fig. 11-14(b).

EXAMPLE 6

Fahrenheit temperature F and celsius temperature C are linearly related. Use the facts that $32° F = 0° C$ *and* $212° F = 100° C$ *to find an equation that relates F and C. Also, find C when* $F = 50$.

Since F and C are linearly related, the graph of the equation is a straight line. In Fig. 11-15 we used F for the horizontal axis and C for the vertical. (We could have just as well reversed our choices.) When $F = 32$, then $C = 0$, and so $(32, 0)$ is on the line. Likewise $(212, 100)$ is on it.

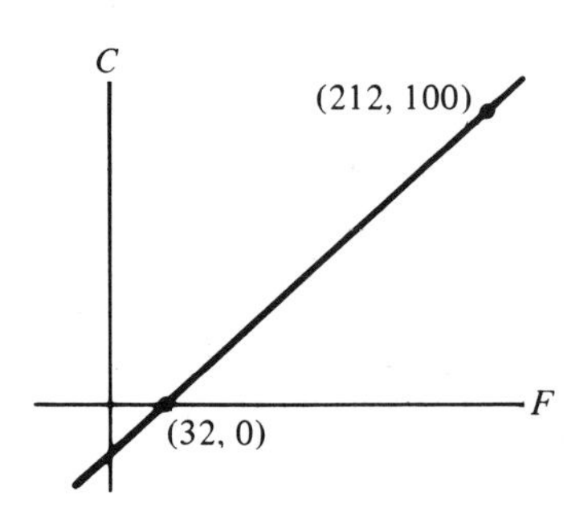

FIG. 11-15

The slope of the line is

$$\frac{100 - 0}{212 - 32} = \frac{100}{180} = \frac{5}{9}.$$

Using a point-slope form with the point (32, 0), we get

$$C - 0 = \frac{5}{9}(F - 32)$$

$$C = \frac{5}{9}(F - 32).$$

We can use this equation to find C when $F = 50$:

$$C = \frac{5}{9}(F - 32) = \frac{5}{9}(50 - 32) = \frac{5}{9}(18) = 10.$$

In the following table is a summary of forms of equations of straight lines.

FORMS OF EQUATIONS OF LINES	
Point-slope form	$y - y_1 = m(x - x_1)$
Slope-intercept form	$y = mx + b$
Vertical line	$x = a$
Horizontal line	$y = b$
General linear form	$Ax + By + C = 0$

Completion Set 11-2

In Problems **1–6**, *fill in the blanks.*

1. An equation of the line with slope 2 and y-intercept 7 is $y =$ ____ $x +$ ____.

2. The line $y = 7x + 3$ has a slope of ____.

3. An equation of the vertical line through (1, 2) is ____________, and an equation of the horizontal line through (1, 2) is ____________.

4. A general linear form of the line $y = -3x + 2$ is ____ $x +$ ____ $y - 2 = 0$.

5. The slope of the line $x = 3$ is ______________________.

6. The slope of the line $y = 7$ is ____.

Problem Set 11-2

In Problems **1–8**, *find an equation of the line having the given properties.*

1. Has slope 2 and y-intercept 4.

2. Has slope 7 and y-intercept -5.

3. Has slope $-\frac{1}{2}$ and y-intercept -3.

4. Has slope 0 and y-intercept $-\frac{1}{2}$.

5. Is horizontal and passes through $(-3, -2)$.

6. Is vertical and passes through $(-1, 4)$.

7. Passes through $(2, -3)$ and is vertical.

8. Passes through the origin and is horizontal.

In Problems **9–18**, *write each line in slope-intercept form and find the slope and y-intercept. Sketch each line.*

9. $y = 2x - 1$

10. $y = -3x + 2$

11. $2y = -6x + 4$

12. $y = x$

13. $y - 4x = 0$

14. $x = 2y + 1$

15. $x + 3y + 3 = 0$

16. $3x - 1 = 5y$

17. $y = 1$

18. $y - 2 = 3(x - 1)$

In Problems **19–22**, *find a general linear form for the given lines.*

19. $y = x - 6$

20. $4x = 5 - 9y$

21. $x = -2y + 2$

22. $\frac{x}{2} + 3 = \frac{y}{3}$

In Problems **23** *and* **24**, *sketch the graph of the given function.*

23. $f(x) = 3x - 2$

24. $f(x) = -2x$

25. A straight line has slope 2 and y-intercept 1. Does the point $(-1, -1)$ lie on the line?

26. Graph the lines $y = x + k$ for $k = -1$, 0, and 1.

27. Suppose that s and t are linearly related so that $s = 40$ when $t = 12$, and $s = 25$ when $t = 18$. Find an equation that relates s and t. Also, find s when $t = 24$.

28. The force F suspended on a spring and the stretch S of the spring that the force produces are linearly related. When $F = 1$ (lb), then $S = .4$ (in.). When $F = 4$, then $S = 1.6$. Find an equation relating F and S. Also, find the stretch when a force of 2 lb is suspended.

29. Suppose that the total cost to produce 10 units of a certain product is $40, and for 20 units the total cost is $70. Assume that total cost is linearly related to the number of units produced. If y represents the total cost of producing x units, find an equation relating x and y. Also, find the total cost of producing 35 units.

30. In a circuit the voltage V and current i are linearly related. When $i = 4$ (amperes), then $V = 2$ (volts); when $i = 12$, then $V = 6$. Find an equation relating V and i. Also, find the voltage when the current is 10.

31. If a ball is thrown straight up in the air with an initial velocity of 160 ft/sec, the velocity at the end of t seconds is given by the function $v(t) = 160 - 32t$. Sketch the graph for values of t between 0 and 5, including 0 and 5.

11-3 PARALLEL AND PERPENDICULAR LINES

You can tell when two lines are parallel or perpendicular by comparing their slopes.

> Two lines having slopes m_1 and m_2 are parallel if, and only if, $m_1 = m_2$.

EXAMPLE 1

Show that the lines $y = 2x - 3$ *and* $y = 2x + 2$ *are parallel.*

Both lines have a slope of 2, so they are parallel. See Fig. 11-16.

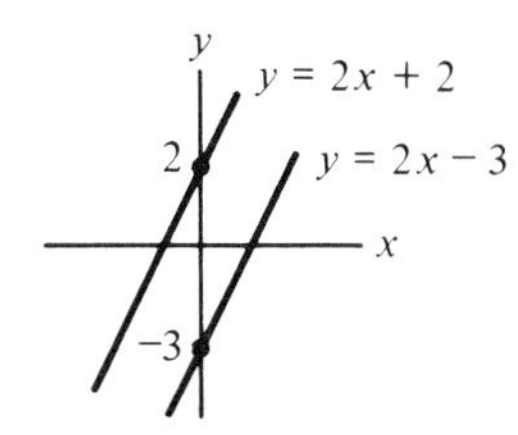

FIG. 11-16

Here's a rule for perpendicular lines.

> Two lines having slopes m_1 and m_2 are perpendicular if, and only if, $m_1 = -\dfrac{1}{m_2}$.

Thus, two lines are perpendicular when the slope of one line is the negative reciprocal of the slope of the other.

EXAMPLE 2

Show that the line $y = \frac{1}{2}x + 3$ *is perpendicular to the line* $y = -2x + 7$.

The line $y = \frac{1}{2}x + 3$ has slope $m_1 = \frac{1}{2}$, while $y = -2x + 7$ has slope $m_2 = -2$. Now, $\frac{1}{2}$ is the negative reciprocal of -2 [that is, $\frac{1}{2} = -\frac{1}{(-2)}$]. Thus the lines are perpendicular. See Fig. 11-17.

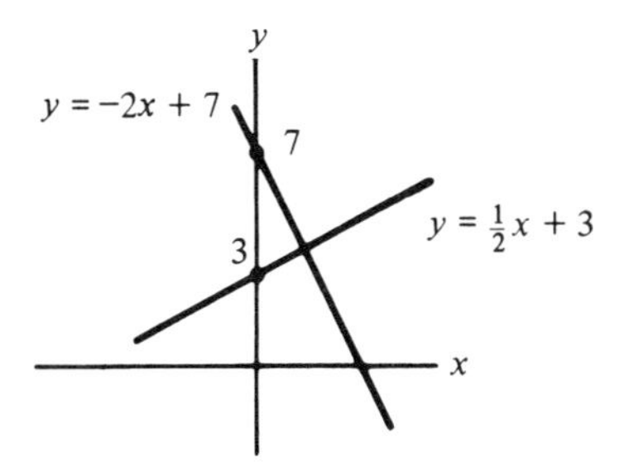

FIG. 11-17

EXAMPLE 3

Find an equation of the line that passes through $(3, -2)$ *and that is*

a. *parallel*

b. *perpendicular*

to the line $y = 3x + 1$. *See Fig.* 11-18.

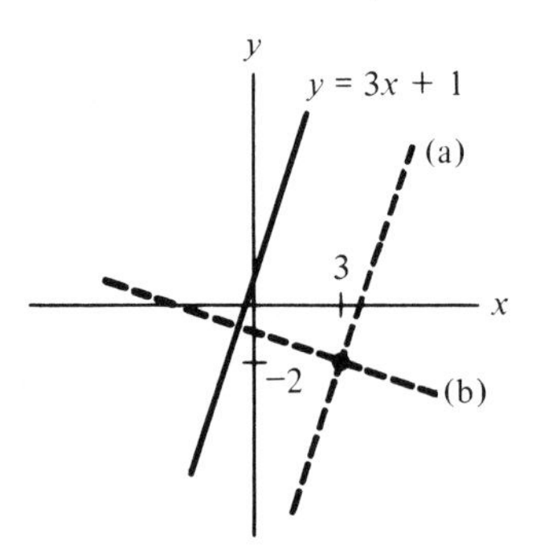

FIG. 11-18

a. The slope of $y = 3x + 1$ is 3. Thus the line through $(3, -2)$ that is *parallel* to $y = 3x + 1$ also has slope 3. Using a point-slope form, we get

$$\begin{aligned} y - (-2) &= 3(x - 3) \\ y + 2 &= 3x - 9 \\ y &= 3x - 11. \end{aligned}$$

b. The slope of a line *perpendicular* to $y = 3x + 1$ must be $-\frac{1}{3}$ (= the negative reciprocal of 3). Using a point-slope form, we get

$$\begin{aligned} y - (-2) &= -\tfrac{1}{3}(x - 3) \\ y + 2 &= -\tfrac{1}{3}x + 1 \\ y &= -\tfrac{1}{3}x - 1. \end{aligned}$$

For the two rules concerning parallel and perpendicular lines, we assumed that each line had a slope. But what happens with vertical lines (they have no slope)?

Two vertical lines are always parallel.

Any vertical line and any horizontal line are perpendicular.

EXAMPLE 4

a. The lines $x = 2$ and $x = 3$ are both vertical. Thus they are parallel. See Fig. 11-19(a).

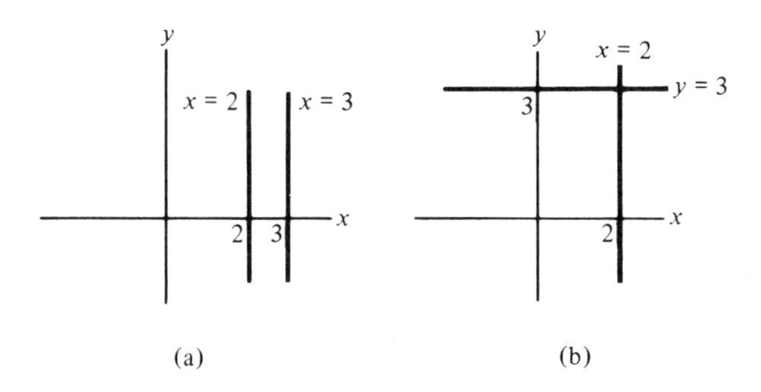

FIG. 11-19

b. The line $x = 2$ is vertical and the line $y = 3$ is horizontal. Thus the lines are perpendicular. See Fig. 11-19(b).

Completion Set 11-3

In Problems **1** *and* **2**, *fill in the blanks.*

1. The slope of $y = 2x + 3$ is ____, so any line parallel to this one must also have a slope of ____. But a line perpendicular to this one must have a slope of _____.

2. Insert *parallel* or *perpendicular*. The lines $y = x + 3$ and $y = -x + 3$ are ______________________________.

Problem Set 11-3

In Problems **1–10**, *determine if the lines are parallel, perpendicular, or neither.*

1. $y = 7x + 2, \quad y = 7x - 3$

2. $y = 4x + 3, \quad y = 5 + 4x$

3. $y = 5x + 2, \quad -5x + y - 3 = 0$

4. $y = x, \quad y = -x$

5. $x + 2y + 1 = 0, \quad y = 2x$

6. $x + 2y = 0, \quad x + y - 4 = 0$

7. $y = 3, \quad y = -\frac{1}{3}$

8. $x = 3, \quad x = -4$

9. $3x + y = 4, \quad 3x - y + 1 = 0$

10. $x - 1 = 0, \quad y = 0$

In Problems **11–20**, *find an equation of the line satisfying the given conditions. Give answer in slope-intercept form if possible.*

11. Passing through $(-1, 3)$ and parallel to $y = 4x - 5$.

12. Passing through $(2, -8)$ and parallel to $x = -4$.

13. Passing through $(2, 1)$ and parallel to $y = 2$.

14. Passing through $(3, -4)$ and parallel to $y = 3 + 2x$.

15. Perpendicular to $y = 3x - 5$ and passing through $(3, 4)$.

16. Perpendicular to $y = -4$ and passing through $(1, 1)$.

17. Passing through $(7, 4)$ and perpendicular to $y = -4$.

18. Passing through $(-5, 4)$ and perpendicular to $2y = -x + 1$.

19. Passing through $(-7, -5)$ and parallel to $2x + 3y + 6 = 0$.

20. Passing through $(-2, 1)$ and parallel to y-axis.

11-4 REVIEW

IMPORTANT TERMS

slope *(p. 256)*

y-intercept *(p. 263)*

point-slope form *(p. 260)*

slope-intercept form *(p. 264)*

perpendicular *(p. 271)*

parallel *(p. 270)*

linearly related *(p. 266)*

linear function *(p. 264)*

general linear equation *(p. 266)*

REVIEW PROBLEMS

In Problems **1–4**, *find the slope of the line passing through the given points.*

1. $(2, 3), (-1, 4)$

2. $(1, -1), (2, 3)$

3. $(2, 1), (5, 1)$

4. $(-2, 3), (3, -2)$

In Problems **5–18**, *find an equation of the line satisfying the given conditions. Give answer in slope-intercept form if possible.*

5. Passes through $(2, -3)$ and has slope -2.

6. Passes through $(-6, 2)$ and has slope $\frac{1}{3}$.

7. Passes through $(-2, 3)$ and $(4, 5)$.

8. Passes through $(1, -1)$ and $(3, 0)$.

9. Passes through $(-2, 2)$ and has slope 0.

10. Passes through $(1, 2)$ and is vertical.

11. Passes through the origin and is vertical.

12. Passes through $(4, -1)$ and is parallel to the x-axis.

13. Has slope 3 and y-intercept -4.

14. Has slope -1 and passes through $(0, 2)$.

15. Passes through $(1, 2)$ and is perpendicular to $-3y + 5x = 7$.

16. Passes through $(-2, 4)$ and is horizontal.

17. Parallel to $y = 3 - 5x$ and passes through $(1, 2)$.

18. Has slope 1 and passes through $(1, 0)$.

In Problems **19–24**, *determine if the lines are parallel, perpendicular, or neither.*

19. $x + 4y + 2 = 0,\ 8x - 2y - 2 = 0$

20. $y - 2 = 2(x - 1),\ 2x + 4y - 3 = 0$

21. $x - 3 = 2(y + 4),\ y = 4x + 2$

22. $3x + 5y + 4 = 0,\ 6x + 10y = 0$

23. $y = \frac{1}{2}x + 5,\ 2x = 4y - 3$

24. $y = 7x,\ y = 7$

In Problems **25** *and* **26**, *write the given line in slope-intercept form and a general linear form. Find the slope.*

25. $3x - 2y = 4$

26. $x = -3y + 4$

In Problems **27** *and* **28**, *graph the functions.*

27. $f(x) = 2x + 4$

28. $f(x) = 6 - 3x$

29. Suppose a and b are linearly related so that $a = 1$ when $b = 2$, and $a = 2$ when $b = 1$. Find a general linear form of an equation that relates a and b. Also, find a when $b = 3$.

CHAPTER 12

Systems of Linear Equations

12-1 METHODS OF ELIMINATION

The manager of a factory is setting up a production schedule for two models of a new product. Each takes one hour to make. The first model requires 4 widgets and 9 klunkers. The second requires 5 widgets and 14 klunkers. From its suppliers the factory gets 335 widgets and 850 klunkers each day. How many of each model should the manager plan to make each day if he wants to use all the widgets and klunkers?

Suppose we let x be the number of the first model made each day and y be the number of the second model. Then these require a total of $4x + 5y$ widgets and $9x + 14y$ klunkers. Since 335 widgets and 850 klunkers are available, we have

$$\begin{cases} 4x + 5y = 335 & (12\text{-}1) \\ 9x + 14y = 850. & (12\text{-}2) \end{cases}$$

We call this set of equations a **system** of two linear equations in the variables (or unknowns) x and y. The problem is to find values of x and y for which *both* equations are true. These are called *solutions* of the system.

Since Eqs. (12-1) and (12-2) are linear, their graphs are straight lines, say L_1 and L_2. Now, the coordinates of any point on a line satisfy the equation of that line. Thus the coordinates of any point of intersection of L_1 and L_2 will satisfy both equations. This means that a point of intersection will give a solution of the system.

If L_1 and L_2 are drawn on the same plane, they will appear in one of three ways:

1. L_1 and L_2 may meet at exactly one point, say (x_0, y_0). See Fig. 12-1(a). Thus the system has the solution $x = x_0$ and $y = y_0$.

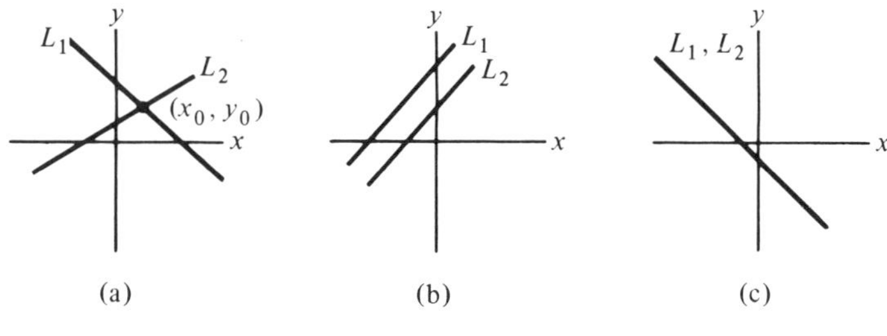

FIG. 12-1

2. L_1 and L_2 may be parallel and have no points in common [see Fig. 12-1(b)]. Thus there is no solution.
3. L_1 and L_2 may be the same line [see Fig. 12-1(c)]. Thus the coordinates of any point on the line are a solution of the system. Consequently there are infinitely many solutions.

Besides graphing, the system can be solved with algebra. To do this we perform operations on the equations until we get one with only one variable in it. That is, one of the variables is *eliminated*. After solving this equation we can easily get the value of the other variable.

Let's solve our widget and klunker problem by eliminating y from the system

$$\begin{cases} 4x + 5y = 335 & (12\text{-}1) \\ 9x + 14y = 850. & (12\text{-}2) \end{cases}$$

One way involves getting the coefficients of the y-terms in each equation to be the same except for sign. We can multiply Eq. (12-1) by 14 [that is, multiply both sides of Eq. (12-1) by 14], and multiply Eq. (12-2) by -5. This gives

$$\begin{cases} 56x + 70y = 4690 & (12\text{-}3) \\ -45x - 70y = -4250. & (12\text{-}4) \end{cases}$$

The left and right sides of Eq. (12-3) are equal, so each can be *added* to the corresponding side of Eq. (12-4). This gives

$$11x = 440,$$

which has only one variable, as planned. Solving this gives

$$x = 40.$$

To find y we'll replace x by 40 in either one of the *original* equations, say Eq. (12-1).

$$\begin{aligned} 4(40) + 5y &= 335 \\ 160 + 5y &= 335 \\ 5y &= 175 \\ y &= 35. \end{aligned}$$

We can check our answer by substituting $x = 40$ and $y = 35$ into *both* of the original equations. In Eq. (12-1) we get $4(40) + 5(35) = 335$, or $335 = 335$. In Eq. (12-2) we get $9(40) + 14(35) = 850$, or $850 = 850$. Thus the solution is

$$x = 40 \quad \text{and} \quad y = 35.$$

Each day the manager should plan to make 40 of the first model and 35 of the second.

The method we used is called **elimination by addition**. Although we chose to eliminate y, you could eliminate x. If you multiply Eq. (12-1) by 9 and Eq. (12-2) by -4, you get

$$\begin{cases} 36x + 45y = 3015 & (12\text{-}5) \\ -36x - 56y = -3400. & (12\text{-}6) \end{cases}$$

Adding Eq. (12-5) to Eq. (12-6) [that is, adding corresponding sides of Eq. (12-5) and Eq. (12-6)], you get

$$\begin{aligned} -11y &= -385 \\ y &= 35. \end{aligned}$$

Replacing y in Eq. (12-1) by 35 gives $x = 40$, as expected.

EXAMPLE 1

Use elimination by addition to solve the system

$$\begin{cases} 3x - 4y = 13 \\ 3y + 2x = 3. \end{cases}$$

Aligning the x- and y-terms gives

$$\begin{cases} 3x - 4y = 13 & (12\text{-}7) \\ 2x + 3y = 3. & (12\text{-}8) \end{cases}$$

Let's eliminate y. We multiply Eq. (12-7) by 3 and Eq. (12-8) by 4:

$$\begin{cases} 9x - 12y = 39 & (12\text{-}9) \\ 8x + 12y = 12. & (12\text{-}10) \end{cases}$$

Adding Eq. (12-9) to Eq. (12-10) gives

$$17x = 51$$

$$\boxed{x = \frac{51}{17} = 3.}$$

Replacing x by 3 in Eq. (12-7) gives

$$3(3) - 4y = 13$$

$$-4y = 4$$

$$\boxed{y = -1.}$$

The solution is $x = 3$ and $y = -1$. Figure 12-2 shows a picture of the system.

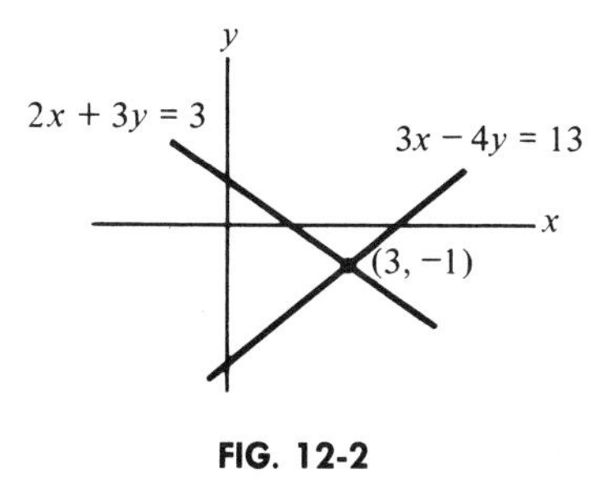

FIG. 12-2

There's another way to solve the system in Example 1:

$$\begin{cases} 3x - 4y = 13 & (12\text{-}11) \\ 3y + 2x = 3. & (12\text{-}12) \end{cases}$$

We first choose one of the equations and solve it for one variable in terms of the other. Let's solve Eq. (12-11) for x in terms of y:

$$\begin{aligned} 3x - 4y &= 13 \\ 3x &= 13 + 4y \\ x &= \frac{13}{3} + \frac{4}{3}y. \end{aligned} \tag{12-13}$$

Next we *substitute* the right side of Eq. (12-13) for x in Eq. (12-12):

$$3y + 2\left(\frac{13}{3} + \frac{4}{3}y\right) = 3. \tag{12-14}$$

Thus x has been eliminated. Solving Eq. (12-14), we have

$$\begin{aligned} 3y + \frac{26}{3} + \frac{8}{3}y &= 3 \\ 9y + 26 + 8y &= 9 \qquad \text{[clearing of fractions]} \\ 17y &= -17 \\ y &= -1. \end{aligned}$$

Replacing y in Eq. (12-13) by -1 gives $x = 3$, as before. This method is called **elimination by substitution**.

EXAMPLE 2

Use the method of elimination by substitution to solve

$$\begin{cases} 4x + 2y + 4 = 0 & (12\text{-}15) \\ 2x + y - 8 = 0. & (12\text{-}16) \end{cases}$$

The variable easiest to solve for is y in Eq. (12-16):

$$y = -2x + 8.$$

Substituting $-2x + 8$ for y in Eq. (12-15) gives

$$\begin{aligned} 4x + 2(-2x + 8) + 4 &= 0 \\ 4x - 4x + 16 + 4 &= 0 \\ 20 &= 0. \end{aligned} \tag{12-17}$$

Equation (12-17) is *never* true. When an impossible equation like this occurs, there is

no solution to the system. The reason in our case is clear from Fig. 12-3. The graphs

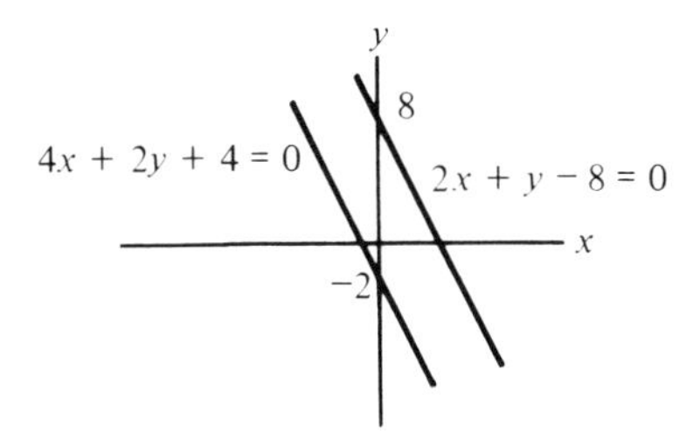

FIG. 12-3

of Eqs. (12-15) and (12-16) are different parallel lines.

EXAMPLE 3

Solve the system

$$\begin{cases} 2x + y = 1 & (12\text{-}18) \\ 4x + 2y = 2. & (12\text{-}19) \end{cases}$$

We'll eliminate y by addition. Multiplying Eq. (12-18) by -2 gives

$$\begin{cases} -4x - 2y = -2 & (12\text{-}20) \\ 4x + 2y = 2. & (12\text{-}21) \end{cases}$$

Adding Eq. (12-20) to Eq. (12-21), we have

$$0 = 0,$$

which is always true. We might have expected this result, since each term in Eq. (12-19) is two times the corresponding term in Eq. (12-18). Equation (12-19) is said to be a *multiple* of Eq. (12-18). This means that the graphs of Eqs. (12-18) and (12-19) are the same line (see Fig. 12-4). Thus, any solution to $2x + y = 1$ is a

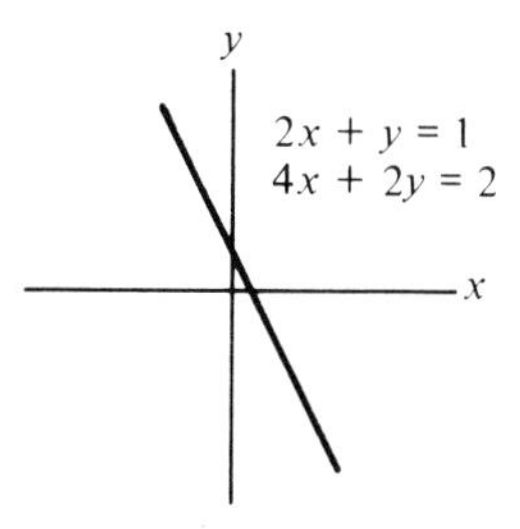

FIG. 12-4

solution to the system. The solution is

the coordinates of any point on the line $2x + y = 1$.

Since we can write that equation as $y = 1 - 2x$, we can easily find some solutions: $x = 0, y = 1$; $x = 1, y = -1$; $x = -3, y = 7$, etc.

EXAMPLE 4

In a laboratory a student is to combine a 25 percent hydrogen peroxide solution (25 percent by volume is hydrogen peroxide) with a 40 percent hydrogen peroxide solution to get 2 liters of a 30 percent solution. How many liters of each solution should the student mix?

Let x be the number of liters of the 25 percent solution, and y be the number of liters of the 40 percent solution that should be mixed. Then

$$x + y = 2. \tag{12-22}$$

See Fig. 12-5. In 2 liters of a 30 percent solution there is $.30(2) = .6$ liter of hydrogen

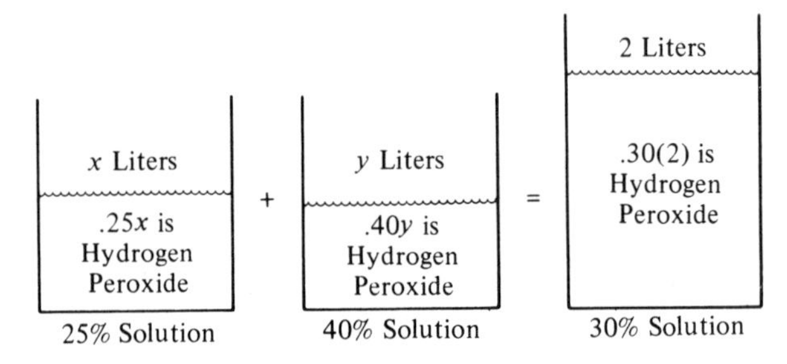

FIG. 12-5

peroxide. This hydrogen peroxide comes from two places: $.25x$ liter of it comes from the 25 percent solution, and $.40y$ liter comes from the 40 percent solution. Thus,

$$.25x + .40y = .6 \tag{12-23}$$

Equations (12-22) and (12-23) form a system. Solving Eq. (12-22) for x give $x = 2 - y$. Substituting $2 - y$ for x in Eq. (12-23) gives

$$\begin{aligned} .25(2 - y) + .40y &= .6 \\ .5 - .25y + .40y &= .6 \\ .15y &= .1 \\ y = \frac{.1}{.15} = \frac{10}{15} &= \frac{2}{3}. \end{aligned}$$

From Eq. (12-22), $x = 2 - y = 2 - \frac{2}{3} = \frac{4}{3}$. Thus $\frac{4}{3}$ liters of the 25 percent solution and $\frac{2}{3}$ liter of the 40 percent solution must be mixed.

Completion Set 12-1

In Problems **1** *and* **2**, *fill in the blanks.*

1. Two ways of solving a system of two linear equations in two variables is by elimination by ____________________ and elimination by ____________________.

2. To solve the system

$$\begin{cases} 3x - 4y = 10 \\ 8x + 5y = -40 \end{cases}$$

by eliminating y by addition, you can first multiply the top equation by 5 and the bottom equation by ____.

Problem Set 12-1

In Problems **1–4**, *use elimination by addition to solve the systems.*

1. $\begin{cases} x + 2y = 1 \\ 3x + y = -2 \end{cases}$

2. $\begin{cases} 2x + 3y = -10 \\ 3x - 2y = -2 \end{cases}$

3. $\begin{cases} 4x - 3y = 6 \\ 3x + 2y = 13 \end{cases}$

4. $\begin{cases} 5x + 3y = 10 \\ 3x + 4y = 6 \end{cases}$

In Problems **5–8**, *use elimination by substitution to solve the system.*

5. $\begin{cases} 4x + y = 6 \\ 3x + 2y = 2 \end{cases}$

6. $\begin{cases} x + 5y = -14 \\ -2x - 7y = 16 \end{cases}$

7. $\begin{cases} 5x + 7y = 13 \\ -x + 5y = 7 \end{cases}$

8. $\begin{cases} 4x - 3y = 13 \\ 3x + y = 0 \end{cases}$

In Problems **9–20**, *solve the systems by addition or substitution.*

9. $\begin{cases} 3x + y = 7 \\ 2x + 2y = -2 \end{cases}$

10. $\begin{cases} 2x - y = -11 \\ y + 5x = -7 \end{cases}$

11. $\begin{cases} 3x - 4y - 13 = 0 \\ 2x + 3y - 3 = 0 \end{cases}$

12. $\begin{cases} 5x - 3y = 2 \\ -10x + 6y = 4 \end{cases}$

13. $\begin{cases} 2q = 36 - 5p \\ 8p = 3q - 54 \end{cases}$

14. $\begin{cases} x = 3 - y \\ 3x + 2y = 19 \end{cases}$

15. $\begin{cases} 4x + 12y = 6 \\ 2x + 6y = 3 \end{cases}$

16. $\begin{cases} 2u - v - 1 = 0 \\ -u + 2v - 7 = 0 \end{cases}$

17. $\begin{cases} 2x + y = 4 \\ 10y - 41 = -20x \end{cases}$

18. $\begin{cases} 3y = 4x - 2 \\ y + \frac{2}{3} = \frac{4}{3}x \end{cases}$

19. $\begin{cases} \dfrac{2}{3}x + \dfrac{1}{2}y = 2 \\ \dfrac{3}{8}x + \dfrac{5}{6}y = -\dfrac{11}{2} \end{cases}$

20. $\begin{cases} 3x + 2y - 3 = x \\ y - 3x = -2 \end{cases}$

21. A chemical manufacturer wishes to fill an order for 700 gallons of a 24 percent acid solution (24 percent by volume is acid). Solutions of 20 percent and 30 percent are in stock. How many gallons of each must be mixed to fill the order?

22. A chemical manufacturer wishes to obtain 500 gallons of a 25 percent acid solution by mixing a 30 percent solution with an 18 percent solution. How many gallons of each should be mixed?

23. Universal Control Co. makes industrial control units. Their new models are the Argon I and the Argon II. To make each Argon I unit, they use 6 doodles and 3 skeeters. To make each Argon II unit, they use 10 doodles and 8 skeeters. The company receives a total of 760 doodles and 500 skeeters each day from its supplier. How many units of each model of the Argon can the company make each day? Assume that all the parts are used.

24. On a trip on a raft, it took $\frac{3}{4}$ hr to travel 12 mi downstream. The return trip took $1\frac{1}{2}$ hr. Find the speed of the raft in still water and the speed of the current.

25. An airplane travels 900 mi in 3 hr with the aid of a tail wind. It takes 3 hr and 36 min for the return trip flying against the same wind. Find the speed of the airplane in still air and the speed of the wind.

26. A private parcel service charges a certain rate per pound for the first 5 lb of a package's weight. For each additional pound it charges another rate. An 8-lb package will be delivered for $3.55. A 15-lb package will be delivered for $6.00. Find the two rates.

27. National Wigwam Co. manufactures prefabricated houses. It makes two models: Early American and Little Big Horn. The company knows that 20 percent more Early American models can be sold than Little Big Horn models. The profit on each Early American sold is $250. On each Little Big Horn sold, the profit is $350. This year the company wants a total profit of $130,000. How many units of each model must be sold?

28. A 10,000-gal railroad tank car is to be filled with solvent from two storage tanks, A and B. Solvent from A is pumped at the rate of 20 gal/min. Solvent from B is pumped at 30 gal/min. Usually both pumps operate at the same time. However, because of a blown fuse the pump on A is delayed 10 min. How many gallons from each storage tank will be used to fill the car?

*A **supply equation** for a product relates the number of units x that manufacturers will sell when the price is p (in dollars) per unit. A **demand equation** for a product relates the number of units x that consumers will buy when the price is p (in dollars) per unit. Suppose you form a system using the supply and demand equations for a product. The values of x and p that satisfy the system are given special names. The x-value is the **equilibrium quantity**. The p-value is the **equilibrium price**. In Problems **29** and **30**, find the equilibrium quantities and prices.*

29. Supply equation: $3x - 200p = -1800$
Demand equation: $3x + 100p = 1800$

30. Supply equation: $p = \frac{3}{100}x + 2$

Demand equation: $p = -\frac{7}{100}x + 12$

*A **total revenue equation** for a manufacturer relates the total amount of money y (in dollars) that the manufacturer receives when he sells x units of his product. His **total cost equation** relates the total amount of money y (in dollars) that it costs him when he produces x units. Suppose you form a system using the total revenue and total cost equations for a manufacturer. The x-value of a solution to this system is called the **break-even quantity**. It corresponds to the point at which the manufacturer has neither a profit nor a loss. In Problems **31** and **32**, find the break-even quantities.*

31. Total revenue equation: $y = 14x$

Total cost equation: $y = \frac{40}{3}x + 1200$

32. Total revenue equation: $y = 3x$
Total cost equation: $y = 2x + 4500$

12-2 SYSTEMS IN THREE UNKNOWNS

A linear equation in three unknowns, say x, y, and z, is an equation that can be put in the form $Ax + By + Cz = D$, where A, B, C, and D are fixed numbers. To solve a system of three linear equations in three unknowns, we use the methods of elimination in Sec. 12-1. For example, we might perform the following steps:

1. Select two *pairs* of the given equations and eliminate the *same* variable from each.
2. From step 1 you have two equations in the same two variables. Solve these by elimination.
3. In step 2 you found values of two of the variables. Substitute these in one of the original equations to find the value of the third variable.

EXAMPLE 1

Solve the system

$$\begin{cases} 4x - y - 3z = 1 & (12\text{-}24) \\ 2x + y + 2z = 5 & (12\text{-}25) \\ 8x + y - z = 5. & (12\text{-}26) \end{cases}$$

To solve this system in three unknowns, we follow the steps on page 285.

1. As our first pair of equations, let's select Eqs. (12-24) and (12-25):

$$\begin{cases} 4x - y - 3z = 1 \\ 2x + y + 2z = 5. \end{cases}$$

Adding these to eliminate y gives

$$6x - z = 6. \qquad (12\text{-}27)$$

As our second pair of equations, let's select Eqs. (12-25) and (12-26):

$$\begin{cases} 2x + y + 2z = 5 \\ 8x + y - z = 5. \end{cases}$$

We must also eliminate y from these. Multiplying the bottom equation by -1 gives

$$\begin{cases} 2x + y + 2z = 5 \\ -8x - y + z = -5. \end{cases}$$

Adding these, we have

$$-6x + 3z = 0. \qquad (12\text{-}28)$$

2. From step 1 we have two equations in two variables, Eqs. (12-27) and (12-28):

$$\begin{cases} 6x - z = 6 & (12\text{-}27) \\ -6x + 3z = 0. & (12\text{-}28) \end{cases}$$

Adding these and solving for z, we obtain

$$2z = 6$$

$$\boxed{z = 3.}$$

Substituting $z = 3$ in Eq. (12-27) gives

$$6x - 3 = 6$$
$$6x = 9$$
$$\boxed{x = \frac{9}{6} = \frac{3}{2}.}$$

3. In step 2 we found $z = 3$ and $x = \frac{3}{2}$. Substituting these in Eq. (12-24), we find y:

$$4x - y - 3z = 1$$
$$4\left(\frac{3}{2}\right) - y - 3(3) = 1$$
$$-y = 4$$
$$\boxed{y = -4.}$$

The solution is $x = \frac{3}{2}$, $y = -4$, and $z = 3$.

Problem Set 12-2

In Problems **1–8**, *solve the systems.*

1. $\begin{cases} x + y + z = 6 \\ x - y + z = 2 \\ 2x - y + 3z = 6 \end{cases}$

2. $\begin{cases} 2x - y + 3z = 12 \\ x + 2y - 3z = -10 \\ x + y - z = -3 \end{cases}$

3. $\begin{cases} x - z = 14 \\ y + z = 21 \\ x - y + z = -10 \end{cases}$

4. $\begin{cases} x + y = -6 \\ z = 4 \\ -x + y + 2z = 16 \end{cases}$

5. $\begin{cases} 2x + y + 6z = 3 \\ x - y + 4z = 1 \\ 3x + 2y - 2z = 2 \end{cases}$

6. $\begin{cases} 5x - 7y + 4z = 2 \\ 3x + 2y - 2z = 3 \\ 2x - y + 3z = 4 \end{cases}$

7. $\begin{cases} 2x - 3y + z = -2 \\ 3x + 3y - z = 2 \\ x - 6y + 3z = -2 \end{cases}$

8. $\begin{cases} x + y + z = -1 \\ 3x + y + z = 1 \\ 4x - 2y + 2z = 0 \end{cases}$

9. The Flip-Flop Co. makes three types of patio furniture: chairs, rockers, and chaise lounges. Each requires wood, plastic, and aluminum, as given in Table 12-1. The company has in stock 400 units of wood, 600 units of plastic, and 1500 units of aluminum. For its end-of-the-season production run, the company wants to use up all the stock. To do this, how many chairs, rockers, and chaise lounges should it make?

TABLE 12-1

	WOOD	PLASTIC	ALUMINUM
Chair	1 unit	1 unit	2 units
Rocker	1 unit	1 unit	3 units
Chaise lounge	1 unit	2 units	5 units

10. Table 12-2 shows how alloys A, B, and C are composed (by weight). How much of A, B, and C must be mixed to produce 100 kilograms of an alloy that is 53 percent copper and 19 percent zinc?

TABLE 12-2

	A	B	C
Copper	50%	60%	40%
Zinc	30%	20%	
Nickel	20%	20%	60%

11. The graph of $y = ax^2 + bx + c$ passes through the points (2, 0), (0, 0), and $(-1, 3)$. Find a, b, and c. *Hint*: The coordinates of each point must satisfy the equation, so replace x by 2 and y by 0 to get an equation in a, b, and c. Do the same for the other two points. In this way obtain three equations in the unknowns a, b, and c.

12. Repeat Problem 11 if the graph passes through the points (2, 5), $(-3, 5)$, and (1, 1).

In studies of the electrical circuit shown in Fig. 12-6, Kirchhoff's laws lead to the following system:

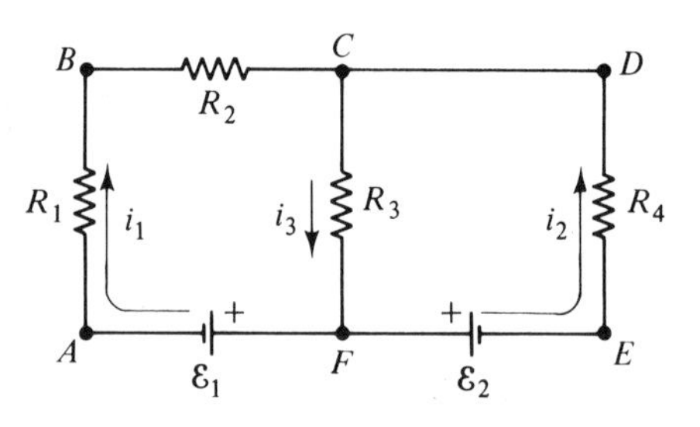

FIG. 12-6

$$\begin{cases} R_1 i_1 + R_2 i_1 + R_3 i_3 + \mathcal{E}_1 = 0 \\ R_3 i_3 + R_4 i_2 + \mathcal{E}_2 = 0 \\ i_1 + i_2 = i_3. \end{cases}$$

Here the R's are resistances (in ohms), the i's are currents (in amperes), and the $\mathcal{E}$'s are potential differences (in volts). Problems **13** *and* **14** *relate to this system.*

13. Find i_1, i_2, and i_3 (in amperes) if $R_1 = 2$, $R_2 = 3$, $R_3 = 5$, and $R_4 = 3$ ohms and $\mathcal{E}_1 = 15$ and $\mathcal{E}_2 = 2$ volts. (It is possible to get negative values of i in your answer.)

14. Find i_1, i_2, and i_3 (in amperes) if $R_1 = 1$, $R_2 = 1$, $R_3 = 4$, and $R_4 = 6$ ohms and $\mathcal{E}_1 = 2$ and $\mathcal{E}_2 = 60$ volts. (It is possible to get negative values of i in your answer.)

12-3 REVIEW

IMPORTANT TERMS

system of equations *(p. 276)*
elimination by addition *(p. 278)*
elimination by substitution *(p. 280)*

REVIEW PROBLEMS

In Problems **1–12**, *solve the systems.*

1. $\begin{cases} 4x + 5y = 3 \\ 3x + 4y = 2 \end{cases}$

2. $\begin{cases} 2x + 3y = 5 \\ x - 2y = -1 \end{cases}$

3. $\begin{cases} 3s + t - 4 = 0 \\ 12s + 4t - 2 = 0 \end{cases}$

4. $\begin{cases} u - 3v + 11 = 0 \\ 4u + 3v - 9 = 0 \end{cases}$

5. $\begin{cases} 3x + \frac{1}{2}y = 2 \\ \frac{1}{2}x - \frac{1}{4}y = 0 \end{cases}$

6. $\begin{cases} \frac{1}{3}x - \frac{1}{2}y = 4 \\ \frac{1}{4}x - \frac{3}{8}y = 3 \end{cases}$

7. $\begin{cases} 6x = 3 - 9y \\ 12y = 4 - 8x \end{cases}$

8. $\begin{cases} 4x = 7 + 12y \\ 5x = 15y - 2 \end{cases}$

9. $\begin{cases} x - y = 2 \\ x + z = 1 \\ y - z = 3 \end{cases}$

10. $\begin{cases} 2x - 4z = 8 \\ x - 2y - 2z = 14 \\ 3x + y + z = 0 \end{cases}$

11. $\begin{cases} 4r - s + 2t = 2 \\ 8r - 3s + 4t = 1 \\ r + 2s + 2t = 8 \end{cases}$

12. $\begin{cases} 3u - 2v + w = -2 \\ 2u + v + w = 1 \\ u + 3v - w = 3 \end{cases}$

13. Two alloys of copper are to be mixed so that the result is 15 kilograms (kg) of a 45 percent alloy (by weight). One alloy is 20 percent copper, and the other is 50 percent copper. How many kg of each should be used?

14. Snoopy Surveys was given a contract to perform a product rating survey for Moldy Crackers. A total of 250 people were interviewed. Snoopy Surveys reported that 62.5 percent more people liked Moldy Crackers than disliked them. The report did not indicate that 16 percent of those interviewed had no comment. How many of those surveyed liked Moldy Crackers? How many disliked them? How many had no comment?

15. A company pays skilled workers in its assembly department \$8 per hour. Semiskilled workers in that department are paid \$4 per hour. Shipping clerks are paid \$5 per hour. Because of an increase in orders, the company wants to employ a total of 70 workers in the assembly and shipping departments. It will pay a total of \$370 per hour to these employees. Because of a union contract, twice as many semiskilled workers as skilled workers must be employed. How many semiskilled workers, skilled workers, and shipping clerks should the company hire?

CHAPTER 13

Exponents and Radicals

13-1 ZERO AND NEGATIVE EXPONENTS

In Chapter 3 the basic rules for exponents were given. They are (assuming no division by 0):

$$a^m a^n = a^{m+n} \tag{13-1}$$

$$(a^m)^n = a^{mn} \tag{13-2}$$

$$\frac{a^m}{a^n} = a^{m-n}, \quad m > n \tag{13-3}$$

$$\frac{a^m}{a^n} = \frac{1}{a^{n-m}}, \quad n > m \tag{13-4}$$

$$\frac{a^n}{a^n} = 1 \tag{13-5}$$

$$(ab)^n = a^n b^n \tag{13-6}$$

$$\left(\frac{a}{b}\right)^n = \frac{a^n}{b^n}. \tag{13-7}$$

Up to now our exponents have been positive integers. In this section we'll give a meaning to the zero exponent and negative exponents.

Let's begin with a^0, where $a \neq 0$.† We want to define a^0 so that rules (13-1)–(13-7) hold if m or n is 0. For instance, by rule (13-1) we must have

$$a^m \cdot a^0 = a^{m+0} = a^m.$$

Dividing both sides of $a^m a^0 = a^m$ by a^m gives $a^0 = 1$. Thus we define

$$\boxed{a^0 = 1.} \tag{13-8}$$

With this definition the other rules also hold. For example, for rule (13-6) we have

$$(ab)^0 = 1 = 1 \cdot 1 = a^0 b^0.$$

EXAMPLE 1

Zero exponents.

a. $2^0 = 1$.

b. $(-5)^0 = 1$.

c. $3\left(\frac{4}{5}\right)^0 = 3(1) = 3$.

d. $(x^2 + 2)^0 = 1$.

We now give a meaning to negative exponents, as in a^{-n}. Again we want rules (13-1)–(13-7) to hold. From rule (13-1) we must have

$$a^n a^{-n} = a^{n-n} = a^0 = 1.$$

Dividing both sides of $a^n a^{-n} = 1$ by a^n gives $a^{-n} = \frac{1}{a^n}$. Thus we have the following rule (actually a definition) for a^{-n}:

$$\boxed{a^{-n} = \frac{1}{a^n}.} \tag{13-9}$$

This definition allows the other rules to hold for negative exponents. For instance, for rule (13-6) we have

$$(ab)^{-n} = \frac{1}{(ab)^n} = \frac{1}{a^n b^n} = \frac{1}{a^n} \cdot \frac{1}{b^n} = a^{-n} b^{-n}.$$

---

†0^0 has no meaning.

From (13-9) we can get another rule. Since

$$\frac{1}{a^{-n}} = \frac{1}{\frac{1}{a^n}} = 1 \cdot \frac{a^n}{1} = a^n,$$

then

$$\boxed{\frac{1}{a^{-n}} = a^n.} \tag{13-10}$$

EXAMPLE 2

Negative exponents.

a. $5^{-2} = \frac{1}{5^2} = \frac{1}{25}$.

b. $-x^{-6} = -(x^{-6}) = -\frac{1}{x^6}$.

c. $\frac{1}{2^{-4}} = 2^4 = 16$.

d. $\frac{1}{x^2} = x^{-2}$.

e. $\left(\frac{a}{b}\right)^{-1} = \frac{1}{\frac{a}{b}} = \frac{b}{a}$.

f. $\frac{1}{(x+2)^{-2}} = (x+2)^2$.

g. $2x^{-3} + (2x)^{-3} = 2\left(\frac{1}{x^3}\right) + \frac{1}{(2x)^3} = \frac{2}{x^3} + \frac{1}{8x^3}$.

a. $\frac{1}{(-2)^{-3}} = (-2)^3 = -8$. *We changed the sign of the exponent only, not the sign of the base.* ***Don't*** *write* $\frac{1}{(-2)^{-3}} = (2)^3$.

b. $x^{-2} + y^{-2} = \frac{1}{x^2} + \frac{1}{y^2}$. ***Don't*** *write* $x^{-2} + y^{-2} = \frac{1}{x^2 + y^2}$. *Note that*

$$x^{-2} + y^{-2} = \frac{1}{x^2} + \frac{1}{y^2} = \frac{y^2}{x^2y^2} + \frac{x^2}{x^2y^2} = \frac{y^2 + x^2}{x^2y^2}.$$

Using rules (13-8)–(13-10), we can replace rules (13-3)–(13-5) by a single rule:

$$\boxed{\frac{a^m}{a^n} = a^{m-n},}$$

regardless of what m and n are. For example,

$$\frac{x^3}{x^6} = x^{3-6} = x^{-3} = \frac{1}{x^3}.$$

Watch how we manipulate factors in the following fraction.

$$\frac{x^3y^{-4}}{z^{-5}} = \frac{x^3\left(\frac{1}{y^4}\right)}{\frac{1}{z^5}} = x^3\left(\frac{1}{y^4}\right)\cdot\frac{z^5}{1} = \frac{x^3z^5}{y^4}.$$

Compare the first and last fractions. *A **factor** of the numerator can be moved to the denominator if you **change** the sign of its exponent.* Similarly, *a **factor** of the denominator can be moved to the numerator by **changing** the sign of its exponent. Thus,*

$$\frac{x^{-7}y^6}{3z^2} = \frac{y^6}{3x^7z^2}.$$

EXAMPLE 3

Simplify and give all answers with positive exponents.

a. $x^{-2}y^{-2} = \dfrac{1}{x^2y^2}$.

b. $\dfrac{16x^{-4}}{32x^7} = \dfrac{1}{2x^7x^4} = \dfrac{1}{2x^{11}}$.

c. $\dfrac{-x^{-2}}{y^{-3}z^2} = -\dfrac{y^3}{x^2z^2}$.

d. $\dfrac{x^{-7}y^6}{x^9y^{-2}} = \dfrac{y^6y^2}{x^9x^7} = \dfrac{y^8}{x^{16}}$.

e. $\dfrac{4+y}{y^{-6}} = (4+y)(y^6) = 4y^6 + y^7$.

The next example shows how rules of exponents are used to simplify expressions involving negative exponents. Remember: *The rules of exponents are true for negative, positive, and zero exponents.*

EXAMPLE 4

Perform the operations and write the answers with positive exponents only.

a. $x^{11}x^{-5} = x^{11+(-5)} = x^6$.

b. $2x^{-7}x^{-6} = 2x^{-7-6} = 2x^{-13} = \dfrac{2}{x^{13}}$.

c. $(3x^7y^{-4})^2 = 3^2(x^7)^2(y^{-4})^2 = 9x^{14}y^{-8} = \dfrac{9x^{14}}{y^8}$.

d. $(x^{-3}y^5)^{-4} = (x^{-3})^{-4}(y^5)^{-4} = x^{12}y^{-20} = \dfrac{x^{12}}{y^{20}}$.

e. $\left(\dfrac{x^{-2}y^3}{z^{-4}}\right)^{-6} = \dfrac{(x^{-2}y^3)^{-6}}{(z^{-4})^{-6}} = \dfrac{x^{12}y^{-18}}{z^{24}} = \dfrac{x^{12}}{y^{18}z^{24}}$.

f. $(x^{-1} - y^{-1})^2 = \left(\dfrac{1}{x} - \dfrac{1}{y}\right)^2 = \left(\dfrac{y-x}{xy}\right)^2 = \dfrac{(y-x)^2}{(xy)^2} = \dfrac{y^2 - 2xy + x^2}{x^2y^2}$.

EXAMPLE 5

When a number is written as a product of a number between 1 *and* 10 *and some integral power of* 10, *then the number is said to be in* ***scientific notation***. *For example,* 832 *in scientific notation is* 8.32×10^2. *Given the formula (from physics)*

$$F = \frac{(9 \times 10^9)q_1q_2}{r^2},$$

find F *if* $q_1 = 1.6 \times 10^{-19}$, $q_2 = 1.6 \times 10^{-19}$, *and* $r = 5.3 \times 10^{-11}$. *Give answer in scientific notation.*

Let's handle the powers of 10 separately.

$$\begin{aligned} F &= \frac{(9 \times 10^9)(1.6 \times 10^{-19})(1.6 \times 10^{-19})}{(5.3 \times 10^{-11})^2} \\ &= \frac{(9 \times 1.6 \times 1.6)(10^9 \times 10^{-19} \times 10^{-19})}{(5.3)^2(10^{-22})} \\ &= \left[\frac{9 \times 1.6 \times 1.6}{(5.3)^2}\right]10^{9-19-19+22}. \end{aligned}$$

The first factor is approximately .82. Thus

$$F = .82 \times 10^{-7}.$$

To write the answer in scientific notation, we want the .82 to be 8.2. To do this we multiply by 10×10^{-1} (which is 1):

$$\begin{aligned} F &= .82 \times 10^{-7} = (.82 \times 10) \times (10^{-1} \times 10^{-7}) \\ &= 8.2 \times 10^{-8}. \end{aligned}$$

Completion Set 13-1

Fill in the blanks.

1. $6^{-2} = \dfrac{1}{(\_\_\_)^2} = \_\_\_\_\_.$

2. $\dfrac{1}{4^{-3}} = 4^{(\_\_)} = \_\_\_\_\_.$

3. $(2^{-3})^2 = 2^{(\_\_\_)} = \dfrac{1}{2^{(\_\_)}} = \dfrac{1}{(\_\_\_\_)}.$

4. $x^{-3}x^5 = x^{(\_\_)}.$

5. $\dfrac{x^3}{x^{-2}} = x^{(\_\_)}.$

6. $(x^{-5})^{-4} = x^{(\_\_\_)}.$

Problem Set 13-1

In Problems **1–15**, *find the values of the numbers.*

1. 7^0

2. $\left(\dfrac{3}{5}\right)^0$

3. 2^{-3}

4. 3^{-2}

5. $\dfrac{1}{3^{-3}}$

6. $\dfrac{2}{4^{-2}}$

7. $2x^0 + (2x)^0$

8. $2(-3)^0$

9. $\dfrac{-1^0}{4^{-1}}$

10. $-5^{-2}(25)$

11. $\dfrac{1}{(-3)^{-3}}$

12. $2^{-1} + 3^{-1}$

13. $\dfrac{2^{-2} + 3^{-2}}{13}$

14. $\dfrac{1 + 2^{-1}}{2^{-1}}$

15. $\dfrac{(3^{-2})^0}{1^{-1}}$

In Problems **16–30**, *write the expression by using positive exponents only. Simplify.*

16. x^{-2}

17. x^{-6}

18. $2^{-1}x$

19. $\dfrac{1}{x^{-3}}$

20. $\dfrac{1}{3x^{-2}}$

21. $3y^{-4}$

22. $2^{-2}x^{-4}$

23. $\dfrac{x}{4^{-2}}$

24. $\dfrac{7^0}{x^{-1}yz^{-2}}$

25. $x^{-5}y^{-7}$

26. $x^{-1}y^{-2}z^4$

27. $\dfrac{2a^2b^{-4}}{c^{-5}}$

28. $\dfrac{a^5b^{-4}}{c^{-3}d}$

29. $\dfrac{x^9y^{-12}}{w^2z^{-4}}$

30. $\dfrac{(x^2 + 4x^4)^0}{x^{-2}}$

In Problems **31–64**, *perform the operations and simplify. Give all answers with positive exponents only.*

31. x^8x^{-7}

32. $x^{-7}x$

33. $x^{-2}x^{-3}$

34. $x^2x^{-4}x^9$

35. $(3x^{-4}y^5)(2xy^{-1})$

36. $(3x^{-7}y^{-3})(x^2y^2)(4x)$

37. $(xy^{-5})^{-4}$

38. $(2x^2y^{-1})^2$

39. $2(x^{-1}y^2)^2$

40. $(x^{-5}y^6)^{-1}$

41. $(3t)^{-2}$

42. $(x^{-4}y^{-4})^4$

43. $(x^{-5}y^5z)^{-3}$

44. $\dfrac{10x^6}{5x^{-2}}$

45. $\dfrac{t^{-8}}{t^{-12}}$

46. $\dfrac{-2b^{-30}}{4b^{-5}}$

47. $\dfrac{x^{-2}y^4}{x^6y^{-1}}$

48. $\dfrac{(x^{-6}y^2)xy^{-2}}{xy}$

49. $\dfrac{x^{-2}(yz)^2w}{x^{-3}y^5}$

50. $\dfrac{x^6}{(x^{-4}y^8)y^{-8}}$

51. $\dfrac{2(2x^2y)^2}{3y^{-13}z^{-2}}$

52. $\dfrac{x^2y^{-5}}{(x^{-8}y^6)^{-3}}$

53. $\dfrac{2^0}{(xy)^{-4}(x^2y)}$

54. $\left(\dfrac{y}{z^{-1}}\right)^{-1}$

55. $\left(\dfrac{8x^2}{5y^2}\right)^{-1}$

56. $\dfrac{1}{(3x^{-1})^{-1}}$

57. $\left(-\dfrac{z^{-1}}{x}\right)^{-1}$

58. $\dfrac{xyz^{-1}}{(x^2)^{-4}}$

59. $\left(\dfrac{x^{-3}y^{-6}z^2}{2xy^{-1}}\right)^{-2}$

60. $\left[\left(\dfrac{x}{y}\right)^{-2}\right]^{-4}$

61. $\left[\dfrac{x^{-1}y^4}{(z^2)^{-2}}\right]^{-5}$

62. $(x + y^{-1})^2$

63. $(x - y)^{-2}$

64. $x^{-2} - y^{-2}$

In Problems **65–70**, *perform the indicated operations and give your answer in scientific notation.*

65. $(1.3 \times 10^{-4})(2.0 \times 10^{6})$

66. $\dfrac{9.3 \times 10^{-1}}{3.1 \times 10^{5}}$

67. $\dfrac{(3.0 \times 10^{11})(4.2 \times 10^{-4})}{2 \times 10^{13}}$

68. $\dfrac{(4.8 \times 10^{-1})(5.0 \times 10^{-2})}{(3.2 \times 10^{-3})(3.0 \times 10^{-4})}$

69. $\dfrac{(1.0 \times 10^{4})^{3}}{2.5 \times 10^{5}}$

70. $\dfrac{(1.2 \times 10^{-2})^{2}}{2.88 \times 10}$

71. Given the formula $F = IBL$, find F if $I = 10^3$, $B = 5 \times 10^{-5}$, and $L = 4$. Give your answer in scientific notation.

72. Given the formula $F = \dfrac{a}{b}$, find F if $a = 3 \times 10^{10}$ and $b = 1.5 \times 10^{8}$. Give your answer in scientific notation.

73. In a certain chemical solution, the hydrogen ion concentration, written $[H^+]$, in moles per liter is given by

$$1 \times 10^{-4} = [H^+] \times 1 \times 10^{-2}.$$

Solve for $[H^+]$ and give your answer in scientific notation.

74. Solve the equation $x^{-2} = 16$. *Hint*: First write the equation with positive exponents.

13-2 FRACTIONAL EXPONENTS

In this section a meaning will be given to a fractional exponent, such as in $a^{1/n}$ and $a^{m/n}$. We want to define $a^{1/n}$, where n is a positive integer, so that the rules of exponents hold. For instance, by rule (13-2) on p. 291 we must have

$$(a^{1/n})^n = a^{n/n} = a^1 = a.$$

But from Chapter 3 you know that

$$(\sqrt[n]{a}\,)^n = a.$$

Thus the nth power of $a^{1/n}$ is the same as the nth power of $\sqrt[n]{a}$. For this reason we define $a^{1/n}$ to be the principal nth root of a:†

†We want $a^{1/n}$ to be a real number. For this, we must restrict a at times. For example, $(-4)^{1/2} = \sqrt{-4}$ is not a real number. In definition (13-12), we must also restrict a.

$$\boxed{a^{1/n} = \sqrt[n]{a}\ .} \tag{13-11}$$

Now you may write such things as

$$13^{1/5} = \sqrt[5]{13} \quad \text{and} \quad \sqrt{2 + 7x} = (2 + 7x)^{1/2}.$$

Remember: *The rules of exponents are true for all types of exponents.*

EXAMPLE 1

Write each of the following in radical form and find the value if possible.

a. $4^{1/2} = \sqrt[2]{4} = 2.$

b. $27^{1/3} = \sqrt[3]{27} = 3.$

c. $(-32)^{1/5} = \sqrt[5]{-32} = -2.$

d. $16^{-(1/2)} = \dfrac{1}{16^{1/2}} = \dfrac{1}{\sqrt{16}} = \dfrac{1}{4}.$

e. $(3x^2 + 4y)^{1/4} = \sqrt[4]{3x^2 + 4y}\ .$

f. $xy^{1/2} + (xy)^{1/2} = x\sqrt{y} + \sqrt{xy}\ .$

EXAMPLE 2

Write each of the following with fractional exponents.

a. $\sqrt{7} = 7^{1/2}.$

b. $\sqrt[3]{x - y} = (x - y)^{1/3}.$

c. $x\sqrt[4]{y} = xy^{1/4}.$

Definition (13-11) can be made more general. We define $a^{m/n}$, where m and n are integers and n is positive, by

$$\boxed{a^{m/n} = \left(\sqrt[n]{a}\right)^m = \sqrt[n]{a^m}\ .} \tag{13-12}$$

That is, $a^{m/n} = (a^{1/n})^m = (a^m)^{1/n}$. By definition (13-12) you may look at $x^{4/3}$ in two ways: first, as the fourth power of the cube root of x, $(\sqrt[3]{x})^4$; and second, as the cube root of the fourth power of x, $\sqrt[3]{x^4}$. In computations it is often easier to take the nth root first and then raise the result to the mth power.

EXAMPLE 3

Find the value of each of the following.

a. $8^{2/3} = (\sqrt[3]{8})^2 = 2^2 = 4$. On the other hand, we could also write $8^{2/3} = \sqrt[3]{8^2} = \sqrt[3]{64} = 4$.

b. $(-27)^{4/3} = (\sqrt[3]{-27})^4 = (-3)^4 = 81$.

c. $27^{-2/3} = \dfrac{1}{27^{2/3}} = \dfrac{1}{(\sqrt[3]{27})^2} = \dfrac{1}{(3)^2} = \dfrac{1}{9}$.

d. $\left(\dfrac{1}{4}\right)^{3/2} = \left(\sqrt{\dfrac{1}{4}}\right)^3 = \left(\dfrac{1}{2}\right)^3 = \dfrac{1}{8}$.

e. $9^{-1/2} = \dfrac{1}{9^{1/2}} = \dfrac{1}{\sqrt{9}} = \dfrac{1}{3}$.

In definition (13-12) we must at times restrict a. Here's why. By that definition $\sqrt{x^2} = x^{2/2} = x$. But this is not always true. If $x = 6$, then $\sqrt{x^2} = \sqrt{36} = 6 = x$. But if $x = -6$, then $\sqrt{x^2} = \sqrt{36} = 6 = -(-6) = -x$. Thus, $\sqrt{x^2}$ is x if x is positive, and is $-x$ if x is negative. More simply, we can say that $\sqrt{x^2} = |x|$. **Unless we say otherwise, we shall assume that all literal numbers appearing in radicals or bases are positive. As a result, we are free to write** $\sqrt{x^2} = x$.

EXAMPLE 4

Write each of the following with fractional exponents and simplify if possible.

a. $\sqrt[5]{x^4} = x^{4/5}$.

b. $x\sqrt[6]{y^5} = xy^{5/6}$.

c. $(\sqrt{3})^4 = (3^{1/2})^4 = 3^{4/2} = 3^2 = 9$.

d. $\sqrt{a} - \sqrt[4]{b} = a^{1/2} - b^{1/4}$.

e. $\dfrac{4}{\sqrt{x-3}} = \dfrac{4}{(x-3)^{1/2}}$.

f. $\sqrt[4]{a^4b^8} = (a^4b^8)^{1/4} = (a^4)^{1/4}(b^8)^{1/4} = ab^2$.

EXAMPLE 5

Perform the indicated operations and give the answer with positive exponents only.

a. $x^{1/2}x^{1/3} = x^{(1/2)+(1/3)} = x^{5/6}$.

b. $\dfrac{2x}{x^{1/4}} = \dfrac{2x^1}{x^{1/4}} = 2x^{1-(1/4)} = 2x^{3/4}$.

c. $(8a^3)^{2/3} = 8^{2/3}(a^3)^{2/3} = (\sqrt[3]{8})^2\, a^2 = 2^2a^2 = 4a^2$.

d. $(15^{-1/2})^4 = 15^{-4/2} = 15^{-2} = \dfrac{1}{15^2} = \dfrac{1}{225}$.

e. $(x^{2/9}y^{-4/3})^{1/2} = (x^{2/9})^{1/2}(y^{-4/3})^{1/2} = x^{1/9}y^{-2/3} = \dfrac{x^{1/9}}{y^{2/3}}$.

f. $\left(\dfrac{x^{12}}{y^6}\right)^{-1/3} = \dfrac{(x^{12})^{-1/3}}{(y^6)^{-1/3}} = \dfrac{x^{-4}}{y^{-2}} = \dfrac{y^2}{x^4}$.

g. $\left(\dfrac{x^{1/5}y^{6/5}}{z^{2/5}}\right)^{15} = \dfrac{(x^{1/5}y^{6/5})^{15}}{(z^{2/5})^{15}} = \dfrac{x^3y^{18}}{z^6}$.

EXAMPLE 6

Perform the indicated operations.

a. $x^{1/2}(2 - x^{3/2}) = 2x^{1/2} - x^{1/2}x^{3/2} = 2x^{1/2} - x^2$.

b. $2x^{1/2} + (x^{1/6})^3 = 2x^{1/2} + x^{3/6} = 2x^{1/2} + x^{1/2} = 3x^{1/2}$.

c.
$$\begin{aligned}(27x)^{1/3} - (2x^{1/3} - 4x^{1/3}) &= \sqrt[3]{27}\, x^{1/3} - (-2x^{1/3})\\ &= 3x^{1/3} + 2x^{1/3}\\ &= 5x^{1/3}.\end{aligned}$$

d. $\dfrac{3x^4 - 2x^2}{x^{1/3}} = \dfrac{3x^4}{x^{1/3}} - \dfrac{2x^2}{x^{1/3}} = 3x^{4-(1/3)} - 2x^{2-(1/3)} = 3x^{11/3} - 2x^{5/3}$.

Completion Set 13-2

Fill in the blanks.

1. $64^{1/3} = \sqrt[n]{64}$, where $n =$ ____.

2. $x^{5/4} = \sqrt[n]{x^m}$, where $m =$ ____ and $n =$ ____.

3. $x^{2/3} = (\sqrt[n]{x})^m$, where $m =$ ____ and $n =$ ____.

4. $\sqrt[7]{x^5} = x^{m/n}$, where $m =$ ____ and $n =$ ____.

5. $(\sqrt[4]{x})^9 = x^{m/n}$, where $m =$ ____ and $n =$ ____.

6. *Insert* T (= *True*) *or* F (= *False*).

a. $\sqrt[3]{2^4} = (\sqrt[3]{2})^4$. ____

b. $\sqrt[4]{2^3} = \sqrt[3]{2^4}$. ____

Problem Set 13-2

In Problems **1–15**, *find the value of each expression.*

1. $25^{1/2}$

2. $125^{1/3}$

3. $81^{-1/2}$

4. $4^{-1/2}$

5. $27^{2/3}$

6. $\left(\frac{1}{9}\right)^{3/2}$

7. $27^{-2/3}$

8. $9^{3/2}$

9. $(-8)^{4/3}$

10. $(-8)^{-1/3}$

11. $16^{3/4}$

12. $(64^{1/2})^{1/3}$

13. $\left(-\frac{1}{32}\right)^{-1/5}$

14. $\left(\frac{1}{27}\right)^{-2/3}$

15. $(4^3)^{2/3}$

In Problems **16–30**, *rewrite each expression by using positive fractional exponents and simplify where possible.*

16. $\sqrt[3]{y}$

17. $\sqrt{x}$

18. $\sqrt[4]{x^9}$

19. $\sqrt[3]{x^2}$

20. $\dfrac{\sqrt{y}}{\sqrt[4]{x}}$

21. $\sqrt[4]{x^3y^5}$

22. $\sqrt[4]{x}\,\sqrt[5]{y}$

23. $\sqrt[6]{x^5y^{12}}$

24. $\dfrac{1}{\sqrt{2}}$

25. $x^2\sqrt[4]{x}$

26. $\dfrac{2}{\sqrt[3]{x-5}}$

27. $\dfrac{3}{\sqrt{x}}$

28. $\sqrt[3]{7-x}$

29. $\sqrt[3]{(x^2-5x)^2}$

30. $\dfrac{1}{\sqrt[6]{(x^2-2x)^5}}$

In Problems **31–76**, *perform the operations and simplify. Write your answers with positive exponents only.*

31. $x^{1/2}x^{3/2}$

32. $x^{2/3}x^{4/3}$

33. $x^{4/3}x^{-1/3}$

34. $x^{-1/2}x^{3/2}$

35. $x^{-2}x^{7/2}x^{5/2}$

36. $x^3x^{-3/8}x^{-5/8}$

37. $x^{1/2}x^{1/4}$

38. $x^{3/5}x^{1/3}$

39. $x^{1/2}(3x^{1/2}y)$

40. $(xy^{-1/2})y^{5/2}$

41. $(x^{1/2})^3$

42. $(y^{-1/3})^6$

43. $(y^6)^{1/3}$

44. $(t^{10})^{4/5}$

45. $2(x^{-2/3})^6$

46. $(-y^{1/3})^6$

47. $(2x^{-2}y^{1/3})^3$

48. $(3x^3y^{1/2})^4$

49. $(ab^2c^3)^{3/4}(a^{1/4}b^{3/4})^5$

50. $(x^{1/2}y^{2/3})(x^2y)^{1/3}$

51. $x^{1/3}(x^{2/3}+3)$

52. $x^{3/2}(1-x^{1/2})$

53. $(27x^{15})^{-1/3}$

54. $(8x^{-6})^{1/3}$

55. $(-8x^{-6})^{1/3}$

56. $(27^{-1}x^{15})^{-1/3}$

57. $\dfrac{x^{4/9}y^{2/5}}{x^{1/9}y^{7/5}}$

58. $\dfrac{x^{4/7}y^{3/20}}{x^{3/7}y^{17/20}}$

59. $\dfrac{x^{2/3}y^{-9/4}}{x^{-4/3}y^{-1/4}}$

60. $\dfrac{xy^{2/5}}{x^{1/2}y^{2/5}z^{-1/3}}$

61. $\left(\dfrac{x^{-4/3}}{y^{-2/3}}\right)^{-3}$

62. $\left(\dfrac{3^{1/4}x^{3/4}}{y^{1/2}}\right)^4$

63. $\left(\dfrac{x^{-1}}{x^{1/3}}\right)^2$

64. $\left(\dfrac{y^{2/3}}{x^{3/4}}\right)^{2/3}$

65. $\left(\dfrac{2^{1/3}x^{2/3}}{x^{1/3}}\right)^6$

66. $\left(\dfrac{x^{3/2}}{x^{-1}}\right)^{-4}$

67. $2x^{1/2}-5x^{1/2}$

68. $7x^{3/2}+2x^{3/2}-5x^{3/2}$

69. $4x^{-1/3}-2(x^2)^{-1/6}$

70. $3x^{1/4}-(x^{-1/2})^{-1/2}$

71. $\dfrac{x^{4/3}+x^2-x^4}{x}$

72. $\dfrac{x^{-1/2}y^{-1/3}}{2}$

73. $\dfrac{x^{1/2} - 3x^{1/3}}{x^{1/4}}$

74. $\dfrac{2x^{1/3} - y^3 - y^{1/4}}{y^{1/4}}$

75. $\dfrac{(2x^{1/2}y)^3(4x)^{-1/2}}{x}$

76. $\dfrac{(x^{1/3}y^{1/6})^3(2x^{-2})}{xy^{1/2}}$

13-3 CHANGING THE FORM OF A RADICAL

The basic rules for radicals were given in Chapter 3. They are:

$$\sqrt[n]{ab} = \sqrt[n]{a}\,\sqrt[n]{b} \tag{13-13}$$

$$\sqrt[n]{\frac{a}{b}} = \frac{\sqrt[n]{a}}{\sqrt[n]{b}} \tag{13-14}$$

$$(\sqrt[n]{a}\,)^n = a. \tag{13-15}$$

Rule (13-15) may also be written as

$$\sqrt[n]{a^n} = a. \tag{13-16}$$

At times you may replace a radical by an expression that does not have a radical. For example, just as $\sqrt[3]{x^3} = x$ [rule (13-16)], we have

$$\sqrt[3]{x^6y^9} = \sqrt[3]{(x^2y^3)^3} = x^2y^3.$$

This shows that you may "remove" a radical of index n when the radicand is an nth power of an expression.

EXAMPLE 1

Removing radicals: $\sqrt[n]{a^n} = a.$

a. $\sqrt{x^6y^8} = \sqrt{(x^3y^4)^2} = x^3y^4.$

b. $\sqrt[4]{\dfrac{x^{16}}{y^8}} = \sqrt[4]{\left(\dfrac{x^4}{y^2}\right)^4} = \dfrac{x^4}{y^2}.$

c. $\sqrt[5]{32x^5y^{15}} = \sqrt[5]{(2xy^3)^5} = 2xy^3.$

One way to change the form of a radical with index n is to "remove" from the radicand all factors whose nth roots can easily be found. Rule (13-13) is used: $\sqrt[n]{ab} = \sqrt[n]{a}\,\sqrt[n]{b}$. For example,

$$\sqrt{50} = \sqrt{25 \cdot 2} = \sqrt{25}\,\sqrt{2} = 5\sqrt{2}\,.$$

EXAMPLE 2

Remove as many factors as possible from the radicand.

a. $\sqrt{8} = \sqrt{4 \cdot 2} = \sqrt{4}\,\sqrt{2} = 2\sqrt{2}\,.$

b. $\sqrt[4]{48} = \sqrt[4]{16 \cdot 3} = \sqrt[4]{16}\,\sqrt[4]{3} = 2\sqrt[4]{3}\,.$

c. $\sqrt[3]{x^3y} = \sqrt[3]{x^3}\,\sqrt[3]{y} = x\sqrt[3]{y}\,.$

d. $\sqrt{25x^7} = \sqrt{25x^6x} = \sqrt{25x^6}\,\sqrt{x} = 5x^3\sqrt{x}\,.$

e. $\sqrt[5]{64x^6y^{14}z^2} = \sqrt[5]{32x^5y^{10}(2xy^4z^2)} = 2xy^2\sqrt[5]{2xy^4z^2}$, since $\sqrt[5]{32x^5y^{10}}$ is $2xy^2$.

When a radicand is a fraction, you may use the rule $\sqrt[n]{\dfrac{a}{b}} = \dfrac{\sqrt[n]{a}}{\sqrt[n]{b}}$ to get an equal expression in which the radicand is not a fraction. This is called **rationalizing the denominator**. For example,

$$\sqrt{\frac{7}{16}} = \frac{\sqrt{7}}{\sqrt{16}} = \frac{\sqrt{7}}{4}\,.$$

Sometimes you have to rewrite the radicand before using the rule. This is the case in

$$\sqrt{\frac{3}{7}}\,.$$

Since the index is 2, we want the denominator to be a square of a number. We can make the denominator 7^2 if we multiply numerator and denominator by 7. Thus,

$$\sqrt{\frac{3}{7}} = \sqrt{\frac{3}{7} \cdot \frac{7}{7}} = \sqrt{\frac{21}{7^2}} = \frac{\sqrt{21}}{\sqrt{7^2}} = \frac{\sqrt{21}}{7}\,.$$

EXAMPLE 3

Rationalizing the denominator.

a. $\sqrt[4]{\dfrac{y}{x^8}} = \dfrac{\sqrt[4]{y}}{\sqrt[4]{x^8}} = \dfrac{\sqrt[4]{y}}{x^2}$.

b. $\sqrt{\dfrac{21}{x}} = \sqrt{\dfrac{21}{x} \cdot \dfrac{x}{x}} = \dfrac{\sqrt{21x}}{\sqrt{x^2}} = \dfrac{\sqrt{21x}}{x}$.

c. $\sqrt[5]{\dfrac{x}{y^2}}$.

Since the index is 5, we want the denominator to be a fifth power. We'll multiply numerator and denominator by y^3.

$$\sqrt[5]{\frac{x}{y^2}} = \sqrt[5]{\frac{x}{y^2} \cdot \frac{y^3}{y^3}} = \sqrt[5]{\frac{xy^3}{y^5}} = \frac{\sqrt[5]{xy^3}}{\sqrt[5]{y^5}} = \frac{\sqrt[5]{xy^3}}{y}.$$

d. $\sqrt[3]{\dfrac{2}{3x^4y^2}}$.

Since the index is 3, we want each factor in the denominator to be a cube. This will be the case if we multiply the 3 by 3^2, the x^4 by x^2, and the y^2 by y. Thus we multiply both numerator and denominator by 3^2x^2y.

$$\sqrt[3]{\frac{2}{3x^4y^2}} = \sqrt[3]{\frac{2}{3x^4y^2} \cdot \frac{3^2x^2y}{3^2x^2y}} = \sqrt[3]{\frac{2 \cdot 3^2x^2y}{3^3x^6y^3}} = \frac{\sqrt[3]{18x^2y}}{\sqrt[3]{(3x^2y)^3}} = \frac{\sqrt[3]{18x^2y}}{3x^2y}.$$

Another rule for radicals involves a root of a root of a number.

$$\boxed{\sqrt[m]{\sqrt[n]{a}} = \sqrt[mn]{a}.} \qquad (13\text{-}17)$$

For example, let's find $\sqrt[3]{\sqrt{64}}$ in two ways. Without using rule (13-17), we have

$$\sqrt[3]{\sqrt{64}} = \sqrt[3]{8} = 2.$$

By using rule (13-17), we have

$$\sqrt[3]{\sqrt{64}} = \sqrt[6]{64} = 2.$$

EXAMPLE 4

Use of the rule $\sqrt[m]{\sqrt[n]{a}} = \sqrt[mn]{a}$.

a. $\sqrt[3]{\sqrt[4]{2}} = \sqrt[12]{2}$.

b. $\sqrt{\sqrt[3]{x}} = \sqrt[6]{x}$.

c. $\sqrt[4]{81} = \sqrt{\sqrt{81}} = \sqrt{9} = 3$.

Sometimes it is possible to *reduce the index* of a radical, as the following shows.

$$\sqrt[6]{x^3} = x^{3/6} = x^{1/2} = \sqrt{x}.$$

Here the index 6 was reduced to 2. You may also use the rule $\sqrt[mn]{a} = \sqrt[m]{\sqrt[n]{a}}$ to do such a problem.

$$\sqrt[6]{x^3} = \sqrt{\sqrt[3]{x^3}} = \sqrt{x}.$$

EXAMPLE 5

Reducing the index.

a. $\sqrt[4]{25} = \sqrt[4]{5^2} = 5^{2/4} = 5^{1/2} = \sqrt{5}$, or

$$\sqrt[4]{25} = \sqrt{\sqrt{25}} = \sqrt{5}.$$

b. $\sqrt[6]{16x^2}$. We can't conveniently take the sixth root of the radicand, but we can take the square root. Since $6 = 3 \cdot 2$, we write

$$\sqrt[6]{16x^2} = \sqrt[3]{\sqrt{16x^2}} = \sqrt[3]{4x},$$

or

$$\sqrt[6]{16x^2} = \sqrt[6]{(4x)^2} = (4x)^{2/6} = (4x)^{1/3} = \sqrt[3]{4x}.$$

c. $\sqrt[12]{8x^6y^9} = \sqrt[4]{\sqrt[3]{8x^6y^9}} = \sqrt[4]{2x^2y^3}$.

We say that a radical is **simplified** when:

1. As many factors as possible are removed from the radicand (Example 2).
2. The denominator is rationalized (Example 3).
3. The index cannot be reduced (Example 5).

EXAMPLE 6

Simplify the following.

a. $\sqrt{\frac{x^5}{z}} = \sqrt{\frac{x^5}{z} \cdot \frac{z}{z}} = \frac{\sqrt{x^5 z}}{\sqrt{z^2}} = \frac{\sqrt{x^5 z}}{z} = \frac{\sqrt{x^4(xz)}}{z} = \frac{x^2\sqrt{xz}}{z}.$

b. $\sqrt[4]{x^6 y^{10}} = \sqrt[4]{x^4 y^8 (x^2 y^2)} = xy^2\sqrt[4]{x^2 y^2} = xy^2\sqrt[4]{(xy)^2}$

$= xy^2(xy)^{2/4} = xy^2(xy)^{1/2}$

$= xy^2\sqrt{xy}.$

c. $\sqrt[6]{\frac{x^3}{y^9}} = \sqrt[6]{\frac{x^3}{y^9} \cdot \frac{y^3}{y^3}} = \sqrt[6]{\frac{x^3 y^3}{y^{12}}} = \frac{\sqrt[6]{x^3 y^3}}{\sqrt[6]{(y^2)^6}} = \frac{\sqrt[6]{(xy)^3}}{y^2} = \frac{\sqrt{xy}}{y^2}.$

d. $\sqrt[3]{x^{-6} y^6} = \sqrt[3]{\frac{y^6}{x^6}} = \sqrt[3]{\left(\frac{y^2}{x^2}\right)^3} = \frac{y^2}{x^2}.$

Completion Set 13-3

Fill in the blanks.

1. $\sqrt[4]{x^4} =$ ____; $\sqrt[3]{z^6} = \sqrt[3]{(z^2)^3} =$ _____.

2. $\sqrt[5]{x^{10} y^{15}} = \sqrt[5]{(\_\_\_\_\_\_)^5} =$ ________.

3. $\sqrt{32} = \sqrt{(\_\_\_)\cdot 2} = \sqrt{\_\_\_} \cdot \sqrt{2} =$ ___ $\sqrt{2}$.

4. $\sqrt[3]{x^5 y^4} = \sqrt[3]{(\_\_\_\_\_\_)x^2 y} = \sqrt[3]{\_\_\_\_\_\_} \cdot \sqrt[3]{x^2 y} =$ ____ $\sqrt[3]{x^2 y}$.

5. $\sqrt{\dfrac{5}{36}} = \dfrac{\sqrt{5}}{\sqrt{}} = \dfrac{\sqrt{5}}{(\quad)}.$

6. $\sqrt[5]{\dfrac{x^3}{y^2}} = \sqrt[5]{\dfrac{x^3(\quad)}{y^5}} = \dfrac{\sqrt[5]{x^3(\quad)}}{\sqrt[5]{y^5}} = \dfrac{\sqrt[5]{x^3(\quad)}}{y}.$

Problem Set 13-3

In Problems **1–14**, *find the root.*

1. $\sqrt[4]{x^8}$	**2.** $\sqrt[3]{x^9}$	**3.** $\sqrt[3]{8x^{12}}$
4. $\sqrt[6]{9^{12}}$	**5.** $\sqrt{9x^{16}y^{18}}$	**6.** $\sqrt[3]{x^3y^3z^6}$
7. $\sqrt[3]{x^3y^6z^9}$	**8.** $\sqrt{(x^3y^7)(x^7y^{13})}$	**9.** $\sqrt[5]{\dfrac{x^{15}}{y^{20}}}$
10. $\sqrt[4]{\dfrac{16x^8}{y^{16}}}$	**11.** $\sqrt{\sqrt{x^8}}$	**12.** $\sqrt[3]{\sqrt{x^{18}}}$
13. $\sqrt[4]{\sqrt[3]{x^{12}}}$	**14.** $\sqrt{\sqrt[6]{x^{24}}}$	

In Problems **15–72**, *simplify the radicals.*

15. $\sqrt{12}$	**16.** $\sqrt{18}$	**17.** $\sqrt{32}$
18. $\sqrt{20}$	**19.** $\sqrt[3]{16}$	**20.** $\sqrt[3]{54}$
21. $\sqrt{x^7}$	**22.** $\sqrt[3]{x^7}$	**23.** $\sqrt[3]{24x^6}$
24. $\sqrt{8x^3}$	**25.** $\sqrt[4]{x^9y^2}$	**26.** $\sqrt[5]{x^{17}y^{20}}$
27. $\sqrt[3]{x^6yz^4}$	**28.** $\sqrt{x^5y^4z}$	**29.** $\sqrt[3]{8a^3y^5}$
30. $\sqrt[3]{24(a+b)^7}$	**31.** $\sqrt[5]{x^{23}y^{10}z^6}$	**32.** $\sqrt[4]{x^3y^3z^7}$
33. $\sqrt{81xy^2z^3w^4}$	**34.** $\sqrt[4]{32x^{17}}$	**35.** $\sqrt{\dfrac{1}{2}}$
36. $\sqrt{\dfrac{1}{5}}$	**37.** $\sqrt[3]{\dfrac{2}{5}}$	**38.** $\sqrt[3]{\dfrac{1}{3}}$
39. $\sqrt[3]{\dfrac{x^2}{y^3}}$	**40.** $\sqrt[4]{\dfrac{3}{x^4}}$	**41.** $\sqrt{\dfrac{x}{y^4}}$

42. $\sqrt[3]{\dfrac{y^2}{x^{12}}}$ 43. $\sqrt{\dfrac{2x}{y}}$ 44. $\sqrt{\dfrac{y}{x^9}}$

45. $\sqrt[3]{\dfrac{2}{xy^2}}$ 46. $\sqrt[3]{\dfrac{2y}{xz}}$ 47. $\sqrt[4]{\dfrac{3}{2x^7yz^2}}$

48. $\sqrt[4]{\dfrac{1}{xy^3z^5}}$ 49. $\sqrt[6]{x^2}$ 50. $\sqrt[8]{(xy)^4}$

51. $\sqrt[4]{9}$ 52. $\sqrt[6]{8}$ 53. $\sqrt[4]{16x^4y^2}$

54. $\sqrt[6]{27x^3y^3z^3}$ 55. $\sqrt[8]{\dfrac{x^4}{y^4}}$ 56. $\sqrt[6]{\dfrac{16x^2}{y^2}}$

57. $\sqrt[12]{x^2y^2z^{10}}$ 58. $\sqrt[15]{x^3y^{15}z^6}$ 59. $\sqrt[4]{x^{10}y^2}$

60. $\sqrt[12]{x^{15}y^{27}}$ 61. $\sqrt[6]{x^{20}x^{26}}$ 62. $\sqrt[8]{x^2z^{10}}$

63. $\sqrt[3]{\sqrt{x^{12}y^5w^{25}}}$ 64. $\sqrt{\sqrt[3]{64x^{12}y^{11}w^7}}$ 65. $\sqrt[4]{\dfrac{16x^5}{y^8}}$

66. $\sqrt[3]{\dfrac{x^6}{y^{-3}}}$ 67. $\sqrt[4]{\dfrac{1}{4}}$ 68. $\sqrt[6]{\dfrac{x^9}{y^{15}}}$

69. $\sqrt[8]{\dfrac{x^4}{y^{12}}}$ 70. $\sqrt{x^4(2 + x)}$ 71. $\sqrt[3]{\sqrt{x^3}}$

72. $\sqrt[4]{\sqrt[3]{\sqrt{x^{48}}}}$

13-4 OPERATIONS WITH RADICALS

You may combine radicals in addition or subtraction when the radicals have the *same index* and the *same radicand*. Such radicals are called **like radicals**. The distributive law is used just as it was in combining *like terms* in Chapter 4. For example,

$$8\sqrt{xy} - 4\sqrt{xy} + 7\sqrt{xy} = (8 - 4 + 7)\sqrt{xy} = 11\sqrt{xy}\ .$$

Sometimes, radicals that are *unlike* can be combined by simplifying them first. For example, in $\sqrt[3]{81} - \sqrt[3]{24}$ the radicands are different. But we can write

$$\sqrt[3]{81} - \sqrt[3]{24} = \sqrt[3]{27\cdot 3} - \sqrt[3]{8\cdot 3}$$
$$= 3\sqrt[3]{3} - 2\sqrt[3]{3} = \sqrt[3]{3}\,.$$

EXAMPLE 1

Adding and subtracting radicals.

a. $(16\sqrt[3]{x} + 15) - (12 - 2\sqrt[3]{x}) = 16\sqrt[3]{x} + 15 - 12 + 2\sqrt[3]{x}$
$$= 18\sqrt[3]{x} + 3.$$

b. $3\sqrt{75} - 2\sqrt{12} + \sqrt{7} = 3\sqrt{25\cdot 3} - 2\sqrt{4\cdot 3} + \sqrt{7}$
$$= 3(5\sqrt{3}) - 2(2\sqrt{3}) + \sqrt{7}$$
$$= 15\sqrt{3} - 4\sqrt{3} + \sqrt{7} = 11\sqrt{3} + \sqrt{7}\,.$$

c. $\sqrt[3]{\frac{3}{4}} + \sqrt[3]{\frac{2}{9}} - \sqrt[3]{\frac{1}{36}} = \sqrt[3]{\frac{3}{4}\cdot\frac{2}{2}} + \sqrt[3]{\frac{2}{9}\cdot\frac{3}{3}} - \sqrt[3]{\frac{1}{6^2}\cdot\frac{6}{6}}$
$$= \frac{\sqrt[3]{6}}{2} + \frac{\sqrt[3]{6}}{3} - \frac{\sqrt[3]{6}}{6}$$
$$= \left(\frac{1}{2} + \frac{1}{3} - \frac{1}{6}\right)\sqrt[3]{6} = \frac{2}{3}\sqrt[3]{6}\,.$$

Radicals having the same index can be multiplied by means of the rule $\sqrt[n]{a}\cdot\sqrt[n]{b} = \sqrt[n]{ab}$. For example,

$$\sqrt{2x^3}\cdot\sqrt{8x} = \sqrt{(2x^3)(8x)} = \sqrt{16x^4} = 4x^2.$$

EXAMPLE 2

Multiplying radicals with the same index.

a. $\sqrt{yz}\cdot\sqrt{3z} = \sqrt{3yz^2} = z\sqrt{3y}\,.$

b. $\sqrt[4]{\frac{3}{2}}\cdot\sqrt[4]{6} = \sqrt[4]{\frac{3}{2}\cdot 6} = \sqrt[4]{9} = \sqrt[4]{3^2} = 3^{2/4} = 3^{1/2} = \sqrt{3}\,.$

c. $\sqrt{3}\,(\sqrt{3} - \sqrt{6}) = \sqrt{3}\cdot\sqrt{3} - \sqrt{3}\cdot\sqrt{6}$
$$= 3 - \sqrt{18} = 3 - \sqrt{9\cdot 2} = 3 - 3\sqrt{2}\,.$$

d. $\left(a\sqrt[3]{a^2b^3}\right)^4 = a^4\left(\sqrt[3]{a^2b^3}\right)^4$ [rule (13-6)]

$= a^4\sqrt[3]{(a^2b^3)^4}$ [definition (13-12)]

$= a^4\sqrt[3]{a^8b^{12}}$ [rule (13-6)]

$= a^4\left(a^2b^4\sqrt[3]{a^2}\right)$ [simplifying]

$= a^6b^4\sqrt[3]{a^2}$.

EXAMPLE 3

Multiplying radicals with the same index.

a. $(\sqrt{8} - \sqrt{3})(\sqrt{18} - \sqrt{48})$. We'll simplify the radicals before multiplying.

$$\begin{aligned}&(2\sqrt{2} - \sqrt{3})(3\sqrt{2} - 4\sqrt{3})\\ &= 2\sqrt{2}\cdot 3\sqrt{2} - 2\sqrt{2}\cdot 4\sqrt{3} - \sqrt{3}\cdot 3\sqrt{2} + \sqrt{3}\cdot 4\sqrt{3}\\ &= 6(\sqrt{2})^2 - 8\sqrt{6} - 3\sqrt{6} + 4(\sqrt{3})^2\\ &= 12 - 11\sqrt{6} + 12\\ &= 24 - 11\sqrt{6}\,.\end{aligned}$$

b. $(\sqrt{5} - 2\sqrt{7})(\sqrt{5} + 2\sqrt{7})$. This has the form $(a - b)(a + b)$. Thus,

$$\begin{aligned}(\sqrt{5} - 2\sqrt{7})(\sqrt{5} + 2\sqrt{7}) &= (\sqrt{5})^2 - (2\sqrt{7})^2\\ &= 5 - 4(7) = 5 - 28 = -23.\end{aligned}$$

c. $(\sqrt{x} - \sqrt{y})^2$. This is the square of a binomial.

$$\begin{aligned}(\sqrt{x} - \sqrt{y})^2 &= (\sqrt{x})^2 - 2\sqrt{x}\sqrt{y} + (\sqrt{y})^2\\ &= x - 2\sqrt{xy} + y.\end{aligned}$$

In the last section you rationalized the denominator for a radical such as $\sqrt{\frac{3}{7}}$. We also use the phrase "rationalizing the denominator" in another situation: when a fraction that has a radical in its denominator, such as $\dfrac{2}{\sqrt[3]{5}}$, is rewritten so that there is no radical in the denominator.

Let's rationalize the denominator of $\dfrac{2}{\sqrt[3]{5}}$. First we write $\dfrac{2}{\sqrt[3]{5}}$ with fractional exponents.

$$\frac{2}{\sqrt[3]{5}} = \frac{2}{5^{1/3}}.$$

We want to get an integer exponent in the denominator. To do this we multiply numerator and denominator by $5^{2/3}$, since $5^{1/3} \cdot 5^{2/3} = 5^1 = 5$.

$$\frac{2}{\sqrt[3]{5}} = \frac{2}{5^{1/3}} = \frac{2}{5^{1/3}} \cdot \frac{5^{2/3}}{5^{2/3}} = \frac{2 \cdot 5^{2/3}}{5^{3/3}} = \frac{2\sqrt[3]{5^2}}{5^1} = \frac{2\sqrt[3]{25}}{5}.$$

With square roots in a denominator, such as in $\frac{4}{\sqrt{3}}$, we usually avoid fractional exponents and simply write

$$\frac{4}{\sqrt{3}} = \frac{4}{\sqrt{3}} \cdot \frac{\sqrt{3}}{\sqrt{3}} = \frac{4\sqrt{3}}{3}.$$

EXAMPLE 4

Rationalizing the denominator.

a. $\frac{6}{\sqrt[5]{2}} = \frac{6}{2^{1/5}} = \frac{6}{2^{1/5}} \cdot \frac{2^{4/5}}{2^{4/5}} = \frac{6\sqrt[5]{2^4}}{2} = 3\sqrt[5]{16}.$

b. $\frac{3}{\sqrt{2x}} = \frac{3}{\sqrt{2x}} \cdot \frac{\sqrt{2x}}{\sqrt{2x}} = \frac{3\sqrt{2x}}{2x}.$

c. $\frac{7}{2\sqrt[5]{3x^3}} = \frac{7}{2(3x^3)^{1/5}} = \frac{7}{2(3^{1/5}x^{3/5})} \cdot \frac{3^{4/5}x^{2/5}}{3^{4/5}x^{2/5}} = \frac{7(3^4x^2)^{1/5}}{2 \cdot 3 \cdot x} = \frac{7\sqrt[5]{81x^2}}{6x}.$

Division of radicals with the same index may be handled by using the rule $\frac{\sqrt[n]{a}}{\sqrt[n]{b}} = \sqrt[n]{\frac{a}{b}}$. For example,

$$\frac{\sqrt{7}}{\sqrt{3}} = \sqrt{\frac{7}{3}} = \sqrt{\frac{7}{3} \cdot \frac{3}{3}} = \frac{\sqrt{21}}{3}.$$

EXAMPLE 5

Division of radicals.

a. $\frac{\sqrt[3]{3x}}{\sqrt[3]{2x}} = \sqrt[3]{\frac{3x}{2x}} = \sqrt[3]{\frac{3}{2}} = \sqrt[3]{\frac{3}{2} \cdot \frac{2^2}{2^2}} = \frac{\sqrt[3]{12}}{2}.$

b.

$$\frac{\sqrt{5xy}}{\sqrt{18xy^3}} = \frac{\sqrt{5xy}}{3y\sqrt{2xy}} = \frac{1}{3y}\sqrt{\frac{5xy}{2xy}} = \frac{1}{3y}\sqrt{\frac{5}{2}} = \frac{1}{3y}\sqrt{\frac{5}{2}\cdot\frac{2}{2}} = \frac{\sqrt{10}}{6y}.$$

Sometimes a denominator of a fraction has two terms and involves square roots, such as $2 - \sqrt{3}$ or $\sqrt{5} + \sqrt{2}$. The denominator may be rationalized by multiplying by an expression that makes the denominator a difference of two squares. For example,

$$\begin{aligned}\frac{4}{\sqrt{5}+\sqrt{2}} &= \frac{4}{\sqrt{5}+\sqrt{2}}\cdot\frac{\sqrt{5}-\sqrt{2}}{\sqrt{5}-\sqrt{2}}\\ &= \frac{4(\sqrt{5}-\sqrt{2})}{(\sqrt{5})^2-(\sqrt{2})^2} = \frac{4(\sqrt{5}-\sqrt{2})}{5-2}\\ &= \frac{4(\sqrt{5}-\sqrt{2})}{3}.\end{aligned}$$

EXAMPLE 6

a. $$\frac{2}{x-\sqrt{3}} = \frac{2}{x-\sqrt{3}}\cdot\frac{x+\sqrt{3}}{x+\sqrt{3}} = \frac{2(x+\sqrt{3})}{x^2-(\sqrt{3})^2} = \frac{2(x+\sqrt{3})}{x^2-3}.$$

b. $$\begin{aligned}\frac{\sqrt{2}}{\sqrt{2}-\sqrt{3}} &= \frac{\sqrt{2}}{\sqrt{2}-\sqrt{3}}\cdot\frac{\sqrt{2}+\sqrt{3}}{\sqrt{2}+\sqrt{3}} = \frac{\sqrt{2}(\sqrt{2}+\sqrt{3})}{2-3}\\ &= \frac{2+\sqrt{6}}{-1} = -2-\sqrt{6}.\end{aligned}$$

c. $$\begin{aligned}\frac{\sqrt{5}-\sqrt{2}}{\sqrt{5}+\sqrt{2}} &= \frac{\sqrt{5}-\sqrt{2}}{\sqrt{5}+\sqrt{2}}\cdot\frac{\sqrt{5}-\sqrt{2}}{\sqrt{5}-\sqrt{2}}\\ &= \frac{(\sqrt{5}-\sqrt{2})^2}{5-2} = \frac{5-2\sqrt{5}\sqrt{2}+2}{3}\\ &= \frac{7-2\sqrt{10}}{3}.\end{aligned}$$

Completion Set 13-4

Fill in the blanks.

1. $6\sqrt[4]{3x} + 2\sqrt[4]{3x} - 3\sqrt[4]{3x} = \_\_\_\_ \sqrt[4]{3x}$.

2. $\sqrt{x^3} + 2x\sqrt{x} = \sqrt{x^2 \cdot x} + 2x\sqrt{x} = \_\_\_\_ \sqrt{x} + 2x\sqrt{x} = \_\_\_\_\_ \sqrt{x}$.

3. $\sqrt[3]{y^2} \cdot \sqrt[3]{y} = \sqrt[3]{(\_\_\_)} = \_\_\_\_.$

4. $\dfrac{1}{\sqrt[3]{x}} = \dfrac{1}{x^{1/3}} \cdot \dfrac{(\qquad)}{x^{2/3}} = \dfrac{\sqrt[3]{(\qquad)}}{x}.$

5. $\dfrac{\sqrt{10xy}}{\sqrt{5x}} = \sqrt{\dfrac{10xy}{5x}} = \sqrt{(\_\_\_\_)}.$

6. $\dfrac{1}{3+\sqrt{2}} = \dfrac{1}{3+\sqrt{2}} \cdot \dfrac{3-\sqrt{2}}{3-\sqrt{2}} = \dfrac{3-\sqrt{2}}{(\_\_\_)}.$

Problem Set 13-4

*In Problems **1–78**, perform the operations and simplify.*

1. $4\sqrt[3]{3} - 2\sqrt[3]{3} + \sqrt[3]{3}$
2. $6\sqrt{7} - (2\sqrt{7} + 3\sqrt{7})$
3. $x^2\sqrt{2x} - 3x^2\sqrt{2x} + 4x^2\sqrt{2x}$
4. $xy\sqrt{x} + 4xy\sqrt{x} - 2xy\sqrt{x}$
5. $3\sqrt{75} - 2\sqrt{12}$
6. $5\sqrt{18} - (\sqrt{2} + 1)$
7. $5\sqrt{8} - (2\sqrt{18} - 4\sqrt{32})$
8. $\sqrt{75} - (\sqrt{27} - 2\sqrt{3})$
9. $2y\sqrt{16x} - 3y\sqrt{9x}$
10. $3\sqrt[3]{16} - \sqrt[3]{54}$
11. $\sqrt[3]{128} - 6\sqrt[3]{16}$
12. $2(\sqrt[3]{54} - 2\sqrt[3]{128}) - 2\sqrt[3]{16}$
13. $30\sqrt{\frac{1}{15}} - 72\sqrt{\frac{5}{12}} + 50\sqrt{\frac{3}{5}}$
14. $20\sqrt{\frac{2}{5}} - 3\sqrt{40} - 4\sqrt{\frac{5}{2}}$
15. $4\sqrt[3]{\frac{1}{2}} + \sqrt[3]{32} - 3\sqrt[3]{4}$
16. $\sqrt{\frac{1}{6}} + \sqrt{\frac{2}{3}} + \sqrt{\frac{3}{2}}$
17. $\sqrt[4]{4x^4} - 3\sqrt[6]{8x^6}$
18. $\sqrt[4]{4x^2} - 3\sqrt[6]{8x^3}$
19. $\sqrt{2}\sqrt{8}$
20. $\sqrt{12}\sqrt{3}$

21. $\sqrt{3}\,\sqrt{4}$

22. $\sqrt{9}\,\sqrt{2}$

23. $(2\sqrt{6})(3\sqrt{3})$

24. $(5\sqrt{27})(2\sqrt{3})$

25. $\sqrt[3]{3}\,\sqrt[3]{9}\,\sqrt[3]{12}$

26. $\sqrt[3]{4}\,\sqrt[3]{16}\,\sqrt[3]{-1}$

27. $\sqrt{2x}\,\sqrt{x}\,\sqrt{3x}$

28. $\sqrt{5y}\,\sqrt{2y}\,\sqrt{y}$

29. $(-\sqrt{3})^2$

30. $(-\sqrt{5})^3$

31. $\left(2\sqrt[3]{x}\right)^4$

32. $\left(\frac{1}{3}\sqrt{x}\right)^3$

33. $\sqrt{30}\,\sqrt{\frac{2}{3}}$

34. $\sqrt{\frac{5}{8}}\,\sqrt{\frac{16}{3}}$

35. $\sqrt{3}\,(2\sqrt{6} - 4\sqrt{3})$

36. $\sqrt{2}\,(\sqrt{2} + 2\sqrt{18})$

37. $(\sqrt{6} + \sqrt{2})(\sqrt{2} - 2\sqrt{6})$

38. $(\sqrt{3} - 1)(\sqrt{3} + 2)$

39. $(2 + \sqrt{7})(2 - \sqrt{7})$

40. $(\sqrt{7} - \sqrt{2})(\sqrt{7} + \sqrt{2})$

41. $(\sqrt{5} + 2)^2$

42. $(1 - \sqrt{5})^2$

43. $(\sqrt{x} - 1)(2\sqrt{x} + 5)$

44. $\sqrt{ab}\,\sqrt{a^2bc^2}\,\sqrt{abc^2}$

45. $\sqrt{3xy^2}\,\sqrt{2xy}\,\sqrt{3xy^3}$

46. $5\sqrt[4]{ab}\left(1 - 2\sqrt[4]{ab}\right)$

47. $(\sqrt{6} - 5)^2$

48. $(\sqrt{3} + 4)^2$

49. $(3\sqrt{8} + \sqrt{3})(8\sqrt{3} - \sqrt{8})$

50. $(5\sqrt{2} + \sqrt{5})(2\sqrt{5} - \sqrt{2})$

51. $\dfrac{3}{\sqrt{7}}$

52. $\dfrac{5}{\sqrt{11}}$

53. $\dfrac{4}{\sqrt{2x}}$

54. $\dfrac{y}{\sqrt{2y}}$

55. $\dfrac{1}{\sqrt[3]{2}}$

56. $\dfrac{3}{\sqrt[4]{2}}$

57. $\dfrac{1}{\sqrt[3]{3x}}$

58. $\dfrac{4}{3\sqrt[3]{x^2}}$

59. $\dfrac{\sqrt{32}}{\sqrt{2}}$

60. $\dfrac{\sqrt{18}}{\sqrt{2}}$

61. $\dfrac{\sqrt{2a^3}}{\sqrt{a}}$

62. $\dfrac{\sqrt{3x^5}}{\sqrt{x}}$

63. $\dfrac{2x\sqrt[4]{x^7}}{\sqrt[4]{x^{12}}}$

64. $\dfrac{3a^2b\sqrt[5]{a^3}}{\sqrt[5]{a^5}}$

65. $\dfrac{\sqrt[3]{6}}{\sqrt[3]{4x}}$

66. $\dfrac{\sqrt[3]{3y}}{\sqrt[3]{x}}$

67. $\dfrac{\sqrt[3]{2x}}{\sqrt[3]{5xy^2}}$

68. $\dfrac{\sqrt{7xy}}{\sqrt{14xy^3}}$

69. $\dfrac{1}{2+\sqrt{3}}$

70. $\dfrac{1}{1-\sqrt{2}}$

71. $\dfrac{\sqrt{2}}{\sqrt{3}-\sqrt{6}}$

72. $\dfrac{5}{\sqrt{6}+\sqrt{7}}$

73. $\dfrac{2\sqrt{2}}{\sqrt{2}-\sqrt{3}}$

74. $\dfrac{2\sqrt{3}}{\sqrt{5}-\sqrt{2}}$

75. $\dfrac{1}{x+\sqrt{5}}$

76. $\dfrac{3-\sqrt{5}}{\sqrt{4}+\sqrt{2}}$

77. $\dfrac{\sqrt{6}-\sqrt{3}}{\sqrt{6}+\sqrt{3}}$

78. $\dfrac{\sqrt{3}-5\sqrt{2}}{\sqrt{3}+\sqrt{2}}$

13-5 OPERATIONS WITH DIFFERENT INDICES†

If two or more radicals have different indices (plural of "index"), they can be rewritten so that they have the same index. This is usually the first step in multiplying or dividing such radicals. For example,

$$\sqrt{10} \quad \text{and} \quad \sqrt[3]{10}$$

have indices of 2 and 3. Both 2 and 3 divide 6, and 6 is the smallest number that is exactly divisible by 2 and 3. We say that 6 is the **least common multiple** (L.C.M.) of 2 and 3. Finding an L.C.M. is similar to finding an L.C.D., except that no fractions are involved. Now we change the indices to 6 as follows:

$$\sqrt{10} = 10^{1/2} = 10^{3/6} = \sqrt[6]{10^3}$$

$$\text{and} \quad \sqrt[3]{10} = 10^{1/3} = 10^{2/6} = \sqrt[6]{10^2}\,.$$

EXAMPLE 1

Change the following to radicals with the same index.

a. $\sqrt[4]{x}\,,\ \sqrt[6]{2x^3}\,.$

The L.C.M. of the indices 4 and 6 is 12. Thus,

$$\sqrt[4]{x} = x^{1/4} = x^{3/12} = \sqrt[12]{x^3}\,,$$

†The material in this section is not needed for the remaining chapters.

and $\sqrt[6]{2x^3} = (2x^3)^{1/6} = (2x^3)^{2/12} = \sqrt[12]{(2x^3)^2} = \sqrt[12]{4x^6}$.

b. $\sqrt{x}$, $\sqrt[3]{y^2}$, $\sqrt[5]{x^2y}$.

The L.C.M. of 2, 3, and 5 is 30.

$$\sqrt{x} = x^{1/2} = x^{15/30} = \sqrt[30]{x^{15}},$$

$$\sqrt[3]{y^2} = (y^2)^{1/3} = (y^2)^{10/30} = \sqrt[30]{(y^2)^{10}} = \sqrt[30]{y^{20}},$$

and $$\sqrt[5]{x^2y} = (x^2y)^{1/5} = (x^2y)^{6/30} = \sqrt[30]{(x^2y)^6} = \sqrt[30]{x^{12}y^6}.$$

You may multiply radicals that don't have the same index by first rewriting them so that they do.

EXAMPLE 2

Multiplying radicals with different indices.

a. $\sqrt{2}\,\sqrt[3]{3} = 2^{1/2}\cdot 3^{1/3} = 2^{3/6}3^{2/6}$

$$= \sqrt[6]{2^3}\,\sqrt[6]{3^2} = \sqrt[6]{2^3\cdot 3^2} = \sqrt[6]{8\cdot 9} = \sqrt[6]{72}.$$

b. $\sqrt[4]{3x^2}\,\sqrt[3]{2x^2} = (3x^2)^{1/4}(2x^2)^{1/3}$

$$= (3x^2)^{3/12}(2x^2)^{4/12} = \sqrt[12]{(3x^2)^3}\cdot\sqrt[12]{(2x^2)^4}$$

$$= \sqrt[12]{(3x^2)^3(2x^2)^4} = \sqrt[12]{3^3x^6\cdot 2^4x^8}$$

$$= \sqrt[12]{432x^{14}} = x\sqrt[12]{432x^2}.$$

To divide radicals with different indices, first rewrite each radical so that their indices are the same. Then simplify.

EXAMPLE 3

Dividing radicals with different indices.

a. $$\frac{\sqrt[3]{10}}{\sqrt{10}} = \frac{10^{1/3}}{10^{1/2}} = \frac{10^{2/6}}{10^{3/6}} = \frac{\sqrt[6]{10^2}}{\sqrt[6]{10^3}} = \sqrt[6]{\frac{10^2}{10^3}} = \sqrt[6]{\frac{1}{10}\cdot\frac{10^5}{10^5}} = \frac{\sqrt[6]{10^5}}{10}.$$

b. $\dfrac{\sqrt{2}}{\sqrt[3]{3}} = \dfrac{2^{1/2}}{3^{1/3}} = \dfrac{2^{3/6}}{3^{2/6}} = \dfrac{\sqrt[6]{2^3}}{\sqrt[6]{3^2}} = \sqrt[6]{\dfrac{2^3}{3^2} \cdot \dfrac{3^4}{3^4}} = \dfrac{\sqrt[6]{8 \cdot 81}}{3} = \dfrac{\sqrt[6]{648}}{3}.$

c. $\dfrac{\sqrt[4]{2}}{\sqrt[3]{xy^2}} = \dfrac{2^{1/4}}{x^{1/3}y^{2/3}} = \dfrac{2^{3/12}}{x^{4/12}y^{8/12}} = \sqrt[12]{\dfrac{2^3}{x^4y^8} \cdot \dfrac{x^8y^4}{x^8y^4}} = \dfrac{\sqrt[12]{8x^8y^4}}{xy}.$

Problem Set 13-5

In Problems **1–6**, *write the radicals so that they have the same index.*

1. $\sqrt[3]{x}, \sqrt{x}$

2. $\sqrt{3x}, \sqrt[3]{y^2}$

3. $\sqrt[3]{2x}, \sqrt[6]{x^5}$

4. $\sqrt[7]{xy^2}, \sqrt[3]{x^2y^2}$

5. $\sqrt[4]{x^3y}, \sqrt{y}, \sqrt[8]{xy^5}$

6. $\sqrt{x}, \sqrt[3]{y}, \sqrt[4]{x^2y}$

In Problems **7–28**, *perform the operations and simplify.*

7. $\sqrt{5} \cdot \sqrt[4]{5}$

8. $\sqrt[3]{2} \cdot \sqrt[6]{3}$

9. $\sqrt{3x} \cdot \sqrt[3]{x^2}$

10. $\sqrt[4]{x} \cdot \sqrt[3]{x^2}$

11. $\sqrt[3]{9x} \cdot \sqrt{3x}$

12. $\sqrt{8x^2} \cdot \sqrt[3]{4x^3}$

13. $(3\sqrt[3]{x^2y})(2\sqrt{2x})$

14. $5\sqrt{ab^2} \cdot \sqrt[3]{ab}$

15. $\sqrt[5]{x^2y^3} \cdot \sqrt[4]{xy^2}$

16. $\sqrt[6]{xy^5} \cdot \sqrt[3]{x^2y}$

17. $\sqrt{x} \cdot \sqrt[3]{x^2y} \cdot \sqrt[4]{x^4y^2}$

18. $\sqrt{x} \cdot \sqrt[4]{y} \cdot \sqrt[8]{xy}$

19. $\dfrac{\sqrt{6}}{\sqrt[4]{6}}$

20. $\dfrac{\sqrt[3]{5}}{\sqrt[6]{5}}$

21. $\dfrac{\sqrt[3]{3}}{\sqrt{2}}$

22. $\dfrac{\sqrt[8]{3}}{\sqrt[4]{2}}$

23. $\dfrac{\sqrt{2x}}{\sqrt[3]{x}}$

24. $\dfrac{\sqrt{2x}}{\sqrt[3]{y}}$

25. $\dfrac{\sqrt[9]{2y}}{\sqrt[3]{y}}$

26. $\dfrac{\sqrt[3]{3y}}{\sqrt[9]{x}}$

27. $\dfrac{2\sqrt{xy}}{\sqrt[4]{xy}}$

28. $\dfrac{2x\sqrt{x}}{\sqrt[3]{2xy}}$

13-6 RADICAL EQUATIONS

The equation $\sqrt{x-7} = 4$ has the variable x under the radical sign. It is called a **radical equation**. One way to solve it is to raise both sides to the same power so as to get rid of the radical. This step sometimes leads to a value of x that actually doesn't satisfy the given equation (that is, an extraneous root). It is important that all "solutions" be checked.

Let's solve our equation by squaring both sides.

$$\sqrt{x-7} = 4$$
$$(\sqrt{x-7})^2 = 4^2$$
$$x - 7 = 16$$
$$x = 23.$$

Substituting 23 for x in $\sqrt{x-7} = 4$ gives $\sqrt{23-7} = 4$ or $\sqrt{16} = 4$, which is true. Thus the solution is $x = 23$.

EXAMPLE 1

Solving radical equations.

a. $\sqrt{x+2} - x + 4 = 0.$

It helps to rewrite the equation so that the radical is by itself on one side.

$$\sqrt{x+2} = x - 4$$
$$x + 2 = (x-4)^2 \qquad \text{[squaring both sides]}$$
$$x + 2 = x^2 - 8x + 16$$
$$0 = x^2 - 9x + 14$$
$$0 = (x-7)(x-2) \qquad \text{[factoring]}$$
$$x = 7 \quad \text{or} \quad x = 2.$$

Now we check these values in the *original* equation.

Replacing x by 7 gives $\sqrt{7+2} - 7 + 4 = 0$ or $3 - 7 + 4 = 0$, which is true.
Replacing x by 2 gives $\sqrt{2+2} - 2 + 4 = 0$ or $2 - 2 + 4 = 0$, which is *false*.

Thus the only solution is $x = 7$.

b. $\sqrt{y-3} - \sqrt{y} = -3.$

When an equation has two terms involving radicals, you should first write the equation so that one radical is on each side.

$$\sqrt{y-3} = \sqrt{y} - 3$$

$$y - 3 = (\sqrt{y} - 3)^2 \qquad \text{[squaring both sides]}$$

$$y - 3 = y - 6\sqrt{y} + 9$$

$$6\sqrt{y} = 12$$

$$\sqrt{y} = 2$$

$$y = 4. \qquad \text{[squaring both sides]}$$

Replacing y by 4 in the left side of the original equation gives $\sqrt{1} - \sqrt{4}$, which is -1. Since this does not equal the right side, -3, there is **no solution.**

c. $\sqrt[3]{x-4} = 3.$

$$\sqrt[3]{x-4} = 3$$

$$x - 4 = 27 \qquad [\textit{cubing} \text{ both sides}]$$

$$x = 31.$$

You may check that $x = 31$ is indeed the solution.

Problem Set 13-6

Solve the following radical equations.

1. $\sqrt{x-2} = 5$

2. $\sqrt{x+7} = 9$

3. $\sqrt{2y-5} = 6$

4. $\sqrt{2x-6} - 16 = 0$

5. $2\sqrt{2x+1} = 3\sqrt{3x-8}$

6. $4\sqrt{8y+1} = 5\sqrt{5y+1}$

7. $\sqrt{x^2+33} = x + 3$

8. $\sqrt{4x-6} - \sqrt{x} = 0$

9. $\sqrt{x} - \sqrt{x+1} = 1$

10. $\sqrt{x-3} + 4 = 1$

11. $\sqrt{x+2} = x - 4$

12. $3\sqrt{x+4} = x - 6$

13. $z + 2 = 2\sqrt{4z-7}$

14. $x + \sqrt{x} - 2 = 0$

15. $\sqrt{x+7} - \sqrt{2x} = 1$

16. $\sqrt{3x} - \sqrt{5x+1} = -1$

17. $\sqrt[3]{x^2+2} = 3$

18. $\sqrt[4]{2x+1} = 3$

13-7 REVIEW

IMPORTANT TERMS

zero exponent *(p. 292)*

negative exponent *(p. 292)*

least common multiple *(p. 317)*

like radicals *(p. 310)*

fractional exponent *(p. 298)*

rationalizing the denominator *(p. 305)*

radical equation *(p. 320)*

REVIEW PROBLEMS

In Problems **1–14**, *evaluate the expressions.*

1. 3^0 **2.** $2\left(-\frac{2}{3}\right)^0$ **3.** 5^{-1}

4. $(-3)^{-1}$ **5.** $4\left(-\frac{2}{3}\right)^{-2}$ **6.** $\dfrac{2^{-1}}{4^{-2}}$

7. $100^{1/2}$ **8.** $64^{1/3}$ **9.** $4^{3/2}$

10. $(25)^{-3/2}$ **11.** $(32)^{-2/5}$ **12.** $\left(\frac{9}{100}\right)^{3/2}$

13. $\left(\frac{1}{16}\right)^{5/4}$ **14.** $\left(-\frac{27}{64}\right)^{2/3}$

In Problems **15–74**, *perform the operations and simplify. Write all answers in terms of positive exponents only. Rationalize all denominators. Avoid fractional exponents in the final form. For example,* $y^{-1}x^{1/2} = \dfrac{\sqrt{x}}{y}$.

15. $\sqrt{32}$ **16.** $\sqrt[3]{24}$ **17.** $\sqrt[3]{2x^3}$

18. $\sqrt{4x}$ **19.** $\sqrt{16x^4}$ **20.** $\sqrt[4]{\dfrac{x}{16}}$

21. $(9z^6)^{1/2}$ **22.** $(16y^8)^{3/4}$ **23.** $\left(\dfrac{27t^3}{8}\right)^{2/3}$

24. $\left(\dfrac{1000}{a^9}\right)^{-2/3}$ **25.** $\dfrac{x^3y^{-2}}{x^5z^2}$ **26.** $\sqrt[5]{x^2y^3z^{-10}}$

27. $2x^{-1}x^{-3}$ **28.** xy^{-1} **29.** $(4t^2)^{-2}$

30. $(-2z)^{-3}$ **31.** $(x^{-2}y^2)^3$ **32.** $(ab^2c^3)^{3/4}$

33. $\dfrac{x^{-2}y^{-6}z^2}{xy^{-1}}$ **34.** $(2x^{-1}y)^{-2}$

35. $(2x^{3/4}y^{1/2})(xy^{3/2})$ **36.** $(xy^{-2}\sqrt{z})^4$

37. $x^{-3}(2x^4y^{-2})$ **38.** $(x^{1/2} + y^{1/2})(x^{1/2} - y^{1/2})$

39. $(-3x^{1/2}y^{2/3})^3$ **40.** $[(x-4)^{1/5}]^{10}$

41. $2x^{1/2}y^{-3}x^{1/3}$

42. $(4xy^3)^{1/2}(-2x^{3/2}y)^4$

43. $\sqrt{7}\sqrt{4}\sqrt{14}$

44. $\dfrac{2}{\sqrt{x^3}}$

45. $\sqrt{\sqrt[3]{t^4}}$

46. $\dfrac{\sqrt{3}\sqrt{6}}{\sqrt{2}}$

47. $\dfrac{2^0}{(2^{-2}x^{1/2}y^{-2})^3}$

48. $\dfrac{\sqrt[3]{t^5}}{\sqrt[3]{t^2}}$

49. $\dfrac{(x^2y^{-1}z)^{-2}}{(xy^{1/2})^{-4}}$

50. $\left(\dfrac{2x^2y}{8y^3z^{-2}}\right)^{1/2}$

51. $2\sqrt{8} - (5\sqrt{2} - \sqrt{18})$

52. $(\sqrt{3} - \sqrt{2})(\sqrt{3} + 2\sqrt{2})$

53. $\sqrt{2}(1 - \sqrt{6})$

54. $\sqrt{75k^4}$

55. $(\sqrt[5]{2})^{10}$

56. $\sqrt{x}\sqrt{x^2y^3}\sqrt{xy^2}$

57. $\dfrac{2}{\sqrt{7}}$

58. $\dfrac{8}{\sqrt[3]{4}}$

59. $\dfrac{3}{\sqrt[4]{x}}$

60. $\sqrt[5]{\sqrt[3]{x^{10}}}$

61. $\sqrt[4]{81x^6}$

62. $(\sqrt[5]{x^2y})^{10}$

63. $\sqrt[3]{\sqrt{\sqrt[3]{x^{36}}}}$

64. $\sqrt[4]{2}\sqrt[4]{24}$

65. $\sqrt{x}\sqrt{3x}$

66. $\sqrt[6]{x^7y^{13}z^{12}}$

67. $\sqrt[6]{\dfrac{x^6}{y^9}}$

68. $\sqrt[9]{x^3y^6}$

69. $\dfrac{3}{\sqrt[3]{xy^2}}$

70. $\dfrac{1}{\sqrt{x}\sqrt{y}}$

71. $\dfrac{\sqrt[3]{3x^2}}{\sqrt[3]{2x}}$

72. $\sqrt[5]{\dfrac{xy^2}{x^2y}}$

73. $\dfrac{1}{\sqrt{6} - 2}$

74. $\dfrac{4\sqrt{6}}{\sqrt{6} + \sqrt{4}}$

In Problems **75–82**, *solve the equations.*

75. $\sqrt{2x + 5} = 5$

76. $\sqrt{3x - 4} = \sqrt{2x + 5}$

77. $\sqrt[3]{11x + 9} = 4$

78. $\sqrt{x^2 + 5x + 25} = x + 4$

79. $\sqrt{y} + 6 = 5$

80. $\sqrt{z^2 + 9} = 5$

81. $\sqrt{x - 1} + \sqrt{x + 6} = 7$

82. $\sqrt{2x + 1} = x - 7$

CHAPTER 14

Exponential and Logarithmic Functions

14-1 EXPONENTIAL FUNCTIONS

In this chapter we'll look at two special types of functions. The first involves a constant raised to a variable power. An example is $f(x) = 2^x$. We call this an *exponential function*. Notice that the base is a fixed number (2), and the exponent is a variable (x). Another exponential function is $y = g(x) = 3^x$. In general:

> An **exponential function** has the form
>
> $$y = f(x) = b^x,$$
>
> where $b > 0$ and $b \neq 1$,[†] and x is any real number.

[†]If $b = 1$, then $f(x) = 1^x = 1$. This function is so uninteresting that we do not call it an expónential function.

Since the exponent x can be any real number, you may wonder how we assign a value to something like $4^{\sqrt{2}}$. Stated simply, we use approximations. First, $4^{\sqrt{2}}$ is approximately $4^{1.4} = 4^{7/5} = \sqrt[5]{4^7}$, which *is* defined. Better approximations are $4^{1.41} = 4^{141/100} = \sqrt[100]{4^{141}}$ and $4^{1.414}$, etc. In this way a meaning of $4^{\sqrt{2}}$ becomes clear.

In Figure 14-1 are the graphs of the exponential functions $y = 2^x$, $y = 3^x$, and $y = (\frac{1}{2})^x = 2^{-x}$. Notice that

1. The domain of an exponential function is all real numbers.
2. The range is all positive numbers.

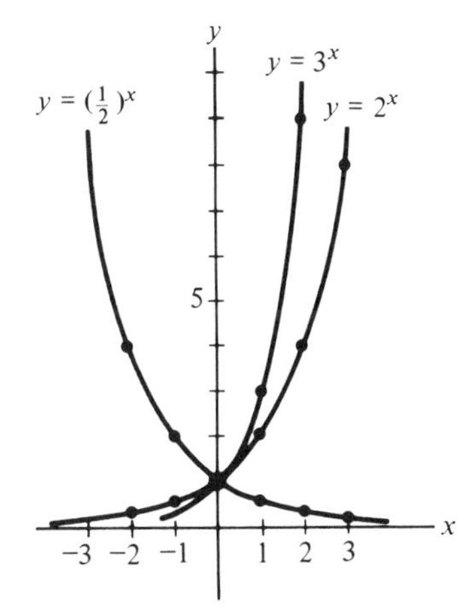

x	2^x	3^x	$(\frac{1}{2})^x$
−2	$\frac{1}{4}$	$\frac{1}{9}$	4
−1	$\frac{1}{2}$	$\frac{1}{3}$	2
0	1	1	1
1	2	3	$\frac{1}{2}$
2	4	9	$\frac{1}{4}$
3	8	27	$\frac{1}{8}$

FIG. 14-1

Since $b^0 = 1$ for every base b, all the graphs pass through (0, 1). Also, the graphs get very close to the x-axis but never touch it.

There are two basic shapes for the graphs of $y = b^x$. These depend on whether $b > 1$ or b is between 0 and 1. (Look at Fig. 14-1 again.)

If $b > 1$, then the graph of $y = b^x$ *rises* from left to right.

If b is between 0 and 1, then the graph of $y = b^x$ *falls* from left to right.

A certain irrational number called e is often used as a base in $y = b^x$.

e is approximately 2.71828 . . . .†

---

†We use the letter e in honor of the mathematician Euler.

This number frequently occurs in studies of growth (populations) and decay (radioactive decay). In Appendix B is a table of (approximate) values of e^x and e^{-x}. The graph of $y = e^x$ is shown in Fig. 14-2.

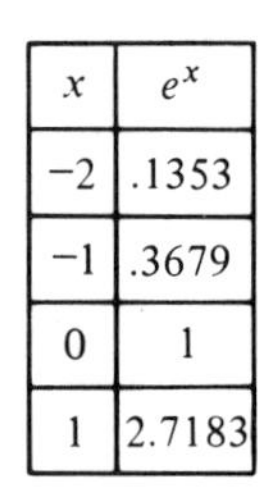

x	e^x
−2	.1353
−1	.3679
0	1
1	2.7183

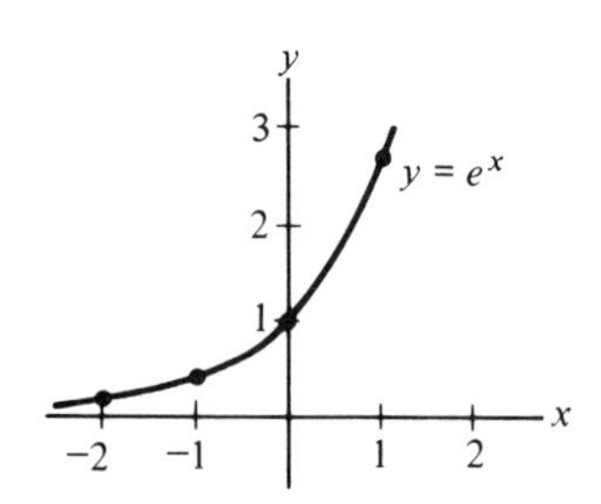

FIG. 14-2

EXAMPLE 1

The predicted population P of a certain city is given by

$$P = 100{,}000e^{.05t},$$

where t is the number of years after 1978. *Predict the population for* 1998.

The number of years from 1978 to 1998 is 20, so let $t = 20$. Then

$$\begin{aligned} P &= 100{,}000e^{.05t} = 100{,}000e^{.05(20)} \\ &= 100{,}000e^{1} = 100{,}000e. \end{aligned}$$

Since $e \approx 2.71828$ ($\approx$ means "is approximately"), then

$$P \approx 100{,}000(2.71828) = 271{,}828.$$

Many economic forecasts are based on population studies.

EXAMPLE 2

A mail order company advertises in a national magazine. The company finds that of all small towns, the percentage P (given as a decimal) of those in which exactly x people respond to an ad is given (approximately) by the formula

$$P = \frac{e^{-.5}(.5)^x}{1 \cdot 2 \cdot 3 \ldots x}.$$

From what percentage of small towns can the company expect exactly 2 *people to respond?*

We want to find P when $x = 2$. For the denominator $1 \cdot 2 \cdot 3 \ldots x$, we must find the product of the integers from 1 to x. Thus,

$$P = \frac{e^{-.5}(.5)^x}{1 \cdot 2 \cdot 3 \ldots x} = \frac{e^{-.5}(.5)^2}{1 \cdot 2}.$$

From the table in Appendix B, $e^{-.5} = .6065$. Thus,

$$P = \frac{.6065(.25)}{2} \approx .0758.$$

Thus, approximately 7.58% of all small towns will have exactly two people respond to the ad.

Completion Set 14-1

Fill in the blanks.

1. A function of the form $f(x) = b^x$ is called an ____________________ function.

2. Is the graph in Fig. 14-3 that of $y = 5^x$ or that of $y = \left(\frac{1}{5}\right)^x$? ____________.

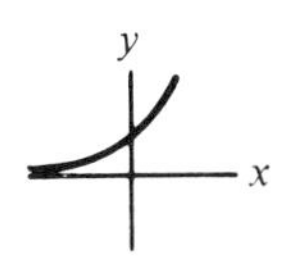

FIG. 14-3

3. Is the value of e closer to 2 or to 3? ____

Problem Set 14-1

In Problems 1–4, graph each function.

1. $y = 4^x$

2. $y = \left(\frac{1}{3}\right)^x$

3. $y = \left(\frac{1}{4}\right)^x$

4. $y = 3^{x/2}$

In Problems 5–8, use the table in Appendix B to find the value of each number.

5. e^3

6. $e^{.5}$

7. $e^{-1.5}$

8. $e^{-.3}$

In Problems **9–14**, *find the given functional values.*

9. $f(x) = 9^x$; $f(2)$, $f(-2)$, $f(\frac{1}{2})$

10. $g(x) = 4(3)^x$; $g(0)$, $g(2)$, $g(-1)$

11. $h(t) = 3(16)^{t/2}$; $h(1)$, $h(\frac{1}{2})$, $h(-\frac{1}{2})$

12. $f(t) = 5 - (\frac{1}{2})^x$; $f(2)$, $f(3)$, $f(-3)$

13. $g(x) = 1 + 2(\frac{1}{8})^{1-x}$; $g(1)$, $g(-1)$, $g(\frac{1}{3})$

14. $h(x) = \dfrac{6(.25)^{(4-x)/2}}{5}$; $h(0)$, $h(2)$, $h(3)$

15. The predicted population P of a city is given by $P = 125{,}000(1.12)^{t/20}$, where t is the number of years after 1978. Predict the population in 1998.

16. For a certain city the population P grows at the rate of 2 percent per year. The formula $P = 1{,}000{,}000(1.02)^t$ gives the population t years after 1977. Find the population in (a) 1977, (b) in 1980.

17. The probability P that a telephone operator will receive exactly x calls during a certain time period is

$$P = \frac{e^{-3}3^x}{1 \cdot 2 \cdot 3 \ldots x}.$$

Find the probability that the operator receives exactly two calls.

18. Repeat Problem **17** for the case that the operator receives exactly one call.

19. At a certain time there are 100 milligrams (mg) of a radioactive substance. It decays so that after t years, the number of mg present, A, is given by $A = 100e^{-.035t}$. How many mg are present after 20 years? Give your answer to the nearest mg.

20. A function used in statistics is

$$y = f(x) = \frac{1}{\sqrt{2\pi}} e^{-x^2/2}.$$

Find $f(0)$, $f(-1)$, and $f(1)$. Assume that $\dfrac{1}{\sqrt{2\pi}} = .399$.

Problems **21** *and* **22** *refer to the following: We say that interest is* ***compounded continuously*** *when it is compounded at every instant of time. The formula* $A = Pe^{rn}$ *gives the amount* A *of a principal* P *after* n *years when the principal is compounded continuously at an annual rate* r *(given as a decimal).*

21. Find the amount that $1000 will become after 8 years with interest compounded continuously at an annual rate of 5 percent.

22. Find the amount at the end of one year for $100 at 5 percent, compounded continuously.

23. P. P. Piddle is president of United Dum Dum Co., a manufacturer of electric forks. In *Prong*, a trade journal, he finds an interesting article. It says that for most electric fork manufacturers, the daily output y of electric forks on the tth day after the start of a production run is given approximately by $y = 500(1 - e^{-.2t})$. Such an equation is called a *learning equation*. It indicates that as time goes on, output per day will increase until a certain level is reached. This may be due to the gain in the workers' skill at their jobs. Suppose Mr. Piddle's company is typical of the industry. Find, to the nearest complete unit, the output on (a) the first day, and (b) the tenth day after the start of a production run.

14-2 LOGARITHMIC FUNCTIONS

Figure 14-4 shows the graph of the exponential function $s = f(t) = 2^t$. The function f sends an input number t into a *positive* output number s.

$$f : t \to s \quad \text{where} \quad s = 2^t.$$

For example, f sends 2 into 4.

$$f : 2 \to 4$$

Now look at the same curve in Fig. 14-5. There you can see that with each

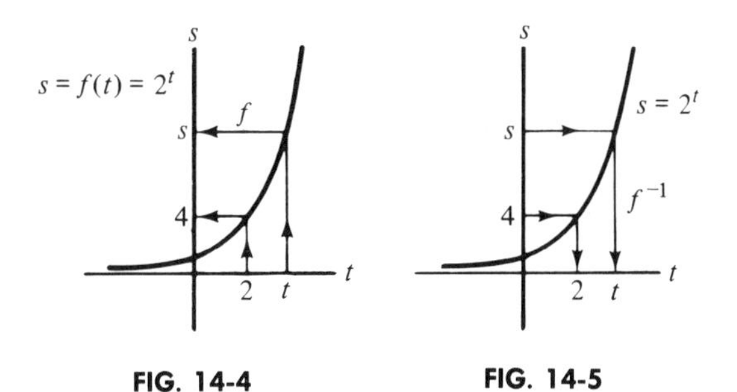

FIG. 14-4 FIG. 14-5

positive number s we can associate exactly one value of t. Notice that with $s = 4$ we associate $t = 2$. Let's think here of s as an input and t as an output. Then we have a function that sends s's into t's. We'll denote this function by the symbol f^{-1} (read "f inverse").†

$$f^{-1} : s \to t \quad \text{where} \quad s = 2^t.$$

Thus, $f^{-1}(s) = t$.

The functions f and f^{-1} are related. In Fig. 14-6 you can see that f^{-1}

---

†f^{-1} stands for a new function. It does not mean $\dfrac{1}{f}$.

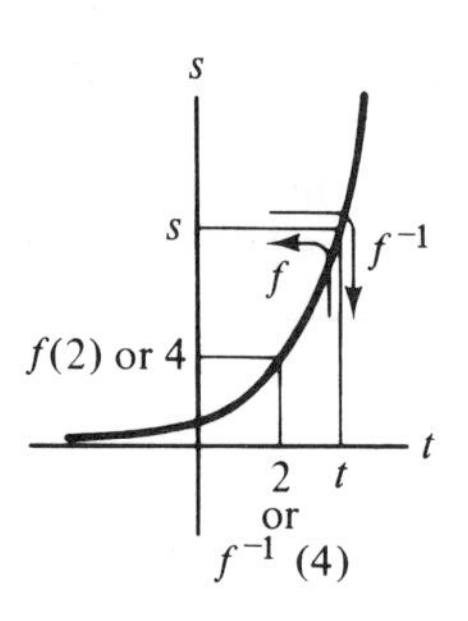

FIG. 14-6

reverses the action of f, and vice versa. For example,

$$f \text{ sends 2 into 4}, \quad \text{and} \quad f^{-1} \text{ sends 4 into 2}.$$

Notice that the domain of f^{-1} is the range of f (all positive numbers) and the range of f^{-1} is the domain of f (all real numbers).

We give a special name to f^{-1}. It is called the **logarithmic** (or log) **function base 2**. Usually we write f^{-1} as $\log_2$ (read "log base 2"). Thus $\log_2$ is just a symbol for a special function. Putting everything together, we have

$$\text{if } s = f(t) = 2^t, \quad \text{then} \quad f^{-1}(s) = \log_2(s) = t. \tag{14-1}$$

Let's generalize to other bases and in (14-1) replace s by x and t by y.

> The **logarithmic function base *b*** is denoted $\log_b$ and
>
> $$y = \log_b x \quad \text{if and only if} \quad b^y = x.$$
>
> The domain of $\log_b$ is all positive numbers, and the range is all real numbers.

The logarithmic function reverses the action of the exponential function. Because of this we sometimes say that the logarithmic function is the *inverse* of the exponential function.

Always remember: when we say that the log base b of x is y, we mean that b raised to the y power is x.

$$\boxed{\log_b x = y \quad \text{means} \quad b^y = x.}$$

For example,

$$\log_2 8 = 3 \text{ because } 2^3 = 8.$$

We say $\log_2 8 = 3$ is the **logarithmic form** of the **exponential form** $2^3 = 8$.

EXAMPLE 1

	Exponential form		*Logarithmic form*
a. Since	$25 = 5^2$,	then	$\log_5 25 = 2$.
b. Since	$3^4 = 81$,	then	$\log_3 81 = 4$.
c. Since	$10^0 = 1$,	then	$\log_{10} 1 = 0$.

EXAMPLE 2

Logarithmic form		*Exponential form*
a. $\log_{10} 1000 = 3$	means	$10^3 = 1000$.
b. $\log_{64} 8 = \frac{1}{2}$	means	$64^{1/2} = 8$.
c. $\log_2 \frac{1}{16} = -4$	means	$2^{-4} = \frac{1}{16}$.

EXAMPLE 3

Graph the function $y = \log_2 x$.

To plot points we'll use the equivalent form $2^y = x$. If $y = 0$, then $x = 1$. Thus (1, 0) lies on the graph. In Fig. 14-7 are other points. You can see that the domain is

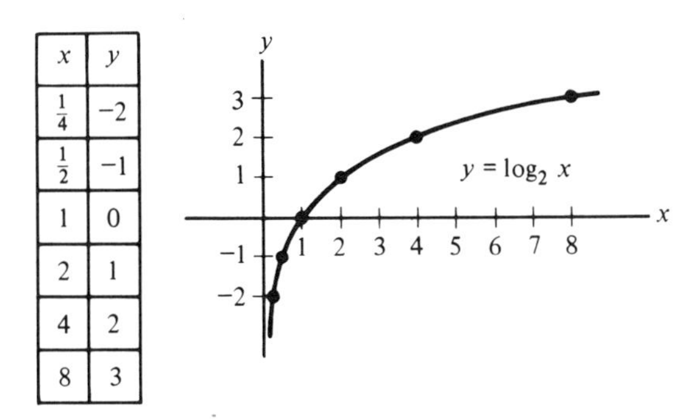

x	y
$\frac{1}{4}$	−2
$\frac{1}{2}$	−1
1	0
2	1
4	2
8	3

FIG. 14-7

all positive numbers and the range is all real numbers.

In Fig. 14-8 is the graph of $y = \log_e x$. Notice that it has the same shape as

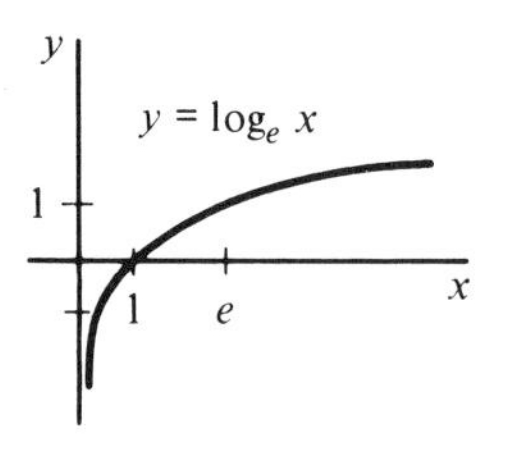

FIG. 14-8

Fig. 14-7.

Logarithms in the base 10 are called **common logarithms**. They were often used to simplify computations before the pocket-calculator age. We usually drop the subscript 10 when writing common logs. Thus,

$$\log x \text{ means } \log_{10} x.$$

Logarithms in the base e are called **natural** (or Naperian) logarithms. We use the symbol "ln" for natural logs. Thus,

$$\ln x \text{ means } \log_e x.$$

EXAMPLE 4

Find

a. log 100.

Here the base is 10. We'll set $\log 100 = y$.

$$\log 100 = y$$
$$10^y = 100 \quad \text{[converting to exponential form].}$$

Clearly, y must be 2. Thus $\log 100 = 2$.

b. ln 1.

Here the base is e. Set $\ln 1 = y$.

$$\ln 1 = y$$
$$e^y = 1 \quad \text{[converting to exponential form].}$$

Clearly, y must be 0. Thus $\ln 1 = 0$.

c. $\log .1$.

$$\log .1 = y$$
$$10^y = .1$$
$$10^y = \frac{1}{10} = 10^{-1}.$$

Thus $y = -1$ and $\log .1 = -1$.

d. $\ln e$.

$$\ln e = y$$
$$e^y = e.$$

Thus $y = 1$ and $\ln e = 1$.

e. $\log_{36} 6$.

$$\log_{36} 6 = y$$
$$36^y = 6.$$

Clearly, y must be $\frac{1}{2}$. Thus $\log_{36} 6 = \frac{1}{2}$.

EXAMPLE 5

Solve for x.

a. $\log_2 x = 4$.

$$\log_2 x = 4$$
$$2^4 = x \quad \text{[converting to exponential form]}$$
$$\boxed{16 = x.}$$

b. $x = \log_5 125$.

$$x = \log_5 125$$
$$5^x = 125 \quad \text{[converting to exponential form].}$$

Clearly, $\boxed{x = 3.}$

c. $x + 1 = \log_4 16$.

$$x + 1 = \log_4 16$$
$$4^{x+1} = 16 \quad \text{[converting to exponential form].}$$

From inspection, the exponent $x + 1$ must be 2. Thus $x + 1 = 2$ or $\boxed{x = 1.}$

d. $\log_x 49 = 2$.

$$\begin{aligned} \log_x 49 &= 2 \\ x^2 &= 49 \\ x &= \pm 7. \end{aligned}$$

A negative number cannot be a base in a logarithm, so we throw out $x = -7$ and choose $\boxed{x = 7.}$

Completion Set 14-2

Fill in the blanks.

1. Insert *logarithmic* or *exponential*. $2^4 = 16$ is the ________________ form of $\log_2 16 = 4$.

2. Since $8^2 = 64$, then $\log_8 64 =$ ____.

3. Since $3^5 = 243$, then $\log_3$ ________ $= 5$.

4. Since $2^7 = 128$, then $\log_2$ ________ $=$ ____.

5. Insert 3^4 or 4^3. If $\log_3 x = 4$, then $x =$ ______.

6. If $\log_b x = y$, then $b^y =$ ______.

7. The base in $\log x$ is ______.

8. The base in $\ln x$ is ______.

Problem Set 14-2

In Problems **1–16**, *write each exponential form logarithmically and each logarithmic form exponentially.*

1. $4^3 = 64$

2. $2 = \log_{12} 144$

3. $10^5 = 100{,}000$

4. $\log_9 3 = \frac{1}{2}$

5. $\log_2 64 = 6$

6. $4^4 = 256$

7. $\log_8 2 = \frac{1}{3}$

8. $8^{2/3} = 4$

9. $6^0 = 1$

10. $\log_{1/2} 4 = -2$

11. $\log_2 x = 14$

12. $10^{.48302} = 3.041$

13. $e^2 = 7.3891$

14. $e^{.33647} = 1.4$

15. $\ln 3 = 1.0986$

16. $\log 5 = .6990$

In Problems **17** *and* **18**, *graph the function.*

17. $y = \log_3 x$

18. $y = \log_4 x$

In Problems **19–32**, *find the values.*

19. $\log_6 36$

20. $\log_2 32$

21. $\log_3 27$

22. $\log_{27} 3$

23. $\log_{16} 4$

24. $\log_7 7$

25. $\log 10$

26. $\log 10{,}000$

27. $\log .01$

28. $\ln (e^4)$

29. $\log_5 1$

30. $\log_2 \sqrt{2}$

31. $\log_2 \frac{1}{8}$

32. $\log_5 \frac{1}{25}$

In Problems **33–58**, *find x.*

33. $\log_2 x = 4$

34. $\log_3 x = 2$

35. $\log_5 x = 3$

36. $\log_4 x = 0$

37. $\log x = -1$

38. $\ln x = 1$

39. $\ln x = 2$

40. $\log_x 100 = 2$

41. $\log_x 8 = 3$

42. $\log_x 3 = \frac{1}{2}$

43. $\log_x \frac{1}{6} = -1$

44. $\log_x y = 1$

45. $\log_4 16 = x$

46. $\log_3 1 = x$

47. $\log 10{,}000 = x$

48. $\log_2 \frac{1}{16} = x$

49. $\log_{25} 5 = x$

50. $\log_9 9 = x$

51. $\log_3 x = -4$

52. $\log_x (2x - 3) = 1$

53. $\log_x 81 = 2$

54. $\log_{1/2} x = -2$

55. $\log_8 64 = x - 1$

56. $\log_x (6 - x) = 2$

57. $\log_3 (x - 2) = 2$

58. $\log_2 4 = 3x - 1$

14-3 PROPERTIES OF LOGARITHMS

Logarithms have many properties. One involves the log of a product, like $\log_b(mn)$. If we let $x = \log_b m$ and $y = \log_b n$, then $b^x = m$ and $b^y = n$. Therefore,

$$mn = b^x b^y = b^{x+y}.$$

Thus, $mn = b^{x+y}$. Converting this to logarithmic form gives $\log_b(mn) = x + y$. But we said $x = \log_b m$ and $y = \log_b n$. Thus we have our first property.

> **Property 1.** $\log_b(mn) = \log_b m + \log_b n$.
> *In short, the log of a product is a sum of logs.*

In some of the examples and exercises, we'll use Table 14-1. It gives the approximate values of a few common logs.† Notice that $\log 4 = .6021$. This means $10^{.6021} = 4$.

TABLE 14-1 COMMON LOGARITHMS

x	$\log x$	x	$\log x$
2	.3010	7	.8451
3	.4771	8	.9031
4	.6021	9	.9542
5	.6990	10	1.0000
6	.7782	e	.4343

EXAMPLE 1

a. *Find* log 15.

Log 15 is not in Table 14-1. But we can write 15 as the product $3 \cdot 5$.

$$\begin{aligned} \log 15 &= \log(3 \cdot 5) \\ &= \log 3 + \log 5 && \text{[Property 1]} \\ &= .4771 + .6990 && \text{[Table 14-1]} \\ \log 15 &= 1.1761. \end{aligned}$$

b. *Find* log 56.

$$\log 56 = \log(8 \cdot 7) = \log 8 + \log 7 = .9031 + .8451 = 1.7482.$$

The next two properties can be proven in the same basic way as Property 1.

> **Property 2.** $\log_b \dfrac{m}{n} = \log_b m - \log_b n$.
> *In short, the log of a quotient is a difference of logs.*

†More complete tables can be found in most libraries.

Property 3. $\log_b (m^n) = n \log_b m.$
In short, the log of a power of a number is the power times a log.

EXAMPLE 2

a. *Find* $\log \frac{9}{2}$.

$$\log \frac{9}{2} = \log 9 - \log 2 \qquad \text{[Property 2]}$$
$$= .9542 - .3010 \qquad \text{[Table 14-1]}$$
$$\log \frac{9}{2} = .6532.$$

b. *Find* $\log 64$.

$$\log 64 = \log (8^2)$$
$$= 2 \log 8 \qquad \text{[Property 3]}$$
$$= 2(.9031) \qquad \text{[Table 14-1]}$$
$$\log 64 = 1.8062.$$

c. *Find* $\log \sqrt{5}$.

$$\log \sqrt{5} = \log (5^{1/2}) = \frac{1}{2} \log 5 = \frac{1}{2}(.6990) = .3495.$$

d. *Find* $\log \frac{15}{7}$.

$$\log \frac{15}{7} = \log 15 - \log 7 = \log(3 \cdot 5) - \log 7$$
$$= [\log 3 + \log 5] - \log 7$$
$$= [.4771 + .6990] - .8451 = .3310.$$

e. *Find* $\log \frac{16}{21}$.

$$\log \frac{16}{21} = \log 16 - \log 21 = \log (4^2) - \log (3 \cdot 7)$$
$$= 2 \log 4 - [\log 3 + \log 7]$$
$$= 2(.6021) - [.4771 + .8451] = -.1180.$$

$$\log_b (m + n) \neq \log_b m + \log_b n.$$
$$\log_b (m - n) \neq \log_b m - \log_b n.$$

EXAMPLE 3

a. *Simplify* $\log_3 \dfrac{1}{x^2}$.

$$\log_3 \frac{1}{x^2} = \log_3 x^{-2} = -2 \log_3 x \qquad \text{[Property 3].}$$

b. *Write* $\log_4 x - \log_4 (x + 3)$ *as a single logarithm.*

$$\log_4 x - \log_4 (x + 3) = \log_4 \frac{x}{x + 3} \qquad \text{[Property 2].}$$

c. *Write* $3 \log_2 10 + \log_2 15$ *as a single logarithm.*

$$\begin{aligned} 3 \log_2 10 + \log_2 15 &= \log_2 (10^3) + \log_2 15 && \text{[Property 3]} \\ &= \log_2 [(10^3)15] && \text{[Property 1]} \\ &= \log_2 15{,}000. \end{aligned}$$

d. *Write* $\ln 3 + \ln 7 - \ln 2 - 2 \ln 4$ *as a single logarithm.*

$$\begin{aligned} &\ln 3 + \ln 7 - \ln 2 - 2 \ln 4 \\ &= \ln 3 + \ln 7 - \ln 2 - \ln (4^2) \\ &= \ln 3 + \ln 7 - \left[\ln 2 + \ln (4^2)\right] \\ &= \ln (3 \cdot 7) - \ln (2 \cdot 4^2) \\ &= \ln 21 - \ln 32 \\ &= \ln \frac{21}{32}. \end{aligned}$$

EXAMPLE 4

a. *Write* $\log_6 \dfrac{x^5 y}{zw}$ *in terms of* $\log_6 x$, $\log_6 y$, $\log_6 z$, *and* $\log_6 w$.

$$\begin{aligned} \log_6 \frac{x^5 y}{zw} &= \log_6 (x^5 y) - \log_6 (zw) \\ &= \log_6 (x^5) + \log_6 y - [\log_6 z + \log_6 w] \\ &= 5 \log_6 x + \log_6 y - \log_6 z - \log_6 w. \end{aligned}$$

b. *Write* $\log \sqrt[3]{\dfrac{x^2}{x - 4}}$ *in terms of* $\log x$ *and* $\log (x - 4)$.

$$\log \sqrt[3]{\frac{x^2}{x-4}} = \log \left(\frac{x^2}{x-4}\right)^{1/3} = \frac{1}{3} \log \frac{x^2}{x-4}$$
$$= \frac{1}{3}\left[\log (x^2) - \log (x-4)\right]$$
$$= \frac{1}{3}[2 \log x - \log (x-4)].$$

Let's look at two more properties of logs. Since $b^0 = 1$ and $b^1 = b$, then by converting to logarithmic forms we get

Property 4. $\log_b 1 = 0.$
Property 5. $\log_b b = 1.$

EXAMPLE 5

a. *Find* log 1.

$$\log 1 = \log_{10} 1 = 0 \qquad \text{[Property 4].}$$

b. *Find* ln e.

$$\ln e = \log_e e = 1 \qquad \text{[Property 5].}$$

c. *Find* log 1000.

$$\log 1000 = \log_{10} 10^3 = 3 \log_{10} 10$$
$$= 3 \cdot 1 \qquad \text{[Property 5]}$$
$$= 3.$$

d. *Find* log (10^n).

$$\log (10^n) = n \log_{10} 10 = n \cdot 1 = n.$$

EXAMPLE 6

a. *Find* $\log_7 \sqrt[9]{7^8}$.

$$\log_7 \sqrt[9]{7^8} = \log_7 (7^{8/9}) = \frac{8}{9} \log_7 7$$
$$= \frac{8}{9} \cdot 1 = \frac{8}{9} \qquad \text{[Property 5].}$$

b. *Find* $\log_3 \frac{27}{81}$.

$$\log_3 \frac{27}{81} = \log_3 \frac{3^3}{3^4} = \log_3 \frac{1}{3}$$
$$= \log_3(3^{-1}) = (-1)\log_3 3 = -1 \cdot 1 = -1.$$

c. *Find* $\ln (e^2) + \log \frac{1}{10}$.

$$\ln (e^2) + \log \frac{1}{10} = 2 \ln e + \log 10^{-1}$$
$$= 2 \log_e e + (-1)\log_{10} 10$$
$$= 2(1) + (-1)(1) = 1.$$

d. *Find* $\log \frac{200}{21}$.

$$\log \frac{200}{21} = \log 200 - \log 21 = \log (2 \cdot 10^2) - \log (3 \cdot 7)$$
$$= \log 2 + \log (10^2) - [\log 3 + \log 7]$$
$$= \log 2 + 2 \log 10 - \log 3 - \log 7$$
$$= .3010 + 2(1) - .4771 - .8451$$
$$= .9788.$$

Completion Set 14-3

In Problems **1–3**, *insert* "+" *or* "−."

1. $\log_5 xy = \log_5 x$ ____ $\log_5 y$.

2. $\log_4 \frac{x}{y} = \log_4 x$ ____ $\log_4 y$.

3. $\ln \frac{xy}{z} = \ln x$ ____ $\ln y$ ____ $\ln z$.

In Problems **4–10**, *fill in the blanks.*

4. $\log_3 (x^5) =$ ____ $\cdot \log_3 x$.

5. $\ln \sqrt{x} =$ ____ $\cdot \ln x$.

6. $\log \frac{x^2 y^3}{z^4} =$ ____ $\cdot \log x +$ ____ $\cdot \log y -$ ____ $\cdot \log z$.

7. $\log 5 + \log 9 = \log[5(\_\_\_\_)] = \log \_\_\_\_\_\_$.

8. $\log_3 8 - \log_3 4 = \log_3 \frac{(\quad)}{4} = \log_3 \_\_\_\_$.

9. $\log_5 5 = \_\_\_\_$.

10. $\log_3 1 = \_\_\_\_$.

Problem Set 14-3

In Problems **1–48**, *use Table* 14-1 *on p.* 336 *and properties of logs to find the values.*

1. $\log 5$ **2.** $\log 3$ **3.** $\log 21$

4. $\log 14$ **5.** $\log 35$ **6.** $\log 12$

7. $\log 25$ **8.** $\log 49$ **9.** $\log \frac{9}{4}$

10. $\log \frac{7}{3}$ **11.** $\log \frac{1}{2}$ **12.** $\log \frac{7}{9}$

13. $\log (8^5)$ **14.** $\log (4^{-3})$ **15.** $\log \frac{1}{5^4}$

16. $\log 2^{.01}$ **17.** $\log \sqrt{2}$ **18.** $\log \sqrt[3]{4}$

19. $\log \sqrt[3]{6^2}$ **20.** $\log \frac{1}{\sqrt[3]{5}}$ **21.** $\log \frac{14}{5}$

22. $\log \frac{3}{35}$ **23.** $\log \frac{15}{28}$ **24.** $\log \frac{81}{25}$

25. $\log 10{,}000$ **26.** $\log 10^{10}$ **27.** $\log .01$

28. $\log (.1)^3$ **29.** $\log 300$ **30.** $\log \sqrt[3]{400}$

31. $\log \frac{100}{9}$ **32.** $\log \frac{1}{600}$ **33.** $\log \frac{27}{5000}$

34. $\log \sqrt{\frac{8}{3}}$ **35.** $\log_7 7^{48}$ **36.** $\log_4 \sqrt[5]{4}$

37. $\log_5 \sqrt[4]{5^3}$ **38.** $\log_2 \frac{1}{\sqrt{2}}$ **39.** $\ln e^4$

40. $\ln \frac{1}{e}$ **41.** $\log_2 \frac{2^6}{2^{10}}$ **42.** $\log_3 [(3^5 \cdot 3^4)^6]$

43. $\log [(10\sqrt{10}\,)^3]$ **44.** $\log_5 \frac{25}{\sqrt{5}}$ **45.** $\log_7 \frac{\sqrt[3]{49}}{7}$

46. $\log_8 (64\sqrt[4]{8})$ **47.** $\log 10 + \ln (e^2)$ **48.** $(\log 100)(\ln \sqrt{e})$

In Problems **49–62**, *write each expression as a single logarithm.*

49. $\log 7 + \log 4$

50. $\log_3 10 - \log_3 5$

51. $\log_2 (x + 2) - \log_2 (x + 1)$

52. $\log (x^2) + \log x$

53. $2 \ln 5 - 3 \ln 4$

54. $3 \log_7 2 + 2 \log_7 3$

55. $\frac{1}{2}\log_4 2 + 3 \log_4 3$

56. $\frac{1}{3}\ln x - 5 \ln (x - 2)$

57. $\log_3 x + \log_3 y - \log_3 z$

58. $\ln x + \ln(x - 1) - 2 \ln y$

59. $\log_2 5 + 2 \log_2 3 - \log_2 7$

60. $\frac{1}{2}(\log_3 x + \log_3 y)$

61. $2 \ln x + 3 \ln y - 4 \ln z - 2 \ln w$

62. $\log_6 5 - 2 \log_6 \frac{1}{5} - 2 \log_6 4$

In Problems **63–68**, *write the expressions in terms of* $\log x$ *and* $\log (x + 1)$.

63. $\log [x(x + 1)^2]$

64. $\log \dfrac{\sqrt{x}}{x + 1}$

65. $\log \dfrac{x^2}{(x + 1)^3}$

66. $\log [x(x + 1)]^3$

67. $\log \left(\dfrac{x}{x + 1}\right)^3$

68. $\log \sqrt{x(x + 1)}$

In Problems **69–74**, *write the expressions in terms of* $\ln x$, $\ln (x + 1)$, *and* $\ln (x + 2)$.

69. $\ln \dfrac{x}{(x + 1)(x + 2)}$

70. $\ln \dfrac{x^2(x + 1)}{x + 2}$

71. $\ln \dfrac{\sqrt{x}}{(x + 1)^2(x + 2)^3}$

72. $\ln \dfrac{1}{x(x + 1)(x + 2)}$

73. $\ln\left[\dfrac{1}{x + 2}\sqrt[5]{\dfrac{x^2}{x + 1}}\right]$

74. $\ln \sqrt{\dfrac{x^4(x + 1)^3}{x + 2}}$

A chemist can determine the acidity or basicity of an aqueous solution at room temperature by finding the "pH" of the solution. To do this, he can first find the hydrogen-ion concentration (in moles per liter). The symbol $[H^+]$ *stands for this concentration. The* pH *is then given by*

$$\text{pH} = -\log [H^+].$$

If $\text{pH} < 7$, *the solution is acidic. If* $\text{pH} > 7$, *it is basic. If* $\text{pH} = 7$, *we say the solution is neutral.*

In Problems **75** *and* **76**, *find the* pH *of a solution with the given* $[H^+]$.

75. $[H^+] = 10^{-9}$

76. $[H^+] = 2 \times 10^{-3}$

For an aqueous solution at room temperature, the product of the hydrogen-ion concentration, $[H^+]$, *and the hydroxide-ion concentration,* $[OH^-]$, *is* 10^{-14} (*where the concentrations are in moles per liter*).

$$[H^+][OH^-] = 10^{-14}.$$

In Problems **77** *and* **78**, *find the* pH *of a solution* (*see explanation above Problem* **75**) *with the given* $[OH^-]$.

77. $[OH^-] = 10^{-4}$

78. $[OH^-] = 3 \times 10^{-2}$

14-4 CHANGE OF BASE FORMULAS

Suppose you want to find $\log_5 2$. This would be easy if you had a table of logs in base 5. However, you can find $\log_5 2$ by just using the table of common logs (base 10) on p. 336. Here's how.

Let $x = \log_5 2$. Then $5^x = 2$. Now take the common log of each side.

$$\log (5^x) = \log 2$$

$$x \log 5 = \log 2$$

$$x = \frac{\log 2}{\log 5} = \frac{.3010}{.6990} = .4306.$$

Thus $\log_5 2 = .4306$.

If we generalize this procedure, we get a so-called *change of base formula*.

A CHANGE OF BASE FORMULA

$$\log_b N = \frac{\log N}{\log b}.$$

This formula lets us find logs in base b by using common logs.

EXAMPLE 1

a. *Find* $\log_3 7$.

Using the change of base formula above with $b = 3$ and $N = 7$ gives

$$\log_3 7 = \frac{\log 7}{\log 3}$$

$$= \frac{.8451}{.4771} = 1.771 \qquad \text{[Table 14-1]}.$$

b. *Find* ln 6.

Here $b = e$ and $N = 6$. Consequently

$$\ln 6 = \frac{\log 6}{\log e} = \frac{.7782}{.4343} = 1.792.$$

Thus $e^{1.792} = 6$.

A more general form of the change of base formula above is

MORE GENERAL CHANGE OF BASE FORMULA

$$\log_b N = \frac{\log_a N}{\log_a b}.$$

This formula lets us find logs in base b by using logs in base a.

EXAMPLE 2

Suppose $\ln 20 = 2.9957$ *and* $\ln 4 = 1.3863$. *Find* $\log_4 20$.

We'll use the more general change of base formula with $b = 4$, $N = 20$, and $a = e$.

$$\log_4 20 = \frac{\log_e 20}{\log_e 4} = \frac{\ln 20}{\ln 4} = \frac{2.9957}{1.3863} = 2.1609.$$

Problem Set 14-4

In Problems **1–8**, *find the values.*

1. $\log_5 3$ **2.** $\log_7 4$ **3.** $\log_4 9$ **4.** $\log_2 3$

5. $\log_8 10$ **6.** $\log_3 20$ **7.** $\ln 2$ **8.** $\ln 10$

9. Suppose $\ln 59 = 4.0775$ and $\ln 15 = 2.7080$. Find $\log_{15} 59$.

10. Suppose $\log_4 13 = 1.8502$ and $\log_4 7 = 1.4037$. Find $\log_7 13$.

11. Suppose $\log_3 11 = 2.1826$. Find $\log_9 11$.

14-5 REVIEW

IMPORTANT TERMS AND SYMBOLS

exponential function *(p. 324)*
$y = b^x$ *(p. 324)*
e *(p. 325)*
logarithmic function *(p. 330)*
$y = \log_b x$ *(p. 330)*
exponential form *(p. 331)*
logarithmic form *(p. 331)*
common logarithm *(p. 332)*
$\log x$ *(p. 332)*
natural logarithm *(p. 332)*
$\ln x$ *(p. 332)*

REVIEW PROBLEMS

In Problems **1–4**, *find the given functional values.*

1. $f(x) = 4 + 3^{2x}$; $f(0)$, $f(1)$, $f\left(-\frac{1}{2}\right)$

2. $g(t) = 4(9)^{t/2}$; $g(1)$, $g(-1)$, $g(3)$

3. $h(s) = 100e^{(s+3)/2}$; $h(-3)$, $h(5)$, $h(-6)$

4. $F(x) = 6 - 3\left(\frac{1}{2}\right)^{x+4}$; $F(-3)$, $F(-1)$, $F(-6)$

In Problems **5** *and* **6**, *graph.*

5. $y = \log_5 x$

6. $y = 5^x$

In Problems **7–12**, *write each exponential form logarithmically and each logarithmic form exponentially.*

7. $3^5 = 243$

8. $\log_7 343 = 3$

9. $\log_{16} 2 = \frac{1}{4}$

10. $10^5 = 100{,}000$

11. $e^4 = 54.598$

12. $\log_9 9 = 1$

In Problems **13–18**, *find the values.*

13. $\log_5 125$

14. $\log_4 16$

15. $\log_2 \frac{1}{16}$

16. $\log_{1/3} \frac{1}{9}$

17. $\log_{1/3} 9$

18. $\log_4 2$

In Problems **19–26**, *find* x.

19. $\log_5 \frac{1}{25} = x$

20. $\log_x 1000 = 3$

21. $\log x = -2$

22. $\log_8 64 = x$

23. $\log_x 81 = 2$

24. $\log_3 x = \frac{1}{2}$

25. $\log_x (4x - 9) = 1$

26. $\log_4 (x - 6) = 2$

In Problems **27–44**, *use Table* 14-1 *on p.* 336 *and properties of logs to find the values.*

27. $\log 16$ **28.** $\log (12^3)$ **29.** $\log \frac{8}{3}$

30. $\log \frac{10}{7}$ **31.** $\log \sqrt[4]{8}$ **32.** $\log \sqrt[3]{2^2}$

33. $\log 500$ **34.** $\log \frac{1}{200}$ **35.** $\log \frac{24}{35}$

36. $\log \frac{800}{21}$ **37.** $\log \sqrt{\frac{5}{2}}$ **38.** $\log (10 \sqrt[4]{10})$

39. $\log_9 1$ **40.** $\log_3 3^{10}$ **41.** $\ln \sqrt[3]{e}$

42. $\log_5 \sqrt[7]{5^4}$ **43.** $\log_7 (49\sqrt{7})$ **44.** $\log_3 \dfrac{\sqrt{3}}{27}$

In Problems **45–50**, *write each expression as a single logarithm.*

45. $2 \log 5 - 3 \log 3$

46. $6 \ln x + 4 \ln y$

47. $2 \ln x + \ln y - 3 \ln z$

48. $\log_6 2 - \log_6 4 - 2 \log_6 3$

49. $\frac{1}{2} \log_2 x + 2 \log_2 (x^2) - 3 \log_2 (x + 1) - 4 \log_2 (x + 2)$

50. $3 \log x + \log y - 2(\log z + \log w)$

In Problems **51–56**, *write the expressions in terms of* $\ln x$, $\ln y$, *and* $\ln z$.

51. $\ln \dfrac{x^2 y}{z^3}$ **52.** $\ln \dfrac{\sqrt{x}}{(yz)^2}$ **53.** $\ln \sqrt[3]{xyz}$

54. $\ln \left[\dfrac{xy^3}{z^2} \right]^4$ **55.** $\ln \left[\dfrac{1}{x} \sqrt{\dfrac{y}{z}} \right]$ **56.** $\ln \left[\left(\dfrac{x}{y} \right)^2 \left(\dfrac{x}{z} \right)^3 \right]$

57. Because of poor advertising, the Kleen-Kut Razor Co. finds that its annual revenues have been cut sharply. The annual revenue R (in dollars) at the end of t years of business is given by the equation $R = 200{,}000e^{-.2t}$. Find the annual revenue at the end of (a) 2 years, (b) 3 years.

In Problems **58–61**, *use the change of base formulas in Sec.* 14-4 *to find the given values.*

58. $\log_6 9$ **59.** $\log_3 5$

60. Suppose that $\log_2 19 = 4.2479$ and $\log_2 5 = 2.3219$. Find $\log_5 19$.

61. Suppose that $\ln 11 = 2.3979$ and $\ln 4 = 1.3863$. Find $\log_4 11$.

CHAPTER 15

INEQUALITIES

15-1 LINEAR INEQUALITIES

An inequality looks something like an equation, except that the symbol "=" is replaced by either "<," ">," "⩽," or "⩾."

We'll look at inequalities involving one variable: for example, $2x + 3 < 13$. Our goal is to find the values of the variable that make the inequality a true statement.

There are certain rules that we use to solve inequalities.

Rule 1. If $a < b$, then $a + c < b + c$ and $a - c < b - c$.

Rule 1 says that you can add (or subtract) the same number to (or from) both sides of an inequality and the inequality symbol remains the same. This rule is also true if you replace "<" by "⩽," ">," or "⩾."

EXAMPLE 1

Use of Rule 1.

a. Since $7 < 10$, then $7 + 3 < 10 + 3$. Thus, $10 < 13$.

b. Since $8 \geqslant 4$, then $8 - 5 \geqslant 4 - 5$. Thus, $3 \geqslant -1$.

c. Since $-4 \leqslant -4$, then $-4 + 4 \leqslant -4 + 4$. Thus, $0 \leqslant 0$.

d. If $x + 2 > 9$, then $x + 2 - 2 > 9 - 2$. Thus, $x > 7$.

Rule 2. If $a < b$ and c is a positive number, then $ac < bc$ and $\frac{a}{c} < \frac{b}{c}$.

Rule 2 says that if you multiply (or divide) both sides of an inequality by a *positive* number, the inequality symbol remains the same. This rule is also true for the other inequality symbols.

EXAMPLE 2

Use of Rule 2.

a. Since $4 < 9$, then $4 \cdot 2 < 9 \cdot 2$. Thus, $8 < 18$.

b. Since $1 > -10$, then $\frac{1}{5} > \frac{-10}{5}$. Thus, $\frac{1}{5} > -2$.

c. Since $0 \geqslant -8$, then $0 \cdot 3 \geqslant -8 \cdot 3$. Thus, $0 \geqslant -24$.

d. If $2x \leqslant -4$, then $\frac{2x}{2} \leqslant \frac{-4}{2}$. Thus, $x \leqslant -2$.

Rule 3. If $a < b$ and c is a positive number, then $a(-c) > b(-c)$ and $\frac{a}{-c} > \frac{b}{-c}$.

Rule 3 says that if you multiply (or divide) both sides of an inequality by a **negative** number, then you must **change** the direction of the inequality symbol. Thus,

"$<$" changes to "$>$."

If you started with the symbol "$\leqslant$," it would change to "$\geqslant$":

"$\leqslant$" changes to "$\geqslant$."

Similarly,

">" changes to "<" and "$\geqslant$" changes to "$\leqslant$."

EXAMPLE 3

Use of Rule 3.

a. Since $-2 < 100$, then $-2(-1) > 100(-1)$. Thus, $2 > -100$.

b. Since $0 > -8$, then $\frac{0}{-2} < \frac{-8}{-2}$. Thus, $0 < 4$.

c. Since $-10 \leqslant -\frac{1}{100}$, then $(-10)(-10) \geqslant (-\frac{1}{100})(-10)$. Thus, $100 \geqslant \frac{1}{10}$.

d. If $-3x < 6$, then $\frac{-3x}{-3} > \frac{6}{-3}$. Thus, $x > -2$.

To solve an inequality, we apply rules 1–3 until the values of the variable are obvious.

EXAMPLE 4

Solve $2x + 3 < 13$.

$$2x + 3 < 13$$

$$2x + 3 - 3 < 13 - 3 \quad \text{[subtracting 3 from both sides]}$$

$$2x < 10$$

$$\frac{2x}{2} < \frac{10}{2} \quad \text{[dividing both sides by 2]}$$

$$\boxed{x < 5.}$$

Thus *all* numbers less than 5 are solutions to $2x + 3 < 13$. For example, since $1 < 5$, then 1 is a solution. Let's check it:

$$2x + 3 < 13$$

$$2(1) + 3 < 13$$

$$5 < 13. \quad \checkmark$$

We can geometrically represent all the solutions by a bold line segment on a number line. See Fig. 15-1. The hollow dot means that 5 is **not included**. Notice that

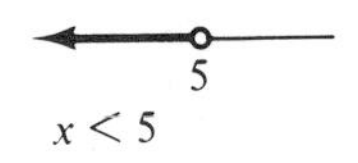

FIG. 15-1

infinitely many values make the inequality true.

In Example 4, we call $2x + 3 < 13$ a *linear inequality*. In fact, any inequality that can be put in the form $ax + b < 0$ is called a **linear inequality**.[†] You can replace "$<$" in $ax + b < 0$ by "$\leqslant$," "$>$," or "$\geqslant$" and still have a linear inequality.

EXAMPLE 5

Solve $5x - 2(7x + 3) \leqslant 12$.

$$5x - 2(7x + 3) \leqslant 12$$

$$5x - 14x - 6 \leqslant 12$$

$$-9x - 6 \leqslant 12$$

$$-9x \leqslant 18 \qquad \text{[adding 6 to both sides]}$$

$$\frac{-9x}{-9} \geqslant \frac{18}{-9} \qquad \text{[dividing both sides by } -9 \text{ and changing direction of inequality]}$$

$$\boxed{x \geqslant -2.}$$

Thus, *all* numbers greater than or equal to -2 are solutions. We can represent them geometrically by the bold line segment in Fig. 15-2. The solid dot means that -2 **is**

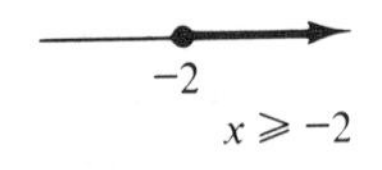

FIG. 15-2

included.

[†]We assume $a \neq 0$.

EXAMPLE 6

Solve $\frac{3}{2}t - 1 < \frac{5}{3}(-3 + t)$.

To clear of fractions, multiply both sides by the L.C.D., 6.

$$6\left(\frac{3}{2}t - 1\right) < 6 \cdot \frac{5}{3}(-3 + t)$$

$$9t - 6 < 10(-3 + t)$$

$$9t - 6 < -30 + 10t$$

$$-6 < -30 + t \qquad \text{[subtracting } 9t \text{ from both sides]}$$

$$24 < t \qquad \text{[adding 30 to both sides]}$$

$$\boxed{t > 24} \qquad \text{[rewriting].}$$

See Fig. 15-3.

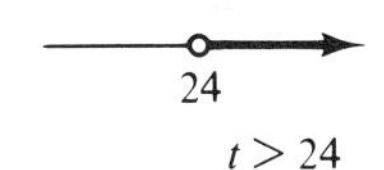

FIG. 15-3

EXAMPLE 7

Solve $2(x - 4) - 3 \geqslant 2x + 1$.

$$2(x - 4) - 3 \geqslant 2x + 1$$

$$2x - 8 - 3 \geqslant 2x + 1$$

$$-11 \geqslant 1 \qquad \text{[subtracting } 2x \text{ from both sides].}$$

Since $-11 \geqslant 1$ is never true, there is **no solution.**

EXAMPLE 8

Solve $3z + 7 > 8z - (5z - 2)$.

$$3z + 7 > 8z - (5z - 2)$$

$$3z + 7 > 3z + 2$$

$$7 > 2 \qquad \text{[subtracting } 3z \text{ from both sides].}$$

Since $7 > 2$ is always true (true for any value of z), the solution is **all real numbers**. See Fig. 15-4.

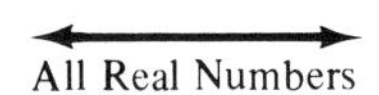

FIG. 15-4

EXAMPLE 9

Determine the domain of the function $f(x) = \sqrt{2x+3}$.

Recall from Sec. 9-3 that the domain consists of all real numbers x except those for which the functional values would involve either division by 0 or imaginary numbers. To avoid imaginary numbers for $f(x) = \sqrt{2x+3}$, we want $2x + 3 \geqslant 0$.

$$2x + 3 \geqslant 0$$
$$2x \geqslant -3$$
$$x \geqslant -\frac{3}{2}.$$

Thus, the domain of $f(x) = \sqrt{2x+3}$ is all x such that $x \geqslant -\frac{3}{2}$.

Completion Set 15-1

In Problems **1–3**, *circle the given values of x that make the inequalities true.*

1. $x > 5$: 4, 10, -3, 3, 5.1, 4.9

2. $x < -2$: -1, 0, 8, -8, -2, $\frac{15}{-2}$

3. $x \geqslant 9.1$: 9.1, -9.1, $\frac{10}{3}$, 9^{-1}, $|-12|$, 3^3

In Problems **4–8**, *insert* "$<$," "$\leqslant$," "$>$," *or* "$\geqslant$."

4. If $x + 2 > 5$, then $x + 2 - 2$ ____ $5 - 2$.

5. If $-x \leqslant 7$, then $(-1)(-x)$ ____ $(-1)(7)$.

6. If $4w \geqslant 8$, then $\frac{4w}{4}$ ____ $\frac{8}{4}$.

7. If $6t < 3t$, then $6t - 3t$ ____ $3t - 3t$.

8. If $-3x > 5$, then $\frac{-3x}{-3}$ ____ $\frac{5}{-3}$.

In Problems **9–13**, *fill in the blanks.*

9. If $3x > 12$, then $x > \frac{12}{(\_\_\_)} =$ ____.

10. If $y - 5 < 2$, then $y < 2 +$ ____ and so $y <$ ____.

11. If $4x + 2 \geqslant 6$, then $4x \geqslant$ ____ and so $x \geqslant$ ____.

12. If $2 - 7x \leqslant 9$, then $-7x \leqslant$ ____ and so $x \geqslant$ ______.

13. In Fig. 15-5, which of (a) or (b) represents all values of x such that $x \geqslant 3$? ____

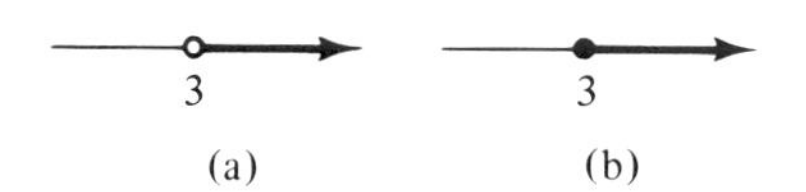

FIG. 15-5

Problem Set 15-1

In Problems **1–32**, *solve the inequalities and indicate your answers on a number line.*

1. $4x > 8$
2. $8x < -2$
3. $3x - 4 \leqslant 5$
4. $5x \geqslant 0$
5. $-4x \geqslant 2$
6. $6 \leqslant 5 - 3y$
7. $3 - 5s > 5$
8. $4s - 1 < -5$
9. $6x - 15 < 2x - 3$
10. $3x + 1 > 2(x + 4)$
11. $2x - 3 \leqslant 4 + 7x$
12. $-3 \geqslant 8(2 - x)$
13. $3 < 2y + 3$
14. $2y + 3 \leqslant 1 + 2y$
15. $3(2 - 3x) > 4(1 - 4x)$
16. $8(x + 1) + 1 < 3(2x) + 1$
17. $2(3x - 2) > 3(2x - 1)$
18. $3 - 2(x - 1) \leqslant 2(4 + x)$
19. $\frac{5}{3}x < 10$
20. $-\frac{1}{2}x > 6$

21. $\dfrac{9y+1}{4} \leqslant 2y - 1$

22. $\dfrac{4y-3}{2} \geqslant \dfrac{1}{3}$

23. $4x - 1 \geqslant 4(x-2) + 7$

24. $0x \leqslant 0$

25. $\dfrac{1-t}{2} < \dfrac{3t-7}{3}$

26. $\dfrac{3(2t-2)}{2} > \dfrac{6t-3}{5} + \dfrac{t}{10}$

27. $2x + 3 \geqslant \frac{1}{2}x - 4$

28. $4x - \frac{1}{2} \leqslant \frac{3}{2}x$

29. $\frac{2}{3}r < \frac{5}{6}r$

30. $\frac{7}{4}t > -\frac{2}{3}t$

31. $\dfrac{y}{2} + \dfrac{y}{3} > y + \dfrac{y}{5}$

32. $\dfrac{5y-1}{-3} < \dfrac{7(y+1)}{-2}$

In Problems **33–36**, *determine the domain of each function.*

33. $f(x) = \sqrt{3x - 6}$

34. $f(x) = \sqrt{7 - 4x}$

35. $g(t) = \sqrt{\dfrac{2t+1}{3}}$

36. $g(t) = \sqrt{\dfrac{3(t+5)}{8}}$

15-2 NONLINEAR INEQUALITIES

Now we'll show you how to solve inequalities that are not linear. Let's begin with $x^2 + 3x - 4 > 0$. First we factor the left side:

$$(x+4)(x-1) > 0.$$

Here we have a product of two factors that is positive. This can happen only if either both factors are positive *or* both factors are negative. Thus there are two cases to look at.

Case 1. *Both factors positive.*

Then $x + 4 > 0$ *and* $x - 1 > 0$. Thus, $x > -4$ *and* $x > 1$. This means that x must lie not only to the right of -4, but also to the right of 1. See Fig. 15-6. Both conditions are met when $x > 1$. Thus the

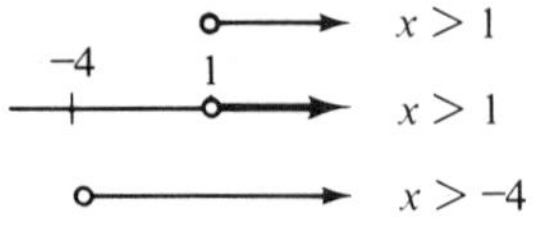

$x > 1$

-4 1 $x > 1$

$x > -4$

FIG. 15-6

solution in Case 1 is $x > 1$.

Case 2. *Both factors negative.*

Then $x + 4 < 0$ *and* $x - 1 < 0$. Thus, $x < -4$ *and* $x < 1$. Thus x must lie to the left of both -4 and 1. See Fig. 15-7. Both conditions

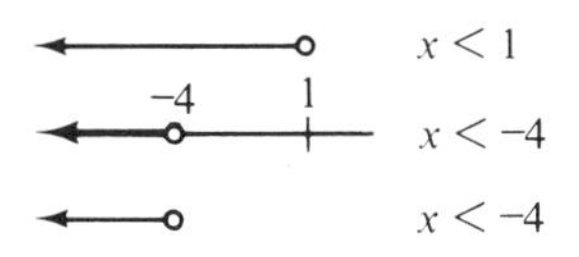

FIG. 15-7

are met when $x < -4$. The solution in Case 2 is $x < -4$.

Since either Case 1 or Case 2 can occur, the original inequality is true when $x > 1$ or $x < -4$. You can write the solution as

$$\boxed{x < -4 \quad \text{or} \quad x > 1.}$$

We can relate solving our inequality $x^2 + 3x - 4 > 0$ to solving the corresponding *equation* $x^2 + 3x - 4 = 0$. The roots of $x^2 + 3x - 4 = 0$ are -4 and 1 [since $x^2 + 3x - 4 = (x + 4)(x - 1) = 0$]. In Fig. 15-8(a), the roots are marked on a number line. These two roots determine three sets of numbers,

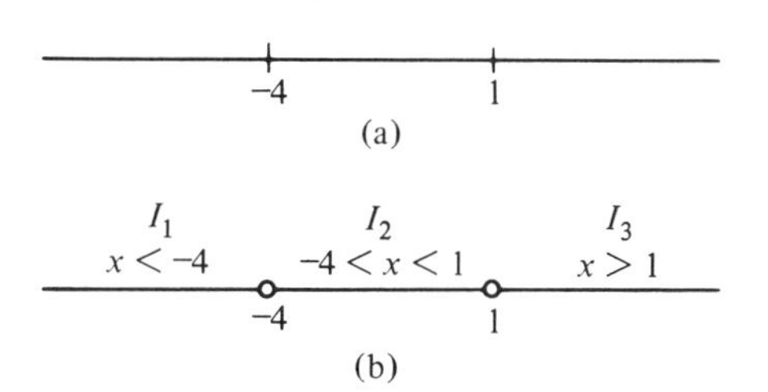

FIG. 15-8

called *intervals*, which are labeled I_1, I_2, and I_3 in Fig. 15-8(b).

I_1: all values of x such that $x < -4$.
I_2: all values of x such that $x > -4$ and $x < 1$.
I_3: all values of x such that $x > 1$.

The conditions on x in I_2 are usually written $-4 < x < 1$, which means that $x > -4$ *and* $x < 1$ simultaneously.

As we found above, $x^2 + 3x - 4$ is positive on I_1 and I_3. It can be shown that $x^2 + 3x - 4$ is negative on I_2. Thus, on each of the intervals determined by the roots of $x^2 + 3x - 4 = 0$, the polynomial $x^2 + 3x - 4$ is either strictly positive or strictly negative.

In general, for any polynomial P, the roots of $P = 0$ give rise to certain intervals.† On each of these intervals, P is always positive or always negative. Thus if P is positive (or negative) at one point on an interval, then P must be positive (or negative) on the *entire* interval. This fact lets us solve inequalities without the bother of going into different cases. Example 1 will show you how.

EXAMPLE 1

Solve $x^2 - 4x - 12 < 0$.

First, factor the left side.

$$x^2 - 4x - 12 < 0$$
$$(x + 2)(x - 6) < 0.$$

Next, find the roots of $(x + 2)(x - 6) = 0$.

$$\text{Roots of } (x + 2)(x - 6) = 0 \text{ are } -2, 6.$$

These two roots give rise to three intervals. See Fig. 15-9. On each interval the product $(x + 2)(x - 6)$ is always positive or always negative. Let's find the sign of

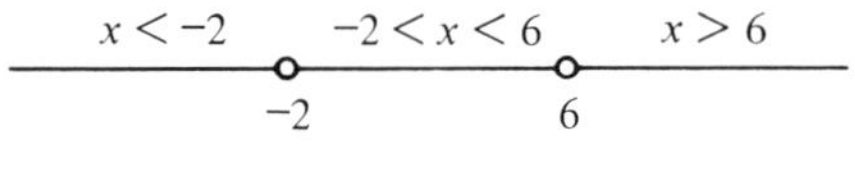

$x < -2$ $\quad -2 < x < 6 \quad$ $x > 6$

$-2 \qquad 6$

FIG. 15-9

the product for $x < -2$. To do this we pick *any* value of x in the interval and see if $(x + 2)(x - 6)$ is positive or negative for that value. Let's choose $x = -8$.

$$\text{If } x = -8, \text{ then } (x + 2)(x - 6) = (-8 + 2)(-8 - 6)$$
$$= (-)(-) = (+).$$

Since the product $(x + 2)(x - 6)$ is positive (+) for $x = -8$, it is positive on the entire interval where $x < -2$. To find the sign of the product on the other intervals, we choose a value of x in each of them.

†We are concerned only with those roots that are real numbers.

For $-2 < x < 6$, let's choose $x = 0$.

$$(x + 2)(x - 6) = (0 + 2)(0 - 6) = (+)(-) = (-).$$

Thus, $\quad (x + 2)(x - 6) < 0 \quad$ for $-2 < x < 6$.

For $x > 6$, let's choose $x = 10$.

$$(x + 2)(x - 6) = (10 + 2)(10 - 6) = (+)(+) = (+).$$

Thus, $\quad (x + 2)(x - 6) > 0 \quad$ for $x > 6$.

Figure 15-10 gives a summary of the signs. Thus the solution of $(x + 2)(x - 6) < 0$

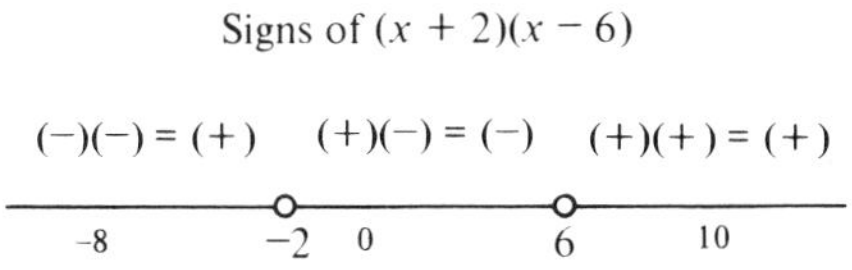

Signs of $(x + 2)(x - 6)$

$(-)(-) = (+)$ $\quad (+)(-) = (-)$ $\quad (+)(+) = (+)$

-8 $\quad -2 \quad 0 \quad 6 \quad 10$

FIG. 15-10

is

$$\boxed{-2 < x < 6.}$$

EXAMPLE 2

Solve $x^2 - 3x \geqslant 6x - 20$.

First we write the inequality so that one side is 0.

$$x^2 - 9x + 20 \geqslant 0.$$

Factoring the left side gives

$$(x - 4)(x - 5) \geqslant 0.$$

Roots of $(x - 4)(x - 5) = 0$ are 4, 5.

These roots give us the intervals in Fig. 15-11. We now choose a point in each

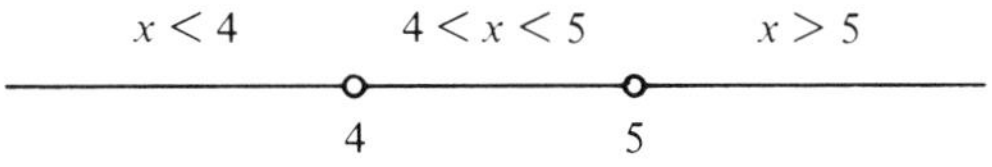

$x < 4$ $\quad 4 < x < 5$ $\quad x > 5$

4 $\quad$ 5

FIG. 15-11

interval and find the sign of the product $(x - 4)(x - 5)$ at those points. See Fig. 15-12. Notice that the product is positive when $x = 3$ and when $x = 6$. Thus,

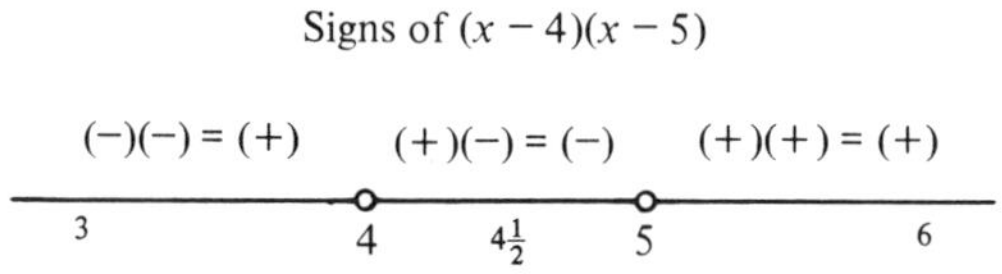

Signs of $(x-4)(x-5)$

$(-)(-) = (+)$	$(+)(-) = (-)$	$(+)(+) = (+)$

3 4 $4\frac{1}{2}$ 5 6

FIG. 15-12

$(x - 4)(x - 5) > 0$ for $x < 4$ or for $x > 5$. But we want the solution of the inequality $(x - 4)(x - 5) \geqslant 0$. Since $(x - 4)(x - 5) = 0$ when $x = 4$ and when $x = 5$, the solution of $(x - 4)(x - 5) \geqslant 0$ is

$$\boxed{x \leqslant 4 \quad \text{or} \quad x \geqslant 5.}$$

*In Example 2, **don't** write the answer as $4 \geqslant x \geqslant 5$. These symbols mean that $x \leqslant 4$ **and** $x \geqslant 5$. But there is no value of x that meets both conditions.*

EXAMPLE 3

Solve $t^3 - 8t^2 + 16t > 0$.

Factoring the left side gives

$$t(t^2 - 8t + 16) > 0$$

$$t(t - 4)^2 > 0.$$

Roots of $t(t - 4)^2 = 0$ are 0, 4.

These roots determine the intervals in Fig. 15-13. At some point in each interval we

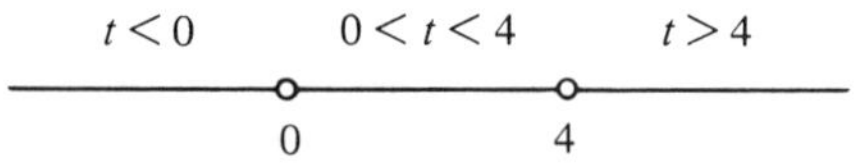

$t<0$	$0<t<4$	$t>4$

0 4

FIG. 15-13

find the sign of $t(t-4)^2$. From Fig. 15-14 we see that $t(t-4)^2 > 0$ for

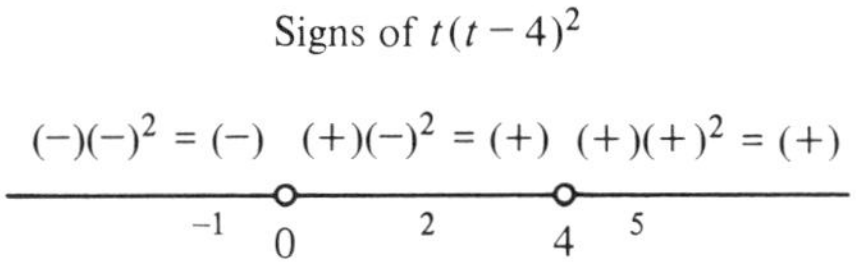

Signs of $t(t-4)^2$

$(-)(-)^2 = (-)$ $(+)(-)^2 = (+)$ $(+)(+)^2 = (+)$

−1 0 2 4 5

FIG. 15-14

$$\boxed{0 < t < 4 \quad \text{or} \quad t > 4.}$$

The next inequality involves a quotient of polynomials. Here we'll find the sign of the quotient on intervals determined by the roots of *two* equations. We get these equations by setting the numerator and denominator equal to 0.

EXAMPLE 4

Solve $\dfrac{x^2 - 4x + 3}{x + 2} > 0.$

Factoring the numerator gives

$$\frac{(x-1)(x-3)}{x+2} > 0.$$

Setting the numerator and denominator equal to 0 gives the roots 1, 3, and -2.

$(x-1)(x-3) = 0$	$x + 2 = 0$
Roots are 1, 3.	Root is -2.

These three roots give rise to four intervals. See Fig. 15-15.

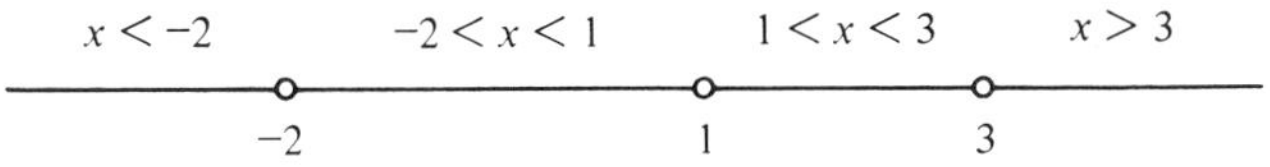

$x < -2$ $-2 < x < 1$ $1 < x < 3$ $x > 3$

−2 1 3

FIG. 15-15

Now we choose a point in each of the intervals (see Fig. 15-16) and find the sign of $\dfrac{(x-1)(x-3)}{x+2}$ at each point. Since this quotient is positive for $x = 0$ and $x = 4$,

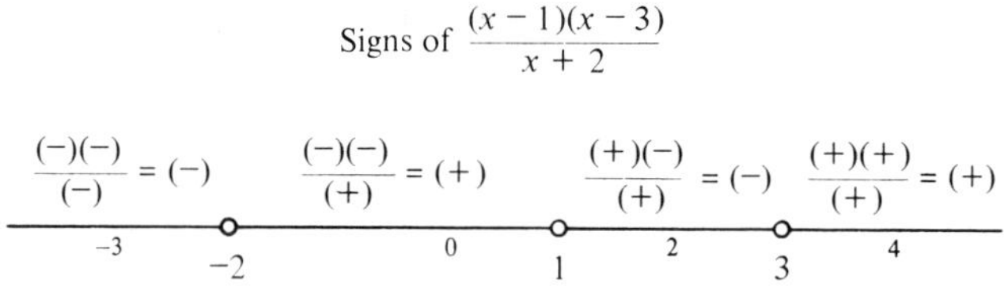

Signs of $\frac{(x-1)(x-3)}{x+2}$

$\frac{(-)(-)}{(-)} = (-)$ $\frac{(-)(-)}{(+)} = (+)$ $\frac{(+)(-)}{(+)} = (-)$ $\frac{(+)(+)}{(+)} = (+)$

−3 −2 0 1 2 3 4

FIG. 15-16

then $\frac{(x-1)(x-3)}{x+2} > 0$ for

$$\boxed{-2 < x < 1 \quad \text{or} \quad x > 3.}$$

EXAMPLE 5

Solve $\frac{x^2-9}{x^2-6x} \leqslant 0.$

Factoring the numerator and denominator gives

$$\frac{(x+3)(x-3)}{x(x-6)} \leqslant 0.$$

Setting both the numerator and denominator equal to 0, we get

$(x+3)(x-3)=0$	$x(x-6)=0$
Roots are $-3, 3$.	Roots are 0, 6.

These four roots give rise to the five intervals in Fig. 15-17. From Fig. 15-18 we see

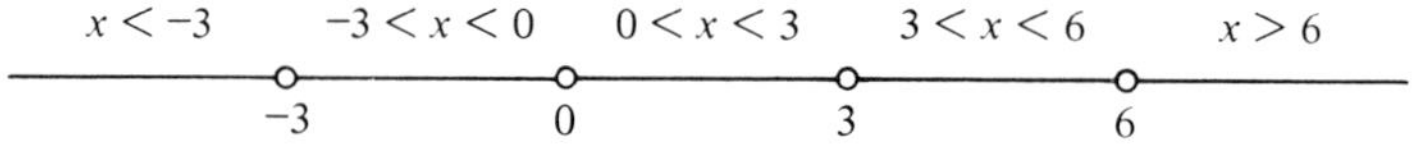

$x<-3$ $-3<x<0$ $0<x<3$ $3<x<6$ $x>6$

−3 0 3 6

FIG. 15-17

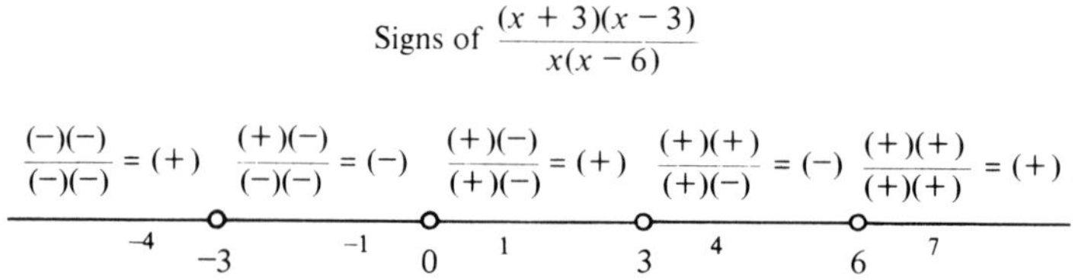

Signs of $\frac{(x+3)(x-3)}{x(x-6)}$

$\frac{(-)(-)}{(-)(-)} = (+)$ $\frac{(+)(-)}{(-)(-)} = (-)$ $\frac{(+)(-)}{(+)(-)} = (+)$ $\frac{(+)(+)}{(+)(-)} = (-)$ $\frac{(+)(+)}{(+)(+)} = (+)$

−4 −3 −1 0 1 3 4 6 7

FIG. 15-18

that $\frac{(x+3)(x-3)}{x(x-6)} < 0$ for $-3 < x < 0$ or $3 < x < 6$. Now $\frac{(x+3)(x-3)}{x(x-6)}$ is 0 only if the numerator is 0, that is, when $x = -3$ or $x = 3$. The solution of $\frac{(x+3)(x-3)}{x(x-6)} \leqslant 0$ must include -3 and 3, and thus is

$$\boxed{-3 \leqslant x < 0 \quad \text{or} \quad 3 \leqslant x < 6.}$$

Completion Set 15-2

In Problems **1–3**, *circle the given values of x that make the inequalities true.*

1. $4 < x < 7$: 7, 5, 3, 6.03, -6, 4, 9.

2. $-3 \leqslant x \leqslant 2$: 1, 6, 2, 0, -2, -3, -5.

3. $-5 < x \leqslant 3$: 4, 3, -3, -5, 5, 0, -6.

In Problems **4–9**, *fill in the blanks.*

4. To solve $(x+3)(x-6) > 0$, you can look at three intervals determined by the numbers -3 and ____.

5. To solve $(x-1)(x-5) < 0$, you can look at three intervals determined by the numbers ____ and ____. These intervals are $x <$ ____, ____ $< x <$ ____, and $x >$ ____.

6. *Insert* "+" *or* "−." To find the signs of $x(x-2)$, you can look at the three intervals in Fig. 15-19. Since the sign of $x(x-2)$ at $x = -1$ is (____)(____) = (____), then

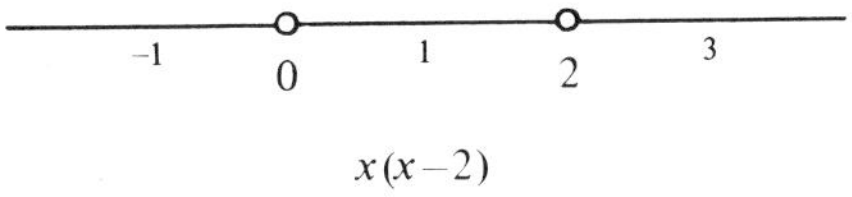

FIG. 15-19

its sign for $x < 0$ is (____). Since the sign of $x(x-2)$ at $x = 1$ is (____)(____) = (____), then its sign for $0 < x < 2$ is (____). Since the sign of $x(x-2)$ at $x = 3$ is (____)(____) = (____), then its sign for $x > 2$ is (____).

7. From Fig. 15-20, the solution of $(x - 4)(x - 8) < 0$ is ______________.

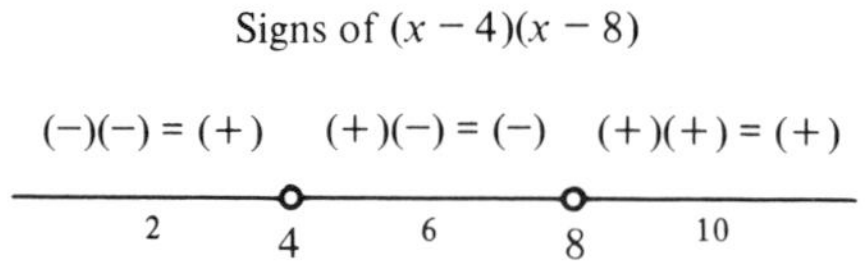

Signs of $(x - 4)(x - 8)$

FIG. 15-20

The solution of $(x - 4)(x - 8) \leqslant 0$ is ______________.

8. From Fig. 15-20, the solution of $(x - 4)(x - 8) > 0$ is ______________.

The solution of $(x - 4)(x - 8) \geqslant 0$ is ______________.

9. To solve $\dfrac{x}{x - 2} \geqslant 0$, you can look at the intervals determined by the numbers ____ and ____.

In Problems **10** *and* **11**, *insert* T (= *True*) *or* F (= *False*).

10. The solution of $\dfrac{x}{x - 2} \geqslant 0$ includes $x = 0$. ____

11. The solution of $\dfrac{x}{x - 2} \geqslant 0$ includes $x = 2$. ____

Problem Set 15-2

In Problems **1–38**, *solve the inequalities.*

1. $(x - 1)(x - 5) < 0$

2. $(x + 3)(x - 8) > 0$

3. $(x + 1)(x - 3) > 0$

4. $(x + 4)(x + 5) < 0$

5. $x^2 - 1 < 0$

6. $x^2 > 9$

7. $x^2 - x - 6 > 0$

8. $x^2 - 2x - 3 \leqslant 0$

9. $5s - s^2 \leqslant 0$

10. $t^2 + 9t + 18 \geqslant 0$

11. $x^2 + 5x < -6$

12. $x^2 - 12 > x$

13. $x^2 + 4x - 5 \geqslant 3x + 15$

14. $x^2 + 9x + 9 \leqslant 2 - x^2$

15. $2z^2 - 5z - 12 < 0$

16. $5t^2 - 1 > 4t$

17. $x^2 + 2x + 1 > 0$

18. $x^2 + 9 \leqslant 6x$

19. $4(t^2 + t) - 1 \geqslant 2$

20. $3s^2 \leqslant 11s - 10$

21. $(x + 2)(x - 1)(x - 4) > 0$

22. $x(x + 3)(5 - x) < 0$

23. $(x + 5)(x - 3)^2 < 0$

24. $(x + 1)^2(x - 4)^2 > 0$

25. $y^3 - y \leqslant 0$

26. $x^3 + 8x^2 + 15x \geqslant 0$

27. $x^3 - 2x^2 \geqslant 0$

28. $p^4 - 2p^3 - 3p^2 \leqslant 0$

29. $\dfrac{x - 4}{x + 8} > 0$

30. $\dfrac{x + 3}{x + 5} < 0$

31. $\dfrac{t}{3 - t} \leqslant 0$

32. $\dfrac{s - 1}{2s - 1} > 0$

33. $\dfrac{5}{x + 2} > 0$

34. $\dfrac{3}{9(x - 8)} < 0$

35. $\dfrac{x + 3}{x^2 - 1} < 0$

36. $\dfrac{x^2 - x - 6}{x + 4} > 0$

37. $\dfrac{x^2 - 5x + 4}{x^2 + 5x + 4} \geqslant 0$

38. $\dfrac{x^2 - 4}{x^2 + 2x + 1} \leqslant 0$

In Problems **39–42**, *determine the domain of each function.*

39. $f(x) = \sqrt{x^2 + 4x - 12}$

40. $g(x) = \sqrt{x^4 - x^2}$

41. $g(t) = \sqrt{\dfrac{t - 1}{t + 5}}$

42. $F(p) = \sqrt{\dfrac{p^2 - 4}{p + 1}}$

15-3 ABSOLUTE VALUE INEQUALITIES

Sometimes it is useful to work with inequalities involving absolute value. Recall that the absolute value of a number is the distance of the number from 0.

If $|x| < 3$, then x is less than 3 units from 0. Thus, x must lie between -3 and 3. That is, $-3 < x < 3$ [Fig. 15-21(a)]. However, if $|x| > 3$, then x must be more than 3 units from 0. Thus, one of two things must be true: either $x > 3$ *or* $x < -3$ [Fig. 15-21(b)]. We can extend these ideas. If $|x| \leqslant 3$, then we have $-3 \leqslant x \leqslant 3$. If $|x| \geqslant 3$, then $x \geqslant 3$ or $x \leqslant -3$.

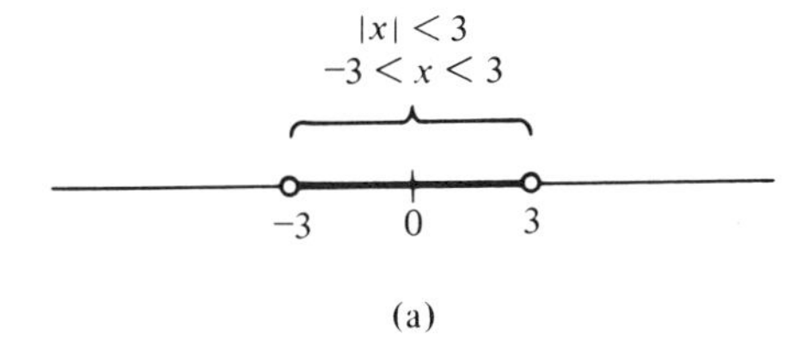

(a)

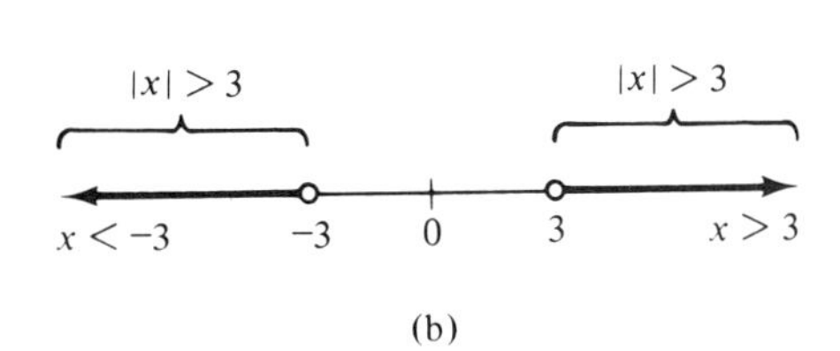

(b)

FIG. 15-21

In general, the solution of $|x| < d$ or $|x| \leqslant d$, where d is a positive number, consists of one interval. However, when $|x| > d$ or $|x| \geqslant d$, you will have two intervals in your solution.

EXAMPLE 1

Solve the following absolute value inequalities.

a. $|x - 2| < 4$.

The number $x - 2$ must be less than 4 units from 0. This means that $x - 2$ lies between -4 and 4. Thus, $-4 < x - 2 < 4$. We solve this as follows.

$$-4 < x - 2 < 4$$

$$-4 + 2 < x < 4 + 2 \qquad \text{[adding 2 to each member]}$$

$$\boxed{-2 < x < 6.}$$

b. $|3 - 2x| \leqslant 5$.

$$-5 \leqslant 3 - 2x \leqslant 5$$

$$-5 - 3 \leqslant -2x \leqslant 5 - 3 \qquad \text{[subtracting 3 from each member]}$$

$$-8 \leqslant -2x \leqslant 2$$

$$4 \geqslant x \geqslant -1 \qquad \text{[dividing each member by } -2 \text{ and changing direction of inequalities]}$$

$$\boxed{-1 \leqslant x \leqslant 4} \qquad \text{[rewriting].}$$

EXAMPLE 2

Solve the following inequalities.

a. $|x + 5| \geqslant 7$.

The number $x + 5$ must be *at least* 7 units from 0. Thus, either $x + 5 \geqslant 7$ *or* $x + 5 \leqslant -7$. This means that

$$\boxed{x \geqslant 2 \quad \text{or} \quad x \leqslant -12.}$$

b. $|3x - 4| > 1$.

Either $3x - 4 > 1$ or $3x - 4 < -1$. Thus, either $3x > 5$ or $3x < 3$. Therefore

$$\boxed{x > \tfrac{5}{3} \quad \text{or} \quad x < 1.}$$

The numbers 5 and 9 are 4 units apart. Also,

$$|9 - 5| = |4| = 4,$$
$$|5 - 9| = |-4| = 4.$$

In general, you may think of $|a - b|$ or $|b - a|$ as the distance between a and b.

EXAMPLE 3

Use absolute value notation to express the fact that

a. *x is less than 3 units from 5.*

Since the distance between x and 5 is less than 3, we must have

$$|x - 5| < 3.$$

b. *x differs from 6 by at least 7.*

$$|x - 6| \geqslant 7.$$

c. *$x < 9$ and $x > -9$ simultaneously.*

$$|x| < 9.$$

d. *x is more than 2 units from -3.*

$$|x - (-3)| > 2$$
$$|x + 3| > 2.$$

e. *$f(x)$ is less than ϵ (epsilon) units from L.*

$$|f(x) - L| < \epsilon.$$

Three basic properties of absolute value are

1. $|ab| = |a| \cdot |b|$.
2. $\left| \dfrac{a}{b} \right| = \dfrac{|a|}{|b|}$.
3. $|a - b| = |b - a|$.

EXAMPLE 4

a. $|(-7)\cdot 3| = |-7|\cdot|3| = 21$; $|(-7)(-3)| = |-7|\cdot|-3| = 21$.

b. $|4-2| = |2-4| = 2$.

c. $|7-x| = |x-7|$.

d. $\left|\frac{-7}{3}\right| = \frac{|-7|}{|3|} = \frac{7}{3}$; $\left|\frac{-7}{-3}\right| = \frac{|-7|}{|-3|} = \frac{7}{3}$.

e. $\left|\frac{x-3}{-5}\right| = \frac{|x-3|}{|-5|} = \frac{|x-3|}{5}$.

Completion Set 15-3

In Problems **1** *and* **2**, *circle the given values of x that make the inequalities true.*

1. $|x-7| < 4$: 3, -1, -3, 5, 7, $\frac{37}{6}$

2. $|4-3x| \geqslant 12$: 0, 6, $-\frac{8}{3}$, $\frac{1}{2}$ $|-10|$, $\frac{16}{3}$

In Problems **3–6**, *fill in the blanks.*

3. If $|x| < 4$, then x is less than ____ units from ______. Thus, x lies between -4 and ____. So $-4 < x <$____.

4. If $|x-1| > 2$, then $x-1$ is more than ____ units from ____. Thus, $x-1 >$____ or $x-1 <$______. This means that $x >$____ or $x <$______.

5. In absolute value notation, the fact that $2x$ is less than 5 units from 6 is written $|2x -$____$| <$____.

6. $|3-2x| = |2x -$____$|$ and $|(x+2)(x-2)| = |x+2|\cdot|$______________$|$.

Problem Set 15-3

In Problems **1–20**, *solve the inequalities.*

1. $|x| < 3$

2. $|x| < 10$

3. $|x| > 6$

4. $|x| > 3$

5. $|2x| \leqslant 2$

6. $|4x| \leqslant 3$

7. $|3x| \geqslant 9$

8. $\left|\frac{x}{2}\right| \geqslant 3$

9. $|x - 4| < 16$

10. $|y + 5| \leqslant 6$

11. $|y + 1| \geqslant 6$

12. $|x - 2| \geqslant 4$

13. $|3x - 5| \leqslant 1$

14. $\left|\frac{x}{3} - 5\right| < 4$

15. $|1 - 3x| < 2$

16. $|4x - 1| > 7$

17. $\left|\frac{1}{2} - t\right| > \frac{1}{2}$

18. $|5 - 2x| > 1$

19. $\left|\frac{3x - 8}{2}\right| \geqslant 4$

20. $\left|\frac{x - 8}{4}\right| \leqslant 2$

21. Why does $|x - 5| < -12$ have no solution?

22. Why does every value of x satisfy $|2x + 1| > -1$?

23. Use absolute value notation to express the fact that

a. x is less than 3 units from 7.

b. x differs from 2 by less than 3.

c. x is no more than 5 units from -7.

d. The distance between 7 and x is 4.

e. $x + 4$ is strictly within 2 units of 0.

f. x is strictly between 3 and -3.

g. $x < -6$ or $x > 6$.

h. $x - 6 \geqslant 4$ or $x - 6 \leqslant -4$.

i. The number x of hours that a machine will operate efficiently differs from 105 by less than 3.

j. The average monthly income x (in dollars) of a family differs from 600 by less than 100.

24. Use absolute value notation to indicate that the prices p and q of two products may differ by no more than 2 (dollars).

25. Show that if $|f(x) - L| < \epsilon$, then $L - \epsilon < f(x) < L + \epsilon$.

26. Show that if $|x - \mu| \leqslant 2\sigma$, where μ is read "mu" and σ is read "sigma," then $-2\sigma + \mu \leqslant x \leqslant 2\sigma + \mu$.

27. In the manufacture of widgets, the average length of a part is .01 centimeter (cm). Use absolute value notation to express the fact that an individual measurement x of a part does not differ from the average by more than .005 cm.

15-4 REVIEW

IMPORTANT TERMS AND SYMBOLS

linear inequality *(p. 350)*

interval *(p. 355)*

$a < x < b$ *(p. 355)*

$a \leqslant x < b$ *(p. 361)*

REVIEW PROBLEMS

In Problems **1–32**, *solve the inequalities.*

1. $3x + 5 < 6$

2. $3(x + 1) > 9$

3. $4 - 2x \geqslant 8$

4. $-2(x + 6) \leqslant x + 4$

5. $3(t + 4) < 9 + 6t$

6. $5(s + 3) > 2(s - 1)$

7. $\frac{1}{3}(x + 2) \geqslant \frac{1}{4}x + 4$

8. $\dfrac{x + 1}{5} \leqslant \dfrac{x}{10} + 2$

9. $x^2 + 4x - 12 < 0$

10. $x^2 + 11x + 28 > 0$

11. $y^2 > 6y$

12. $z^2 + 6z < -5$

13. $2x^2 + 5x \geqslant x^2 - 4x - 20$

14. $3x(x - 4) \leqslant 2x^2 - 27$

15. $x(3x + 2) < 1$

16. $6x(x - 1) > 1 - 5x$

17. $(x + 4)(x - 5)(x - 9) > 0$

18. $(x + 2)(x + 4)^2 < 0$

19. $p^3 - 8p^2 \leqslant 0$

20. $r^3 - 9r \geqslant 0$

21. $\dfrac{x + 9}{x + 2} \geqslant 0$

22. $\dfrac{x + 3}{x^2 - 3x + 2} \leqslant 0$

23. $\dfrac{x^2 - 6x + 9}{x^2 + 7x + 10} > 0$

24. $\dfrac{x^2 + 2x - 8}{x^2 - 25} < 0$

25. $|3x| > 6$

26. $|x + 4| \leqslant 6$

27. $|4x - 1| < 1$

28. $\left|\dfrac{5x - 8}{12}\right| \geqslant 1$

29. $|4 - 2x| \geqslant 4$

30. $|-1 - x| < 1$

31. $|x + \frac{1}{2}| \leqslant \frac{3}{2}$

32. $|\frac{2}{3}x - 5| > 4$

In Problems **33–36**, *determine the domain of each function.*

33. $f(x) = \sqrt{4x + 10}$

34. $F(t) = \sqrt{\dfrac{2t - 6}{3}}$

35. $g(z) = \sqrt{2z - z(z - 1)}$

36. $G(s) = \sqrt{\dfrac{s + 8}{s}}$

CHAPTER 16

CONIC SECTIONS

16-1 THE DISTANCE FORMULA

We can find a formula that gives the distance between two points in a plane. Suppose the points are (x_1, y_1) and (x_2, y_2), as in Fig. 16-1. The length of the

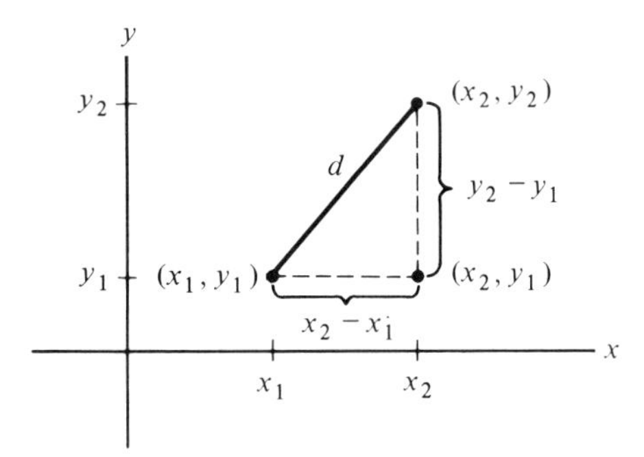

FIG. 16-1

segment joining them is d. By drawing segments parallel to the axes we have a right triangle with d as the length of the hypotenuse. The lengths of the other sides are $x_2 - x_1$ and $y_2 - y_1$. By the Pythagorean theorem,†

$$d^2 = (x_2 - x_1)^2 + (y_2 - y_1)^2.$$

Since d cannot be negative, we must have

$$\boxed{d = \sqrt{(x_2 - x_1)^2 + (y_2 - y_1)^2}\ ,}$$

which is called the **distance formula.**

EXAMPLE 1

Find the distance between the given points.

a. (5, − 2) and (−3, 4).

Suppose $(x_1, y_1) = (5, -2)$ and $(x_2, y_2) = (-3, 4)$. Then

$$\begin{aligned} d &= \sqrt{(x_2 - x_1)^2 + (y_2 - y_1)^2} \\ &= \sqrt{(-3 - 5)^2 + [4 - (-2)]^2} \\ &= \sqrt{(-8)^2 + (6)^2} = \sqrt{64 + 36} \\ &= \sqrt{100} = 10. \end{aligned}$$

If we chose (−3, 4) as (x_1, y_1), we would get the same result.

b. (3, −6) and the origin.

Let $(x_1, y_1) = (0, 0)$ and $(x_2, y_2) = (3, -6)$. Then

$$\begin{aligned} d &= \sqrt{(x_2 - x_1)^2 + (y_2 - y_1)^2} \\ &= \sqrt{(3 - 0)^2 + (-6 - 0)^2} \\ &= \sqrt{9 + 36} = \sqrt{45} = 3\sqrt{5}\ . \end{aligned}$$

---

†The Pythagorean theorem says that the square of the length of the hypotenuse of a right triangle is equal to the sum of the squares of the lengths of the other two sides.

Problem Set 16-1

Find the distance between the given points.

1. (2, 3), (5, 7)

2. (4, 5), (12, 11)

3. (0, 5), (2, −2)

4. (1, 3), (1, 4)

5. (−1, −2), (−3, −4)

6. (−4, 4), origin

7. (2, −3), (−5, −3)

8. $\left(-\frac{3}{2}, \frac{1}{2}\right), \left(\frac{1}{2}, \frac{3}{2}\right)$

9. $\left(1, -\frac{1}{2}\right)$, origin

10. $(4, 0), (-1, \sqrt{11})$

11. (−5, 0), (0, −12)

12. (0, −2), (−5, −2)

16-2 THE CIRCLE

In the remainder of this chapter we'll consider four curves that may be formed by the intersection of a plane with a right circular cone. These so-called **conic sections**, or **conics**, are the *circle*, the *parabola*, the *ellipse*, and the *hyperbola*. They are shown in Fig. 16-2, where the type of curve obtained depends on how

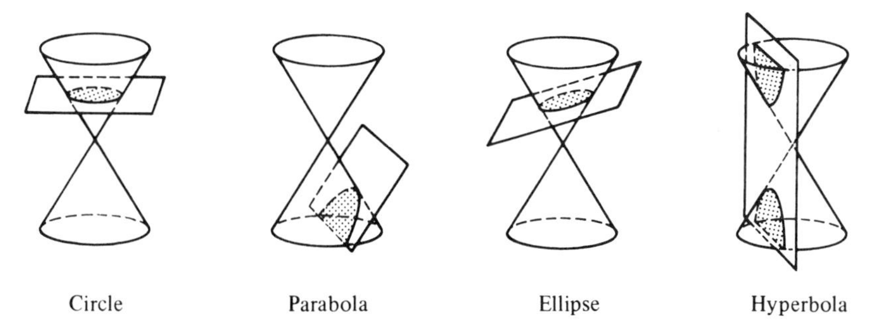

FIG. 16-2

the plane cuts the cone. In this section we'll look at the circle.

A **circle** consists of all points in a plane that are at a given distance from a fixed point in the plane. The fixed point is the **center**, and the given distance is the **radius** of the circle.

To be general, suppose that a circle has a radius of r and has its center at

(h, k), as in Fig. 16-3. Let (x, y) be any point on the circle. Then its distance

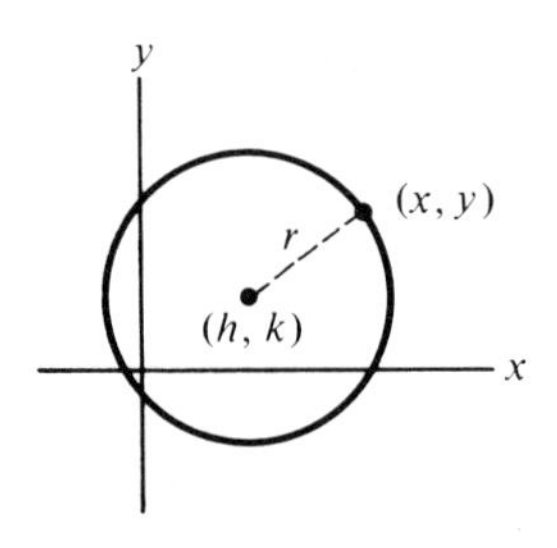

FIG. 16-3

from (h, k) must be r. Applying the distance formula to (h, k) and (x, y) gives

$$\sqrt{(x-h)^2+(y-k)^2} = r.$$

Squaring both sides, we have

$$(x-h)^2+(y-k)^2 = r^2.$$

Any point on the circle must satisfy this equation. Also, it can be shown that all points satisfying the equation lie on the circle. Thus we say that

$$\boldsymbol{(x-h)^2+(y-k)^2 = r^2} \qquad (16\text{-}1)$$

is the **standard form** of an equation of the circle with center (h, k) and radius r.

EXAMPLE 1

Find the standard form of an equation of the circle having the given center and radius.

a. Center $(-3, 2)$ and radius 4.

We use Eq. (16-1) with $h = -3$, $k = 2$, and $r = 4$.

$$(x-h)^2+(y-k)^2 = r^2$$
$$[x-(-3)]^2+(y-2)^2 = 4^2$$
$$(x+3)^2+(y-2)^2 = 16.$$

b. Center $(0, 0)$ and radius $\sqrt{3}$.

Here $h = 0$, $k = 0$, and $r = \sqrt{3}$.

$$(x - h)^2 + (y - k)^2 = r^2$$
$$(x - 0)^2 + (y - 0)^2 = (\sqrt{3})^2$$
$$x^2 + y^2 = 3.$$

In general,

$$x^2 + y^2 = r^2 \qquad (16\text{-}2)$$

is the standard form of an equation of the circle with center at the origin and radius r.

EXAMPLE 2

Describe and draw the graph of the given equation.

a. $(x - 1)^2 + (y + 4)^2 = 9$.

This has the form of Eq. (16-1). By writing this equation as

$$(x - 1)^2 + [y - (-4)]^2 = (3)^2,$$

we see that $h = 1$, $k = -4$, and $r = 3$. Thus the graph is a circle with center $(1, -4)$ and radius 3. See Fig. 16-4(a).

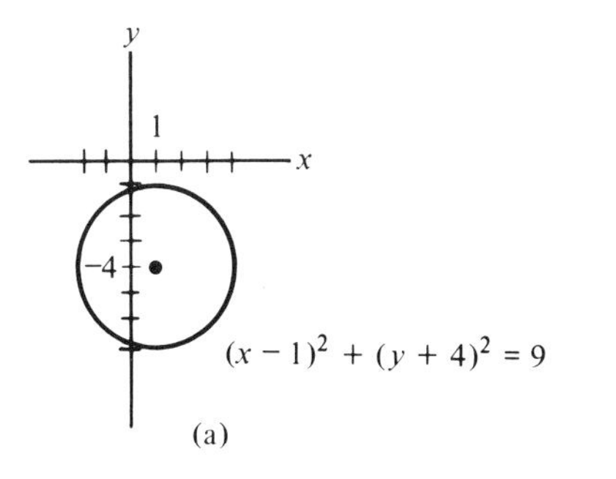

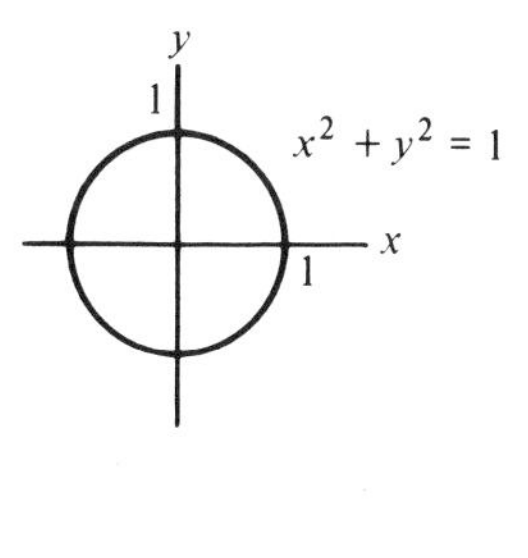

FIG. 16-4

b. $x^2 + y^2 = 1$.

This has the form of Eq. (16-2) with $r^2 = 1$. Thus $r = 1$. The graph is a circle with center at the origin and radius 1. See Fig. 16-4(b).

If we expand the terms in the standard form of a circle, we get

$$(x - h)^2 + (y - k)^2 = r^2$$

$$x^2 - 2hx + h^2 + y^2 - 2ky + k^2 = r^2$$

$$x^2 + y^2 + (-2h)x + (-2k)y + (h^2 + k^2 - r^2) = 0 \qquad \text{[rearranging]}.$$

Since h, k, and r are constants, this equation has the form

$$\mathbf{x^2 + y^2 + Dx + Ey + F = 0}, \qquad (16\text{-}3)$$

where D, E, and F are constants. This is called the **general form** for an equation of a circle.

EXAMPLE 3

Find the general form of an equation of the circle with center (1, 4) *and radius* 2.

The standard form is

$$(x - 1)^2 + (y - 4)^2 = 2^2.$$

Expanding gives

$$x^2 - 2x + 1 + y^2 - 8y + 16 = 4$$

$$x^2 + y^2 - 2x - 8y + 13 = 0,$$

which is the general form.

When a circle is given in general form, we can find the center and radius by going to standard form. This involves "completing the square."† For example,

$$x^2 + y^2 - 6x + 10y + 5 = 0 \qquad (16\text{-}4)$$

is in general form. We now regroup:

$$(x^2 - 6x \quad) + (y^2 + 10y \quad) = -5.$$

To complete the square in x we add $[\frac{1}{2}(-6)]^2$ or 9 to *both* sides; to complete the square in y we add $[\frac{1}{2}(10)]^2$ or 25 to *both* sides. This gives

$$(x^2 - 6x + 9) + (y^2 + 10y + 25) = -5 + 9 + 25,$$

†Completing the square was discussed on p. 157.

or the standard form

$$(x - 3)^2 + (y + 5)^2 = 29.$$

Thus the graph of Eq. (16-4) is a circle with center $(3, -5)$ and radius $\sqrt{29}$.

EXAMPLE 4

Describe the graphs of the following equations.

a. $x^2 + y^2 - y = 0.$

$$x^2 + y^2 - y = 0$$

$$x^2 + (y^2 - y \quad) = 0$$

$$x^2 + \left(y^2 - y + \tfrac{1}{4}\right) = \tfrac{1}{4} \qquad \text{[completing the square]}$$

$$(x - 0)^2 + \left(y - \tfrac{1}{2}\right)^2 = \left(\tfrac{1}{2}\right)^2 \qquad \text{[standard form].}$$

The graph is a circle with center $(0, \frac{1}{2})$ and radius $\frac{1}{2}$.

b. $2x^2 + 2y^2 + 8x - 3y + 5 = 0.$

To complete the squares, we first get the coefficients of the x^2- and y^2-terms to be 1. Thus we divide both sides by 2.

$$x^2 + y^2 + 4x - \tfrac{3}{2}y + \tfrac{5}{2} = 0$$

$$(x^2 + 4x) + (y^2 - \tfrac{3}{2}y) = -\tfrac{5}{2} \qquad \text{[regrouping]}$$

$$(x^2 + 4x + 4) + (y^2 - \tfrac{3}{2}y + \tfrac{9}{16}) = -\tfrac{5}{2} + 4 + \tfrac{9}{16} \qquad \text{[completing the square]}$$

$$(x + 2)^2 + (y - \tfrac{3}{4})^2 = \tfrac{33}{16}.$$

Thus the graph is a circle with center $(-2, \frac{3}{4})$ and radius $\sqrt{\frac{33}{16}}$ or $\frac{\sqrt{33}}{4}$.

Every circle has an equation of the general form $x^2 + y^2 + Dx + Ey + F = 0$. But an equation of this form does not always have a circle for its graph. The graph could be a point, or there may not be any graph at all. You will know what it is after completing the square.

For example, the equation

$$x^2 + y^2 - 2x - 6y + 10 = 0 \qquad (16\text{-}5)$$

may be written

$$(x^2 - 2x + 1) + (y^2 - 6y + 9) = -10 + 1 + 9$$

$$(x - 1)^2 + (y - 3)^2 = 0.$$

This implies a center at $(1, 3)$, but a radius of 0. Thus the graph is the single point (1, 3). Also, if the 10 in Eq. (16-5) were 11, then we would get

$$(x-1)^2+(y-3)^2=-1.$$

This implies that r^2 is -1, a negative number. But for a circle we must have $r \geqslant 0$, so r^2 cannot be negative. Thus the equation does not define a circle. In fact, since the left side of the above equation is a sum of squares, it can never equal -1. Therefore the equation has no graph.

Completion Set 16-2

Fill in the blanks.

1. The equation $(x-3)^2+(y+5)^2=49$ is in standard form. Here $h =$ ____, $k =$ ______, and $r^2 =$ ______. Thus the graph is a circle with center at ____________ and a radius of ____.

2. An equation of the circle with center $(-3, 5)$ and radius 4 is $(x +$ ____$)^2 +$ (____________$)^2 =$ ____^{2}.

3. The graph of $(x-1)^2+(y-1)^2=0$ is the point ______________.

4. The graph of $x^2+y^2=5$ is a circle with center at the point ______________ and a radius of ____.

5. Completing the square in the equation

$$x^2+y^2+16x+12y=0$$

gives $(x^2+16x+$ ______$) + (y^2+12y+$ ______$) =$ ______ $+$ ______. This can be written $(x +$ ____$)^2 + (y +$ ____$)^2 = ($______$)^2$. Thus the graph is a circle with center ________________ and radius ______.

Problem Set 16-2

In Problems **1–6**, *find the standard and general forms of an equation of the circle having the given center C and radius r.*

1. $C = (2, 3)$, $r = 6$

2. $C = (4, -5)$, $r = 2$

3. $C = (-1, 6)$, $r = 4$

4. $C = (-2, -3)$, $r = 1$

5. $C = (0, 0)$, $r = \frac{1}{2}$

6. $C = (3, 0)$, $r = \sqrt{3}$

In Problems **7–12**, *give the center C and radius r of the circle having the given equation. Also, draw the circle.*

7. $x^2 + y^2 = 9$

8. $(x - 1)^2 + (y - 2)^2 = 1$

9. $(x - 3)^2 + (y + 4)^2 = 4$

10. $(x + 6)^2 + (y + 1)^2 = 3^2$

11. $(x + 2)^2 + y^2 = 1$

12. $x^2 + (y - 3)^2 = 16$

In Problems **13–24**, *describe the graphs of the equations.*

13. $x^2 + y^2 - 2x - 4y - 4 = 0$

14. $x^2 + y^2 + 4x - 6y + 9 = 0$

15. $x^2 + y^2 + 6y + 5 = 0$

16. $x^2 + y^2 - 12x + 27 = 0$

17. $x^2 + y^2 + 2x - 2y + 3 = 0$

18. $x^2 + y^2 + 6x - 2y - 15 = 0$

19. $x^2 + y^2 - 14x + 4y + 37 = 0$

20. $x^2 + y^2 + 4y - 8x + 21 = 0$

21. $x^2 + y^2 - 3x + y + \frac{5}{2} = 0$

22. $9x^2 + 9y^2 - 6x + 18y + 9 = 0$

23. $2x^2 + 2y^2 - 4x + 7y + 2 = 0$

24. $16x^2 + 16y^2 + 24x - 48y - 3 = 0$

16-3 PARABOLAS, ELLIPSES, AND HYPERBOLAS

We'll now look at three more conic sections—the *parabola*, the *ellipse*, and the *hyperbola*.

In Fig. 16-5 is the graph of the equation $y = 2x^2$. This curve is called a

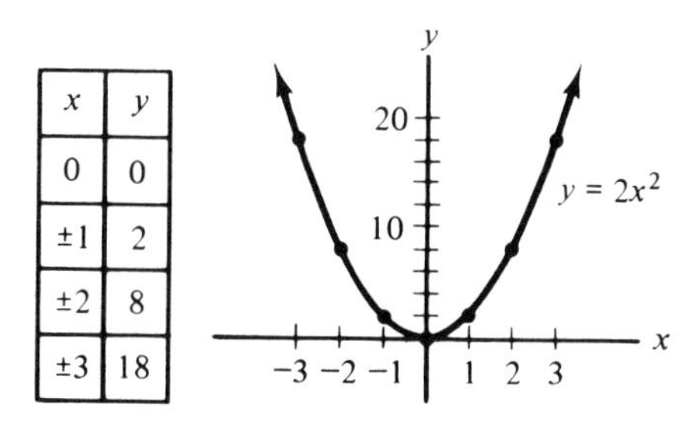

x	y
0	0
±1	2
±2	8
±3	18

Parabola (Opens Upward)

FIG. 16-5

parabola. Notice that it passes through the origin but does not intersect the axes at any other point. Because of the appearance of its graph, we say that the parabola $y = 2x^2$ *opens upward*.

Figure 16-6 shows the graphs of $y = -2x^2$, $x = 2y^2$, and $x = -2y^2$, which

PARABOLAS

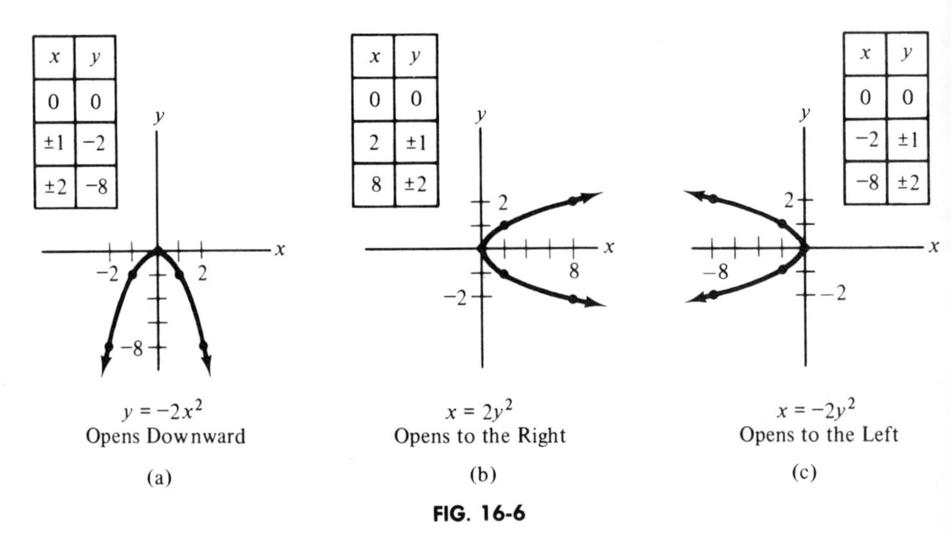

FIG. 16-6

are also parabolas. Again notice that each curve passes through the origin but does not intersect the axes at any other point. For the equation $y = -2x^2$, since $-2x^2$ is either negative or 0, y must always be nonpositive. Thus the parabola $y = -2x^2$ *opens downward* [Fig. 16-6(a)]. The parabola $x = 2y^2$ *opens to the right*, since x (or $2y^2$) is nonnegative [Fig. 16-6(b)]. Finally, the parabola $x = -2y^2$ *opens to the left*, since x (or $-2y^2$) is nonpositive [Fig. 16-6(c)].

In general, the graph of any equation of the form

$$y = ax^2,$$

where a is a constant different from 0, is a *parabola* that intersects the axes only at the origin. If a is positive, the parabola *opens upward* [Fig. 16-7(a)]. If a is

PARABOLAS

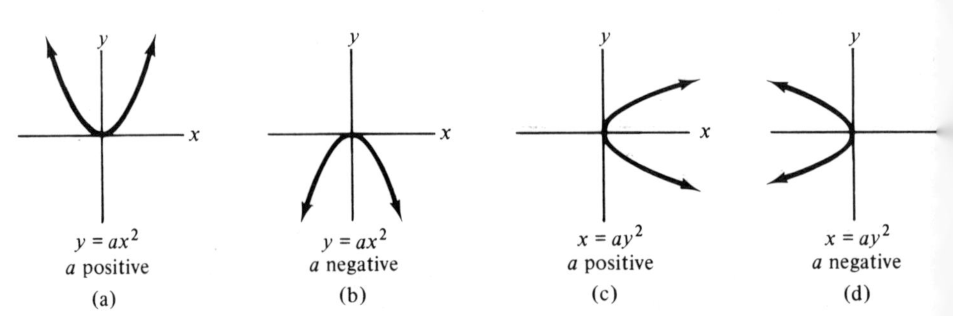

FIG. 16-7

negative, the parabola *opens downward* [Fig. 16-7(b)]. Also, the graph of any equation of the form

$$x = ay^2,$$

where a is a constant different from 0, is a *parabola* that intersects the axes only at the origin. If a is positive, the parabola *opens to the right* [Fig. 16-7(c)]. If a is negative, it *opens to the left* [Fig. 16-7(d)].

Now you should be able to identify and quickly sketch the graph of an equation that has the form $y = ax^2$ or $x = ay^2$.

EXAMPLE 1

Identify and sketch the graph of each equation.

a. $y = -\frac{1}{2}x^2$.

This equation has the form $y = ax^2$, where a is negative $\left(a = -\frac{1}{2}\right)$. Thus its graph is a *parabola opening downward.* By plotting a few points we can sketch the graph as in Fig. 16-8(a).

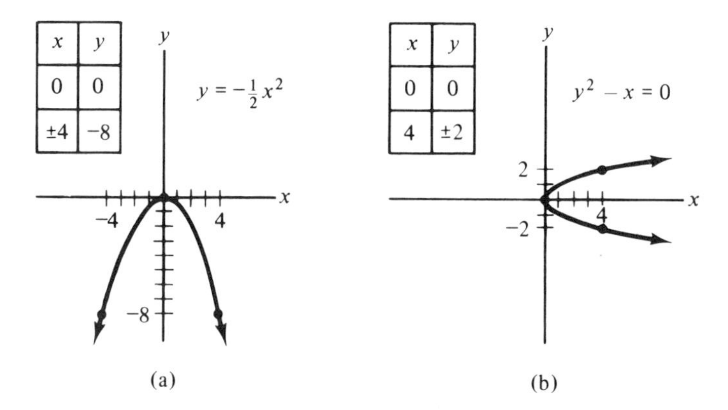

FIG. 16-8

b. $y^2 - x = 0$.

We first rewrite the equation to see if we can get a familiar form. We can write it as $y^2 = x$, which means that

$$x = y^2.$$

This equation has the form $x = ay^2$, where $a = 1$ (a positive number). Thus its graph is a *parabola opening to the right*. By plotting a few points, we get the graph in Fig. 16-8(b).

The graph of $\dfrac{x^2}{3^2} + \dfrac{y^2}{2^2} = 1$ is shown in Fig. 16-9 and is called an **ellipse**.

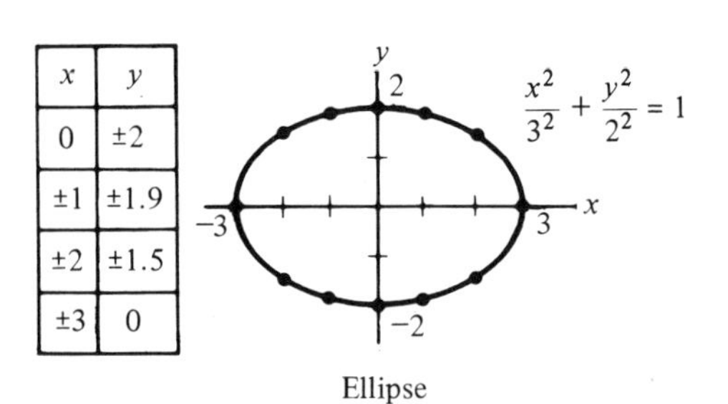

x	y
0	±2
±1	±1.9
±2	±1.5
±3	0

Ellipse

FIG. 16-9

The ellipse cuts the x-axis at the points $(-3, 0)$ and $(3, 0)$. It cuts the y-axis at $(0, -2)$ and $(0, 2)$. We say that the ***x*-intercepts** of the ellipse are -3 and 3, and its ***y*-intercepts** are -2 and 2.

More generally, the graph of any equation of the form

$$\frac{x^2}{a^2} + \frac{y^2}{b^2} = 1,$$

where a and b are positive constants and $a \neq b$, is an *ellipse* with x-intercepts $-a$ and a, and y-intercepts $-b$ and b. In Fig. 16-10(a) we have $a > b$, and the

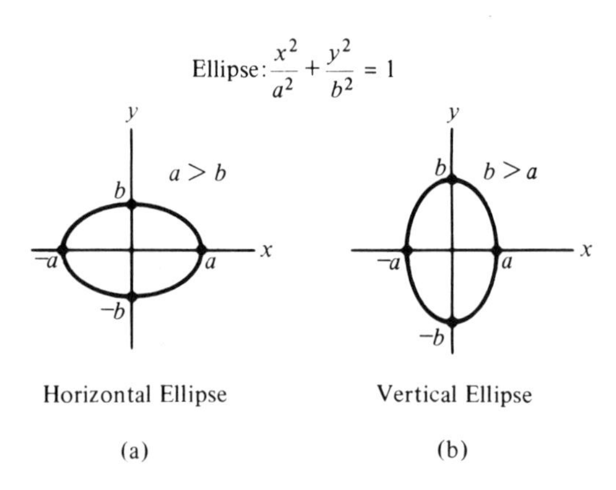

Horizontal Ellipse

(a)

Vertical Ellipse

(b)

FIG. 16-10

ellipse is said to be *horizontal*. In Fig. 16-10(b) we have $b > a$, and the graph is a *vertical* ellipse. By plotting only the intercepts of an ellipse, you should be able to quickly sketch its graph.

EXAMPLE 2

Identify and sketch the graph of each equation.

a. $\frac{x^2}{25} + \frac{y^2}{9} = 1.$

This equation has the form $\frac{x^2}{a^2} + \frac{y^2}{b^2} = 1$, where $a = 5$, $b = 3$, and $a > b$. Thus the graph is a *horizontal ellipse* having x-intercepts -5 and 5 and y-intercepts -3 and 3. See Fig. 16-11(a).

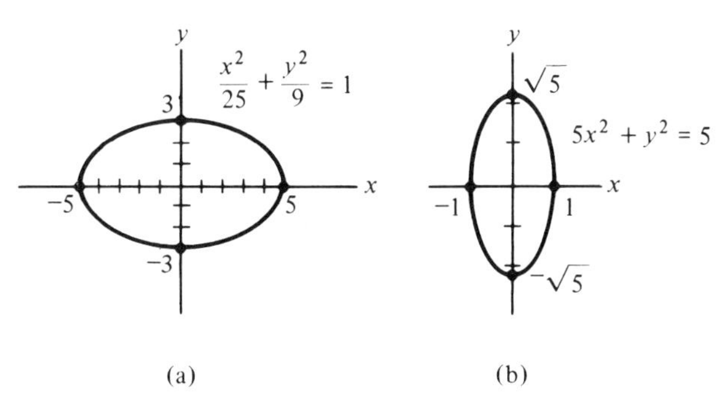

FIG. 16-11

b. $5x^2 + y^2 = 5.$

We can write this equation in a familiar form if we first divide both sides by 5.

$$5x^2 + y^2 = 5$$

$$\frac{5x^2}{5} + \frac{y^2}{5} = \frac{5}{5}$$

$$x^2 + \frac{y^2}{5} = 1,$$

which can be written as

$$\frac{x^2}{1^2} + \frac{y^2}{(\sqrt{5})^2} = 1.$$

This equation has the form $\frac{x^2}{a^2} + \frac{y^2}{b^2} = 1$, where $a = 1$, $b = \sqrt{5}$, and $b > a$. Thus the graph is a *vertical ellipse* whose x-intercepts are -1 and 1, and y-intercepts are $-\sqrt{5}$ and $\sqrt{5}$. See Fig. 16-11(b) above.

The graph of $\dfrac{x^2}{4^2} - \dfrac{y^2}{2^2} = 1$ is shown in Fig. 16-12 and is called a

x	y
±4	0
±5	±1.5
±6	±2.2
±7	±2.9

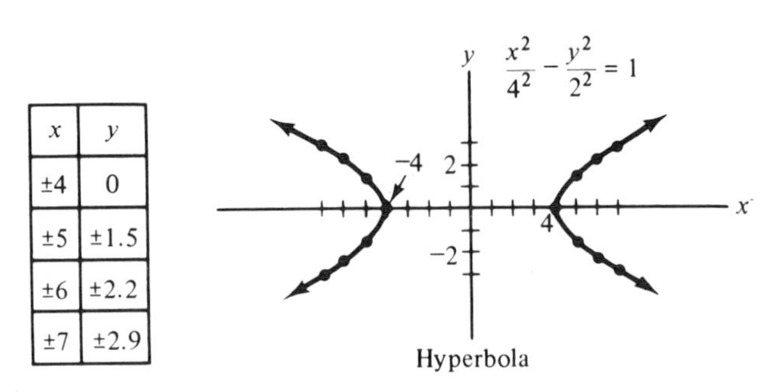

Hyperbola

FIG. 16-12

hyperbola. The x-intercepts are -4 and 4, and there are no y-intercepts.

More generally, the graph of any equation of the form

$$\frac{x^2}{a^2} - \frac{y^2}{b^2} = 1,$$

where a and b are positive constants, is a (*horizontal*) *hyperbola*, as shown in Fig. 16-13(a). The x-intercepts are $-a$ and a, and there are no y-intercepts. In Fig.

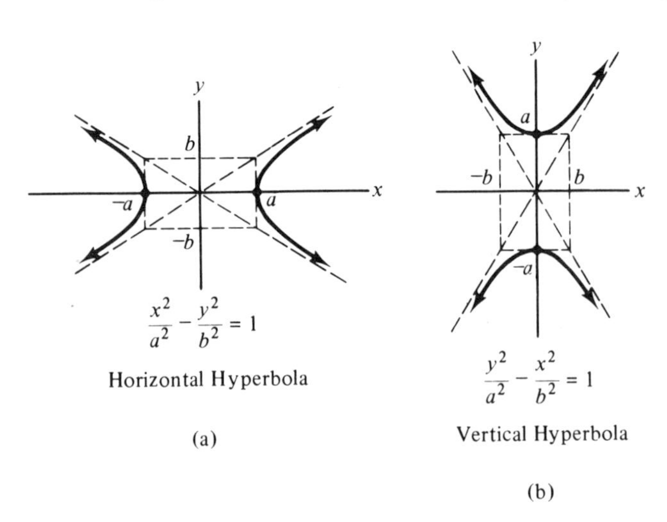

Horizontal Hyperbola

(a)

Vertical Hyperbola

(b)

FIG. 16-13

16-13(a), a broken rectangle of dimensions $2a$ by $2b$ and centered at the origin has been drawn. The diagonals of this rectangle have been extended (broken lines). Notice that as the points on the hyperbola get farther from the origin, they get very close to the extended diagonal lines. Such lines are called **asymptotes** and can be used as a guide for sketching the hyperbola. The rectangle and asymptotes are *not* part of the graph.

The graph of an equation having the form

$$\frac{y^2}{a^2} - \frac{x^2}{b^2} = 1,$$

where a and b are positive constants, is shown in Fig. 16-13(b) above and is also a *hyperbola*. However, this hyperbola has y-intercepts of $-a$ and a, but no x-intercepts. It is called a *vertical* hyperbola. As before, the extended diagonals of the rectangle determined by a and b are asymptotes.

EXAMPLE 3

Identify and sketch the graph of each equation.

a. $\dfrac{x^2}{4} - \dfrac{y^2}{4} = 1.$

This equation has the form $\dfrac{x^2}{a^2} - \dfrac{y^2}{b^2} = 1$, where $a = 2$ and $b = 2$. Thus the graph is a *horizontal hyperbola* with x-intercepts -2 and 2. By first plotting the intercepts and drawing the asymptotes, we then sketch the hyperbola as in Fig. 16-14(a).

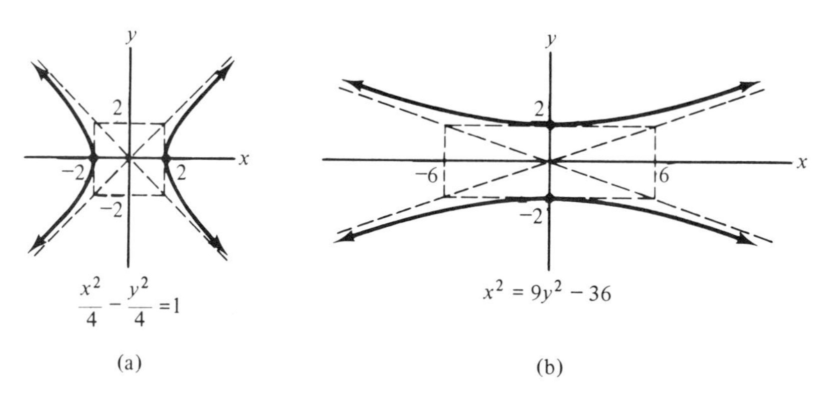

FIG. 16-14

b. $x^2 = 9y^2 - 36.$

We'll write the equation in a familiar form. Since the squares of both x and y are involved, let's first get those terms on one side and the constant term on the other side.

$$x^2 = 9y^2 - 36$$

$$x^2 - 9y^2 = -36. \qquad \text{[subtracting } 9y^2 \text{ from both sides]}$$

Now we divide both sides by -36 in order to get the right side equal to 1.

$$\frac{x^2}{-36} - \frac{9y^2}{-36} = \frac{-36}{-36}$$

$$-\frac{x^2}{36} + \frac{y^2}{4} = 1$$

$$\frac{y^2}{4} - \frac{x^2}{36} = 1$$

$$\frac{y^2}{2^2} - \frac{x^2}{6^2} = 1.$$

The last equation has the form $\frac{y^2}{a^2} - \frac{x^2}{b^2} = 1$, where $a = 2$ and $b = 6$. Thus its graph is a *vertical hyperbola* with y-intercepts -2 and 2. After plotting the intercepts and drawing the asymptotes, we sketch the hyperbola as in Fig. 16-14(b) above.

EXAMPLE 4

Identify and sketch the graph of each equation.

a. $\frac{x^2}{16} + \frac{y^2}{16} = 1.$

You may be tempted to say that this equation fits a form of an ellipse. But in each form of ellipse that was given before, the denominators are different. If we multiply both sides of the given equation by 16, we get

$$16\left[\frac{x^2}{16} + \frac{y^2}{16}\right] = 16 \cdot 1$$

$$x^2 + y^2 = 16.$$

This is an equation of a *circle* with center at the origin and radius 4. See Fig. 16-15(a).

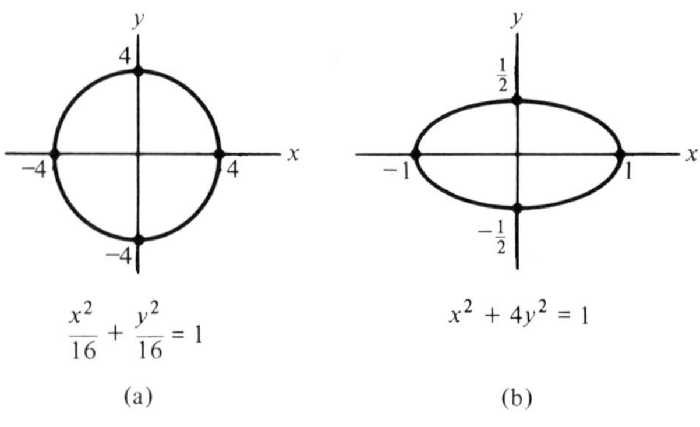

FIG. 16-15

b. $x^2 + 4y^2 = 1$.

We can write $4y^2$ as $\dfrac{y^2}{\frac{1}{4}}$. Thus,

$$x^2 + 4y^2 = 1$$

$$x^2 + \frac{y^2}{\frac{1}{4}} = 1$$

$$\frac{x^2}{1^2} + \frac{y^2}{\left(\frac{1}{2}\right)^2} = 1.$$

This equation has the form $\dfrac{x^2}{a^2} + \dfrac{y^2}{b^2} = 1$, where $a = 1$, $b = \frac{1}{2}$, and $a > b$. Thus its graph is a *horizontal ellipse* with x-intercepts -1 and 1, and y-intercepts $-\frac{1}{2}$ and $\frac{1}{2}$. See Fig. 16-15(b).

c. $3x + 2y^2 = 0$.

An equation with just an x-term and a y^2-term may be a parabola. Solving for x gives

$$3x = -2y^2$$

$$x = -\frac{2}{3}y^2.$$

This equation has the form $x = ay^2$, where $a = -\frac{2}{3}$ (a negative number). Thus its graph is a *parabola opening to the left*. See Fig. 16-16(a).

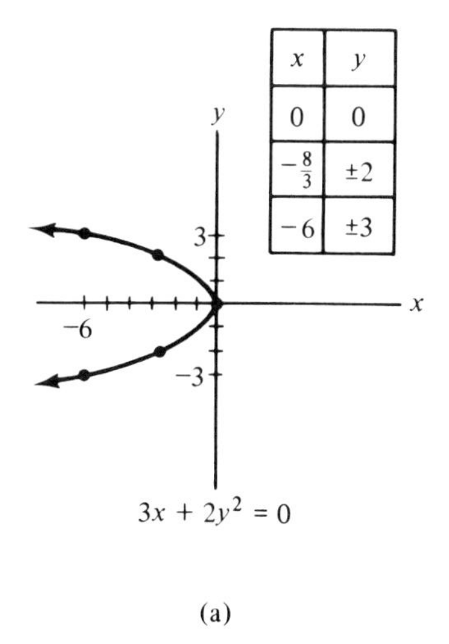

(a)

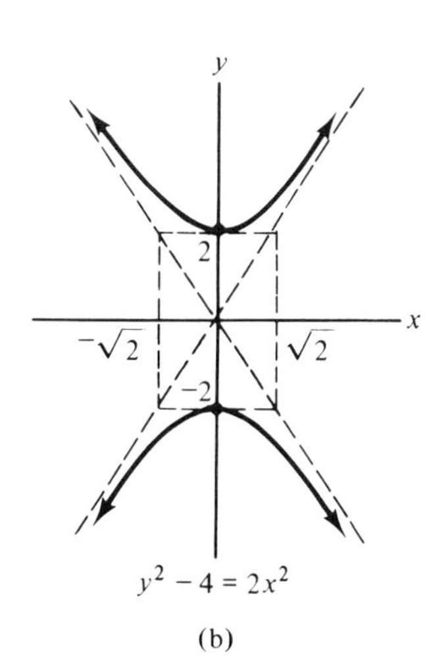

(b)

FIG. 16-16

d. $y^2 - 4 = 2x^2$.

Writing the equation so that the x^2- and y^2-terms are on the left side and the constant is on the right side, we have

$$y^2 - 4 = 2x^2$$
$$y^2 - 2x^2 = 4.$$

Dividing both sides by 4 gives

$$\frac{y^2}{4} - \frac{2x^2}{4} = \frac{4}{4}$$
$$\frac{y^2}{4} - \frac{x^2}{2} = 1$$
$$\frac{y^2}{2^2} - \frac{x^2}{(\sqrt{2}\,)^2} = 1.$$

This equation has the form $\frac{y^2}{a^2} - \frac{x^2}{b^2} = 1$, where $a = 2$ and $b = \sqrt{2}$. Thus its graph is a *vertical hyperbola* with y-intercepts 2 and -2. See Fig. 16-16(b) above.

Here is a summary of forms of equations of parabolas, ellipses, hyperbolas, and circles that have been considered in this section.

Parabola

$y = ax^2$	$x = ay^2$
$a > 0$, opens upward	$a > 0$, opens to the right
$a < 0$, opens downward	$a < 0$, opens to the left

Ellipse

$$\frac{x^2}{a^2} + \frac{y^2}{b^2} = 1 \qquad \begin{cases} \text{horizontal if } a > b \\ \text{vertical if } b > a \end{cases}$$

Hyperbola

Horizontal: $\frac{x^2}{a^2} - \frac{y^2}{b^2} = 1$ Vertical: $\frac{y^2}{a^2} - \frac{x^2}{b^2} = 1$

Circle (center at origin)

$$x^2 + y^2 = r^2 \quad \text{or} \quad \frac{x^2}{r^2} + \frac{y^2}{r^2} = 1.$$

Completion Set 16-3

Fill in the blanks.

1. The graph of $x = y^2$ is called a(n) ____________________.

2. A conic section having asymptotes is the ____________________.

3. The graph of $\frac{x^2}{4} - \frac{y^2}{9} = 1$ is called a(n) ____________________, and the graph of $\frac{x^2}{4} + \frac{y^2}{9} = 1$ is called a(n) ____________________.

4. Insert *downward* or *upward*. The parabola $y = -4x^2$ opens ____________________.

5. Insert *horizontal* or *vertical*. The graph of $\frac{x^2}{16} + \frac{y^2}{25} = 1$ is a ____________________ ellipse.

6. Insert *horizontal* or *vertical*. The graph of $\frac{x^2}{16} - \frac{y^2}{25} = 1$ is a ____________________ hyperbola.

Problem Set 16-3

In Problems **1–24**, *identify and sketch the graph of each equation.*

1. $y = 3x^2$

2. $x = -y^2$

3. $\frac{x^2}{9} + \frac{y^2}{25} = 1$

4. $\frac{y^2}{16} - \frac{x^2}{9} = 1$

5. $\frac{x^2}{9} - \frac{y^2}{25} = 1$

6. $\frac{x^2}{16} + \frac{y^2}{9} = 1$

7. $x^2 + 4y^2 = 16$

8. $2y + 3x^2 = 0$

9. $4x^2 + 4y^2 = 1$

10. $\frac{x^2}{36} - y = 0$

11. $y^2 + 6x = 0$

12. $\frac{x^2}{36} + \frac{y^2}{36} = 1$

13. $y^2 - x^2 = 1$

14. $\frac{x^2}{4} - 1 = \frac{y^2}{4}$

15. $2x = 3y^2$

16. $25x^2 - y^2 = 25$

17. $9x^2 = 1 - y^2$

18. $\frac{x^2}{3} + \frac{y^2}{4} = 3$

19. $25x^2 - 2y^2 = -100$

20. $y^2 - 16x = 0$

21. $\frac{x^2}{4} = -\frac{y}{6}$

22. $9y^2 - 16x^2 - 144 = 0$

23. $x^2 = 5 - 20y^2$

24. $\frac{x^2}{\sqrt{16}} + \frac{y^2}{\sqrt{36}} = 1$

16-4 REVIEW

IMPORTANT TERMS

distance formula *(p. 370)*
circle *(p. 371)*
center *(p. 371)*
radius *(p. 371)*
standard form *(p. 372)*
general form *(p. 374)*
parabola *(p. 377)*
ellipse *(p. 380)*
hyperbola *(p. 382)*
asymptote *(p. 383)*
conic sections *(p. 371)*

REVIEW PROBLEMS

In Problems **1–4**, *find the distance between the given points.*

1. $(1, 4), (-3, 2)$

2. $(-1, -1), (-6, -1)$

3. $(-8, 2), (0, -4)$

4. $(-3, -4), (-1, 1)$

In Problems **5–8**, *find the standard and general forms of an equation of a circle having the given center C and radius r.*

5. $C = (0, 0), r = 5$

6. $C = (0, -2), r = \sqrt{2}$

7. $C = (1, -1), r = \frac{1}{2}$

8. $C = \left(\frac{1}{2}, \frac{1}{4}\right), r = 1$

In Problems **9–16**, *describe the graph of the given equation. For those that are circles, give the center C and radius r.*

9. $x^2 + y^2 = 16$

10. $(x - 4)^2 + (y - 3)^2 = 1$

11. $(x + 2)^2 + (y - 1)^2 = 7$

12. $x^2 + (y - 1)^2 - 12 = 0$

13. $x^2 + y^2 + 5x - 6y + 2 = 0$

14. $x^2 + y^2 - x - 1 = 0$

15. $9x^2 + 9y^2 - 18x - 6y + 10 = 0$

16. $3x^2 + 3y^2 - 6x + 12y + 16 = 0$

In Problems **17–24**, *identify and sketch the graph of each equation.*

17. $9x^2 - 100y^2 = 900$

18. $3x - 4y^2 = 0$

19. $36x^2 + y^2 = 36$

20. $36y^2 - x^2 = 4$

21. $5x^2 + 2y = 0$

22. $x^2 + \frac{1}{2}y^2 = 1$

23. $6x^2 - 1 = -6y^2$

24. $\frac{1}{49}x^2 - \frac{1}{49}y^2 = 1$

CHAPTER 17

TRIGONOMETRY

Trigonometry deals with the study of six special functions, called *trigonometric functions*. These are used in areas such as surveying, measuring distances, solving triangles, mechanics, electricity, and the study of vibrating objects like springs, strings, and membranes.

The inputs for the trigonometric functions are angles. We'll look at angles first and then define the functions in Sec. 17-2.

17-1 ANGLES

The angle in Fig. 17-1(a) can be thought of as generated by rotating the half-line OA about its endpoint O to the position of OB. We call OA the **initial side** of the angle, OB the **terminal side**, and O the **vertex**.

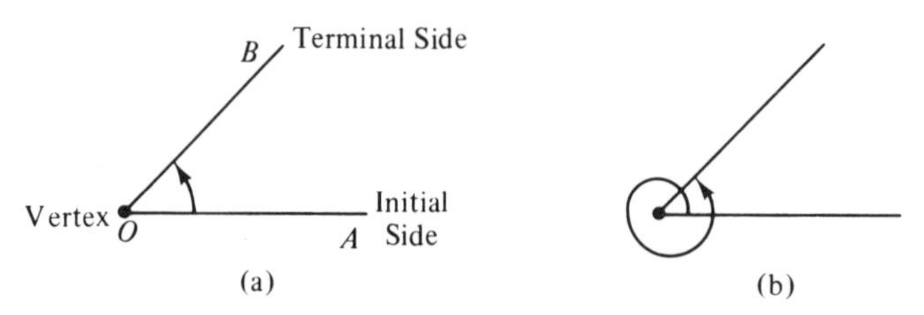

FIG. 17-1

The angles in Fig. 17-1(a, b) may look the same, but they are not. In (b) the rotation is one revolution more than in (a).

We often represent angles by Greek letters such as θ (theta), α (alpha), β (beta). If an angle, say θ, is generated by a counterclockwise rotation, then θ is called a **positive angle** [see Fig. 17-2(a)]. When the rotation is clockwise, θ is a **negative angle** [see Fig. 17-2(b)]. We use arrows to indicate the direction of rotation.

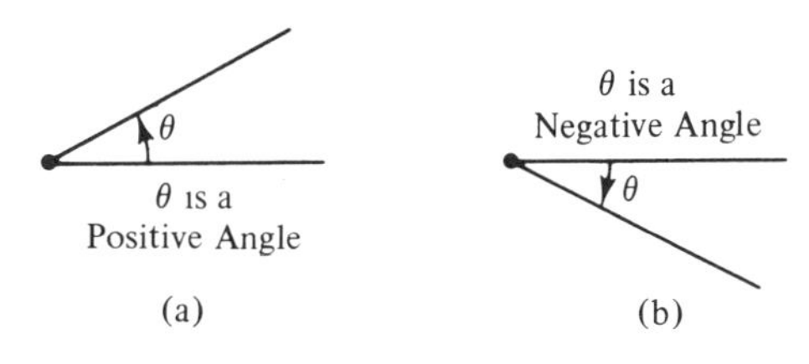

FIG. 17-2

An angle θ is said to be in **standard position** when its vertex is at the origin of a rectangular coordinate plane and its initial side lies on the positive horizontal axis. See Fig. 17-3. If the terminal side of θ lies in Quadrant I, then θ is called a **first-quadrant angle**. Similarly, there are **second-, third-,** and **fourth-quadrant angles**. An angle whose terminal side lies on an axis is called a **quadrantal angle**.

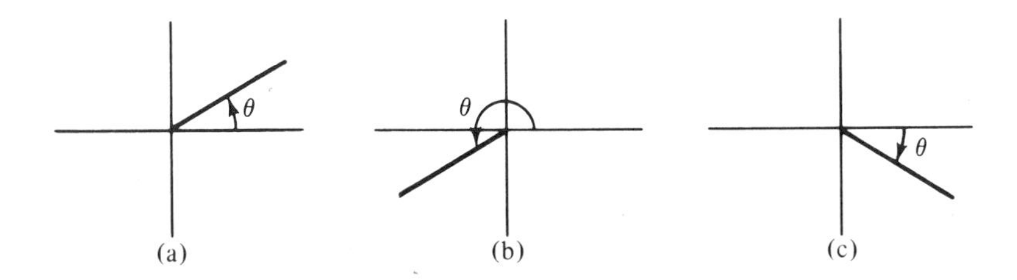

Angles in Standard Position

FIG. 17-3

EXAMPLE 1

a. Figure 17-3 shows a first-quadrant angle in (a), a third-quadrant angle in (b), and a fourth-quadrant angle in (c).

b. The angles in Fig. 17-4 are quadrantal angles. For the first angle, the initial and terminal sides are the same.

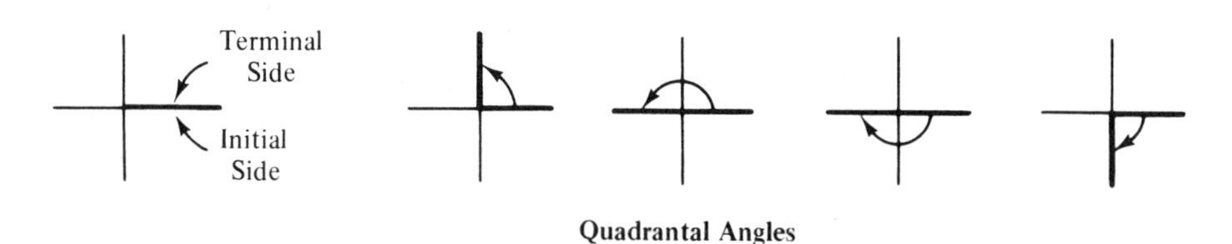

FIG. 17-4

Two common units for measuring the amount of rotation of an angle are *degrees* and *radians*. We'll first look at degrees.

An angle generated by one counterclockwise revolution has a measure of 360 degrees, written 360°. Also, 1° is the measure of an angle generated by $\frac{1}{360}$ of a revolution.

Figure 17-5 gives the degree measures of some angles. Notice that the measure of a negative angle is preceded by a minus sign.

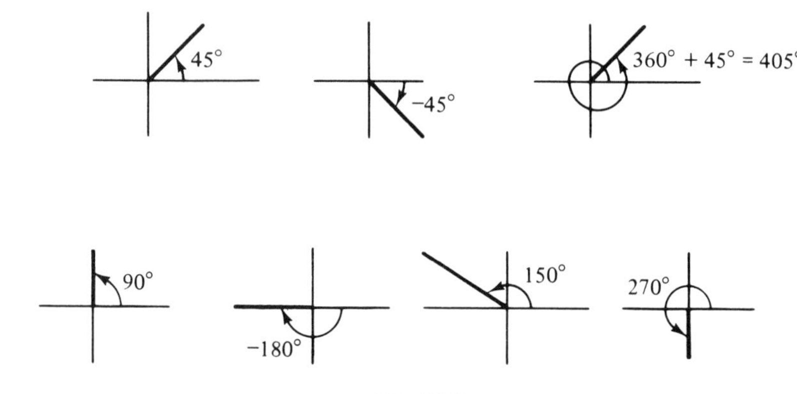

FIG. 17-5

In the top-left diagram in Fig. 17-5 is an angle whose measure is 45°. Usually we avoid the phrase "whose measure is" and simply say the angle is 45°. If θ represents this angle, then we write $\theta = 45°$.

An angle α greater than 0° but less than 90° (that is, $0° < \alpha < 90°$) is called an **acute angle**. One between 90° and 180° (that is, $90° < \alpha < 180°$) is an **obtuse angle**. A 90° angle is called a **right angle**. See Fig. 17-6.

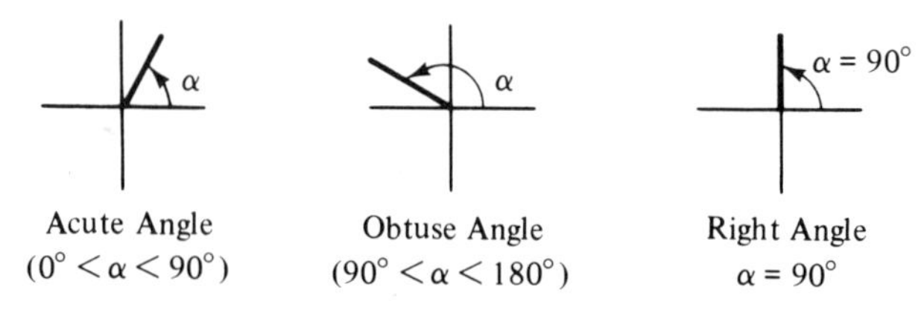

FIG. 17-6

Just as we divide a meter into centimeters and millimeters, we divide a degree into **minutes** and **seconds**.

$$1^\circ = 60 \text{ minutes} \quad (\text{written } 60')$$
$$1' = 60 \text{ seconds} \quad (\text{written } 60'').$$

EXAMPLE 2

a. *In Fig. 17-7(a), find the acute angle β if $\alpha = 321^\circ\ 15'\ 30''$.*

We see that $\alpha + \beta = 360^\circ$. Thus, $\beta = 360^\circ - \alpha$. Since α involves minutes and seconds, we rewrite 360° using minutes and seconds.

$$\begin{aligned} 360^\circ &= 359^\circ + 1^\circ = 359^\circ\ 60' \\ &= 359^\circ\ 59' + 1' = 359^\circ\ 59'\ 60''. \end{aligned}$$

Now, $\beta = 360^\circ - \alpha$:

$$\begin{array}{rrrr} & 359^\circ & 59' & 60'' \\ - & 321^\circ & 15' & 30'' \\ \hline \beta = & 38^\circ & 44' & 30''. \end{array}$$

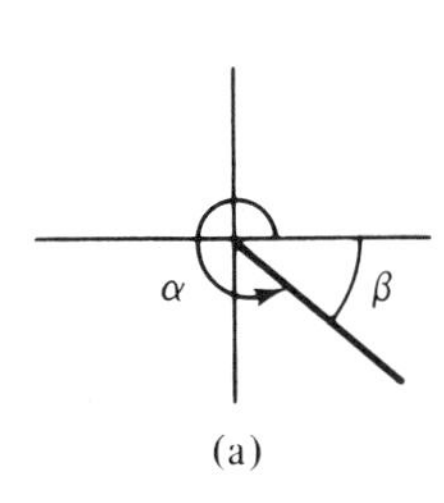

(a)

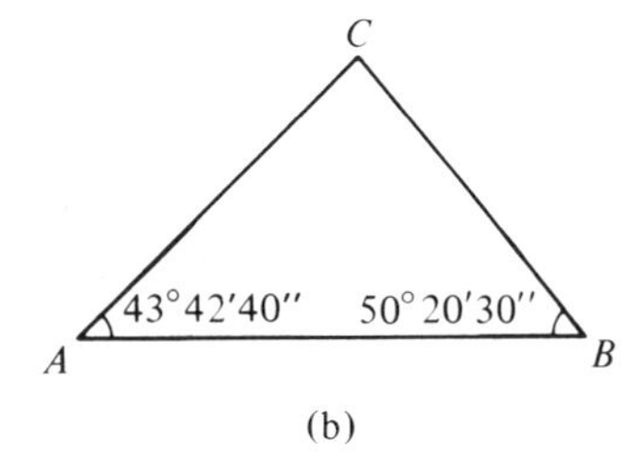

(b)

FIG. 17-7

b. *In Fig. 17-7(b) is triangle ABC. The angle at vertex A is called angle A. Similarly, there are angles B and C.*[†] *For any triangle, the sum of its angles is 180°. Given $A = 43^\circ\ 42'\ 40''$ and $B = 50^\circ\ 20'\ 30''$, find C.*

Since $A + B + C = 180^\circ$, then $C = 180^\circ - (A + B)$. We'll find $A + B$ and then subtract it from 180°.

$$A + B: \qquad \begin{array}{rrrr} & 43^\circ & 42' & 40'' \\ + & 50^\circ & 20' & 30'' \\ \hline A + B = & 93^\circ & 62' & 70''. \end{array}$$

[†]An angle here can also be named by three letters, where the letter associated with the vertex is the middle letter. Thus, angle A can also be called angle CAB or angle BAC.

Now we simplify $A + B$ as follows. Since $70'' = 60'' + 10'' = 1'\ 10''$, then

$$A + B = 93^\circ\ (62 + 1)'\ 10'' = 93^\circ\ 63'\ 10''.$$

But $63' = 60' + 3' = 1^\circ\ 3'$. Thus,

$$A + B = (93 + 1)^\circ\ 3'\ 10'' = 94^\circ\ 3'\ 10''.$$

Since $A + B$ involves minutes and seconds, to subtract $A + B$ from 180° we'll first rewrite 180°, using minutes and seconds.

$$180^\circ = 179^\circ\ 60' = 179^\circ\ 59'\ 60''.$$

Now, $C = 180^\circ - (A + B)$:

$$\begin{array}{r} 179^\circ\ \ 59'\ \ 60'' \\ -\ \ 94^\circ\ \ \ 3'\ \ 10'' \\ \hline C = 85^\circ\ \ 56'\ \ 50''. \end{array}$$

Another unit for measuring an angle is the **radian**. Here's how to find an angle of 1 radian. First, take a circle of radius r and mark off an arc of length r, as in Fig. 17-8(a). The angle determined by the center and the endpoints of the

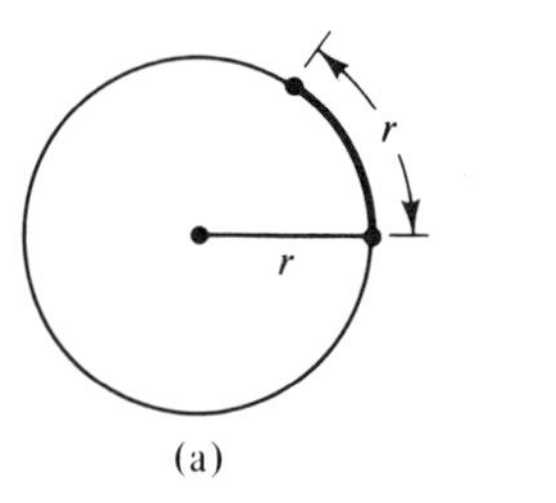

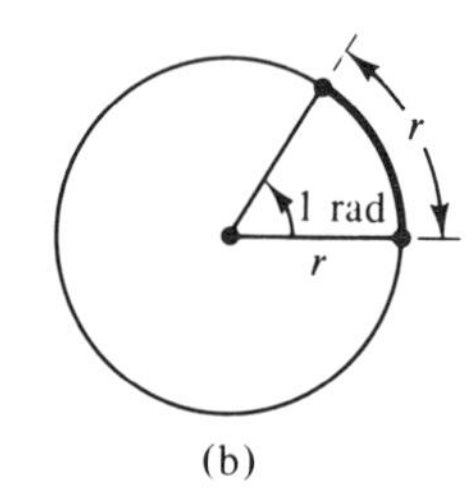

FIG. 17-8

arc is 1 radian (rad) [see Fig. 17-8(b)]. If we mark off an arc of length $2r$, then the angle generated is 2 rad (see Fig. 17-9).

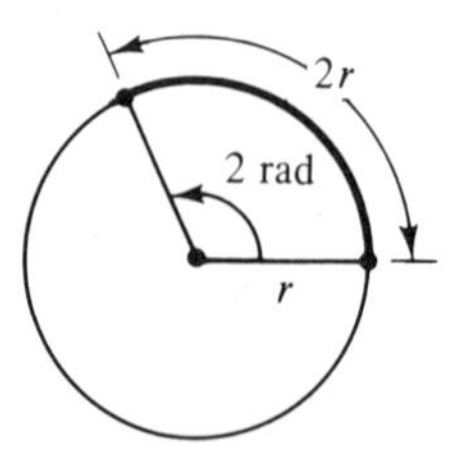

FIG. 17-9

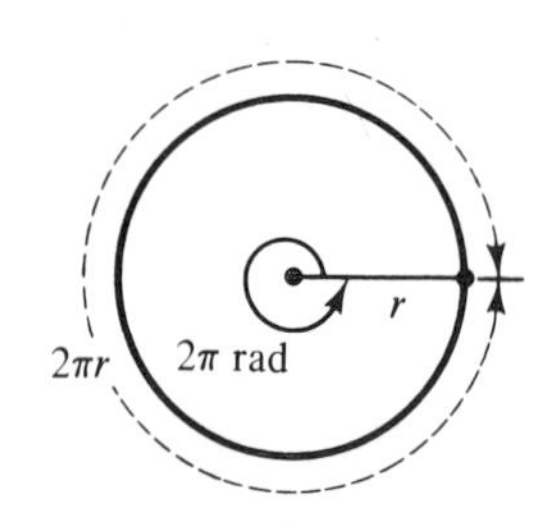

FIG. 17-10

Since the circumference of a circle (the distance around a circle) of radius r is $2\pi r$, it follows that an angle of one complete revolution is 2π rad (see Fig. 17-10). Thus,

$$2\pi \text{ rad} = 360°.$$

Dividing both sides by 2, we have

$$\boxed{\pi \text{ rad} = 180°.}$$

This formula lets you convert from radians to degrees and vice versa.

EXAMPLE 3

a. *Convert 2 rad to degrees.*

Since π rad = 180°, then dividing both sides by π rad gives $1 = \dfrac{180°}{\pi \text{ rad}}$. Thus,

$$2 \text{ rad} = 2 \text{ rad}\cdot(1) = 2 \text{ rad}\cdot \frac{180°}{\pi \text{ rad}}.$$

On the right side the radian units will cancel and leave us with degrees.

$$2 \text{ rad} = 2 \cancel{\text{rad}}\cdot \frac{180°}{\pi \cancel{\text{rad}}} = \left(\frac{360}{\pi}\right)^{\circ} \approx 114.59°.$$

b. *Convert* $\frac{\pi}{4}$ *rad to degrees.*

$$\frac{\pi}{4} \text{ rad} = \frac{\pi}{4} \text{ rad}\cdot \frac{180°}{\pi \text{ rad}} = \frac{\cancel{\pi}}{\cancel{4}} \cancel{\text{rad}}\cdot \frac{\overset{45°}{\cancel{180°}}}{\cancel{\pi} \cancel{\text{rad}}} = 45°.$$

c. *Convert* $-\frac{2\pi}{3}$ *rad to degrees.*

$$-\frac{2\pi}{3} \text{ rad} = -\frac{2\cancel{\pi}}{\cancel{3}} \cancel{\text{rad}}\cdot \frac{\overset{60°}{\cancel{180°}}}{\cancel{\pi} \cancel{\text{rad}}} = -120°.$$

EXAMPLE 4

a. *Convert 90° to radians.*

Since π rad = 180°, then $\dfrac{\pi \text{ rad}}{180°} = 1$. Thus,

$$90° = 90°\cdot(1) = 90°\cdot \frac{\pi \text{ rad}}{180°}.$$

On the right side the degree units will cancel and leave us with radians.

$$90° = 90° \cdot \frac{\pi \text{ rad}}{180°} = \frac{\pi}{2} \text{ rad.}$$

b. *Convert 30° to radians.*

$$30° = 30° \cdot \frac{\pi \text{ rad}}{180°} = \frac{\pi}{6} \text{ rad.}$$

Usually we drop the symbol "rad" on radian measure. Thus an angle of 2 means 2 rad and an angle of π means π rad.

Figure 17-11 shows a circle of radius r. An arc of length s is cut off by the

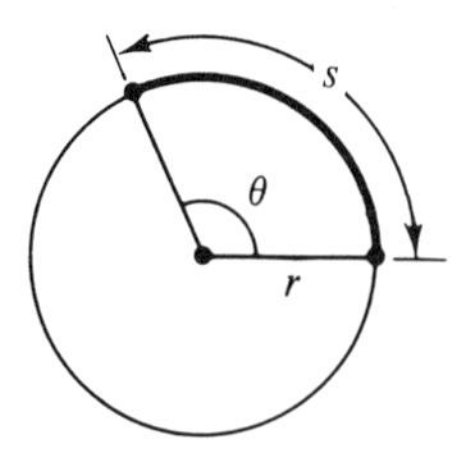

FIG. 17-11

central angle θ.[†] If θ is in radians, then the number of r's that make up s is θ. Thus, $s = \theta r$.

> The arc length s cut off on a circle of radius r by a central angle θ is
>
> $$s = r\theta,$$
>
> where θ is in **radians.**

EXAMPLE 5

Find the arc length cut off on a circle of radius 5 by a central angle of 60°. See Fig. 17-12.

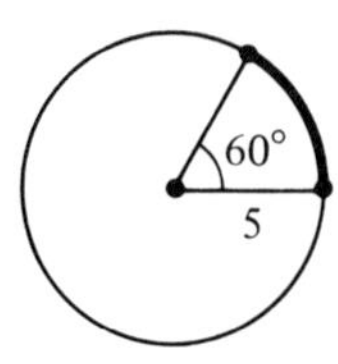

FIG. 17-12

[†]A *central angle* of a circle is a nonnegative angle formed by two radii.

To use the formula for arc length, we must first convert 60° to radians.

$$60^\circ = 60^\circ \cdot \frac{\pi \text{ rad}}{180^\circ} = \frac{\pi}{3} \text{ (rad)}.$$

The arc length is

$$s = r\theta = 5 \cdot \frac{\pi}{3} = \frac{5\pi}{3}.$$

Figure 17-13 shows a circle of radius r. The shaded region cut by the central

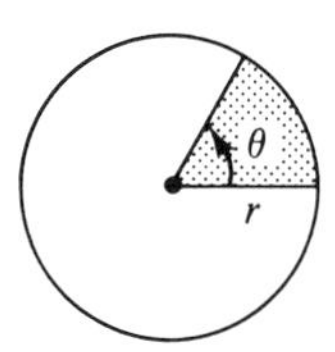

FIG. 17-13

angle θ is called a **sector** of the circle. The area of this sector is given by a formula:

> For a circle of radius r, the area A of a sector with central angle θ is
>
> $$A = \tfrac{1}{2} r^2 \theta,$$
>
> where θ is in **radians.**

EXAMPLE 6

For a circle of radius 12, *find the area of the sector with central angle* 210°.

We first convert 210° to radians.

$$210^\circ = 210^\circ \cdot \frac{\pi \text{ rad}}{180^\circ} = \frac{7\pi}{6} \text{ (rad)}.$$

By the formula for the area of a sector,

$$A = \tfrac{1}{2} r^2 \theta = \tfrac{1}{2} \cdot 12^2 \cdot \frac{7\pi}{6} = 12 \cdot 7\pi = 84\pi \text{ (sq units)}.$$

When you use the formulas for arc length and area of a sector, make sure θ is in ***radians***.

Figure 17-14(a) shows angles of 30° and −330°. Since these angles (when in standard position) have the same terminal side, we call them **coterminal angles**. In Fig. 17-14(b) you can see that 30° and 390° are also coterminal.

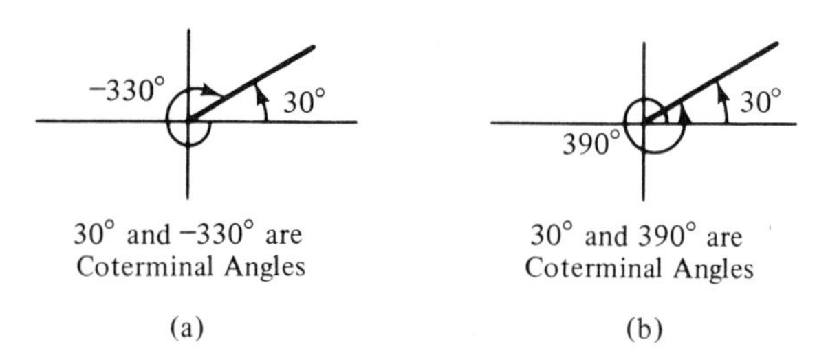

30° and −330° are Coterminal Angles

(a)

30° and 390° are Coterminal Angles

(b)

FIG. 17-14

Clearly, if you add or subtract a multiple of 360° (or 2π) to or from an angle θ, then the resulting angle is coterminal with θ.

EXAMPLE 7

a. *Find two positive and two negative angles that are coterminal with* 135°.

Two such positive angles are

$$135° + 360° = 495°$$

and $$135° + 2(360°) = 855°.$$

Two such negative angles are

$$135° - 360° = -225°$$

and $$135° - 2(360°) = -585°.$$

See Fig. 17-15.

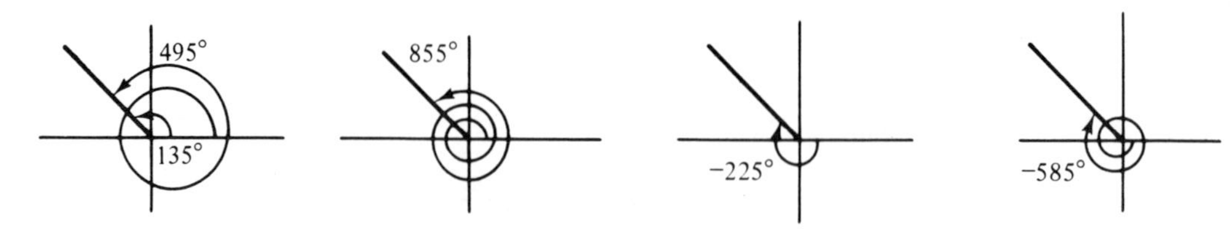

FIG. 17-15

b. *Find the angle θ that is coterminal with $\frac{16\pi}{3}$ and such that $0 \leqslant \theta < 2\pi$.*

Since $\frac{16\pi}{3} \geqslant 2\pi$, we'll keep subtracting 2π from $\frac{16\pi}{3}$ until we reach an angle between 0 and 2π.

$$\frac{16\pi}{3} - 2\pi = \frac{16\pi}{3} - \frac{6\pi}{3} = \frac{10\pi}{3}.$$

Since $\frac{10\pi}{3} \geqslant 2\pi$, we continue.

$$\frac{10\pi}{3} - 2\pi = \frac{10\pi}{3} - \frac{6\pi}{3} = \frac{4\pi}{3}.$$

Now, $0 \leqslant \frac{4\pi}{3} < 2\pi$ and $\frac{4\pi}{3}$ is coterminal with $\frac{16\pi}{3}$. Thus, $\theta = \frac{4\pi}{3}$.

Completion Set 17-1

In Problems **1–10**, *fill in the blanks.*

1. An angle in standard position has its initial side along the positive horizontal axis and its vertex at the point ______________.

2. An angle in standard position that has its terminal side in Quadrant III is called a ______________-quadrant angle.

3. A right angle has a measure of ______ degrees.

4. Insert *acute* or *obtuse*. A 75° angle is an ______________ angle.

5. Insert *positive* or *negative*. An angle generated by a clockwise rotation is a ____________________ angle.

6. Two angles in standard position that have the same terminal side are called ________________________ angles.

7. $60'' =$ ____$'$ and $60' =$ ____°.

8. $180° =$ ____ rad.

9. Insert *degrees* or *radians*. In using the formulas for arc length ($s = r\theta$) and area of a sector $\left(A = \frac{1}{2}r^2\theta\right)$, the angle θ must be in ________________.

10. In sixty minutes the *minute* hand of a clock goes through an angle of ______ rad.

In Problems **11** *and* **12**, *insert* $\dfrac{\pi \text{ rad}}{180°}$ *or* $\dfrac{180°}{\pi \text{ rad}}$.

11. To convert 10° to radians, multiply 10° by ____________.

12. To convert 3 rad to degrees, multiply 3 rad by ____________.

Problem Set 17-1

In Problems **1–20**, *convert each degree measurement to radians, and each radian measurement to degrees.*

1. $0°$ **2.** $180°$ **3.** $\dfrac{\pi}{6}$ **4.** $\dfrac{\pi}{3}$

5. $\dfrac{\pi}{2}$ **6.** $-\dfrac{3\pi}{4}$ **7.** $45°$ **8.** $720°$

9. $-120°$ **10.** $150°$ **11.** $\dfrac{7\pi}{6}$ **12.** $-315°$

13. $\dfrac{5\pi}{3}$ **14.** -4π **15.** $-270°$ **16.** $330°$

17. $\dfrac{5\pi}{4}$ **18.** $4°$ **19.** $-\dfrac{7\pi}{3}$ **20.** 4

In Problems **21–36**, *determine whether the angle is quadrantal, or a first-, second-, third-, or fourth-quadrant angle.*

21. $130°$ **22.** $\dfrac{\pi}{3}$ **23.** $-45°$ **24.** $90°$

25. $\dfrac{5\pi}{6}$ **26.** $250°$ **27.** $-\pi$ **28.** $-\dfrac{\pi}{4}$

29. $370°$ **30.** 0 **31.** $210°$ **32.** $\dfrac{2\pi}{3}$

33. 4π **34.** $\pi°$ **35.** $-\dfrac{\pi}{6}$ **36.** $270°$

In Problems **37–42**, *two angles of triangle ABC are given. Find the third angle.*

37. $A = 40°\ 30'$, $B = 110°\ 10'$ **38.** $B = 62°\ 40'$, $C = 70°\ 40'$

39. $A = 53°\ 24'\ 50''$, $C = 21°\ 10'\ 25''$ **40.** $A = 112°\ 50'\ 22''$, $B = 30°\ 22'\ 28''$

41. $B = 64°\ 36'\ 42''$, $C = 48°\ 39'\ 30''$ **42.** $A = 10°\ 8'\ 15''$, $C = 71°\ 55'\ 55''$

Problems **43–48** *refer to Fig.* 17-16. *Find the acute angle* β.

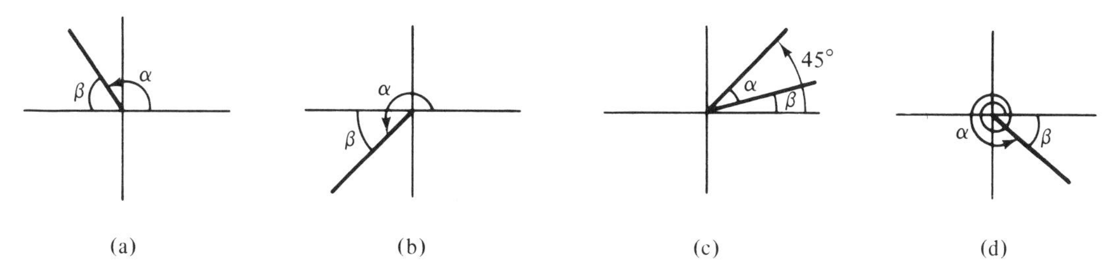

FIG. 17-16

43. In (a), $\alpha = 125°\ 18'$

44. In (b), $\alpha = 223°\ 25'\ 20''$

45. In (c), $\alpha = 31°\ 45'\ 2''$

46. In (d), $\alpha = 680°\ 0'\ 30''$

47. In (b), $\alpha = \dfrac{5\pi}{4}$

48. In (c), $\alpha = \dfrac{\pi}{6}$

In Problems **49–54**, *find the arc length cut off by the given central angle* θ *on a circle with the given radius* r.

49. $\theta = \dfrac{\pi}{4}, r = 8$

50. $\theta = 2, r = 10$

51. $\theta = 30°, r = 4$

52. $\theta = 150°, r = 1$

53. $\theta = 300°, r = 5$

54. $\theta = 135°, r = \frac{1}{2}$

In Problems **55–60**, *for a circle with the given radius* r, *find the area of the sector with the given central angle* θ. *Assume answers are in square units.*

55. $\theta = \dfrac{\pi}{3}, r = 3$

56. $\theta = \dfrac{5\pi}{6}, r = 2$

57. $\theta = 45°, r = 10$

58. $\theta = 225°, r = 5$

59. $\theta = 120°, r = 3$

60. $\theta = 10°, r = 4$

In Problems **61–66**, *find two positive and two negative angles that are coterminal with the given angle.*

61. $20°$

62. $100°$

63. $315°$

64. $90°$

65. $\dfrac{\pi}{6}$

66. $\dfrac{2\pi}{3}$

In Problems **67–72**, *find the angle* θ *that is coterminal with the given angle and that meets the given condition.*

67. $420°, 0° \leqslant \theta < 360°$

68. $1080°, 0° \leqslant \theta < 360°$

69. $\dfrac{13\pi}{4}, 0 \leqslant \theta < 2\pi$

70. $\dfrac{15\pi}{3}, 0 \leqslant \theta < 2\pi$

71. $\frac{17\pi}{2}, 0 \leqslant \theta < 2\pi$

72. $-\frac{5\pi}{6}, 0 \leqslant \theta < 2\pi$

73. An engineer is to design a cloverleaf for an exit on a certain interstate road. A part of the cloverleaf can be thought of as an arc of a circle of radius 300 meters. The central angle of this arc is 120°. Find the length of this part of the cloverleaf.

74. Suppose that a pendulum of length 20 centimeters (cm) swings through an arc of length 12 cm. Through what angle, in *degrees*, does the pendulum swing?

75. A circular road surrounds a lake. The road can be considered as the region between two concentric circles (they have the same center) having radii of 200 ft and 220 ft. Reflective material is to be applied to a portion of the road that is used as a pedestrian crossway. This portion can be considered as the region between the circles and bounded by a central angle of 5°. How many square feet of reflective material are required?

17-2 TRIGONOMETRIC FUNCTIONS

In this section we shall introduce six special functions called *trigonometric functions*. As you will see, their inputs are angles and their outputs are numbers.

We begin with an angle θ in standard position (Fig. 17-17). On the terminal side we pick *any* point (x, y) other than $(0, 0)$. Let r be the length of the line

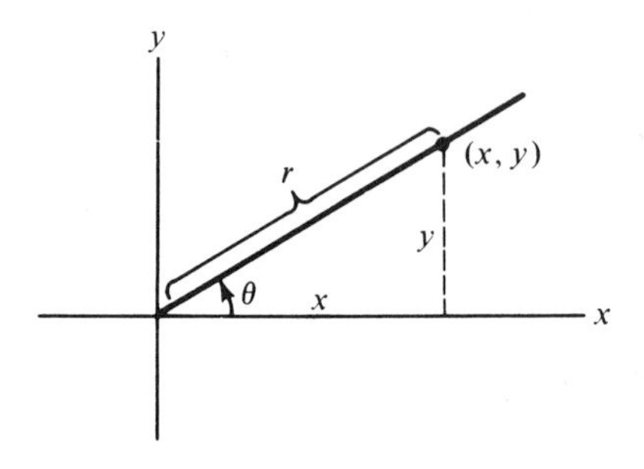

FIG. 17-17

segment from $(0, 0)$ to (x, y). By dropping a perpendicular from (x, y) to the x-axis, we see that r is the hypotenuse of a right triangle with sides x and y. By the Pythagorean theorem,

$$r^2 = x^2 + y^2.$$

Since r cannot be negative (r is a length), then

$$r = \sqrt{x^2 + y^2}.$$

It is in terms of the three numbers x, y, and r that we define the six trigonometric functions of θ. Their names (and abbreviations) are **sine** function (**sin**), **cosine** function (**cos**), **tangent** function (**tan**), **cotangent** function (**cot**), **secant** function (**sec**), and **cosecant** function (**csc**). Just as we take a function f and look at $f(x)$, we'll take, say, the sine function and look at $\sin\theta$.

TRIGONOMETRIC FUNCTIONS

Suppose θ is in standard position. Let (x, y) be any point except (0, 0) on the terminal side, and let $r = \sqrt{x^2 + y^2}$. Then

$$\sin\theta = \frac{y}{r}$$

$$\cos\theta = \frac{x}{r}$$

$$\tan\theta = \frac{y}{x}$$

$$\cot\theta = \frac{x}{y}$$

$$\sec\theta = \frac{r}{x}$$

$$\csc\theta = \frac{r}{y}.$$

Notice that the inputs are angles and the outputs are numbers. We assume that the denominators are not zero.

EXAMPLE 1

a. *Find the six trigonometric (trig) values of θ in Fig.* 17-18(a).

$$(x, y) = (3, 4).$$

$$r = \sqrt{x^2 + y^2} = \sqrt{3^2 + 4^2} = \sqrt{9 + 16} = \sqrt{25} = 5.$$

$$\sin\theta = \frac{y}{r} = \frac{4}{5}$$

$$\cos\theta = \frac{x}{r} = \frac{3}{5}$$

$$\tan\theta = \frac{y}{x} = \frac{4}{3}$$

$$\cot\theta = \frac{x}{y} = \frac{3}{4}$$

$$\sec\theta = \frac{r}{x} = \frac{5}{3}$$

$$\csc\theta = \frac{r}{y} = \frac{5}{4}.$$

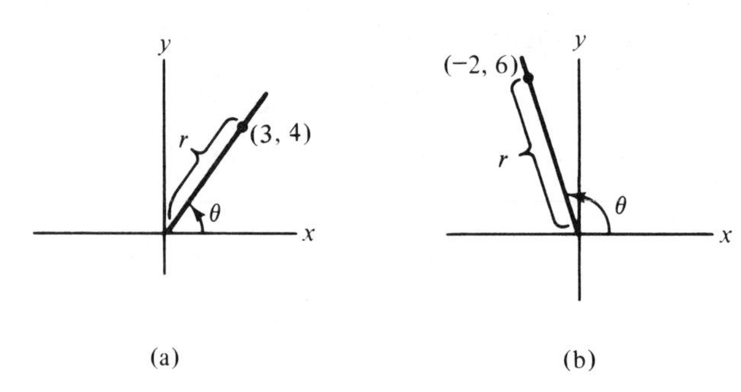

FIG. 17-18

b. *Find the six trig values of θ in Fig.* 17-18(b).

$$(x, y) = (-2, 6).$$

$$r = \sqrt{x^2 + y^2} = \sqrt{(-2)^2 + 6^2} = \sqrt{40}$$
$$= \sqrt{4 \cdot 10} = 2\sqrt{10}\,.$$

Thus, $$\sin\theta = \frac{y}{r} = \frac{6}{2\sqrt{10}} = \frac{3}{\sqrt{10}} = \frac{3}{\sqrt{10}} \cdot \frac{\sqrt{10}}{\sqrt{10}} = \frac{3\sqrt{10}}{10}$$

[rationalizing the denominator].

Also,

$$\cos\theta = \frac{x}{r} = \frac{-2}{2\sqrt{10}} = -\frac{1}{\sqrt{10}} = -\frac{\sqrt{10}}{10},$$

$$\tan\theta = \frac{y}{x} = \frac{6}{-2} = -3,$$

$$\cot\theta = \frac{x}{y} = \frac{-2}{6} = -\frac{1}{3},$$

$$\sec\theta = \frac{r}{x} = \frac{2\sqrt{10}}{-2} = -\sqrt{10},$$

$$\csc\theta = \frac{r}{y} = \frac{2\sqrt{10}}{6} = \frac{\sqrt{10}}{3}.$$

The next example shows how easy it is to find the trig values of quadrantal angles.[†]

[†]Recall that these angles have their terminal sides on an axis.

EXAMPLE 2

a. *Find the six trig values of* $0°$.

Figure 17-19(a) shows the angle $0°$. We must choose a point (x, y) on the terminal side. Let's pick (1, 0). Then $r = 1$, and we have

$$\sin 0° = \frac{y}{r} = \frac{0}{1} = 0,$$

$$\cos 0° = \frac{x}{r} = \frac{1}{1} = 1,$$

$$\tan 0° = \frac{y}{x} = \frac{0}{1} = 0,$$

$$\cot 0° = \frac{x}{y} = \frac{1}{0}, \text{ which is **not defined**,}$$

$$\sec 0° = \frac{r}{x} = \frac{1}{1} = 1,$$

$$\csc 0° = \frac{r}{y} = \frac{1}{0}, \text{ which is **not defined**.}$$

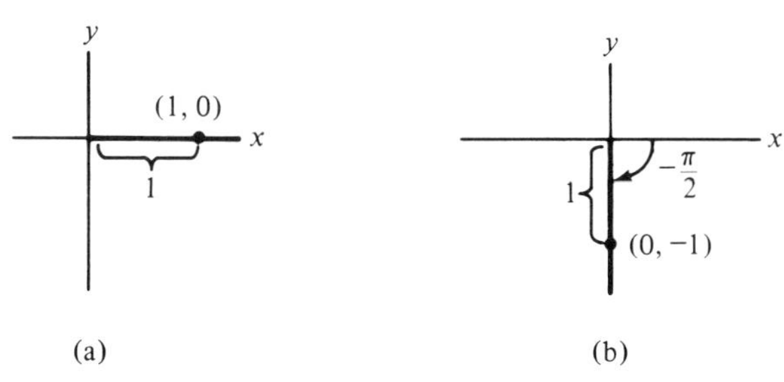

FIG. 17-19

b. *Find the six trig values of* $-\frac{\pi}{2}$.

See Fig. 17-19(b). We'll choose $(0, -1)$ as (x, y). Then $r = 1$. Thus,

$$\sin\left(-\frac{\pi}{2}\right) = \frac{y}{r} = \frac{-1}{1} = -1,$$

$$\cos\left(-\frac{\pi}{2}\right) = \frac{x}{r} = \frac{0}{1} = 0,$$

$$\tan\left(-\frac{\pi}{2}\right) = \frac{y}{x} = \frac{-1}{0}, \text{ which is **not defined**,}$$

$$\cot\left(-\frac{\pi}{2}\right) = \frac{x}{y} = \frac{0}{-1} = 0,$$

$$\sec\left(-\frac{\pi}{2}\right) = \frac{r}{x} = \frac{1}{0}, \text{ which is **not defined**,}$$

$$\csc\left(-\frac{\pi}{2}\right) = \frac{r}{y} = \frac{1}{-1} = -1.$$

Suppose that α and β are coterminal angles, as in Fig. 17-20. The point (x, y) lies on the terminal sides of *both* α and β. Thus, **the trig values of coterminal angles must be the same.**

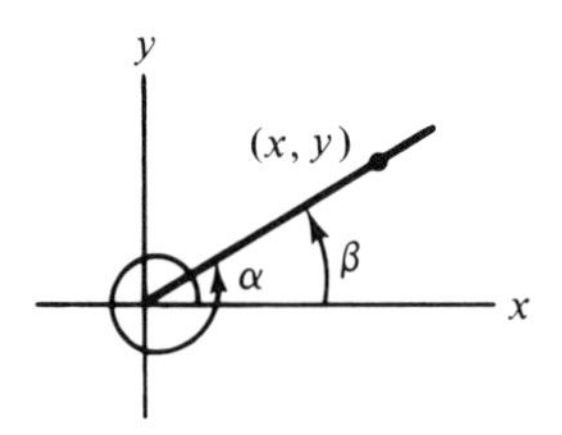

FIG. 17-20

EXAMPLE 3

Find the trig values of $\frac{3\pi}{2}$.

In Example 2(b) we found the trig values of $-\frac{\pi}{2}$. But $\frac{3\pi}{2}$ and $-\frac{\pi}{2}$ are coterminal (see Fig. 17-21). Thus both angles must have the same trig values. From

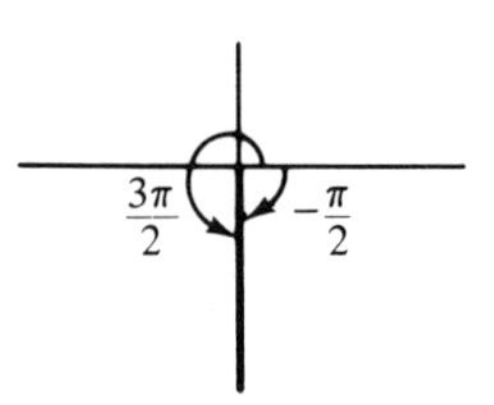

FIG. 17-21

Example 2(b) we have

$$\sin \frac{3\pi}{2} = -1,$$

$$\cos \frac{3\pi}{2} = 0,$$

$$\tan \frac{3\pi}{2} \text{ is not defined,}$$

$$\cot \frac{3\pi}{2} = 0,$$

$$\sec \frac{3\pi}{2} \text{ is not defined,}$$

$$\csc \frac{3\pi}{2} = -1.$$

We know that $\sin \theta = \dfrac{y}{r}$ and $\csc \theta = \dfrac{r}{y}$. Since $\dfrac{y}{r}$ and $\dfrac{r}{y}$ are reciprocals, then

$$\boxed{\sin \theta = \frac{1}{\csc \theta} \quad \text{and} \quad \csc \theta = \frac{1}{\sin \theta}.}$$

Thus if $\sin \theta = \frac{1}{4}$, then $\csc \theta = 4$. The sine and cosecant are called *reciprocal functions*. Similarly,

$$\boxed{\begin{array}{lcl} \cos \theta = \dfrac{1}{\sec \theta} & \text{and} & \sec \theta = \dfrac{1}{\cos \theta} \\ \tan \theta = \dfrac{1}{\cot \theta} & \text{and} & \cot \theta = \dfrac{1}{\tan \theta}. \end{array}}$$

Check these out yourself.

Since $\sin \theta = \dfrac{y}{r}$ and $\cos \theta = \dfrac{x}{r}$, then

$$\frac{\sin \theta}{\cos \theta} = \frac{\frac{y}{r}}{\frac{x}{r}} = \frac{y}{r} \cdot \frac{r}{x} = \frac{y}{x} = \tan \theta.$$

Thus,

$$\boxed{\tan \theta = \frac{\sin \theta}{\cos \theta}.}$$

Similarly,

$$\boxed{\cot \theta = \frac{\cos \theta}{\sin \theta}.}$$

EXAMPLE 4

Suppose $\sin\theta = \frac{1}{3}$ *and* $\cos\theta = -\frac{2\sqrt{2}}{3}$. *Find the other trig values of* θ.

$$\tan\theta = \frac{\sin\theta}{\cos\theta} = \frac{\frac{1}{3}}{-\frac{2\sqrt{2}}{3}} = -\frac{1}{2\sqrt{2}} = -\frac{\sqrt{2}}{4},$$

$$\cot\theta = \frac{1}{\tan\theta} = \frac{1}{-\frac{1}{2\sqrt{2}}} = -2\sqrt{2},$$

$$\sec\theta = \frac{1}{\cos\theta} = \frac{1}{-\frac{2\sqrt{2}}{3}} = -\frac{3}{2\sqrt{2}} = -\frac{3\sqrt{2}}{4},$$

$$\csc\theta = \frac{1}{\sin\theta} = \frac{1}{\frac{1}{3}} = 3.$$

Figure 17-22(a) shows a point (x, y) on the terminal side of a *first-quadrant* angle θ. In Quadrant I, both x and y are positive. Also, r is positive (it's always positive). Thus, $\sin\theta = \frac{y}{r}$ is positive, and $\cos\theta = \frac{x}{r}$ is positive. In fact, all the trig values of θ are positive in the first quadrant.

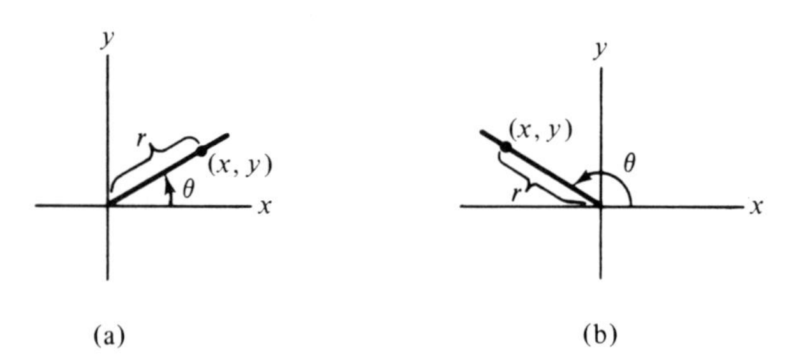

FIG. 17-22

Now, if θ is a *second-quadrant* angle, as in Fig. 17-22(b), then x is negative and y is positive. Thus, $\sin\theta = \frac{y}{r}$ is positive, and its reciprocal, $\csc\theta = \frac{r}{y}$, is positive. The other trig values of θ are negative in the second quadrant.

Similarly, we can find the signs of the trig functions for third- and fourth-quadrant angles. Figure 17-23 gives the trig functions that are positive for angles in the various quadrants.

Positive Trig Functions

sin csc	ALL
tan cot	cos sec

FIG. 17-23

EXAMPLE 5

Use the chart in Fig. 17-23 *to find the signs of the following trig values.*

a. tan 200°. A 200° angle is a third-quadrant angle (see Fig. 17-24). From the chart, we know that tan 200° is positive.

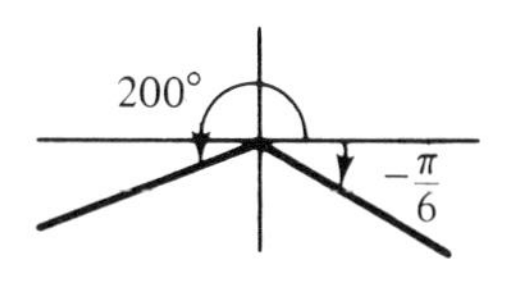

FIG. 17-24

b. $\sin\left(-\frac{\pi}{6}\right)$. An angle of $-\frac{\pi}{6}$ is a fourth-quadrant angle (see Fig. 17-24). From the chart, the only positive functions here are the cosine and secant. Thus, $\sin\left(-\frac{\pi}{6}\right)$ is negative.

EXAMPLE 6

Suppose that $\sin\theta = -\frac{1}{4}$. *Find the other trig values of* θ *if* $\tan\theta$ *is positive.*

First we find the quadrant of θ. Since $\sin\theta$ is negative, then from Fig. 17-23 we see that θ is a third- or fourth-quadrant angle. But since $\tan\theta$ is positive, then θ is a first- or third-quadrant angle. Both conditions are met if θ is a third-quadrant angle. Since $\sin\theta = \frac{y}{r} = -\frac{1}{4}$, we may choose $y = -1$ and $r = 4$† (remember, r must be positive). See Fig. 17-25.

†We could also choose, for example, $y = -2$ and $r = 8$.

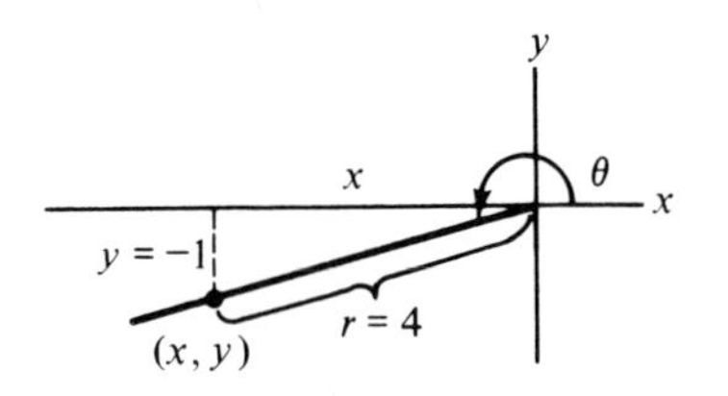

FIG. 17-25

To find x we apply the Pythagorean theorem to the right triangle formed by the terminal side of θ, the x-axis, and the broken line.

$$x^2 + y^2 = r^2$$
$$x^2 + (-1)^2 = 4^2$$
$$x^2 = 16 - 1 = 15$$
$$x = \pm\sqrt{15}\,.$$

Since θ is a third-quadrant angle, x must be negative. Thus,

$$x = -\sqrt{15}\,, \qquad y = -1, \qquad r = 4.$$

The other trig values are

$$\cos\theta = \frac{x}{r} = \frac{-\sqrt{15}}{4} = -\frac{\sqrt{15}}{4}\,,$$

$$\tan\theta = \frac{y}{x} = \frac{-1}{-\sqrt{15}} = \frac{\sqrt{15}}{15}\,,$$

$$\cot\theta = \frac{x}{y} = \frac{-\sqrt{15}}{-1} = \sqrt{15}\,,$$

$$\sec\theta = \frac{r}{x} = \frac{4}{-\sqrt{15}} = -\frac{4\sqrt{15}}{15}\,,$$

$$\csc\theta = \frac{r}{y} = \frac{4}{-1} = -4.$$

Completion Set 17-2

Fill in the blanks.

1. Suppose that (1, 2) lies on the terminal side of an angle θ in standard position. Then $r = \sqrt{(\_\_\_)^2 + (\_\_\_)^2} = \sqrt{\_\_\_}\,.$

2. Suppose that (x, y) lies on the terminal side of an angle θ in standard position. Then

$$\sin\theta = \frac{(\quad)}{r} \text{ and } \cos\theta = \frac{(\quad)}{r}.$$

3. $\tan\theta = \dfrac{(\qquad\qquad)}{\cos\theta}$.

4. If $\sin\theta = \frac{3}{4}$, then $\csc\theta =$ ____.

5. If $\cos 10° = .9848$, then $\cos 370° =$ ____________.

6. The two trig functions that are positive for a third-quadrant angle are the _______ and _______ functions.

7. If $\sin\theta$ is negative and $\cos\theta$ is positive, then θ is a ______________-quadrant angle.

Problem Set 17-2

In Problems **1–12**, *the given point lies on the terminal side of an angle θ in standard position. Find the six trig values of θ.*

1. $(6, 8)$ **2.** $(12, 5)$ **3.** $(-3, 4)$ **4.** $(2, -3)$

5. $(-1, -1)$ **6.** $(-3, -3)$ **7.** $(4, 2)$ **8.** $(-1, 7)$

9. $(-2, -10)$ **10.** $(2, \sqrt{5})$ **11.** $(1, -\sqrt{3})$ **12.** $(-3\sqrt{3}, 3)$

In Problems **13–18**, *find the six trig values of the angles.*

13. $90°$ **14.** $180°$ **15.** $270°$

16. 2π **17.** -3π **18.** $-\dfrac{3\pi}{2}$

In Problems **19–24**, *two trig values of θ are given. Find the other four trig values.*

19. $\sin\theta = \dfrac{1}{4}, \cos\theta = \dfrac{\sqrt{15}}{4}$

20. $\sin\theta = -\dfrac{\sqrt{21}}{5}, \cos\theta = \dfrac{2}{5}$

21. $\sec\theta = -\dfrac{5}{4}, \csc\theta = -\dfrac{5}{3}$

22. $\cos\theta = -\dfrac{\sqrt{5}}{3}, \csc\theta = \dfrac{3}{2}$

23. $\sin\theta = \dfrac{3\sqrt{10}}{10}, \tan\theta = -3$

24. $\cos\theta = -\dfrac{8}{9}, \tan\theta = -\dfrac{\sqrt{17}}{8}$

In Problems **25–32**, *use the facts that*

$$\sin 30° = \frac{1}{2}, \quad \cos 30° = \frac{\sqrt{3}}{2},$$

$$\sin 135° = \frac{\sqrt{2}}{2}, \quad \textit{and} \quad \cos 135° = -\frac{\sqrt{2}}{2}$$

to find the given trig values. Hint: Make use of coterminal angles.

25. $\sin 495°$

26. $\sin 390°$

27. $\cos\left(-\frac{11\pi}{6}\right)$

28. $\cos\left(-\frac{5\pi}{4}\right)$

29. $\tan 750°$

30. $\cot(-330°)$

31. $\csc\left(-\frac{13\pi}{4}\right)$

32. $\sec \frac{11\pi}{4}$

In Problems **33–50**, *give the sign of each given trig value.*

33. $\sin 100°$

34. $\tan 200°$

35. $\csc 280°$

36. $\cos 72°$

37. $\cot 175°$

38. $\sec 240°$

39. $\tan \frac{\pi}{4}$

40. $\sin \frac{7\pi}{6}$

41. $\cos \frac{4\pi}{3}$

42. $\csc \frac{5\pi}{6}$

43. $\sec \frac{11\pi}{6}$

44. $\cot \frac{7\pi}{4}$

45. $\cos 460°$

46. $\sin(-20°)$

47. $\tan\left(-\frac{3\pi}{4}\right)$

48. $\csc \frac{8\pi}{3}$

49. $\sec\left(-\frac{7\pi}{6}\right)$

50. $\cot \frac{25\pi}{6}$

In Problems **51–56**, *a trig value of* θ *and a condition on* θ *is given. Find the five other trig values.*

51. $\sin\theta = \frac{1}{6}$ and $\cos\theta$ is negative

52. $\cos\theta = -\frac{1}{3}$ and $\tan\theta$ is positive

53. $\tan\theta = -\frac{\sqrt{33}}{4}$ and $\sin\theta$ is negative

54. $\cot\theta = \sqrt{15}$ and $\sec\theta$ is positive

55. $\csc\theta = -2$ and $\cos\theta$ is negative

56. $\sec\theta = -\frac{\sqrt{21}}{3}$ and $\cot\theta$ is negative

57. The points (2, 1) and (6, 3) both lie on the terminal side of θ in Fig. 17-26. (a) Choose

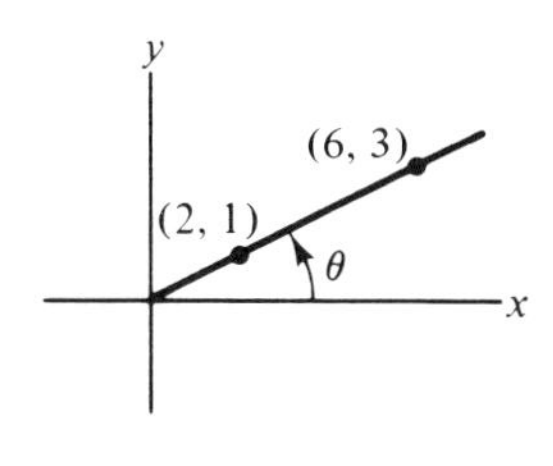

FIG. 17-26

(2, 1) as (x, y) and find the trig values of θ. (b) Do the same for (6, 3). Your answers in (a) and (b) should be the same. This illustrates that it doesn't matter what point you use on the terminal side of θ, except (0, 0), to find the trig values of θ.

17-3 SPECIAL ANGLES

Figure 17-27(a) shows an acute angle θ in standard position. We formed a right triangle by dropping a perpendicular from (x, y) to the x-axis. The symbol "⌈" indicates a 90° angle.

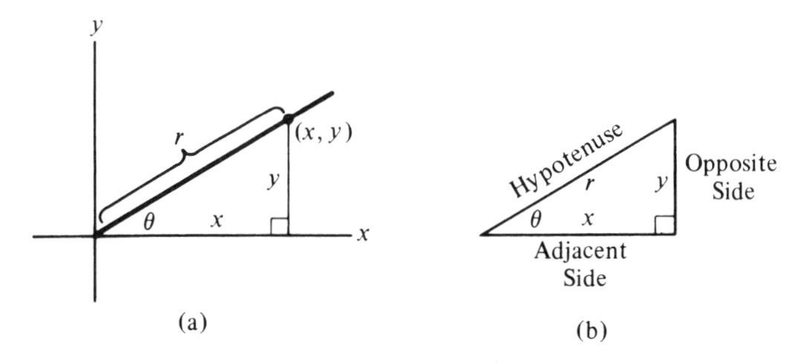

FIG. 17-27

Figure 17-27(b) shows the triangle by itself. We call side y the **opposite side** of θ, and x the **adjacent side**. The **hypotenuse** is r. The trig values of θ can be given in terms of these sides.

$$\sin\theta = \frac{y}{r} = \frac{\text{opposite side}}{\text{hypotenuse}} \qquad \cot\theta = \frac{x}{y} = \frac{\text{adjacent side}}{\text{opposite side}}$$

$$\cos\theta = \frac{x}{r} = \frac{\text{adjacent side}}{\text{hypotenuse}} \qquad \sec\theta = \frac{r}{x} = \frac{\text{hypotenuse}}{\text{adjacent side}}$$

$$\tan\theta = \frac{y}{x} = \frac{\text{opposite side}}{\text{adjacent side}} \qquad \csc\theta = \frac{r}{y} = \frac{\text{hypotenuse}}{\text{opposite side}}$$

EXAMPLE 1

Find the trig values of angles A and B in Fig. 17-28.

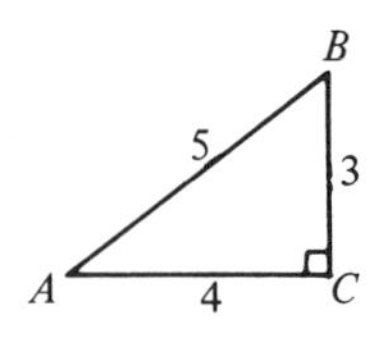

FIG. 17-28

For angle A, the opposite side is 3 and the adjacent side is 4.

$$\sin A = \frac{\text{opp}}{\text{hyp}} = \frac{3}{5} \qquad \cot A = \frac{\text{adj}}{\text{opp}} = \frac{4}{3}$$

$$\cos A = \frac{\text{adj}}{\text{hyp}} = \frac{4}{5} \qquad \sec A = \frac{\text{hyp}}{\text{adj}} = \frac{5}{4}$$

$$\tan A = \frac{\text{opp}}{\text{adj}} = \frac{3}{4} \qquad \csc A = \frac{\text{hyp}}{\text{opp}} = \frac{5}{3}.$$

For angle B, the opposite side is 4 and the adjacent side is 3.

$$\sin B = \frac{\text{opp}}{\text{hyp}} = \frac{4}{5} \qquad \cot B = \frac{\text{adj}}{\text{opp}} = \frac{3}{4}$$

$$\cos B = \frac{\text{adj}}{\text{hyp}} = \frac{3}{5} \qquad \sec B = \frac{\text{hyp}}{\text{adj}} = \frac{5}{3}$$

$$\tan B = \frac{\text{opp}}{\text{adj}} = \frac{4}{3} \qquad \csc B = \frac{\text{hyp}}{\text{opp}} = \frac{5}{4}.$$

For any right triangle, the sum of the two acute angles is 90°. Such angles are called **complementary**. In Fig. 17-28, angles A and B are complementary.

Notice in Example 1 that $\sin A = \cos B$, $\tan A = \cot B$, $\sec A = \csc B$, etc. The sine and cosine are called **cofunctions** of each other. Tangent and cotangent are also cofunctions, as well as secant and cosecant. In general, **cofunctions of complementary angles are equal**. For example, $\sin 20° = \cos 70°$.

By using right triangles you can find the trig values of 30°, 45°, and 60°, which are called **special angles**. To handle 45°, draw a right triangle so that two sides have length 1, as in Fig. 17-29. From geometry, the acute angles must each

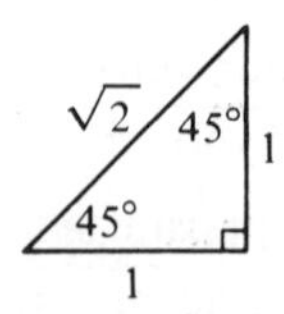

FIG. 17-29

be 45°. By the Pythagorean theorem,

$$(\text{hypotenuse})^2 = 1^2 + 1^2 = 2.$$

Since the hypotenuse must be positive,

$$\text{hypotenuse} = \sqrt{2}\ .$$

From this right triangle we get

$$\sin 45° = \frac{\text{opp}}{\text{hyp}} = \frac{1}{\sqrt{2}} = \frac{\sqrt{2}}{2},$$

$$\cos 45° = \frac{\text{adj}}{\text{hyp}} = \frac{1}{\sqrt{2}} = \frac{\sqrt{2}}{2}, \text{ etc.}$$

You can find the trig values of 30° and 60° by drawing a 30°-60°-90° triangle. One appears in Fig. 17-30 with hypotenuse 2. From geometry, the side

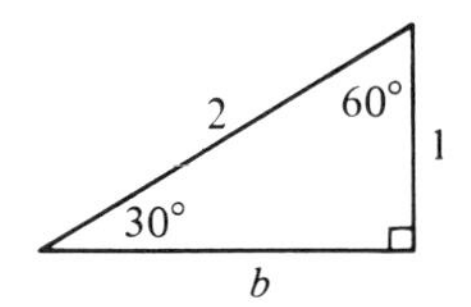

FIG. 17-30

opposite the 30° angle is half the hypotenuse. Thus that side must be 1. By the Pythagorean theorem you can find side b.

$$1^2 + b^2 = 2^2$$
$$b^2 = 3$$
$$b = \sqrt{3}\ . \qquad \left[\text{Ignore } -\sqrt{3}\text{, since you want } b \text{ to be positive.}\right]$$

Thus,

$$\sin 30° = \frac{\text{opp}}{\text{hyp}} = \frac{1}{2}, \text{ etc.}$$

and

$$\sin 60° = \frac{\text{opp}}{\text{hyp}} = \frac{b}{2} = \frac{\sqrt{3}}{2}, \text{ etc.}$$

Here again are the important triangles (Fig. 17-31). From them we get Table 17-1.

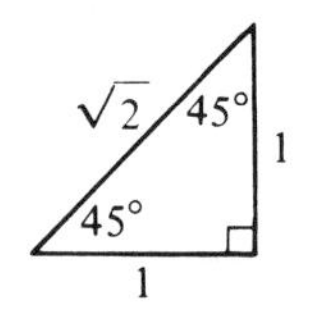

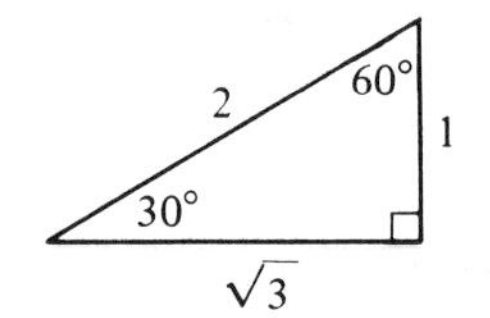

FIG. 17-31

TABLE 17-1

		30° (or $\frac{\pi}{6}$)	45° (or $\frac{\pi}{4}$)	60° (or $\frac{\pi}{3}$)
sin	$\frac{\text{opp}}{\text{hyp}}$	$\frac{1}{2}$	$\frac{1}{\sqrt{2}} = \frac{\sqrt{2}}{2}$	$\frac{\sqrt{3}}{2}$
cos	$\frac{\text{adj}}{\text{hyp}}$	$\frac{\sqrt{3}}{2}$	$\frac{1}{\sqrt{2}} = \frac{\sqrt{2}}{2}$	$\frac{1}{2}$
tan	$\frac{\text{opp}}{\text{adj}}$	$\frac{1}{\sqrt{3}} = \frac{\sqrt{3}}{3}$	$\frac{1}{1} = 1$	$\frac{\sqrt{3}}{1} = \sqrt{3}$
cot	$\frac{\text{adj}}{\text{opp}}$	$\frac{\sqrt{3}}{1} = \sqrt{3}$	$\frac{1}{1} = 1$	$\frac{1}{\sqrt{3}} = \frac{\sqrt{3}}{3}$
sec	$\frac{\text{hyp}}{\text{adj}}$	$\frac{2}{\sqrt{3}} = \frac{2\sqrt{3}}{3}$	$\frac{\sqrt{2}}{1} = \sqrt{2}$	$\frac{2}{1} = 2$
csc	$\frac{\text{hyp}}{\text{opp}}$	$\frac{2}{1} = 2$	$\frac{\sqrt{2}}{1} = \sqrt{2}$	$\frac{2}{\sqrt{3}} = \frac{2\sqrt{3}}{3}$

Don't waste your time memorizing Table 17-1. Your best bet is to memorize the triangles from which it came. Table 17-2 gives the trig values for quadrantal angles. A dash means that the trig value is not defined.

TABLE 17-2

	0° (or 0)	90° $\left(\text{or } \frac{\pi}{2}\right)$	180° (or π)	270° $\left(\text{or } \frac{3\pi}{2}\right)$
sin	0	1	0	−1
cos	1	0	−1	0
tan	0	—	0	—
cot	—	0	—	0
sec	1	—	−1	—
csc	—	1	—	−1

With each angle θ that is not quadrantal, we can associate another angle called the **reference angle** of θ. It is the *acute* angle between the terminal side of θ and the x-axis. Figure 17-32 shows different situations.

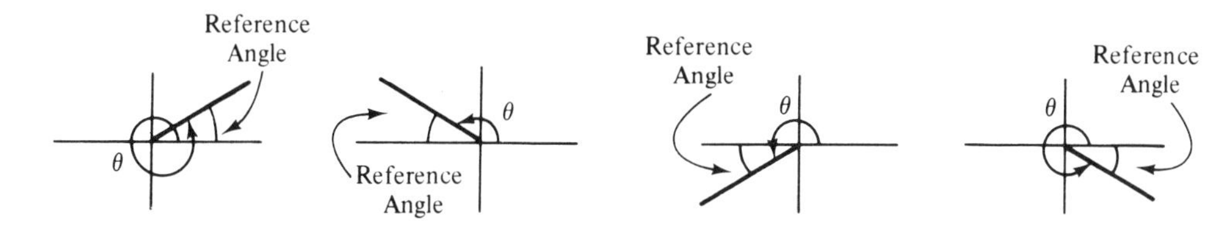

FIG. 17-32

EXAMPLE 2

Find the reference angle for each of the following angles.

a. 400° [see Fig. 17-33(a)].

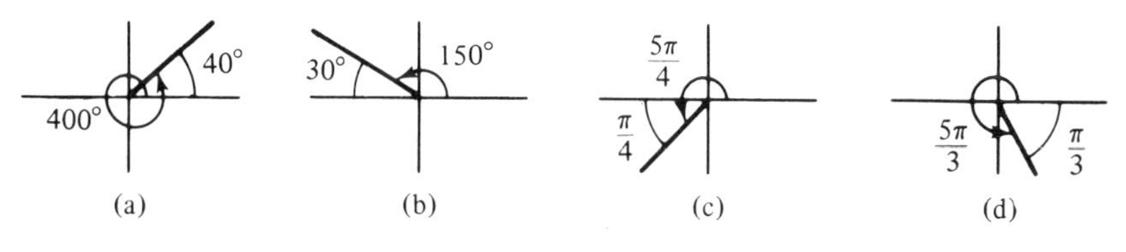

FIG. 17-33

Since

$$360° + (\text{reference angle}) = 400°,$$

then

$$\text{reference angle} = 400° - 360° = 40°.$$

b. 150° [see Fig. 17-33(b)].

$$150° + (\text{reference angle}) = 180°$$
$$\text{reference angle} = 180° - 150° = 30°.$$

c. $\frac{5\pi}{4}$ [see Fig. 17-33(c)].

$$\pi + (\text{reference angle}) = \frac{5\pi}{4}$$
$$\text{reference angle} = \frac{5\pi}{4} - \pi = \frac{\pi}{4}.$$

d. $\frac{5\pi}{3}$ [see Fig. 17-33(d)].

$$\frac{5\pi}{3} + (\text{reference angle}) = 2\pi$$

$$\therefore \quad \text{reference angle} = 2\pi - \frac{5\pi}{3} = \frac{\pi}{3}.$$

If an angle is not quadrantal, we can find its trig values from its reference angle. For example, let's look at 150° [see Fig. 17-34(a)]. Its reference angle is

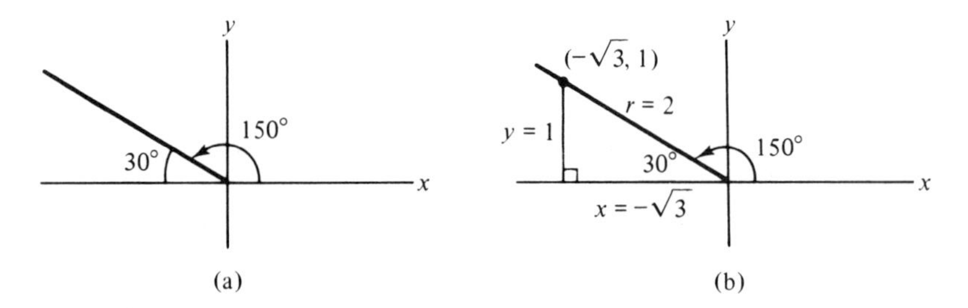

FIG. 17-34

30°. To locate a point on the terminal side of 150°, we'll make use of our 30°-60°-90° triangle. From Fig. 17-34(b), $(-\sqrt{3}, 1)$ lies on the terminal side and $r = 2$. Thus,

$$\sin 150° = \frac{y}{r} = \frac{1}{2},$$

$$\cos 150° = \frac{x}{r} = \frac{-\sqrt{3}}{2} = -\frac{\sqrt{3}}{2}, \text{ etc.}$$

But from Table 17-1,

$$\sin 30° = \frac{1}{2}, \qquad \cos 30° = \frac{\sqrt{3}}{2}, \text{ etc.}$$

Notice that the trig values of 150° are the same as those of its reference angle except, perhaps, for sign.

Thus, to find cos 150° we can find cos 30° and put a minus sign in front of it. The minus sign is needed because the cosine of a second-quadrant angle (150°) is negative. In general,

> To find a trig value of an angle θ, find the same trig value of its reference angle and attach the proper sign to the result. This sign depends on the function involved and the quadrant of θ.
>
> $$\text{trig function of } \theta = \pm \text{trig function of reference angle}$$
>
> ↑ *sign depends on trig function and quadrant of* θ

EXAMPLE 3

a. *Find* sin 135°.

Since 135° is a second-quadrant angle [Fig. 17-35(a)], sin 135° is positive. The reference angle is 180° − 135° = 45°. Thus from the 45° right triangle in Fig. 17-35(b),

$$\sin 135° = +\sin 45° = +\frac{\text{opp}}{\text{hyp}} = +\frac{1}{\sqrt{2}} = \frac{\sqrt{2}}{2}.$$

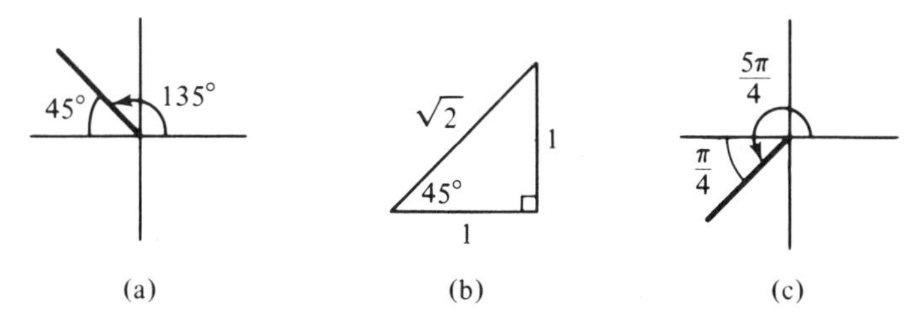

FIG. 17-35

b. *Find* $\cos \frac{5\pi}{4}$.

Since $\frac{5\pi}{4}$ is a third-quadrant angle [Fig. 17-35(c)], $\cos \frac{5\pi}{4}$ is negative. The reference angle is $\frac{5\pi}{4} - \pi = \frac{\pi}{4}$ (or 45°). From Fig. 17-35(b),

$$\cos \frac{5\pi}{4} = -\cos \frac{\pi}{4} = -\frac{\text{adj}}{\text{hyp}} = -\frac{1}{\sqrt{2}} = -\frac{\sqrt{2}}{2}.$$

c. *Find* $\cot\left(-\frac{\pi}{6}\right)$.

Since $-\frac{\pi}{6}$ is a fourth-quadrant angle [Fig. 17-36(a)], $\cot\left(-\frac{\pi}{6}\right)$ is negative. The reference angle is $\frac{\pi}{6}$ (or 30°). From Fig. 17-36(b),

$$\cot\left(-\frac{\pi}{6}\right) = -\cot\frac{\pi}{6} = -\frac{\text{adj}}{\text{opp}} = -\frac{\sqrt{3}}{1} = -\sqrt{3}.$$

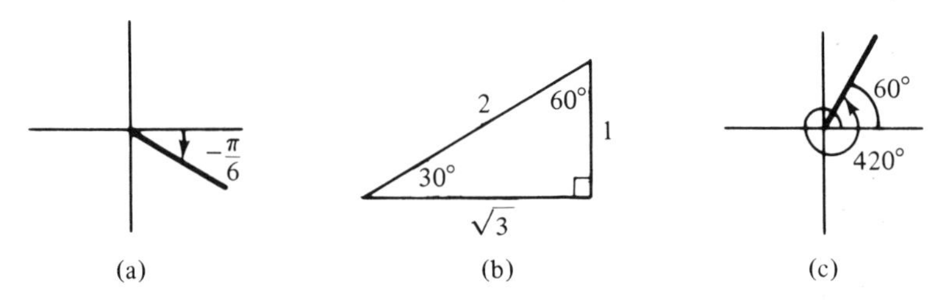

FIG. 17-36

d. *Find* sec 420°.

Since 420° is a first-quadrant angle [Fig. 17-36(c)], sec 420° is positive. The reference angle is 420° − 360° = 60°. From Fig. 17-36(b),

$$\sec 420° = +\sec 60° = +\frac{\text{hyp}}{\text{adj}} = +\frac{2}{1} = 2.$$

Completion Set 17-3

Problems **1** *and* **2** *refer to Fig.* 17-37. *Fill in the blanks.*

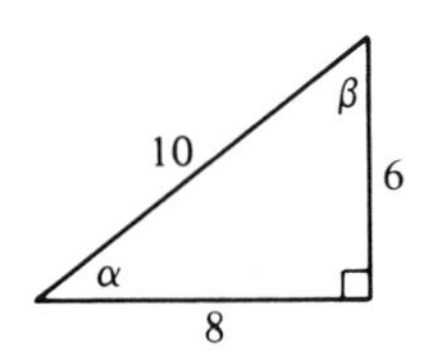

FIG. 17-37

1. The adjacent side of α is ____ and its opposite side is ____.

2. The adjacent side of β is ____ and its opposite side is ____.

Fill in the blanks.

3. If θ is an acute angle of a right triangle, then

a. $\sin\theta = \dfrac{(\qquad)}{\text{hyp}}$ b. $\tan\theta = \dfrac{(\qquad)}{\text{adj}}$

c. $\dfrac{\text{hyp}}{\text{adj}} = \underline{\qquad}\ \theta$ d. $\dfrac{\text{adj}}{\text{opp}} = \underline{\qquad}\ \theta$

4. To find sin 210°, first find the quadrant of 210°. It is a ____________-quadrant angle, so sin 210° is ____________________ (fill in *positive* or *negative*).

The reference angle of 210° is ______°. Thus, sin 210° = (   ) sin ______°. Using Fig. 17-36(b), we have

$$\sin 210^\circ = (\quad)\frac{(\qquad)}{\text{hyp}} = (\quad)\frac{(\quad)}{2}.$$

5. *Insert* T (= *True*) *or* F (= *False*). sin 15° = cos 75°.____

Problem Set 17-3

In Problems **1** *and* **2**, *find the six trig values of angles A and B in the given figure.*

1. Figure 17-38(a) **2.** Figure 17-38(b)

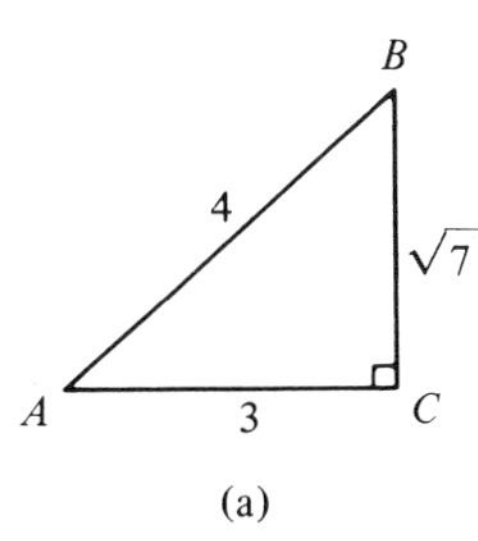

(a)

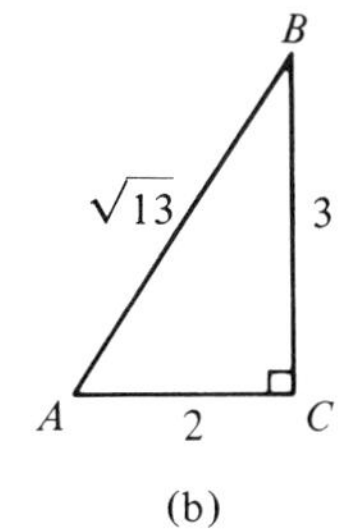

(b)

FIG. 17-38

In Problems **3–18**, *find the value of the given expression. Use the special triangles. Do not use Table* 17-1.

3. sin 30° **4.** sin 60° **5.** tan 45°

6. $\cos \frac{\pi}{4}$ 7. $\cos \frac{\pi}{3}$ 8. $\cos \frac{\pi}{6}$

9. $\cot 30°$ 10. $\tan 60°$ 11. $\sec 45°$

12. $\sec \frac{\pi}{3}$ 13. $\csc \frac{\pi}{3}$ 14. $\csc \frac{\pi}{4}$

15. $(\cot 60°)^2$ 16. $(\cot 45°)(\sec 45°)$

17. $\left(\tan \frac{\pi}{6}\right)\left(\sin \frac{\pi}{3}\right)$ 18. $\left(\sec \frac{\pi}{6}\right)\left(\sin \frac{\pi}{4}\right)$

In Problems **19–26**, *find the reference angle of the given angle.*

19. $100°$ 20. $350°$ 21. $215°$ 22. $375°$

23. $-\frac{\pi}{3}$ 24. $\frac{5\pi}{6}$ 25. $\frac{13\pi}{6}$ 26. $-\frac{3\pi}{4}$

In Problems **27–56**, *find the given trig value.*

27. $\sin 120°$ 28. $\sin 240°$ 29. $\cos 210°$

30. $\cos 300°$ 31. $\sin \frac{5\pi}{4}$ 32. $\tan \frac{7\pi}{4}$

33. $\tan \frac{11\pi}{6}$ 34. $\cos 225°$ 35. $\sec 330°$

36. $\cot \frac{7\pi}{6}$ 37. $\cos 135°$ 38. $\tan \frac{4\pi}{3}$

39. $\csc 150°$ 40. $\sec 315°$ 41. $\cos \frac{2\pi}{3}$

42. $\sin \frac{11\pi}{6}$ 43. $\cot \frac{4\pi}{3}$ 44. $\cos \frac{5\pi}{6}$

45. $\csc(-45°)$ 46. $\tan(-210°)$ 47. $\tan \frac{3\pi}{4}$

48. $\csc 405°$ 49. $\sec \frac{5\pi}{3}$ 50. $\cos \frac{7\pi}{4}$

51. $\sin 390°$ 52. $\cot(-315°)$ 53. $\cot \frac{8\pi}{3}$

54. $\sec \frac{17\pi}{6}$ 55. $\cos\left(-\frac{11\pi}{6}\right)$ 56. $\csc\left(-\frac{2\pi}{3}\right)$

17-4 TRIGONOMETRIC TABLES

By now you should be able to find the trig values of quadrantal angles and angles whose reference angles are 30°, 45°, or 60°. To handle other angles you can use the table of trig functions in Appendix C on p. 503.

In the **left-hand** columns, under the heading *Degrees*, are listed angles from 0° to 45° in changes of 10′. A sample table appears in Table 17-3.

TABLE 17-3

Degrees	Radians	Sin	Cos	Tan	Cot	Sec	Csc		
36° 00′	.6283	.5878	.8090	.7265	1.376	1.236	1.701	.9425	54° 00′
10	312	901	073	310	368	239	695	396	50
20	341	925	056	355	360	241	688	367	40
30	.6370	.5948	.8039	.7400	1.351	1.244	1.681	.9338	30
40	400	972	021	445	343	247	675	308	20
50	429	995	004	490	335	249	668	279	10
37° 00′	.6458	.6018	.7986	.7536	1.327	1.252	1.662	.9250	53° 00′
10	487	041	969	581	319	255	655	221	50
20	516	065	951	627	311	258	649	192	40
30	.6545	.6088	.7934	.7673	1.303	1.260	1.643	.9163	30
40	574	111	916	720	295	263	636	134	20
50	603	134	898	766	288	266	630	105	10
38° 00′	.6632	.6157	.7880	.7813	1.280	1.269	1.624	.9076	52° 00′
		Cos	Sin	Cot	Tan	Csc	Sec	Radians	Degrees

Here's how to find a trig value of a listed angle between 0° and 45°. First, read **down** the left-hand column until you find the angle. Then move to the right to the number in the column headed at the **top** by the name of the function you want. That's all there is to it. Notice that one of the columns gives the radian measure of the angle.

EXAMPLE 1

a. *Find* sin 37°.

Look at Table 17-3.

Read down **left-hand** column to 37°.

Move to the right to column with *Sin* heading at **top**.

Read **.6018.**

This is an approximate value of sin 37°. Most entries are. Also, notice that 37° = .6458 radian. Thus, sin .6458 = .6018.

b. *Find* cos 36° 40′.

Read down **left-hand** column to 36° 40′.

Move to the right to column with *Cos* heading at **top**.

Read **.8021**.

Notice that the ".8" part of the answer was carried over from an earlier entry.

In Appendix C, angles from 45° to 90° are listed in the **right-hand** columns. The names of the functions for these angles are given at the **bottom**. Always *read the angles from* 45° *to* 90° *from bottom to top*. For example, the angle just below 53° is 52° 50′, not 53° 50′.

EXAMPLE 2

Find csc 53° 20′.

Read **up** the **right-hand** column to 53° 20′.

Move to the **left** to column with *Csc* at the **bottom**.

Read **1.247**.

Here again, part of our answer was carried over from a prior entry.

To use the table for angles other than those from 0° to 90°, we use reference angles. Example 3 will show how.

EXAMPLE 3

Find tan 164° 40′.

164° 40′ is a second-quadrant angle (see Fig. 17-39). Thus tan 164° 40′ is negative.

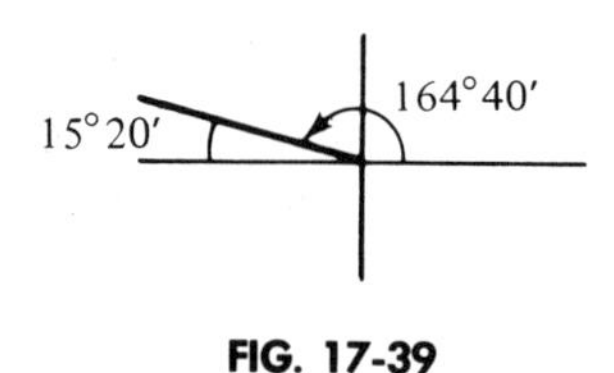

FIG. 17-39

The reference angle is

$$180° - 164° \, 40' = 179° \, 60' - 164° \, 40' = 15° \, 20'.$$

The table in Appendix C gives tan 15° 20′ = .2742. Thus,

$$\tan 164° \, 40' = -\tan 15° \, 20' = -.2742.$$

The next two examples show how to find an angle when you know one of its trig values.

EXAMPLE 4

Find the acute angle θ *given that* $\cos\theta = .8542$.

In Appendix C, look in the Cos columns for .8542. Notice that it is in a column with Cos at the **top**. Go to the **left-hand** column and read 31° 20′, which is θ.

EXAMPLE 5

Find all angles θ *such that* $\tan\theta = -2.747$ *and* $0° \leqslant \theta < 360°$.

Forget about the minus sign for a moment. Look in the Tan columns for 2.747. It is in a column with Tan at the **bottom**. Go to the **right-hand** column and read 70°. This is the reference angle for θ. Now, back to the minus sign. The tangent function is negative for second- and fourth-quadrant angles, so there must be two values of θ. See Fig. 17-40. Thus, $\theta = 110°$ and $\theta = 290°$.

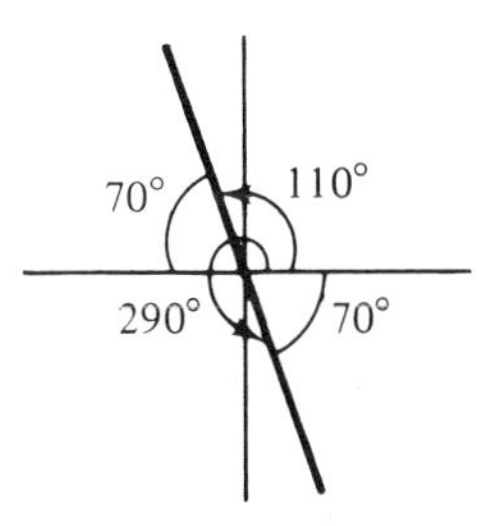

FIG. 17-40

Sometimes you may want a trig value of an angle, such as 27° 13′, which is between two angles in the table. We can estimate the value by using a method called **linear interpolation.**

Linear interpolation works something like this. We know that 15° is halfway between 0° and 30°, so we reason that sin 15° is halfway between sin 0° and sin 30°. Actually, this is not true. Sin 15° = .2588, while linear interpolation gives .25. Nevertheless, we are not too far off. When the angle we're interested in isn't too far away from listed angles, the interpolation method usually gives a good approximation.

EXAMPLE 6

Find sin 27° 13′.

27° 13′ lies between 27° 10′ and 27° 20′, which are in the trig table. We reason that sin 27° 13′ lies between sin 27° 10′ and sin 27° 20′. We look up these two values and write them in a three-line arrangement. *Always write the smallest angle first.*

$$\begin{aligned} \sin 27^\circ\ 10' &= .4566 \\ \sin 27^\circ\ 13' &= \ ? \\ \sin 27^\circ\ 20' &= .4592 \end{aligned}$$

(3 and 10 minutes; 26 **increase**)

Notice that as an angle increases 10 minutes from 27° 10′ to 27° 20′, the sine of the angle *increases* by 26 (ten-thousandths). Since 27° 13′ is $\frac{3}{10}$ "of the way" from 27° 10′ to 27° 20′, we assume that sin 27° 13′ lies approximately $\frac{3}{10}$ "of the way" from .4566 to .4592. Thus $\frac{3}{10}$ of the total increase must be added to .4566.

$$\frac{3}{10}(26) = 7.8 \approx 8.$$

Thus, 4566 + 8 = 4574 and so

$$\sin 27^\circ\ 13' = .4574.$$

EXAMPLE 7

Find cos 258° 46′.

Since 258° 46′ is a third-quadrant angle, cos 258° 46′ is negative. The reference angle is 78° 46′ (check this), and so cos 258° 46′ = − cos 78° 46′. The angle 78° 46′ lies between 78° 40′ and 78° 50′. The information we need is given below. **Remember: Write the smallest angle first.**

$$\begin{aligned} \cos 78^\circ\ 40' &= .1965 \\ \cos 78^\circ\ 46' &= \ ? \\ \cos 78^\circ\ 50' &= .1937 \end{aligned}$$

(6 and 10 minutes; 28 **decrease**)

Assuming that cos 78° 46′ is $\frac{6}{10}$ "of the way" from cos 78° 40′ to cos 78° 50′, we compute $\frac{6}{10}$ of the *decrease*.

$$\frac{6}{10}(28) = 16.8 \approx 17.$$

Since cos θ **decreases** as θ goes from 78° 40′ to 78° 50′, we **subtract** 17 from 1965: 1965 − 17 = 1948. Thus,

$$\cos 78^\circ 46' = .1948$$

and $$\cos 258^\circ\ 46' = -.1948.$$

EXAMPLE 8

Find θ *if* $\tan\theta = 1.168$ *and* θ *is acute.*

Looking at the Tan columns of the trig table, we find that 1.168 lies between the entries 1.164 and 1.171. These correspond to angles of 49° 20′ and 49° 30′, respectively. We write this down, *writing the smallest angle first.*

$$\begin{aligned} \tan 49^\circ 20' &= 1.164 \\ \tan\theta &= 1.168 \\ \tan 49^\circ 30' &= 1.171 \end{aligned}$$

(differences: 10 between the angles; 4 between 1.164 and 1.168; 7 between 1.164 and 1.171)

Tan θ is $\frac{4}{7}$ "of the way" from tan 49° 20′ to tan 49° 30′, so θ is $\frac{4}{7}$ "of the way" from 49° 20′ to 49° 30′. Since these angles differ by 10′,

$$\frac{4}{7}(10) \approx 5.7' \approx 6'.$$

Thus, $\theta = 49^\circ\ 20' + 6' = 49^\circ\ 26'$.

EXAMPLE 9

Find θ *if* $\cos\theta = -.8360$ *and* $0^\circ \leqslant \theta < 360^\circ$.

We ignore the minus sign for now and find the reference angle for θ, say θ_R. Looking at the Cos columns of the trig table, we find that .8360 lies between the entries .8371 and .8355. These correspond to angles of 33° 10′ and 33° 20′, respectively. We write this down, *writing the smallest angle first.*

$$\begin{aligned} \cos 33^\circ 10' &= .8371 \\ \cos\theta_R &= .8360 \\ \cos 33^\circ 20' &= .8355 \end{aligned}$$

(differences: 10 between the angles; 11 between .8371 and .8360; 16 between .8371 and .8355)

We reason that θ_R is $\frac{11}{16}$ "of the way" from 33° 10′ to 33° 20′. Since these angles differ by 10′,

$$\frac{11}{16}(10) \approx 6.9' \approx 7'.$$

Thus, $\theta_R = 33^\circ\ 10' + 7' = 33^\circ\ 17'$. Now, cosine is negative for second- and third-quadrant angles (see Fig. 17-41). Thus,

$$\theta = 180° - 33°\ 17' = 146°\ 43'$$

and $$\theta = 180° + 33°\ 17' = 213°\ 17'.$$

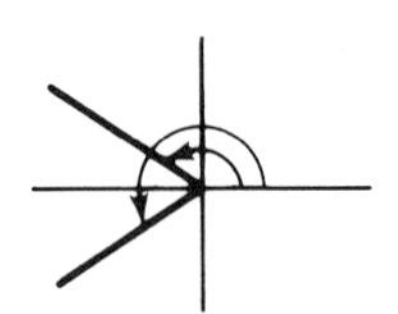

FIG. 17-41

Problem Set 17-4

In Problems **1–26**, *find the given trig values. Use linear interpolation where needed.*

1. $\sin 24°$

2. $\cos 4°$

3. $\tan 41°\ 20'$

4. $\sec 39°\ 10'$

5. $\cos 59°$

6. $\sin 75°$

7. $\cot 67°\ 40'$

8. $\tan 50°\ 50'$

9. $\cos 160°$

10. $\sin 100°$

11. $\sin 230°\ 10'$

12. $\cos 342°\ 20'$

13. $\sec .2094$

14. $\cot 1.0850$

15. $\csc \dfrac{16\pi}{9}$

16. $\cos \dfrac{4\pi}{5}$

17. $\sin 23°\ 12'$

18. $\tan 12°\ 5'$

19. $\sec 43°\ 26'$

20. $\cot 31°\ 37'$

21. $\cos 84°\ 48'$

22. $\cos 79°\ 16'$

23. $\tan 161°\ 35'$

24. $\sin 200°\ 24'$

25. $\cos 310°\ 33'$

26. $\sin(-12°\ 42')$

In Problems **27–36**, *find the acute angle θ that has the given trig value.*

27. $\cos\theta = .7470$

28. $\sin\theta = .3201$

29. $\sin\theta = .9628$

30. $\tan\theta = 2.628$

31. $\sin\theta = .2120$

32. $\sin\theta = .8350$

33. $\cot\theta = 2.000$

34. $\csc\theta = 1.590$

35. $\cos\theta = .4741$

36. $\cot\theta = 1.180$

In Problems **37–42**, *find all angles* θ *that meet the given conditions and such that* $0° \leqslant \theta < 360°$.

37. $\sin\theta = .9032$

38. $\tan\theta = -.4245$

39. $\cos\theta = -.8616$

40. $\sec\theta = 1.466$

41. $\sin\theta = -.6461$ and $\tan\theta$ is negative

42. $\cos\theta = -.8616$ and $\tan\theta$ is negative

43. In doing his physics homework dealing with reflection, a student must find a so-called critical angle θ_c. It is acute and is given by

$$1.5 \sin\theta_c = 1.$$

Find θ_c to the nearest minute. *Hint*: First divide both sides by 1.5.

44. In studies involving the bending of a ray of light that passes from one medium to another, the following formula is used:

$$n = \frac{\sin i}{\sin r},$$

where n is called the index of refraction, i is the angle of incidence, and r is the angle of refraction. For light passing from air to water, suppose that $i = 30°$ and $r = 22°$. Find n.

45. Under certain conditions, the index of refraction n of a prism is given by

$$n = \frac{\sin\frac{1}{2}(A + D)}{\sin\left(\frac{1}{2}A\right)},$$

where A is the angle of the prism and D is the angle of deviation. Find n if $A = 60°$ and $D = 50°$.

17-5 RIGHT TRIANGLES

With trigonometric functions we can *solve a right triangle*. This means finding all unknown sides and angles when you know

any two sides,

or one side and one acute angle.

We'll call the angles A, B, and C. The sides opposite these angles are a, b, and c, respectively, as in Fig. 17-42. We always label the right angle ($= 90°$) as C. The sum of A and B is $90°$.

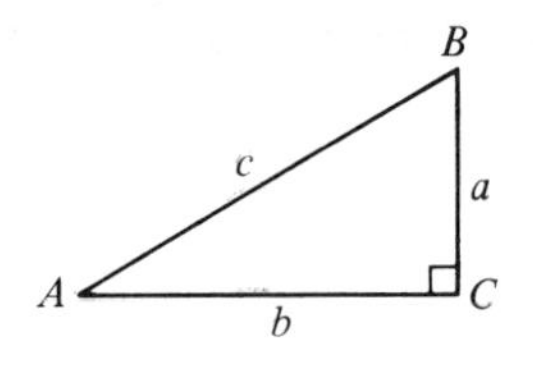

FIG. 17-42

The following examples will show you how to handle different types of right triangle problems.

EXAMPLE 1

Given two sides.

Solve the right triangle ABC given that $c = 13$ *and* $a = 5$ (*see Fig.* 17-43).

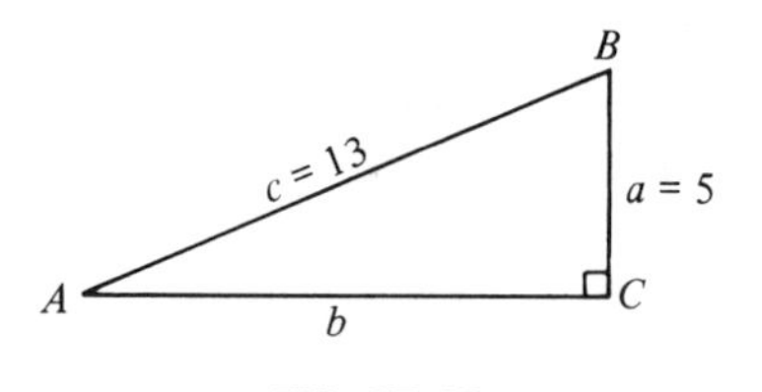

FIG. 17-43

First we use the Pythagorean theorem to find b.

$$a^2 + b^2 = c^2$$

$$5^2 + b^2 = 13^2$$

$$b^2 = 13^2 - 5^2 = 169 - 25 = 144$$

$$b = 12 \qquad \text{[ignore } -12\text{]}.$$

Now we find angle A.

$$\sin A = \frac{\text{opp}}{\text{hyp}} = \frac{5}{13} = .3846$$

$$A = 22° \; 37' \qquad \text{[trig tables and interpolation]}.$$

Since $A + B = 90°$,

$$B = 90° - A = 90° - 22° \; 37' = 67° \; 23'.$$

Thus the missing parts are

$$b = 12, \quad A = 22° \; 37', \quad \text{and} \quad B = 67° \; 23'.$$

To find angle A in Example 1 we used the sine function. Actually, any trig function could be used. For example,

$$\tan A = \frac{\text{opp}}{\text{adj}} = \frac{a}{b} = \frac{5}{12} = .4167.$$

However, if we had made an error in finding b, we would not get a correct value for A. Wherever practical, use the given parts to find the missing parts of a right triangle.

EXAMPLE 2

Given one side and an acute angle.

Solve the right triangle ABC given that $A = 32°$ and $b = 15$ (see Fig. 17-44).

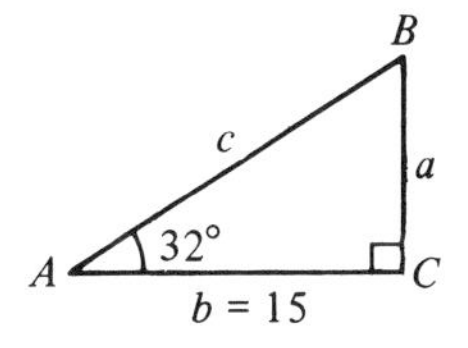

FIG. 17-44

The remaining angle is found easily.

$$B = 90° - A = 90° - 32° = 58°.$$

Let's find a next. Since

$$\tan 32° = \frac{\text{opp}}{\text{adj}} = \frac{a}{15},$$

then
$$\begin{aligned} a &= 15 \tan 32° \\ &= 15(.6249) \quad \text{[from tables]} \\ &= 9.37 \quad \text{[to two decimal places].} \end{aligned}$$

We can find c with the cosine function.

$$\begin{aligned} \cos 32° &= \frac{\text{adj}}{\text{hyp}} = \frac{15}{c} \\ c \cos 32° &= 15 \\ c &= \frac{15}{\cos 32°}. \end{aligned}$$

With no calculator handy, the best bet is to replace this division by a multiplication.

Replacing $\dfrac{1}{\cos 32°}$ by sec 32°, we have

$$\begin{aligned} c &= 15 \sec 32° \\ &= 15(1.179) \qquad \text{[from tables]} \\ &= 17.7 \end{aligned}$$

Thus the missing parts are

$$a = 9.37, \qquad c = 17.7, \quad \text{and} \quad B = 58°.$$

Two types of angles are in Fig. 17-45. Suppose that a person at P looks up at Q [see Fig. 17-45(a)]. Then the angle that PQ makes with the horizontal line

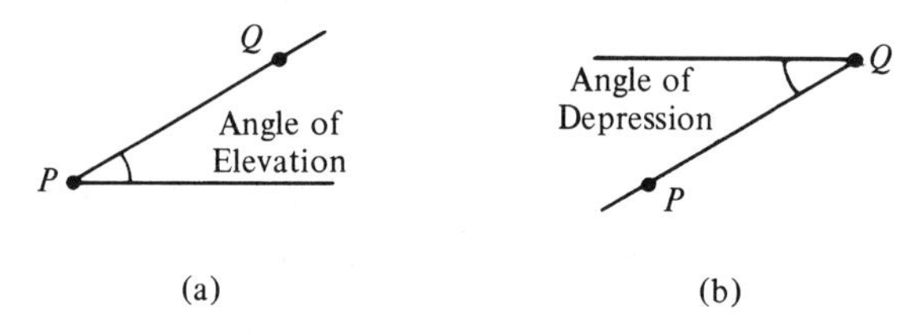

FIG. 17-45

is called the **angle of elevation** of Q from P. On the other hand, suppose that a person at Q looks down at P [see Fig. 17-45(b)]. Then the angle that PQ makes with the horizontal line is the **angle of depression** of P from Q. The angle of elevation of Q from P equals the angle of depression of P from Q.

EXAMPLE 3

The length of a kite string is 228 *ft, and the angle of elevation of the kite is* 60° (*see Fig.* 17-46). *Find the height of the kite.*

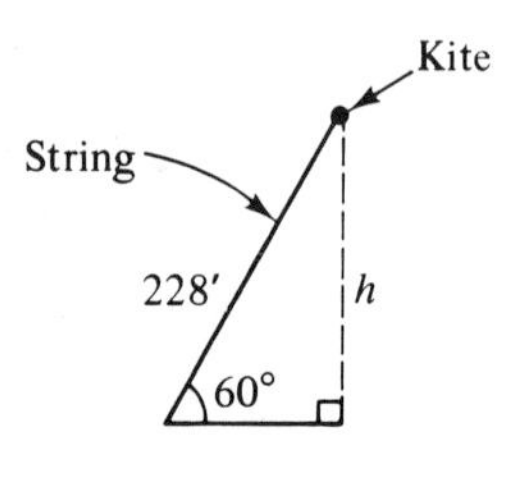

FIG. 17-46

If h is the height, then

$$\sin 60° = \frac{\text{opp}}{\text{hyp}} = \frac{h}{228}$$
$$h = 228 \sin 60°.$$

No trig tables are needed here.

$$h = 228\left(\frac{\sqrt{3}}{2}\right) = 114\sqrt{3} \text{ ft.}$$

Use of tables would have given $h = 228(.8660) = 197.4$ ft.

EXAMPLE 4

A plane flying at an altitude of 1 mile begins to climb at a constant angle of 28°. To the nearest second, how long will it take the plane to reach an altitude of $1\frac{3}{4}$ miles if its constant speed throughout the climb is 200 miles per hour?

The plane must increase its altitude by $\frac{3}{4}$ mile. See the right triangle in Fig. 17-47.

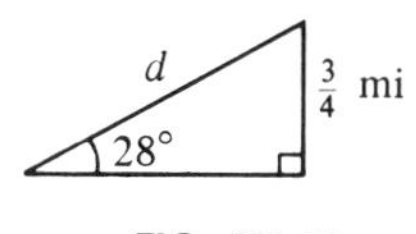

FIG. 17-47

We first find the distance d of the climb.

$$\csc 28° = \frac{\text{hyp}}{\text{opp}} = \frac{d}{\frac{3}{4}}$$
$$d = \frac{3}{4} \csc 28° = \frac{3}{4}(2.130)$$
$$= 1.598 \text{ mi.}$$

We wish to find the time to travel this 1.598 mi. Since distance = (rate)(time), we have

$$\text{time} = \frac{\text{distance}}{\text{rate}} = \frac{1.598}{200} = .008 \text{ hr.}$$

But we want .008 hr expressed in seconds. Since 1 hr = 3600 sec,

$$.008 \text{ hr} = .008(3600 \text{ sec}) = 29 \text{ sec.}$$

Problem Set 17-5

In Problems **1–6**, *ABC is a right triangle with right angle C. Solve for the remaining parts of the triangle in each case.* ***Do not use trig tables.***

1. $B = 60°$, $a = 3$
2. $A = 30°$, $b = 3$
3. $A = 45°$, $c = 6$
4. $B = 60°$, $c = 1$
5. $b = 4$, $c = 8$
6. $a = 2$, $b = 2$

In Problems **7–16**, *solve the right triangle ABC for the remaining parts. Calculate the sides to one decimal place.*

7. $c = 4$, $A = 27°$
8. $a = 10$, $A = 14°$
9. $a = 8$, $b = 6$
10. $a = 3$, $c = 9$
11. $b = 5$, $B = 58°$
12. $a = 20$, $b = 10$
13. $b = 7$, $c = 10$
14. $c = 10$, $B = 40°$
15. $b = 100$, $A = 15°$
16. $a = 4$, $B = 69°$
17. The length of a kite string is 600 ft and the angle of elevation of the kite is 34°. How high is the kite? Give answer to the nearest foot.
18. A wire bracing an antenna is 50 ft long. One end is attached to the top of the antenna, and the other end is attached to the ground at a distance of 40 ft from the base of the antenna. Find the angle that the wire makes with the ground. Give answer to nearest degree.
19. From the top of a 300-ft tower, the angle of depression of a man on the ground is 40°. How far from the base of the tower is the man? Give answer to the nearest foot.
20. From the top of a 50-meter cliff that overlooks a bay, the angle of depression of a buoy is 14°. To the nearest meter, what is the distance along the water of the buoy from the cliff?
21. A bridge will be constructed at a certain point along a river. To gather engineering data, a surveyor is sent out to find the width w of the river (see Fig. 17-48). With his

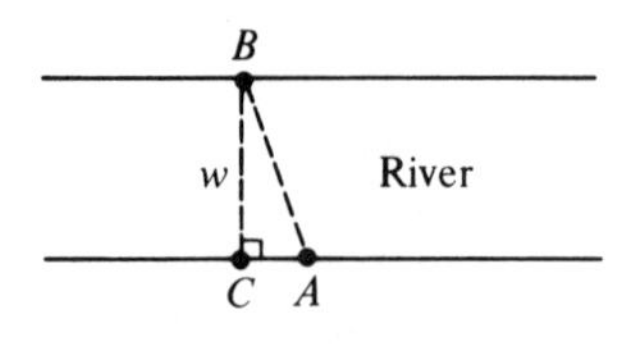

FIG. 17-48

transit at C, the surveyor determines the right angle C. He measures the distance from point C to point A and finds it to be 100 ft. Then, with his transit at point A, he determines that angle A is 69° 40′. What is the width of the river to the nearest foot?

22. A tree casts a shadow 50 ft long when the angle of elevation of the sun is 50°. Find the height of the tree to the nearest foot.

23. An airplane pilot wants to increase his altitude by 3 mi. He plans to climb at a constant angle of 10° and a constant rate of 200 mph. To the nearest minute, how long will it take to reach that altitude?

24. A flagpole is on top of a building [see Fig. 17-49(a)]. From a point 80 ft away from the base of the building, the angle of elevation of the top of the building is 37°, and the angle of elevation of the top of the flagpole is 45°. Find the length of the flagpole to the nearest foot.

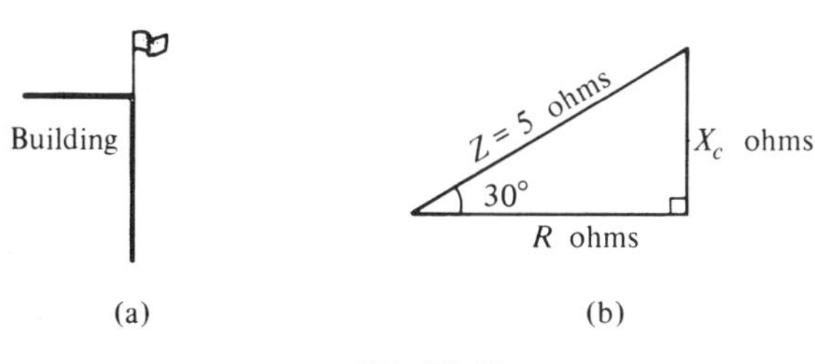

FIG. 17-49

25. In a certain electric circuit, quantities called impedance (Z), reactance (X_c), and resistance (R) can be represented as the sides of a right triangle shown in Fig. 17-49(b). If $Z = 5$ (ohms), find X_c and R.

26. After passing a toll gate, a car travels on a road that rises until it reaches a bridge. If the angle of elevation of the road is 3° 30′ and the length of the road is 170 meters, how high above ground level is the bridge?

17-6 CIRCULAR FUNCTIONS

As you well know, the trig functions have domains that consist of angles. However, in this section you'll see that we can think of these functions as having domains consisting of real numbers (rather than geometric "objects"). The first step is to show you how we can match real numbers with points on a circle.

We begin with a circle of radius 1 (called a **unit circle**) that has its center at the origin (see Fig. 17-50). Its equation is $x^2 + y^2 = 1$. A vertical number line is attached to the circle so that the number 0 on the line touches the circle at the point (1, 0).[†]

[†]The unit distance on the number line is the same as that of the axes.

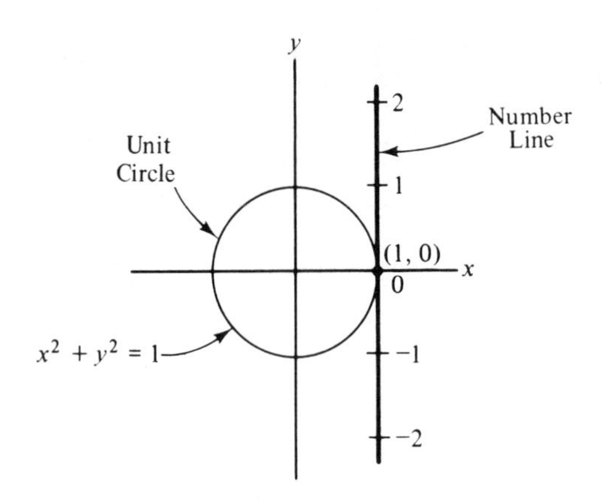

FIG. 17-50

Now, think of the number line as a piece of string. We can continuously wrap the positive part around the circle in a counterclockwise direction [Fig. 17-51(a)]. The negative part can be wrapped around it in a clockwise direction [Fig. 17-51(b)].

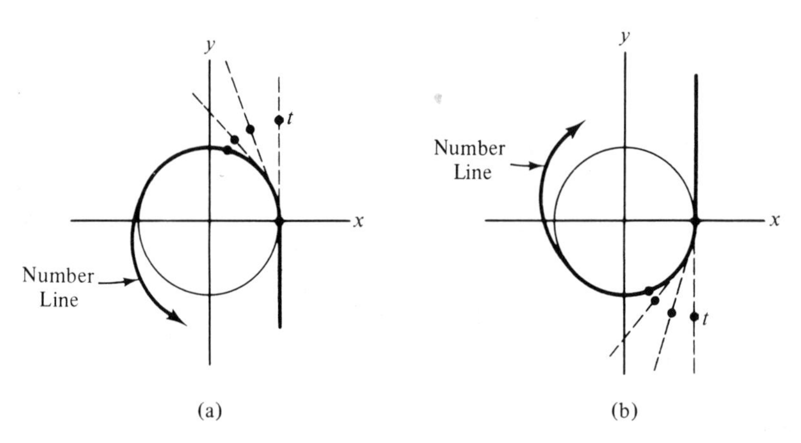

FIG. 17-51

Figure 17-51 shows clearly that each number t on the number line will touch the circle at exactly one point. In Fig. 17-51(a), notice that the distance along the circle from (1, 0) to the point that t touches is t.

Let's look at some examples. We know the real number 0 is wrapped into (1, 0). Since the circumference of the circle is $2\pi r = 2\pi(1) = 2\pi$, the number 2π also wraps into (1, 0). See Fig. 17-52(a). We write this as

$$2\pi \to (1, 0).$$

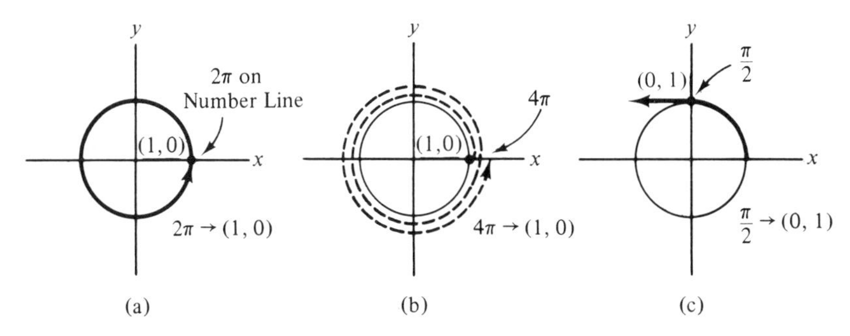

FIG. 17-52

Twice the circumference is $2(2\pi) = 4\pi$. Thus 4π wraps into (1, 0) also. See Fig. 17-52(b). One-quarter of the circumference is $\frac{1}{4}(2\pi) = \frac{\pi}{2}$, and so $\frac{\pi}{2}$ wraps into (0, 1). See Fig. 17-52(c).

Wrapping the negative part of the number line in a clockwise direction gives situations as in Fig. 17-53.

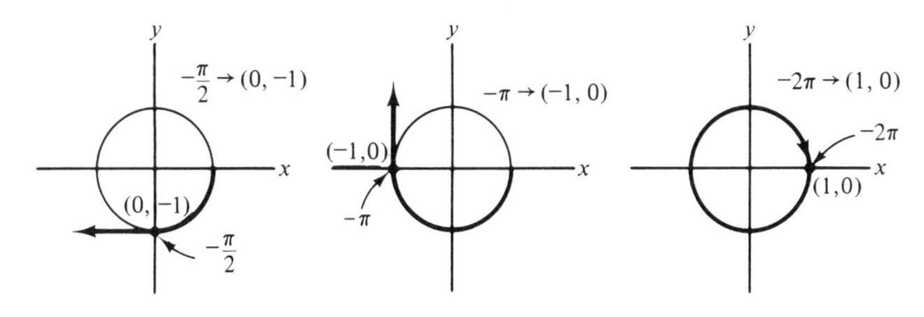

FIG. 17-53

With each number t we can also match an angle. Here's how. Suppose t wraps into the point (x, y) on the circle [see Fig. 17-54(a)]. Then the arc length

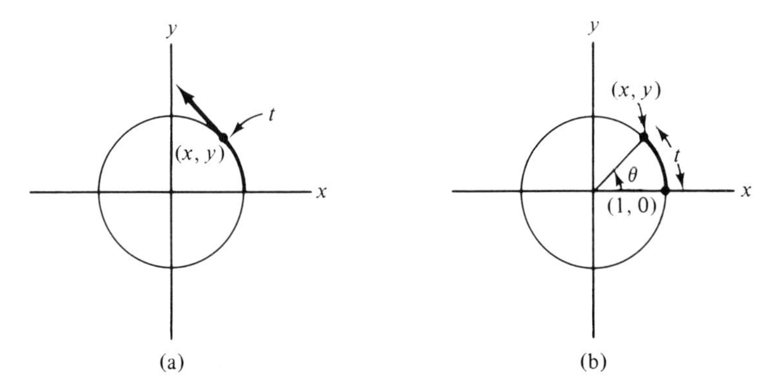

FIG. 17-54

from (1, 0) to (x, y) is t [see Fig. 17-54(b)]. This arc has central angle θ† and radius 1. But recall that the formula for arc length is

$$\text{arc length} = r\theta,$$

where θ is in radians. Thus,

$$t = (1)\theta$$
$$t = \theta$$
$$\text{or} \quad \theta = t \text{ rad}.$$

Therefore, with the number t we can match the angle of t rad in a natural way. See Fig. 17-55.

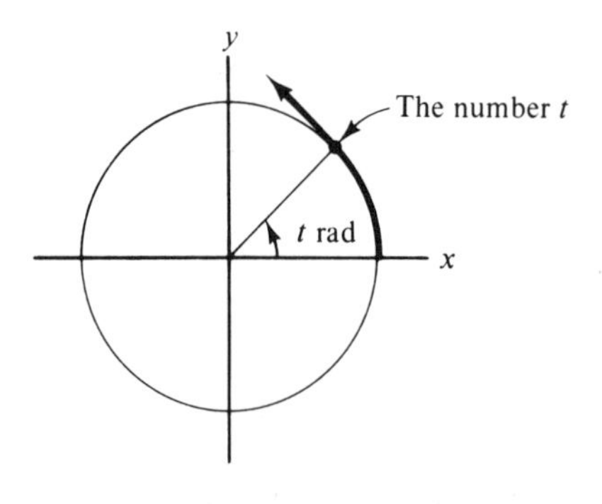

FIG. 17-55

Now we can define the trig functions of *numbers* in a logical way:

$$\begin{pmatrix}\text{sine of the} \\ \text{number } t\end{pmatrix} = \begin{pmatrix}\text{sine of the angle} \\ \text{of } t \text{ radians}\end{pmatrix}$$

$$\text{or} \quad \sin t = \sin(t \text{ rad}).$$

Similarly,

$$\cos t = \cos(t \text{ rad})$$
$$\tan t = \tan(t \text{ rad})$$
$$\cot t = \cot(t \text{ rad})$$
$$\sec t = \sec(t \text{ rad})$$
$$\csc t = \csc(t \text{ rad}).$$

---

†θ is positive if t is positive and θ is negative if t is negative.

Since we used a circle to develop these functions of numbers, we sometimes call them **circular functions**. Thus,

trig functions have domains consisting of *angles*,

and circular functions have domains consisting of *numbers*.

EXAMPLE 1

Find $\sin \frac{\pi}{4}$.

$$\sin \frac{\pi}{4} = \sin\left(\frac{\pi}{4} \text{ rad}\right) = \sin 45° = \frac{\sqrt{2}}{2}.$$

In previous sections, when we wrote $\sin \frac{\pi}{4}$ we meant $\sin\left(\frac{\pi}{4} \text{ rad}\right)$. You now may be asking yourself, "Well, if I see '$\sin \frac{\pi}{4}$,' is it the value of a trig function or of a circular function?" The answer is that it really doesn't matter. In either case the value of $\sin \frac{\pi}{4}$ is $\frac{\sqrt{2}}{2}$. In short, $\sin t$ can be thought of as the sine of the number t, or the sine of an angle of t rad. Similar comments apply to cosine, tangent, etc.

7-7 REVIEW

MPORTANT TERMS AND SYMBOLS

initial side *(p. 390)*
terminal side *(p. 390)*
vertex *(p. 390)*
positive angle *(p. 391)*
negative angle *(p. 391)*
standard position *(p. 391)*
quadrant of an angle *(p. 391)*
quadrantal angle *(p. 391)*
degree *(p. 392)*
minute *(p. 393)*
second *(p. 393)*
sector *(p. 397)*
coterminal angles *(p. 398)*
trigonometric functions *(p. 403)*
$\sin \theta$, $\cos \theta$ *(p. 403)*
$\tan \theta$, $\cot \theta$ *(p. 403)*
$\sec \theta$, $\csc \theta$ *(p. 403)*
reciprocal functions *(p. 407)*
opposite side *(p. 413)*
adjacent side *(p. 413)*
complementary angles *(p. 414)*
cofunctions *(p. 414)*

acute angle *(p. 392)*
obtuse angle *(p. 392)*
right angle *(p. 392)*
radian *(p. 394)*
arc length *(p. 396)*
special angles *(p. 414)*
reference angle *(p. 417)*
angle of elevation *(p. 432)*
angle of depression *(p. 432)*
circular functions *(p. 439)*

REVIEW PROBLEMS

In Problems **1–8**, *convert each degree measurement to radians, and each radian measurement to degrees.*

1. $300°$ **2.** $750°$ **3.** $\frac{5\pi}{6}$ **4.** $-\frac{9\pi}{4}$

5. $-50°$ **6.** $36°$ **7.** $\frac{\pi}{18}$ **8.** $\frac{3\pi}{10}$

In Problems **9–12**, *find the quadrants of the angles.*

9. $224°$ **10.** $-370°$ **11.** $\frac{7\pi}{3}$ **12.** $\frac{9\pi}{10}$

In Problems **13** *and* **14**, *find* $\alpha + \beta$ *and* $\alpha - \beta$.

13. $\alpha = 83°\ 20'\ 40''$, $\beta = 10°\ 40'\ 50''$

14. $\alpha = 150°\ 15'\ 20''$, $\beta = 42°\ 15'\ 45''$

In Problems **15–18**, *find the arc length cut off by the given central angle* θ *on a circle with the given radius* r.

15. $\theta = \frac{7\pi}{6}$, $r = 12$ **16.** $\theta = \frac{5\pi}{4}$, $r = 10$

17. $\theta = 135°$, $r = 2$ **18.** $\theta = 60°$, $r = 3$

In Problems **19** *and* **20**, *find the area of the sector of the circle with the given central angle* θ *and radius* r. *Assume answers are in square units.*

19. $\theta = 120°$, $r = 9$ **20.** $\theta = 5°$, $r = 2$

21. Find an angle θ that is coterminal with $500°$ and such that $0° \leqslant \theta < 360°$.

22. Find an angle θ that is coterminal with $-\frac{5\pi}{4}$ and such that $0 \leqslant \theta < 2\pi$.

In Problems **23–26**, *the given point lies on the terminal side of an angle* θ *in standard position. Find the trig values of* θ.

23. $(1, -6)$ **24.** $(-2, 0)$ **25.** $(-2, -3\sqrt{5})$ **26.** $(\sqrt{11}, -5)$

27. If $\sin\theta = \frac{1}{5}$ and $\cos\theta = -\frac{2\sqrt{6}}{5}$, find the other trig values of θ.

28. If $\tan\theta = -\dfrac{3}{5}$ and $\sec\theta = -\dfrac{\sqrt{34}}{5}$, find the other trig values of θ.

29. If $\cos\theta = \frac{3}{7}$ and $\tan\theta$ is negative, find the other trig values of θ.

30. If $\tan\theta = -\dfrac{\sqrt{51}}{7}$ and $\sin\theta$ is positive, find the other trig values of θ.

In Problems **31–48**, *find the given trig values. Do not use tables.*

31. $\sin 60°$
32. $\cos 30°$
33. $\tan 0°$
34. $\sec \dfrac{\pi}{4}$
35. $\cot \dfrac{\pi}{6}$
36. $\csc \dfrac{\pi}{2}$
37. $\cos \dfrac{2\pi}{3}$
38. $\tan \dfrac{5\pi}{6}$
39. $\sec \pi$
40. $\sin(-120°)$
41. $\csc 135°$
42. $\cot 270°$
43. $\tan 210°$
44. $\sec 405°$
45. $\sin(-180°)$
46. $\cos \dfrac{11\pi}{6}$
47. $\csc \dfrac{4\pi}{3}$
48. $\cot\left(-\dfrac{3\pi}{2}\right)$

In Problems **49–56**, *use trig tables to find the given trig value. Use linear interpolation where needed.*

49. $\cos 155°\ 40'$
50. $\sec 372°\ 30'$
51. $\tan(-100°\ 10')$
52. $\csc(-208°\ 50')$
53. $\sin 58°\ 14'$
54. $\tan 74°\ 48'$
55. $\csc 215°\ 33'$
56. $\cot 313°\ 26'$

In Problems **57–62**, *use trig tables and linear interpolation to find all angles θ that meet the given conditions and such that $0° \leqslant \theta < 360°$.*

57. $\sin\theta = .5503$
58. $\cos\theta = .4100$
59. $\cot\theta = -.8065$
60. $\sec\theta = -1.698$
61. $\cos\theta = -.9235$ and $\tan\theta$ is positive
62. $\tan\theta = -7.450$ and $\sin\theta$ is negative

In Problems **63–68**, *solve the right triangle ABC for the remaining parts. Give the sides to one decimal place.*

63. $b = 3,\ A = 42°$
64. $b = 2,\ B = 25°$
65. $c = 10,\ B = 76°$
66. $c = 4,\ A = 50°$
67. $a = 2,\ c = 5$
68. $a = 1,\ b = 4$

69. A wrecking company wants to make a bid for demolishing a smokestack. To do this they need to know its height. At a point 100 ft from the base of the stack, they find that the angle of elevation of its top is 32°. To the nearest foot, what is the height of the smokestack?

CHAPTER 18

Graphs of Trigonometric Functions

18-1 GRAPHS OF TRIGONOMETRIC FUNCTIONS

At times it is convenient to consider the behavior of the six trigonometric functions by analyzing their graphs. In this way we can discover certain features of these functions.

GRAPH OF SINE FUNCTION

To graph the sine function $y = \sin x$, we'll first find some points that lie on the sine curve. Table 18-1 lists some values of x between 0 and 2π radians along with the corresponding y-values. The points (x, y) are plotted in Fig. 18-1 and then connected by a smooth curve.

Since the angles x, $x + 2\pi$, and $x - 2\pi$ are coterminal angles, then we have $\sin x = \sin(x + 2\pi) = \sin(x - 2\pi)$. Thus the graph of $y = \sin x$ repeats itself every 2π radians. This is indicated by the broken curve in Fig. 18-1. To

TABLE 18-1

x (rad)	$y = \sin x$	x (rad)	$y = \sin x$
0	0	$\frac{7\pi}{6}$	$-\frac{1}{2} = -.5$
$\frac{\pi}{6}$	$\frac{1}{2} = .5$	$\frac{4\pi}{3}$	$-\frac{\sqrt{3}}{2} \approx -.87$
$\frac{\pi}{3}$	$\frac{\sqrt{3}}{2} \approx .87$	$\frac{3\pi}{2}$	-1
$\frac{\pi}{2}$	1	$\frac{5\pi}{3}$	$-\frac{\sqrt{3}}{2} \approx -.87$
$\frac{2\pi}{3}$	$\frac{\sqrt{3}}{2} \approx .87$	$\frac{11\pi}{6}$	$-\frac{1}{2} = -.5$
$\frac{5\pi}{6}$	$\frac{1}{2} = .5$	2π	0
π	0		

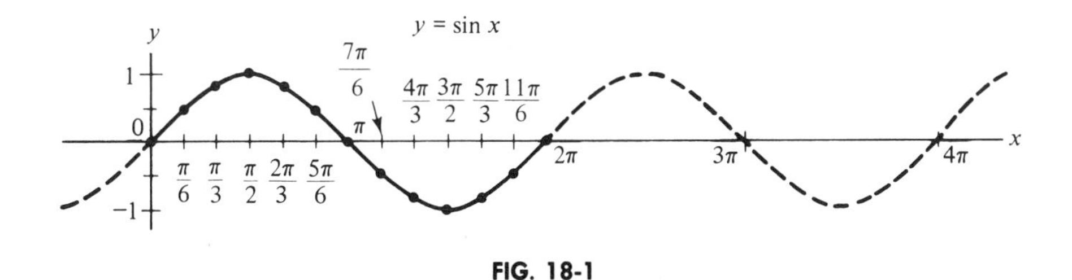

FIG. 18-1

describe this repetition we say that the sine function is **periodic** with **period** 2π.

$$y = \sin x \text{ has period } 2\pi.$$

The graph of $y = \sin x$ over one period is called a **cycle** of the sine curve.

Notice in Fig. 18-1 that the cycle from $x = 0$ to $x = 2\pi$ has four basic parts:

1. From $x = 0$ to $x = \frac{\pi}{2}$, the curve rises from $y = 0$ to $y = 1$.
2. From $x = \frac{\pi}{2}$ to $x = \pi$, it falls from $y = 1$ to $y = 0$.
3. From $x = \pi$ to $x = \frac{3\pi}{2}$, it continues to fall from $y = 0$ to $y = -1$.
4. From $x = \frac{3\pi}{2}$ to $x = 2\pi$, it rises from $y = -1$ to $y = 0$.

The maximum value of $\sin x$ is 1 and its minimum value is -1. For any periodic function, the expression

$$\frac{1}{2}\,(\text{maximum value} - \text{minimum value})$$

is called the **amplitude** of the function. Thus the amplitude of $y = \sin x$ is $\frac{1}{2}[1 - (-1)] = \frac{1}{2}[2] = 1$.

Amplitude of $y = \sin x$ is 1.

GRAPH OF COSINE FUNCTION

Figure 18-2 shows the graph of $y = \cos x$. It too has period 2π and amplitude 1.

$y = \cos x$ has period 2π and amplitude 1.

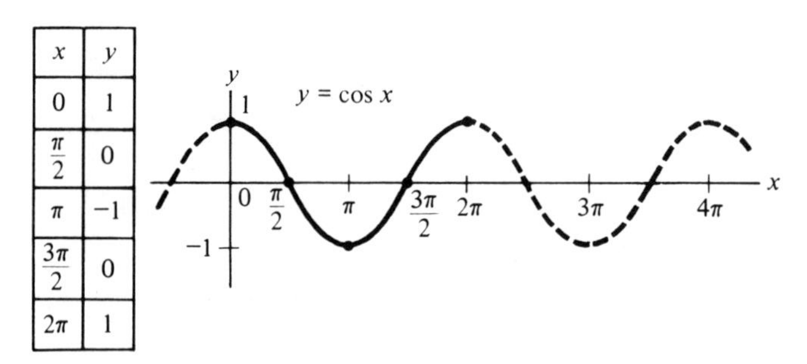

x	y
0	1
$\frac{\pi}{2}$	0
π	-1
$\frac{3\pi}{2}$	0
2π	1

FIG. 18-2

For the cycle from $x = 0$ to $x = 2\pi$, there are four basic parts:

1. From $x = 0$ to $x = \frac{\pi}{2}$, the curve falls from $y = 1$ to $y = 0$.
2. From $x = \frac{\pi}{2}$ to $x = \pi$, it continues to fall from $y = 0$ to $y = -1$.
3. From $x = \pi$ to $x = \frac{3\pi}{2}$, it rises from $y = -1$ to $y = 0$.
4. From $x = \frac{3\pi}{2}$ to $x = 2\pi$, it continues to rise from $y = 0$ to $y = 1$.

GRAPH OF TANGENT FUNCTION

In Table 18-2 are x- and y-values for $y = \tan x$ (some y-values are approximate). The dashes when $x = \pm \frac{\pi}{2}$ and $x = \pm \frac{3\pi}{2}$ mean that $\tan x$ is not defined there.

TABLE 18-2

x	0	$\frac{\pi}{6}$	$\frac{\pi}{3}$	$\frac{\pi}{2}$	$\frac{2\pi}{3}$	$\frac{5\pi}{6}$	π	$\frac{7\pi}{6}$	$\frac{4\pi}{3}$	$\frac{3\pi}{2}$
$y = \tan x$	0	.6	1.7	—	−1.7	−.6	0	.6	1.7	—

x	$-\frac{\pi}{6}$	$-\frac{\pi}{3}$	$-\frac{\pi}{2}$	$-\frac{2\pi}{3}$	$-\frac{5\pi}{6}$	$-\pi$	$-\frac{7\pi}{6}$	$-\frac{4\pi}{3}$	$-\frac{3\pi}{2}$
$y = \tan x$	−.6	−1.7	—	1.7	.6	0	−.6	−1.7	—

Figure 18-3 shows the graph of $y = \tan x$. As expected, there are no

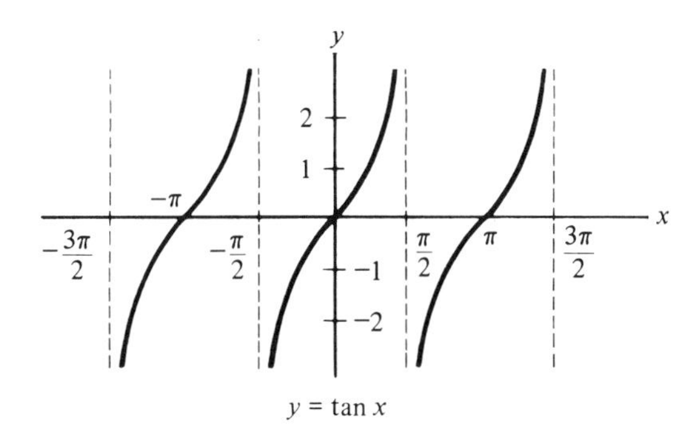

FIG. 18-3

points when $x = \pm \frac{\pi}{2}, \pm \frac{3\pi}{2}$. But for x near these values, $\tan x$ increases or decreases without bound. Notice that the graph gets very close to the lines $x = \pm \frac{\pi}{2}, \pm \frac{3\pi}{2}$, but does not touch them. These lines are called **vertical asymptotes** of the graph. However, they are *not* part of the graph.

From Fig. 18-3 we see that the portion of the graph between $x = -\frac{\pi}{2}$ and $x = \frac{\pi}{2}$ repeats itself. Thus,

> $y = \tan x$ has period π.

Only three cycles of $y = \tan x$ are shown in Fig. 18-3. But it should be clear that the graph continues both to the left and right.

Since $\tan x$ has no maximum or minimum values, we don't assign an amplitude to it.

GRAPHS OF COTANGENT, SECANT, AND COSECANT FUNCTIONS

To graph $y = \csc x$, we first sketch the sine function (see broken curve in Fig. 18-4). Notice that when $x = \frac{\pi}{6}$, then $\sin x = \frac{1}{2}$. Since the sine and cosecant are reciprocal functions, then $\csc \frac{\pi}{6} = 2$. Thus, $\left(\frac{\pi}{6}, 2\right)$ lies on the graph

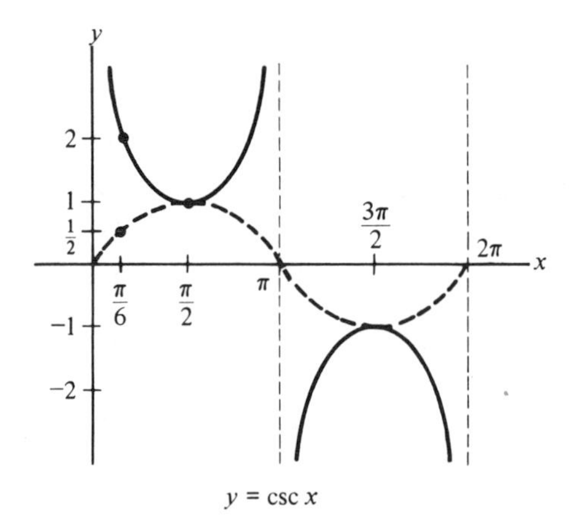

FIG. 18-4

of $y = \csc x$. Similarly, since $\sin \frac{\pi}{2} = 1$, then $\csc \frac{\pi}{2} = 1$. This gives us th point $\left(\frac{\pi}{2}, 1\right)$. In this way we can get the entire graph of $y = \csc x$. The soli curve in Fig. 18-4 shows one cycle. Thus,

$y = \csc x$ has period 2π.

For $x = 0, \pi, 2\pi$, etc., we see that $\sin x = 0$. Thus $\csc x$ is not defined fo these values. The lines $x = 0, \pi, 2\pi$, etc. are vertical asymptotes. No amplitud is assigned to $y = \csc x$.

In a similar way we can graph the cotangent and secant with the aid of their reciprocal functions—tangent and cosine.

$$\cot x = \frac{1}{\tan x}, \qquad \sec x = \frac{1}{\cos x}.$$

Figures 18-5 and 18-6 show the graphs of the tangent and cosine functions (broken curves). Using them, we drew the graphs of $y = \cot x$ and $y = \sec x$ (solid curves). Notice that

> $y = \cot x$ has period π,
> and $y = \sec x$ has period 2π.

No amplitudes are assigned to these functions.

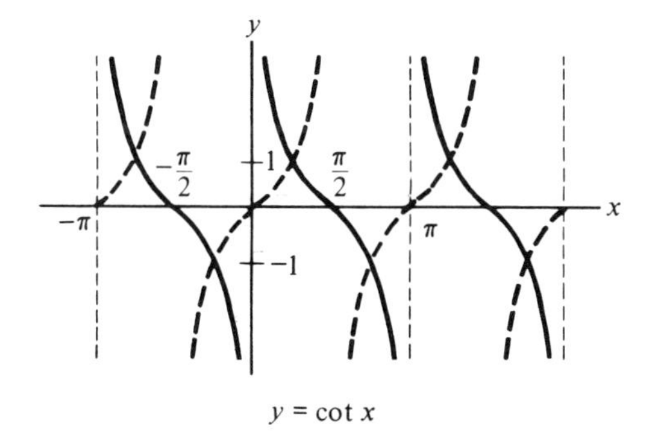

FIG. 18-5

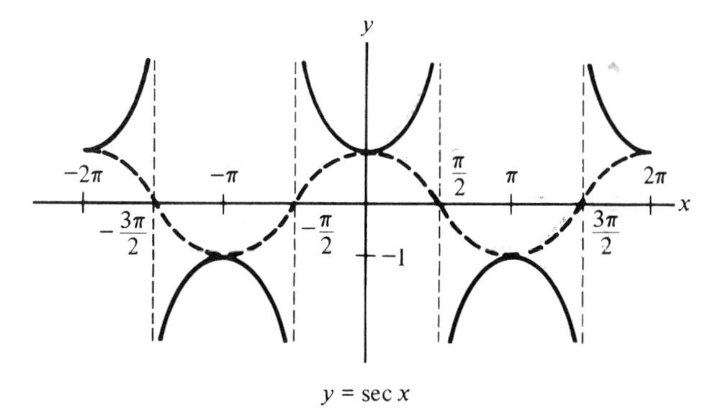

FIG. 18-6

Completion Set 18-1

In Problems **1–6**, *fill in the period for each function.*

1. $y = \sin x$, period is ______.

2. $y = \tan x$, period is ______.

3. $y = \csc x$, period is ______.

4. $y = \cos x$, period is ______.

5. $y = \cot x$, period is ______.

6. $y = \sec x$, period is ______.

In Problems **7** *and* **8**, *fill in the blanks.*

7. The amplitude of $y = \sin x$ is ____.

8. The amplitude of $y = \cos x$ is ____.

Problem Set 18-1

1. Draw the basic shapes of the six trig functions, $y = \sin x$, $y = \cos x$, etc., for x between 0 and 2π.

2. Draw two cycles of $y = \sin x$ beginning at $x = 0$. Use this graph to find all angles x between 0 and 4π for which $\sin x = 1$.

3. Draw two cycles of $y = \cos x$ beginning at $x = -2\pi$. Use this graph to find all angles x between -2π and 2π for which $\cos x = 0$.

4. Draw the graph of $y = \cot x$. Use this graph to find all angles x between 0 and 2π for which $\cot x = 0$.

5. Draw the graph of $y = \csc x$. Use this graph to find all angles x between 0 and 4π for which $\csc x = -1$.

6. Draw the graph of $y = \sec x$. Use this graph to find all angles x between $-\frac{5\pi}{2}$ and $\frac{5\pi}{2}$ for which $\sec x = 1$.

7. Draw the graph of $y = \tan x$. Use this graph to find all angles x between $-\frac{\pi}{2}$ and $\frac{5\pi}{2}$ for which $\tan x = 0$.

18-2 VARIATIONS OF THE SINE AND COSINE

In this section you will learn how to draw the graphs of variations of the sine and cosine functions. The simplest ones have the form $y = a \sin x$ or $y = a \cos x$, where a is a constant.

EXAMPLE 1

Draw the graph of $y = 2 \sin x$.

The values of $\sin x$ vary between -1 and 1 (that is, $-1 \leqslant \sin x \leqslant 1$). Thus the values of two times $\sin x$ vary between -2 and 2. In fact, the y-value of each point on the curve $y = 2 \sin x$ is just two times the corresponding y-value on the curve $y = \sin x$. This means that the graphs of $y = 2 \sin x$ and $y = \sin x$ differ in a vertical sense, but not in a horizontal sense. The amplitude of $y = 2 \sin x$ is 2, but its period is still 2π. To draw the graph, we draw a basic sine curve and label the axes to indicate the amplitude 2 and period 2π. See Fig. 18-7.

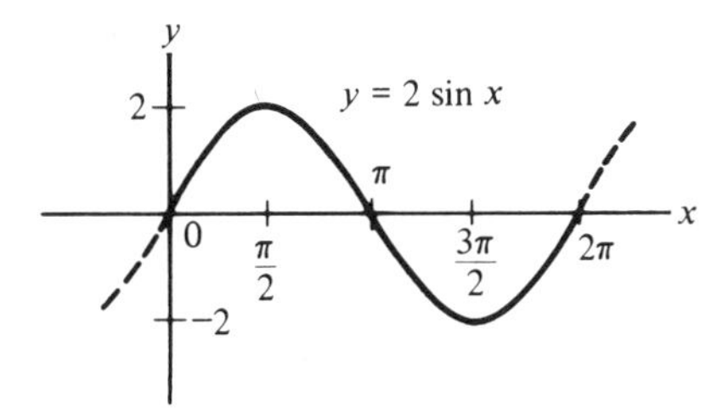

FIG. 18-7

EXAMPLE 2

Draw the graph of $y = -2 \sin x$.

The values of $y = -2 \sin x$ are related to those of $y = 2 \sin x$ in Example 1. Whenever $2 \sin x$ is positive, then $-2 \sin x$ is negative; whenever $2 \sin x$ is negative, then $-2 \sin x$ is positive. The effect is to *invert* the curve in Fig. 18-7. Thus we get the graph of $y = -2 \sin x$ which is given in Fig. 18-8. It shows that the amplitude of $y = -2 \sin x$ is 2 (which is $|-2|$) and the period is 2π. We say that the graph of $y = -2 \sin x$ is the *reflection* in the x-axis of the graph of $y = 2 \sin x$.

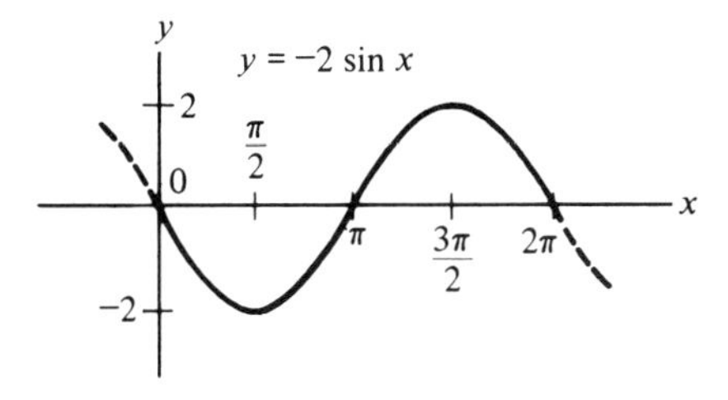

FIG. 18-8

From Examples 1 and 2, we can say that

the amplitude of $y = a \sin x$ is $|a|$ and its period is 2π. The basic shape is that of a sine curve ($a > 0$) or inverted sine curve ($a < 0$). These facts are also true if "sine" is replaced by "cosine."

EXAMPLE 3

Draw the graph of $y = 4 \cos x$.

Here $a = 4$ (> 0). Thus the graph is a cosine curve with amplitude 4 and period 2π (see Fig. 18-9).

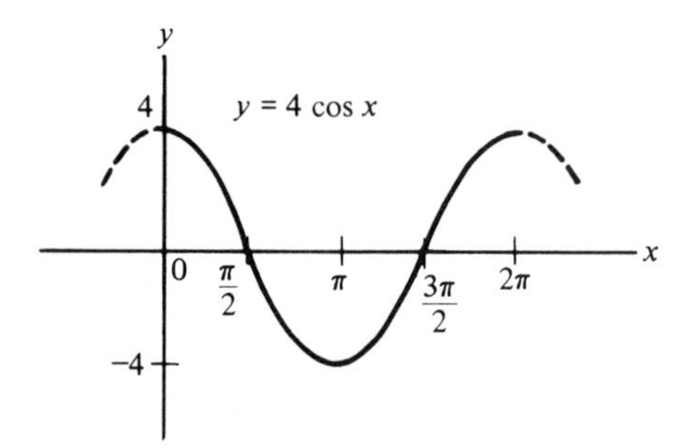

FIG. 18-9

A more general variation of the sine curve is

$$y = a \sin(bx + c) \qquad a, b, c \text{ constants}, b > 0.$$

As before, the amplitude is $|a|$. The basic shape is still a sine curve if $a > 0$, or an inverted sine curve if $a < 0$. But the graph may be affected in a horizontal way, depending on the angle $bx + c$.

To see why, let's graph $y = 4 \sin\left(2x + \frac{\pi}{2}\right)$. The amplitude is 4, and the basic shape is a sine curve. Now, remember that we get a cycle of the curve $y = \sin \theta$ whenever θ ranges over an interval of 2π, say from 0 to 2π. Thus, we'll get a cycle of $y = 4 \sin\left(2x + \frac{\pi}{2}\right)$ whenever the angle $2x + \frac{\pi}{2}$ ranges over an interval of 2π. Let's have $2x + \frac{\pi}{2}$ range from 0 to 2π. Thus a cycle begins when

$$2x + \frac{\pi}{2} = 0$$

$$2x = -\frac{\pi}{2}$$

$$x = -\frac{\pi}{4}.$$

It ends when

$$2x + \frac{\pi}{2} = 2\pi$$

$$2x = 2\pi - \frac{\pi}{2} = \frac{3\pi}{2}$$

$$x = \frac{3\pi}{4}.$$

Now we can sketch a cycle of $y = 4 \sin\left(2x + \frac{\pi}{2}\right)$. First we draw a cycle of the basic sine curve beginning at $-\frac{\pi}{4}$ and ending at $\frac{3\pi}{4}$. See Fig. 18-10. Then

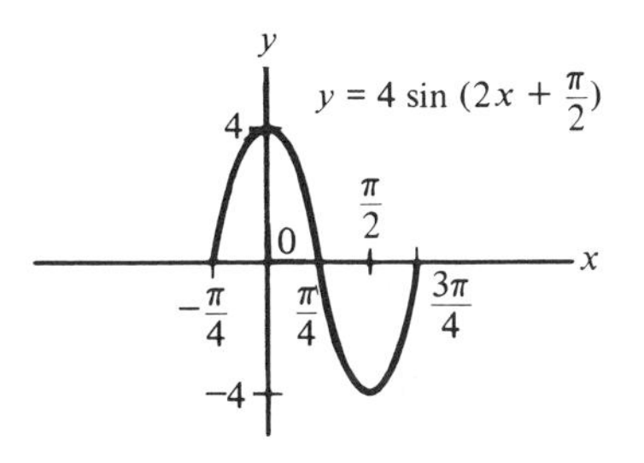

FIG. 18-10

we label the y-axis with a 4 and -4 so that the amplitude is clear. Of course, the curve repeats itself indefinitely to the right and left.

The period is easy to find. The length of the interval from $-\frac{\pi}{4}$ to $\frac{3\pi}{4}$ is

$$\frac{3\pi}{4} - \left(-\frac{\pi}{4}\right) = \pi.$$

Thus the period is π.

On the x-axis we also label the values of x where $4 \sin\left(2x + \frac{\pi}{2}\right)$ has a maximum value (4), or minimum value (-4), or is 0. We find these by dividing the period p by 4,

$$\frac{p}{4} = \frac{\pi}{4},$$

and successively adding the result to the beginning x-value, $-\frac{\pi}{4}$.

$$-\frac{\pi}{4} + \frac{\pi}{4} = 0; \quad 0 + \frac{\pi}{4} = \frac{\pi}{4}; \quad \frac{\pi}{4} + \frac{\pi}{4} = \frac{2\pi}{4} = \frac{\pi}{2}; \quad \frac{\pi}{2} + \frac{\pi}{4} = \frac{3\pi}{4}.$$

These are shown in the diagram.

Our entire discussion above on amplitude, reflection, and period applies equally well to the curve $y = a\cos(bx + c)$.

To draw the graphs of $y = a\sin(bx + c)$ or $y = a\cos(bx + c)$, follow these steps.

a. *Find the amplitude. It is $|a|$.*

b. *Find the basic shape of the curve, that is, regular $(a > 0)$ or inverted $(a < 0)$.*

c. *Solve $bx + c = 0$ to find a value of x when a cycle begins. Call this value x_{begin}.*

d. *Solve $bx + c = 2\pi$ to find the value of x when the cycle in step c ends. Call this x_{end}.*

e. *Find the period p from the results of steps c and d:*

$$p = x_{end} - x_{begin}.$$

f. *Find the x-values of "important" points. By "important" we mean those points where the function has a maximum, minimum, or 0-value. One of these is x_{begin}. To find the others, divide p by 4 and successively add this result to x_{begin}. Also label the y-axis to show the maximum and minimum values.*

EXAMPLE 4

Draw the graph of $y = -\cos\left(\pi x - \frac{\pi}{4}\right)$.

a. The amplitude is $|a| = |-1| = 1$.

b. Since $a < 0$, the basic shape is an *inverted* cosine curve.

c. If $\pi x - \frac{\pi}{4} = 0$, then $\pi x = \frac{\pi}{4}$ and so $x = \frac{1}{4}$. Thus a cycle begins when $x_{\text{begin}} = \frac{1}{4}$.

d. If $\pi x - \frac{\pi}{4} = 2\pi$, then $\pi x = \frac{9\pi}{4}$, and so $x = \frac{9}{4}$. The cycle above ends when $x_{\text{end}} = \frac{9}{4}$.

e. The period p is

$$p = x_{\text{end}} - x_{\text{begin}} = \frac{9}{4} - \frac{1}{4} = 2.$$

f. $\frac{p}{4}$ is $\frac{2}{4}$.

Important points: when $x = \frac{1}{4}, \frac{3}{4}, \frac{5}{4}, \frac{7}{4}$, and $\frac{9}{4}$.

One cycle of the graph is drawn in Fig. 18-11.

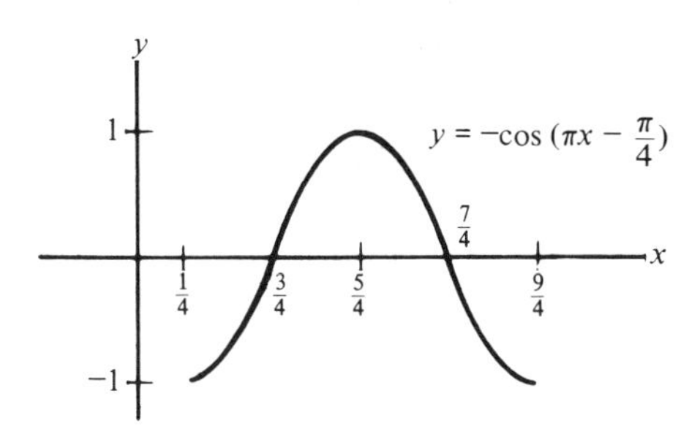

FIG. 18-11

EXAMPLE 5

Draw the graph of $y = -2\sin(\frac{1}{2}x)$.

a. The amplitude is $|a| = |-2| = 2$.

b. Since $a < 0$, the basic shape is an *inverted* sine curve.

c. If $\frac{1}{2}x = 0$, then $x = 0$. A cycle begins when $x_{\text{begin}} = 0$.

d. If $\frac{1}{2}x = 2\pi$, then $x = 4\pi$. The cycle above ends when $x_{\text{end}} = 4\pi$.

e. The period p is

$$p = x_{\text{end}} - x_{\text{begin}} = 4\pi - 0 = 4\pi.$$

f. Important points: when $x = 0, \pi, 2\pi, 3\pi$, and 4π.

One cycle of the graph is drawn in Fig. 18-12.

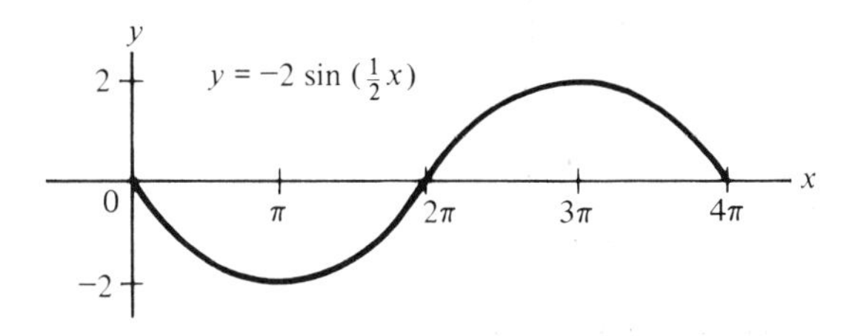

FIG. 18-12

EXAMPLE 6

Draw the graph of $y = 3 \cos 6x$.

a. The amplitude is $|a| = |3| = 3$.

b. Since $a > 0$, the basic shape is a cosine curve.

c. If $6x = 0$, then $x = 0$. A cycle begins when $x_{\text{begin}} = 0$.

d. If $6x = 2\pi$, then $x = \frac{\pi}{3}$. The cycle above ends when $x_{\text{end}} = \frac{\pi}{3}$.

e. The period p is

$$p = x_{\text{end}} - x_{\text{begin}} = \frac{\pi}{3} - 0 = \frac{\pi}{3}.$$

f. Important points: when $x = 0, \frac{\pi}{12}, \frac{\pi}{6}, \frac{\pi}{4}$, and $\frac{\pi}{3}$.

One cycle of the graph is drawn in Fig. 18-13.

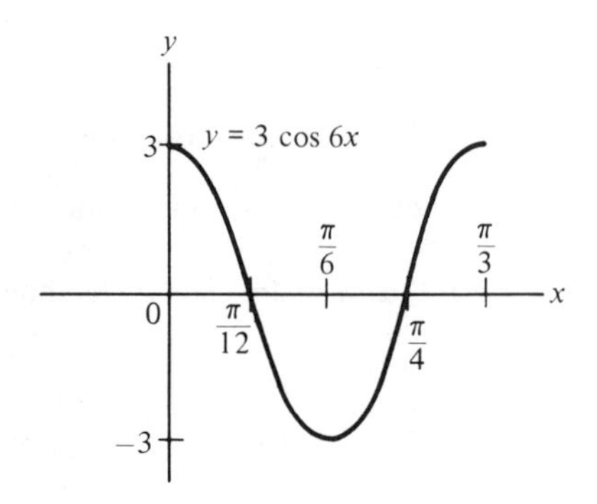

FIG. 18-13

Completion Set 18-2

Fill in the blanks.

1. The maximum value of $y = 3 \sin\left(\frac{6}{97}x - \frac{2\pi}{63}\right)$ is ____.

2. The minimum value of $y = 4 \cos (\pi x - 3)$ is ______.

3. The curve $y = 2 \sin (3x)$ has amplitude ____. A cycle begins when $3x =$ ____. Thus, $x_{\text{begin}} =$ ______. This cycle ends when $3x =$ ______. Thus, $x_{\text{end}} =$ ______. The period is ______ − ____ = ______.

Problem Set 18-2

In Problems **1–24**, *draw one cycle of each curve. Give the amplitude A and the period p.*

1. $y = 3 \sin x$

2. $y = \frac{1}{2} \sin x$

3. $y = -4 \sin x$

4. $y = -\sin x$

5. $y = \frac{1}{2} \cos x$

6. $y = 5 \cos x$

7. $y = -\cos x$

8. $y = -\frac{2}{3} \cos x$

9. $y = \sin 2x$

10. $y = \sin \dfrac{x}{2}$

11. $y = -4 \cos 3x$

12. $y = \frac{1}{2} \cos 5x$

13. $y = 2 \sin \frac{1}{3} x$

14. $y = -2 \sin 4x$

15. $y = \sin \left(x + \dfrac{\pi}{4}\right)$

16. $y = -\dfrac{1}{4} \sin \left(x + \dfrac{\pi}{2}\right)$

17. $y = 4 \sin \left(2x - \dfrac{\pi}{2}\right)$

18. $y = 2 \sin \left(2x - \dfrac{2\pi}{3}\right)$

19. $y = 2 \cos \left(x + \dfrac{\pi}{3}\right)$

20. $y = -\cos \left(4x - \dfrac{\pi}{2}\right)$

21. $y = -2 \cos \left(\dfrac{x}{2} - \dfrac{2\pi}{3}\right)$

22. $y = -3 \cos \left(x - \dfrac{\pi}{6}\right)$

23. $y = \sin \left(\pi x - \dfrac{\pi}{2}\right)$

24. $y = 3 \sin \left(\dfrac{2}{3} \pi x - \dfrac{3\pi}{2}\right)$

25. Draw the graphs of $y = \sin x$ and $y = \sin 2x$ on the same coordinate plane. How many cycles of $y = \sin 2x$ are there for each cycle of $y = \sin x$?

26. Draw the graphs of $y = \cos x$ and $y = \cos \dfrac{x}{2}$ on the same coordinate plane. How many cycles of $y = \cos x$ are there for each cycle of $y = \cos \dfrac{x}{2}$?

27. Find the period of $y = 1.2 \sin \left(\dfrac{3\pi x}{2}\right)$.

28. Find the period of $y = 2\cos(3x + 1)$.

29. Draw the graph of $y = 2\sin\dfrac{x}{4}$ from $x = -2\pi$ to $x = 4\pi$.

30. Draw the graph of $y = \dfrac{1}{4}\cos 4x$ from $x = -\dfrac{\pi}{2}$ to $x = \pi$.

31. Find the period of $y = a\sin(bx + c)$.

32. Find the period of $y = a\cos(bx + c)$.

18-3 INVERSE TRIGONOMETRIC FUNCTIONS

For the function $w = f(t) = \sin t$, the inputs are t's and the outputs are w's. Let's reverse these roles. Think of the number w as an *input* and the angle t as an *output*. For example, let the input be $w = \frac{1}{2}$. Then the output is an angle t such that

$$\frac{1}{2} = \sin t.$$

There are many angles whose sine is $\frac{1}{2}$. Some are $-\dfrac{7\pi}{6}, \dfrac{\pi}{6}, \dfrac{5\pi}{6}$, and $\dfrac{13\pi}{6}$, as shown in Fig. 18-14.

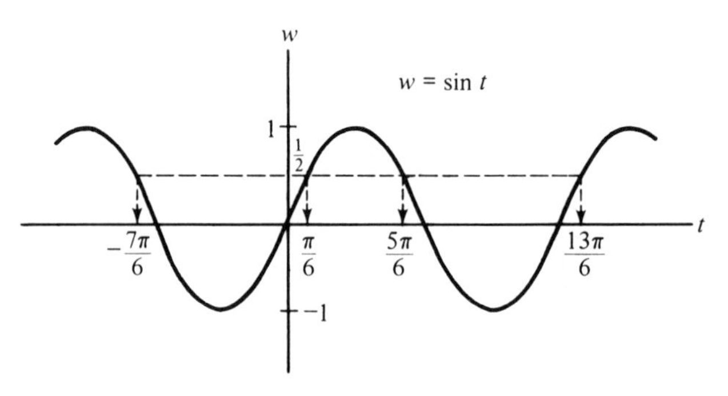

FIG. 18-14

Since there is more than one output for one input, the equation $w = \sin t$ does *not* define t as a function of w. However, if we restrict t so that

$$-\frac{\pi}{2} \leqslant t \leqslant \frac{\pi}{2},$$

then we *do* have a function of w (see Fig. 18-15). Thus, if $\frac{1}{2}$ is the input, then

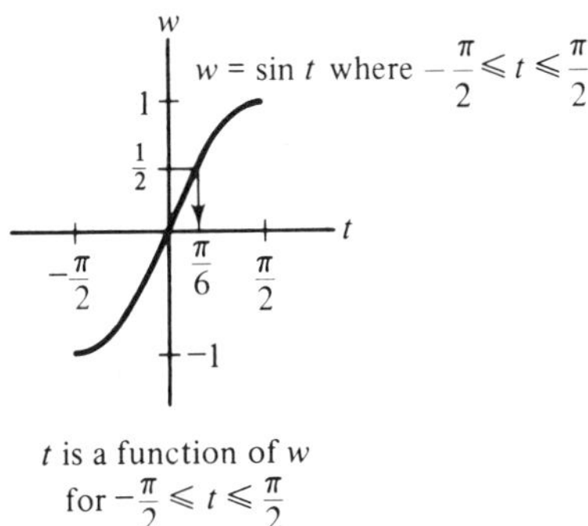

FIG. 18-15

the output is $\frac{\pi}{6}$ because

$$\sin \frac{\pi}{6} = \frac{1}{2}$$

$$\text{and} \qquad -\frac{\pi}{2} \leqslant \frac{\pi}{6} \leqslant \frac{\pi}{2}.$$

Similarly, if -1 is the input, then $-\frac{\pi}{2}$ is the output because

$$\sin\left(-\frac{\pi}{2}\right) = -1$$

$$\text{and} \qquad -\frac{\pi}{2} \leqslant -\frac{\pi}{2} \leqslant \frac{\pi}{2}.$$

The name of this function which reverses the action of the sine function and sends numbers into angles is the **arcsine function** or **inverse sine function**. It is written Arcsin or Sin^{-1}. Thus

$$\text{Arcsin } w = t.$$

In the discussion above, let's replace the input number w by x, and the output angle t by y. Then to say

$$y = \text{Arcsin } x$$

means that y is an angle whose sine is x (that is, $\sin y = x$) and such that

$$-\frac{\pi}{2} \leqslant y \leqslant \frac{\pi}{2}.$$

Note that the domain is all x such that

$$-1 \leqslant x \leqslant 1.$$

$y = \text{Arcsin}\, x$ if and only if $x = \sin y$,
where $-\frac{\pi}{2} \leqslant y \leqslant \frac{\pi}{2}$ and $-1 \leqslant x \leqslant 1$.

We can also define inverse trigonometric functions for the cosine and tangent.

$y = \text{Arccos}\, x$ if and only if $x = \cos y$,
where $0 \leqslant y \leqslant \pi$ and $-1 \leqslant x \leqslant 1$.

$y = \text{Arctan}\, x$ if and only if $x = \tan y$, where
$-\frac{\pi}{2} < y < \frac{\pi}{2}$ and x is any real number.

There are three other inverse trig functions: Arccot, Arcsec, and Arccsc. However, we shall not consider these since they are not used very often.

EXAMPLE 1

Find each of the following.

a. Arcsin $\frac{\sqrt{2}}{2}$.

Arcsin $\frac{\sqrt{2}}{2}$ is *the* angle between $-\frac{\pi}{2}$ and $\frac{\pi}{2}$ whose sine is $\frac{\sqrt{2}}{2}$. Clearly,
Arcsin $\frac{\sqrt{2}}{2} = \frac{\pi}{4}$.

b. Arccos $\frac{1}{2}$.

Arccos $\frac{1}{2}$ is *the* angle between 0 and π whose cosine is $\frac{1}{2}$. Thus, Arccos $\frac{1}{2} = \frac{\pi}{3}$.

c. Arccos $\left(-\frac{\sqrt{3}}{2}\right)$.

Arccos $\left(-\frac{\sqrt{3}}{2}\right)$ is *the* angle, call it y, between 0 and π such that

$\cos y = -\frac{\sqrt{3}}{2}$. Since $\cos y$ is negative, then y must be a second-quadrant angle. If its reference angle is y_{ref}, then $\cos y_{\text{ref}} = \frac{\sqrt{3}}{2}$. Clearly, y_{ref} must be $\frac{\pi}{6}$, and so $y = \pi - \frac{\pi}{6} = \frac{5\pi}{6}$. Thus, $y = \text{Arccos}\left(-\frac{\sqrt{3}}{2}\right) = \frac{5\pi}{6}$.

d. Arctan (-1).

Arctan (-1) is *the* angle between $-\frac{\pi}{2}$ and $\frac{\pi}{2}$ whose tangent is -1. Therefore, Arctan $(-1) = -\frac{\pi}{4}$.

EXAMPLE 2

Find each of the following.

a. $\sin\left(\text{Arcsin}\,\frac{1}{4}\right)$.

Arcsin $\frac{1}{4}$ is a certain angle whose sine is $\frac{1}{4}$. Thus, the sine of this angle is $\frac{1}{4}$, so

$$\sin\left(\text{Arcsin}\,\frac{1}{4}\right) = \frac{1}{4}.$$

In general, $\sin(\text{Arcsin}\,x) = x$.

b. $\cos(\text{Arcsin}\,0)$.

Since Arcsin $0 = 0$, then

$$\cos(\text{Arcsin}\,0) = \cos(0) = 1.$$

c. $\text{Arcsin}\left(\sin\frac{5\pi}{6}\right)$.

$$\text{Arcsin}\left(\sin\frac{5\pi}{6}\right) = \text{Arcsin}\left(\frac{1}{2}\right) = \frac{\pi}{6}.$$

Note that $\text{Arcsin}\left(\sin\frac{5\pi}{6}\right) \neq \frac{5\pi}{6}$.

Completion Set 18-3

Fill in the blanks.

1. Arcsin x is the angle between $-\frac{\pi}{2}$ and $\frac{\pi}{2}$ whose sine is ____.

2. Arccos x is the angle between ____ and ____ whose cosine is ____.

3. Arctan x is the angle between ______ and ____ whose ____________________ is ____.

4. The angles 0, $\pm 2\pi$, $\pm 4\pi$, etc. all have their cosine equal to 1. But Arccos 1 = ____.

Problem Set 18-3

Find the value of each of the following.

1. Arcsin 1
2. Arcsin $\frac{\sqrt{3}}{2}$
3. Arccos $\frac{\sqrt{3}}{2}$
4. Arccos (-1)
5. Arctan 0
6. Arcsin $\left(-\frac{\sqrt{2}}{2}\right)$
7. Arcsin $\left(-\frac{1}{2}\right)$
8. Arccos 0
9. Arctan 1
10. Arctan $(-\sqrt{3})$
11. Arccos $\left(-\frac{\sqrt{2}}{2}\right)$
12. Arccos $\left(-\frac{1}{2}\right)$
13. sin $\left(\text{Arcsin}\,\frac{1}{3}\right)$
14. cos $\left[\text{Arccos}\left(-\frac{1}{5}\right)\right]$
15. cos $\left(\text{Arcsin}\,\frac{1}{2}\right)$
16. sin [Arccos (-1)]
17. cot (Arccos 0)
18. tan (Arccos 1)
19. cos [Arctan (-1)]
20. sin $\left(\text{Arctan}\,\frac{\sqrt{3}}{3}\right)$
21. Arcsin $\left(\sin\frac{2\pi}{3}\right)$
22. Arcsin $\left(\sin\frac{\pi}{3}\right)$

18-4 REVIEW

IMPORTANT TERMS AND SYMBOLS

cycle *(p. 443)*

period *(p. 443)*

amplitude *(p. 444)*

y = Arcsin x *(p. 457)*

y = Arccos x *(p. 458)*

y = Arctan x *(p. 458)*

REVIEW PROBLEMS

In Problems **1–6**, *draw one cycle of each curve. Give the amplitude A and the period p.*

1. $y = -\sin 3x$

2. $y = 3\cos 6x$

3. $y = 4\cos\left(x - \frac{\pi}{2}\right)$

4. $y = 2\sin\left(3x + \frac{\pi}{2}\right)$

5. $y = 3\cos\left(2x + \frac{\pi}{6}\right)$

6. $y = -\cos\left(\frac{x}{2} - \frac{\pi}{3}\right)$

Find the value of each of the following.

7. Arcsin (-1)

8. Arccos $\frac{1}{2}$

9. Arctan $\frac{\sqrt{3}}{3}$

10. Arcsin $\left(-\frac{\sqrt{3}}{2}\right)$

11. Arccos $\left(-\frac{\sqrt{3}}{2}\right)$

12. Arctan (-1)

13. tan (Arccos 1)

14. cos $\left(\text{Arcsin } \frac{\sqrt{2}}{2}\right)$

15. sin $\left[\text{Arctan }(-\sqrt{3})\right]$

16. csc (Arccos 0)

17. sin (Arcsin .8)

18. Arctan (tan 2π)

19. Arccos (cos 3π)

20. cos $\left[\text{Arccos }\left(-\frac{1}{10}\right)\right]$

CHAPTER 19

Trigonometric Formulas and Identities

19-1 BASIC TRIGONOMETRIC IDENTITIES

In Chapter 17 we gave the following trigonometric relationships:

$$1.\ \csc\theta = \frac{1}{\sin\theta} \qquad \sin\theta = \frac{1}{\csc\theta}$$

$$2.\ \sec\theta = \frac{1}{\cos\theta} \qquad \cos\theta = \frac{1}{\sec\theta}$$

$$3.\ \cot\theta = \frac{1}{\tan\theta} \qquad \tan\theta = \frac{1}{\cot\theta}$$

$$4.\ \tan\theta = \frac{\sin\theta}{\cos\theta}$$

$$5.\ \cot\theta = \frac{\cos\theta}{\sin\theta}.$$

Each of Eqs. 1–5 is true for *all* values of θ for which both sides are defined They are called **trigonometric identities**. There are three more basic identities each involving a power of a trig function. In stating them, we shall abbreviat

$(\sin \theta)^2$ by writing $\sin^2 \theta$ (read "sine squared theta"), and similarly with the other trig functions. Do not confuse $\sin^2 \theta$ with $\sin \theta^2$, which means $\sin (\theta^2)$. The three identities in their various forms are:

$$\textbf{6. } \sin^2 \theta + \cos^2 \theta = 1 \quad \text{or} \quad \begin{cases} \sin^2 \theta = 1 - \cos^2 \theta \\ \cos^2 \theta = 1 - \sin^2 \theta \end{cases}$$

$$\textbf{7. } 1 + \tan^2 \theta = \sec^2 \theta \quad \text{or} \quad \begin{cases} \tan^2 \theta = \sec^2 \theta - 1 \\ \sec^2 \theta - \tan^2 \theta = 1 \end{cases}$$

$$\textbf{8. } 1 + \cot^2 \theta = \csc^2 \theta \quad \text{or} \quad \begin{cases} \cot^2 \theta = \csc^2 \theta - 1 \\ \csc^2 \theta - \cot^2 \theta = 1. \end{cases}$$

Identity 6 is simple to derive. If θ is any angle in standard position and (x, y) is a point on its terminal side, then

$$\sin^2 \theta + \cos^2 \theta = \left(\frac{y}{r}\right)^2 + \left(\frac{x}{r}\right)^2 = \frac{y^2}{r^2} + \frac{x^2}{r^2}$$

$$= \frac{y^2 + x^2}{r^2} = \frac{r^2}{r^2} = 1.$$

The proofs of 7 and 8 are similar.

EXAMPLE 1

a. $\sin^2 20° + \cos^2 20° = 1.$

b. $1 + \tan^2 \frac{\pi}{3} = \sec^2 \frac{\pi}{3}.$

c. $1 + \cot^2 4x = \csc^2 4x.$

Identities 1–8 are considered the basic trigonometric identities, and you should become totally familiar with them. They are used to simplify expressions involving trig functions and to prove other identities.

For example, let's prove the identity

$$\cot x \sin x = \cos x.$$

To do this, we pick one side and make substitutions until it is the same as the other side. Usually it is best to pick the more complicated side. In our case the more complicated side is $\cot x \sin x$. We can replace $\cot x$ by $\dfrac{\cos x}{\sin x}$ from identity 5. Thus,

$$\cot x \sin x \text{ becomes } \frac{\cos x}{\sin x} \cdot \sin x.$$

This reduces to

$$\cos x,$$

which is the same as the right side. Thus the identity is proven. We usually write our work in a vertical arrangement as follows:

Left Side	*Right Side*
$\cot x \sin x$	$\cos x.$
$= \dfrac{\cos x}{\sin x} \sin x$	
$= \cos x.$	

EXAMPLE 2

Prove the identity $\sec\theta - \tan\theta \sin\theta = \cos\theta$.

Since the left side is the more complicated side, we'll try to get it to look like the right side. In each step of the proof we'll indicate to you the number of the basic identity being used.

$\sec\theta - \tan\theta \sin\theta$		$\cos\theta.$
$= \dfrac{1}{\cos\theta} - \dfrac{\sin\theta}{\cos\theta}\sin\theta$	[2, 4]	
$= \dfrac{1}{\cos\theta} - \dfrac{\sin^2\theta}{\cos\theta}$		
$= \dfrac{1 - \sin^2\theta}{\cos\theta}$	[combining]	
$= \dfrac{\cos^2\theta}{\cos\theta}$	[6]	
$= \cos\theta.$		

EXAMPLE 3

Prove the identity $\dfrac{\tan^2 x}{1 + \sec x} = \sec x - 1.$

$\dfrac{\tan^2 x}{1 + \sec x}$		$\sec x - 1.$
$= \dfrac{\sec^2 x - 1}{1 + \sec x}$	[7]	
$= \dfrac{(\sec x + 1)(\sec x - 1)}{1 + \sec x}$	[factoring]	
$= \sec x - 1.$		

EXAMPLE 4

Prove $\csc^2 x - \dfrac{\cos^2 x}{\sin^2 x} = 1.$

Here we can write $\dfrac{\cos^2 x}{\sin^2 x}$ as $\left(\dfrac{\cos x}{\sin x}\right)^2$ or $\cot^2 x$ (by 5).

$\csc^2 x - \dfrac{\cos^2 x}{\sin^2 x}$		$1.$
$= \csc^2 x - \cot^2 x$	[5]	
$= 1.$	[8]	

EXAMPLE 5

Prove the identity $1 - \cot^4 x = 2\csc^2 x - \csc^4 x.$

$1 - \cot^4 x.$	$2\csc^2 x - \csc^4 x$	
	$= \csc^2 x\,(2 - \csc^2 x)$	[factoring]
	$= (1 + \cot^2 x)[2 - (1 + \cot^2 x)]$	[8]
	$= (1 + \cot^2 x)[1 - \cot^2 x]$	
	$= 1 - \cot^4 x$	[multiplying].

You may also prove an identity by *separately* manipulating *both* sides until they are the same. Usually we write both sides in terms of sines and cosines when no other approach to the problem is obvious. Example 6 shows this method.

EXAMPLE 6

Prove the identity $\tan x + \cot x = \csc x \sec x$.

Left Side		*Right Side*	
$\tan x + \cot x$		$\csc x \sec x$	
$= \dfrac{\sin x}{\cos x} + \dfrac{\cos x}{\sin x}$	[4, 5]	$= \dfrac{1}{\sin x} \cdot \dfrac{1}{\cos x}$	[1, 2]
$= \dfrac{\sin^2 x + \cos^2 x}{\cos x \sin x}$	[combining]	$= \dfrac{1}{\sin x \cos x}.$	
$= \dfrac{1}{\cos x \sin x}.$	[6]		

In Example 6 there is really no need to manipulate the right side, because we can take the last line on the left and write

$$\frac{1}{\cos x \sin x} = \frac{1}{\cos x} \cdot \frac{1}{\sin x} = \sec x \csc x,$$

which agrees with the right side of the identity.

Completion Set 19-1

In Problems **1–5**, *fill in the blanks.*

1. Since $\tan x = \dfrac{\sin x}{\cos x}$ is true for all values of x for which the equation is defined, this equation is called a trigonometric __________________.

2. $\sin^2 x + \cos^2 x =$ ____.

3. $1 +$ ________________ $= \csc^2 \theta$.

4. $\sec^2 x -$ ________________ $= 1$.

5. $\dfrac{\cos 2x}{\sin 2x} =$ ________________.

6. How many of the following expressions are equal to 1?

$$\sin^2\theta + \cos^2\theta, \quad \csc^2 3° - \cot^2 3°, \quad \sec^2\theta + \tan^2\theta$$

Ans. ________

7. Which of the following statements are identities?

a. $\sin^2\theta = 1 + \cos^2\theta$

b. $\dfrac{1}{\sec\theta} = \cos\theta$

c. $\sin\theta \csc\theta = 1$

d. $1 + \sec^2\theta = \tan\theta$

Ans. ___________

Problem Set 19-1

*In Problems **1–34**, prove the identities.*

1. $\tan x \cos x = \sin x$

2. $\cot x \tan x = 1$

3. $\dfrac{\csc x}{\sec x} = \cot x$

4. $\dfrac{\sin^2 x}{1 - \sin^2 x} = \tan^2 x$

5. $\dfrac{\cos^2\theta}{1 - \sin\theta} = 1 + \sin\theta$

6. $\sec^2\theta - \dfrac{\sin^2\theta}{\cos^2\theta} = 1$

7. $\dfrac{1}{1 - \cos^2\theta} = \csc^2\theta$

8. $\dfrac{1}{1 + \tan^2\theta} = \cos^2\theta$

9. $\dfrac{1 - \cos^2 x}{1 - \sin^2 x} = \tan^2 x$

10. $\dfrac{1}{\csc^2 x - 1} = \tan^2 x$

11. $\dfrac{1 - \sin x}{\cos x} = \sec x - \tan x$

12. $\dfrac{\cos^2 x}{1 - \sin^2 x} = 1$

13. $\dfrac{1 + \tan^2 x}{\csc x} = \tan x \sec x$

14. $\dfrac{1 + \cot^2 x}{\sec x} = \cot x \csc x$

15. $(\sin x)(1 + \cot^2 x) = \csc x$

16. $(1 + \tan^2 x) \cos x = \sec x$

17. $\csc^4\theta - \cot^4\theta = \csc^2\theta + \cot^2\theta$

18. $\dfrac{\csc^4\theta - 1}{\cot^2\theta} = \csc^2\theta + 1$

19. $\dfrac{2}{1 - \cos x} = \dfrac{2 \sec x}{\sec x - 1}$

20. $\dfrac{\sin^2 x + \cos^2 x}{\cos^2 x} = \sec^2 x$

21. $\dfrac{\sin x \cos y}{\sin y \cos x} = \tan x \cot y$

22. $\dfrac{\cot x}{1 + \cot^2 x} = \sin x \cos x$

23. $\dfrac{1}{\tan x + \cot x} = \cos x \sin x$

24. $\dfrac{1 + \cot^2 x}{\cos^2 x \csc^2 x + 1} = 1$

25. $\dfrac{1 + \tan x}{\sec x + \csc x} = \sin x$

26. $\dfrac{\sin^2 x}{\cos^4 x + \cos^2 x \sin^2 x} = \tan^2 x$

27. $\dfrac{1}{\sec x - \tan x} - \dfrac{1}{\sec x + \tan x} = 2 \tan x$

28. $\dfrac{\cos x}{\tan x + \sec x} - \dfrac{\cos x}{\tan x - \sec x} = 2$

29. $\dfrac{\cos^2 x - \cos^2 x \sin^2 x}{\sin^2 x - \cos^2 x \sin^2 x} = \cot^4 x$

30. $\sin^2 x + 1 - \cos^2 x - 2 \sin^2 x \cos^2 x = 2 \sin^4 x$

31. $\dfrac{\sin^2 x}{\cos x}(\tan x - \cos x \cot x) = \sin x(\tan^2 x - \cos x)$

32. $\dfrac{\sin x + \tan x}{1 + \cos x} = \tan x$

33. $\cos^4 x - \sin^4 x + 1 = 2 \cos^2 x$

34. $(1 + \cos x) \csc x + \dfrac{1}{\csc x(1 + \cos x)} = 2 \csc x$

35. Derive identity 7.

36. Derive identity 8.

37. Suppose a ball of mass m is suspended from a string and is pulled aside by a horizontal force F until the string makes an angle θ with the vertical direction. Then three forces are acting on the ball, as shown in Fig. 19-1. The downward force due to

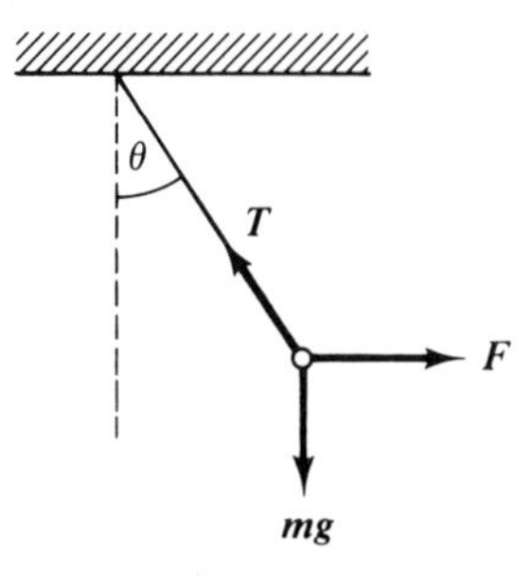

FIG. 19-1

gravity is mg, the tension in the string is T, and F is the supporting force. By applying physical laws it can be shown that

$$T \cos \theta - mg = 0$$

and

$$F - T \sin \theta = 0.$$

Show that $F = mg \tan \theta$. *Hint*: Solve the first equation for T and substitute the result into the second equation.

38. Suppose a weight W slides down a plane, inclined at an angle θ, at constant speed. Such a situation leads to the system of equations

$$\begin{cases} W \sin \theta - \mu N = 0 \\ N - W \cos \theta = 0, \end{cases}$$

where N is a force that the plane exerts on the block and μ (the greek letter "mu") is a constant involved with friction. Solve this system for μ and show that $\mu = \tan \theta$.

39. When a beam of circularly polarized light falls on a polarizing sheet, the resulting amplitude E of the electric field component is given by

$$E = \sqrt{E_x^2 + E_y^2},$$

where $E_x = E_m \sin(\omega t)$ and $E_y = E_m \cos(\omega t)$. Here ω is the greek letter "omega." Prove that $E = E_m$. You may assume that $E_m > 0$.

19-2 FUNCTIONS OF THE SUM AND DIFFERENCE OF ANGLES

The following identities, called the **addition formulas**, express a trig function of the sum of two angles in terms of functions of the individual angles.

$$\textbf{9. } \sin(x + y) = \sin x \cos y + \cos x \sin y$$

$$\textbf{10. } \cos(x + y) = \cos x \cos y - \sin x \sin y$$

$$\textbf{11. } \tan(x + y) = \frac{\tan x + \tan y}{1 - \tan x \tan y}.$$

In words, identity 9 says that *the sine of the sum of two angles is the sine of the first times the cosine of the second, plus the cosine of the first times the sine of the second.*

$\sin(x + y) \neq \sin x + \sin y$, *and similarly for* $\cos(x + y)$ *and* $\tan(x + y)$. *For example,*

$$\sin(30° + 60°) = \sin 90° = 1,$$

but $$\sin 30° + \sin 60° = \frac{1}{2} + \frac{\sqrt{3}}{2} = \frac{1 + \sqrt{3}}{2}.$$

EXAMPLE 1

Find $\sin 75°$ *by using the trig values of* $45°$ *and* $30°$.

Since $75° = 45° + 30°$, we may use identity 9 with $x = 45°$ and $y = 30°$.

$$\begin{aligned}
\sin 75° &= \sin(45° + 30°) \\
&= \sin 45° \cos 30° + \cos 45° \sin 30° \\
&= \frac{\sqrt{2}}{2} \cdot \frac{\sqrt{3}}{2} + \frac{\sqrt{2}}{2} \cdot \frac{1}{2} \\
&= \frac{\sqrt{6}}{4} + \frac{\sqrt{2}}{4} = \frac{\sqrt{6} + \sqrt{2}}{4}.
\end{aligned}$$

EXAMPLE 2

Simplify the equation $y = \tan(x + \pi)$.

We expand $\tan(x + \pi)$ by using identity 11.

$$y = \tan(x + \pi) = \frac{\tan x + \tan \pi}{1 - \tan x \tan \pi}$$

$$= \frac{\tan x + 0}{1 - (\tan x)(0)} = \frac{\tan x}{1 - 0}$$

$$y = \tan x.$$

The next identities express a trig function of $-\theta$ in terms of trig functions of θ. We'll derive these identities for the case where θ is a second-quadrant angle, but they are true for any angle θ.

Figure 19-2 shows the angles θ and $-\theta$, both in standard position, along with points on their terminal sides. These points are chosen so that they give the

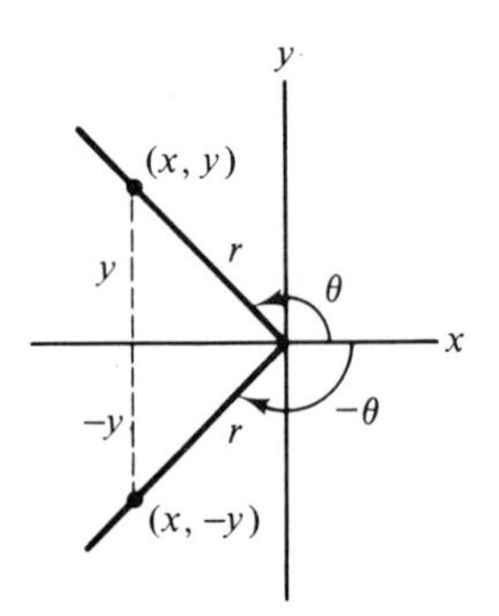

FIG. 19-2

same value of r. Notice that their first coordinates are equal, but their second coordinates differ in sign. We have

$$\sin(-\theta) = \frac{-y}{r} = -\frac{y}{r} = -\sin\theta$$

$$\cos(-\theta) = \frac{x}{r} = \cos\theta$$

$$\tan(-\theta) = \frac{-y}{x} = -\frac{y}{x} = -\tan\theta.$$

Since cosecant and sine are reciprocals, then

$$\csc(-\theta) = \frac{1}{\sin(-\theta)} = \frac{1}{-\sin\theta} = -\csc\theta.$$

We can do the same for secant and cotangent. In summary, we have

$$\textbf{12.}\ \begin{cases} \sin(-\theta) = -\sin\theta & \csc(-\theta) = -\csc\theta \\ \cos(-\theta) = \cos\theta & \sec(-\theta) = \sec\theta \\ \tan(-\theta) = -\tan\theta & \cot(-\theta) = -\cot\theta. \end{cases}$$

For example, $\sin(-30°) = -\sin 30° = -\left(\frac{1}{2}\right) = -\frac{1}{2}$, and $\cos\left(-\frac{\pi}{4}\right) = \cos\frac{\pi}{4} = \frac{\sqrt{2}}{2}$.

With the identities above we may find formulas for the trig functions of the *difference* of two angles. For example, to find $\sin(x - y)$ we have

$$\begin{aligned} \sin(x - y) &= \sin[x + (-y)] \\ &= \sin x \cos(-y) + \cos x \sin(-y) && \text{[from 9]} \\ &= \sin x \cos y + (\cos x)(-\sin y) && \text{[from 12]} \\ &= \sin x \cos y - \cos x \sin y. \end{aligned}$$

We can do the same kind of thing to identities 10 and 11. Thus we have the **subtraction formulas**:

$$\textbf{13.}\ \sin(x - y) = \sin x \cos y - \cos x \sin y$$

$$\textbf{14.}\ \cos(x - y) = \cos x \cos y + \sin x \sin y$$

$$\textbf{15.}\ \tan(x - y) = \frac{\tan x - \tan y}{1 + \tan x \tan y}.$$

EXAMPLE 3

Find cos 15° *by using a subtraction formula.*

Since $15° = 45° - 30°$, using identity 14 we have

$$\begin{aligned} \cos 15° &= \cos(45° - 30°) \\ &= \cos 45° \cos 30° + \sin 45° \sin 30° \\ &= \frac{\sqrt{2}}{2} \cdot \frac{\sqrt{3}}{2} + \frac{\sqrt{2}}{2} \cdot \frac{1}{2} \\ &= \frac{\sqrt{6}}{4} + \frac{\sqrt{2}}{4} = \frac{\sqrt{6} + \sqrt{2}}{4}. \end{aligned}$$

EXAMPLE 4

Show that $\tan(\pi - x) = -\tan x$.

From identity 15 we have

$$\tan(\pi - x) = \frac{\tan\pi - \tan x}{1 + \tan\pi\tan x}$$
$$= \frac{0 - \tan x}{1 + (0)\tan x} = \frac{-\tan x}{1} = -\tan x.$$

EXAMPLE 5

If $\sin\alpha = \frac{1}{3}$, $\cos\beta = \frac{3}{4}$, *and* α *is a first-quadrant angle and* β *is a fourth-quadrant angle, find* $\cos(\alpha - \beta)$.

Angles α and β are sketched in Fig. 19-3. We'll first get values of x, y, and r for each angle. Since $\sin\alpha = \frac{y}{r} = \frac{1}{3}$, then we can choose a point (x, y) on the terminal side

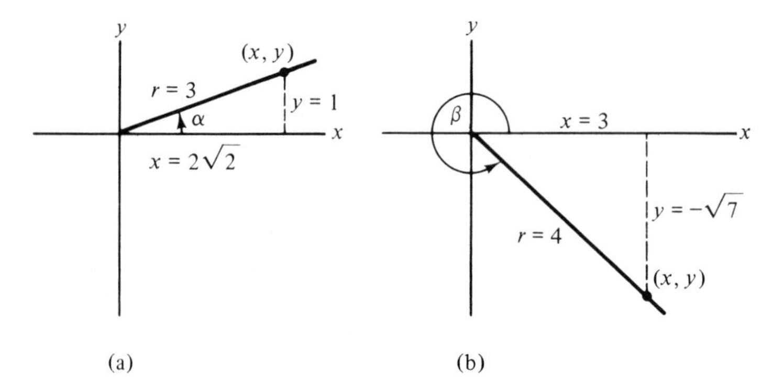

FIG. 19-3

of α so that $y = 1$ and $r = 3$. See Fig. 19-3(a). By the Pythagorean theorem, $x^2 + 1^2 = 3^2$. Thus $x^2 = 3^2 - 1^2 = 8$ and so $x = 2\sqrt{2}$, since x is positive in Quadrant I. Similarly, we can get values for x, y, and r for β. See Fig. 19-3(b). By identity 14 we have

$$\cos(\alpha - \beta) = \cos\alpha\cos\beta + \sin\alpha\sin\beta$$
$$= \left(\frac{2\sqrt{2}}{3}\right)\left(\frac{3}{4}\right) + \left(\frac{1}{3}\right)\left(-\frac{\sqrt{7}}{4}\right)$$
$$= \frac{6\sqrt{2}}{12} - \frac{\sqrt{7}}{12} = \frac{6\sqrt{2} - \sqrt{7}}{12}.$$

Completion Set 19-2

In Problems **1–4**, *fill in* T (= *True*) *or* F (= *False*).

1. $\sin(-13°) = -\sin 13°$. ____

2. $\tan(\alpha - \beta) = \tan\alpha - \tan\beta$. ____

3. $\cos(-214°) = \cos 214°$. ____

4. $\cos(\alpha + \beta) = \cos\alpha\cos\beta + \sin\alpha\sin\beta$. ____

In Problems **5–8**, *fill in the blanks.*

5. $\sin(3x + 2y) = (\sin 3x)($______________$) + (\cos 3x)($______________$)$.

6. $\cos(\pi - x) = ($______________$)(\cos x) + (\sin\pi)($______________$) =$

______________ $+ 0 =$ ______________.

7. $\sin 32° \cos 4° - \cos 32° \sin 4° = \sin($______ $-$ ____$)° = \sin($______$)°$.

8. $\cos 20° \cos 40° - \sin 20° \sin 40° = \cos($______$)° =$ ____.

Problem Set 19-2

1. Find cos 75° by using the trig values of 30° and 45°.

2. Find sin 15° by using the trig values of 30° and 45°.

3. Find sin 195° by using the trig values of 225° and 30°.

4. Find cos 165° by using the trig values of 120° and 45°.

In Problems **5–12**, *use the formulas of this section to find the given trig values. Rationalize your answers.*

5. tan 15°

6. tan 75°

7. cos 105°

8. sin 105°

9. sin 255°

10. cos 255°

11. tan 255°

12. tan 345°

13. If α and β are second-quadrant angles and $\tan\alpha = -\frac{1}{2}$ and $\tan\beta = -\frac{2}{3}$, find (a)

$\sin(\alpha + \beta)$, (b) $\cos(\alpha + \beta)$, (c) $\tan(\alpha + \beta)$, and (d) the quadrant in which $\alpha + \beta$ lies.

14. If α is a first-quadrant angle and $\sin \alpha = \frac{3}{5}$, and β is a second-quadrant angle and $\cos \beta = -\frac{3}{4}$, find (a) $\sin(\alpha - \beta)$, (b) $\cos(\alpha - \beta)$, and (c) $\tan(\alpha - \beta)$.

In Problems **15–20**, *write each expression in terms of* $\sin x$, $\cos x$, *or* $\tan x$, *as in Example* 4.

15. $\sin(x + \pi)$

16. $\cos(x + \pi)$

17. $\cos\left(\frac{\pi}{2} - x\right)$

18. $\sin\left(\frac{\pi}{2} - x\right)$

19. $\tan\left(x + \frac{\pi}{4}\right)$

20. $3 \sin\left(x + \frac{\pi}{2}\right)$

21. Express $\cos 23° \cos 47° - \sin 23° \sin 47°$ as a trig value of one angle only.

22. Express $\sin 18° \cos 10° - \cos 18° \sin 10°$ as a trig value of one angle only.

23. Prove the identity

$$\frac{\sin(\alpha + \beta)}{\cos \alpha \cos \beta} = \tan \alpha + \tan \beta.$$

24. Prove the identity

$$\cos(\alpha + \beta) - \cos(\alpha - \beta) = -2 \sin \alpha \sin \beta.$$

25. Derive identity 14.

26. Derive identity 15.

27. In a certain three-phase electrical generator, the phases are expressed as $I \cos \theta$, $I \cos(\theta + 120°)$, and $I \cos(\theta + 240°)$. Show that

$$I \cos \theta + I \cos(\theta + 120°) + I \cos(\theta + 240°) = 0.$$

28. The displacement x of a certain object, undergoing harmonic motion, as a function of time t is

$$x = 2\sqrt{2} \cos\left(2t - \frac{\pi}{4}\right).$$

a. By expanding the right side, show that the displacement is the sum of two different motions: a sine function and a cosine function.
b. For the two different motions in part a, find the contribution of each to the displacement when $t = \frac{\pi}{4}$.

19-3 DOUBLE- AND HALF-ANGLE FORMULAS

Using the addition formulas, we can derive formulas that express the trig functions of twice an angle in terms of functions of the angle itself. By letting $y = x$ in identity 9, we get

$$\begin{aligned}\sin 2x &= \sin (x + x) = \sin x \cos x + \cos x \sin x \\ &= 2 \sin x \cos x.\end{aligned}$$

Thus,

$$\textbf{16.}\ \ \sin 2x = 2 \sin x \cos x.$$

Letting $y = x$ in identity 10, we have

$$\begin{aligned}\cos 2x &= \cos (x + x) = \cos x \cos x - \sin x \sin x \\ &= \cos^2 x - \sin^2 x.\end{aligned}$$

But since $\sin^2 x + \cos^2 x = 1$, the last expression can be written as either

$$\cos^2 x - \sin^2 x = \cos^2 x - (1 - \cos^2 x) = 2 \cos^2 x - 1,$$

or as

$$\cos^2 x - \sin^2 x = (1 - \sin^2 x) - \sin^2 x = 1 - 2 \sin^2 x.$$

Thus we have

$$\textbf{17.}\ \ \begin{cases}\cos 2x &= \cos^2 x - \sin^2 x \\ &= 2 \cos^2 x - 1 \\ &= 1 - 2 \sin^2 x.\end{cases}$$

Similarly, letting $y = x$ in identity 11 gives

$$\textbf{18.}\ \ \tan 2x = \frac{2 \tan x}{1 - \tan^2 x}.$$

Identities 16–18 are called the **double-angle formulas**.

Don't mix up $\sin 2\theta$ *with* $2 \sin \theta$. *If* $\theta = 30°$, *then* $\sin 2\theta = \sin 60° = \frac{\sqrt{3}}{2}$, *but* $2 \sin \theta = 2 \sin 30° = 2(\frac{1}{2}) = 1$.

EXAMPLE 1

Use a double-angle formula to evaluate sin 60°.

Since we want $\sin 2x = \sin 60°$, we let $x = 30°$. From identity 16 we have

$$\sin 2x = 2 \sin x \cos x$$
$$\sin 60° = \sin(2 \cdot 30°) = 2 \sin 30° \cos 30°$$
$$= 2\left(\frac{1}{2}\right)\left(\frac{\sqrt{3}}{2}\right) = \frac{\sqrt{3}}{2}.$$

EXAMPLE 2

If θ is a second-quadrant angle and $\sin \theta = \frac{3}{5}$, *find* $\tan 2\theta$.

In Fig. 19-4 we've drawn a second-quadrant angle θ such that $\sin \theta = \dfrac{3}{5}$. To find x we solve $x^2 + 3^2 = 5^2$ and get $x = -4$ (negative, since x is negative in the

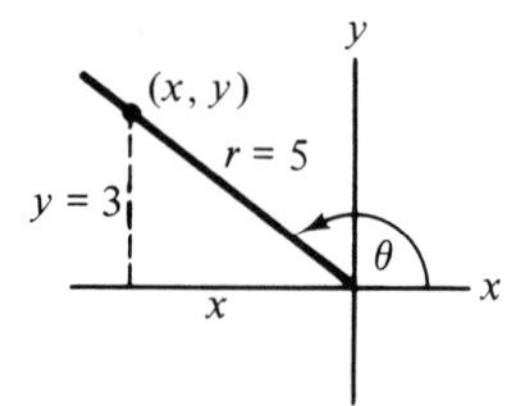

FIG. 19-4

second-quadrant). Using identity 18 with $\tan \theta = \dfrac{3}{x} = \dfrac{3}{-4} = -\dfrac{3}{4}$, we have

$$\tan 2\theta = \frac{2 \tan \theta}{1 - \tan^2 \theta} = \frac{2\left(-\frac{3}{4}\right)}{1 - \left(-\frac{3}{4}\right)^2}$$
$$= \frac{-\frac{3}{2}}{1 - \frac{9}{16}} = \frac{-\frac{3}{2}}{\frac{7}{16}} = -\frac{24}{7}.$$

There are formulas that express a trig function of half an angle $\left(\dfrac{\theta}{2}\right)$ in terms of the angle itself. They are called **half-angle formulas**. To derive one of them we use the double-angle formula

$$\cos 2x = 1 - 2\sin^2 x.$$

Letting $x = \frac{\theta}{2}$ gives

$$\cos\left[2\left(\frac{\theta}{2}\right)\right] = 1 - 2\sin^2\frac{\theta}{2}$$

$$\cos\theta = 1 - 2\sin^2\frac{\theta}{2},$$

which can be written

$$2\sin^2\frac{\theta}{2} = 1 - \cos\theta$$

$$\sin^2\frac{\theta}{2} = \frac{1-\cos\theta}{2}.$$

Solving for $\sin\frac{\theta}{2}$ gives the identity

$$\textbf{19. } \sin\frac{\theta}{2} = \pm\sqrt{\frac{1-\cos\theta}{2}}.$$

The choice of whether to use the (+) or (−) sign depends on the sign of $\sin\frac{\theta}{2}$. If $\frac{\theta}{2}$ is a first- or second-quadrant angle, use the (+) sign. For a third- or fourth-quadrant angle, use the (−) sign.

It can also be shown that

$$\textbf{20. } \cos\frac{\theta}{2} = \pm\sqrt{\frac{1+\cos\theta}{2}}$$

$$\textbf{21. } \tan\frac{\theta}{2} = \frac{\sin\theta}{1+\cos\theta}.$$

In identity 20, use the (+) sign if $\frac{\theta}{2}$ is a first- or fourth-quadrant angle. Use the (−) sign for the other quadrants. Again, the sign depends on $\frac{\theta}{2}$, not θ.

EXAMPLE 3

Use a half-angle formula to find sin 75°.

We want $\frac{\theta}{2} = 75°$ and so $\theta = 150°$. Since sin 75° is positive, we use the (+) sign with identity 19.

$$\sin 75° = \sin \frac{150°}{2} = +\sqrt{\frac{1 - \cos 150°}{2}}$$

$$= \sqrt{\frac{1 - \left(-\frac{\sqrt{3}}{2}\right)}{2}} = \sqrt{\frac{2 + \sqrt{3}}{4}}$$

$$= \frac{\sqrt{2 + \sqrt{3}}}{2}.$$

EXAMPLE 4

Find tan 105° *from the trig values of* 210°.

We use identity 21 with $\theta = 210°$.

$$\tan 105° = \tan \frac{210°}{2} = \frac{\sin 210°}{1 + \cos 210°}$$

$$= \frac{-\frac{1}{2}}{1 + \left(-\frac{\sqrt{3}}{2}\right)} = -\frac{\frac{1}{2}}{\frac{2 - \sqrt{3}}{2}}$$

$$= -\frac{1}{2 - \sqrt{3}} = -\frac{1}{2 - \sqrt{3}} \cdot \frac{2 + \sqrt{3}}{2 + \sqrt{3}}$$

$$= -\frac{2 + \sqrt{3}}{4 - 3} = -(2 + \sqrt{3}).$$

As an aid to you, here is a summary of the identities of this chapter, along with some other identities for future reference.

Trigonometric Identities

1. $\csc \theta = \dfrac{1}{\sin \theta}$

2. $\sec \theta = \dfrac{1}{\cos \theta}$

3. $\cot \theta = \dfrac{1}{\tan \theta}$

4. $\tan \theta = \dfrac{\sin \theta}{\cos \theta}$

5. $\cot\theta = \dfrac{\cos\theta}{\sin\theta}$

6. $\sin^2\theta + \cos^2\theta = 1$

7. $1 + \tan^2\theta = \sec^2\theta$

8. $1 + \cot^2\theta = \csc^2\theta$

9. $\sin(x + y) = \sin x \cos y + \cos x \sin y$

10. $\cos(x + y) = \cos x \cos y - \sin x \sin y$

11. $\tan(x + y) = \dfrac{\tan x + \tan y}{1 - \tan x \tan y}$

12. $\sin(-\theta) = -\sin\theta \qquad \csc(-\theta) = -\csc\theta$
$\cos(-\theta) = \cos\theta \qquad \sec(-\theta) = \sec\theta$
$\tan(-\theta) = -\tan\theta \qquad \cot(-\theta) = -\cot\theta$

13. $\sin(x - y) = \sin x \cos y - \cos x \sin y$

14. $\cos(x - y) = \cos x \cos y + \sin x \sin y$

15. $\tan(x - y) = \dfrac{\tan x - \tan y}{1 + \tan x \tan y}$

16. $\sin 2x = 2 \sin x \cos x$

17. $\cos 2x = \cos^2 x - \sin^2 x$
$= 2\cos^2 x - 1$
$= 1 - 2\sin^2 x$

18. $\tan 2x = \dfrac{2\tan x}{1 - \tan^2 x}$

19. $\sin\dfrac{\theta}{2} = \pm\sqrt{\dfrac{1 - \cos\theta}{2}}$

20. $\cos\dfrac{\theta}{2} = \pm\sqrt{\dfrac{1 + \cos\theta}{2}}$

21. $\tan\dfrac{\theta}{2} = \dfrac{\sin\theta}{1 + \cos\theta}$

22. $\sin(x + y) + \sin(x - y) = 2\sin x \cos y$

23. $\cos(x + y) + \cos(x - y) = 2\cos x \cos y$

24. $\sin(x + y) - \sin(x - y) = 2\cos x \sin y$

25. $\cos(x + y) - \cos(x - y) = -2\sin x \sin y$

26. $\sin x + \sin y = 2\sin\dfrac{x + y}{2}\cos\dfrac{x - y}{2}$

27. $\cos x + \cos y = 2\cos\dfrac{x + y}{2}\cos\dfrac{x - y}{2}$

28. $\sin x - \sin y = 2\cos\dfrac{x + y}{2}\sin\dfrac{x - y}{2}$

29. $\cos x - \cos y = -2\sin\dfrac{x + y}{2}\sin\dfrac{x - y}{2}$

Completion Set 19-3

Fill in the blanks.

1. $\sin 2x = 2 \sin x$ (______________).

2. $\cos 2x = 1 - 2$ (______________).

3. Insert *positive* or *negative*. If $\theta = 130°$, then $\cos \dfrac{\theta}{2}$ is ____________________.

4. By using a half-angle formula, sin 15° can be found from trig values of ______°.

5. $\dfrac{2 \tan 40°}{1 - \tan^2 40°} = \tan$ ______°.

6. $\cos^2 10° - \sin^2 10° =$ ________ 20°.

Problem Set 19-3

In Problems **1–6**, *use a double-angle formula to find the given trig value.*

1. sin 60°

2. cos 60°

3. cos 240°

4. sin 240°

5. tan 120°

6. tan 240°

In Problems **7–12**, *use a half-angle formula to find the given trig value.*

7. sin 15°

8. cos 75°

9. cos 22.5°

10. sin 157.5°

11. $\tan 112\frac{1}{2}°$

12. $\sin 67\frac{1}{2}°$

In Problems **13–16**, *find* sin x, cos x, tan x, sin $2x$, cos $2x$, *and* tan $2x$ *from the given information.*

13. $\cos x = \frac{3}{5}$, $\sin x$ is positive

14. $\sec x = 5$, $\sin x$ is negative

15. $\sin x = -\frac{1}{3}$, $\cot x$ is positive

16. $\cos x = -\frac{1}{4}$, $\tan x$ is negative

In Problems **17–20**, *find* sin x, cos x, tan x, $\sin \dfrac{x}{2}$, $\cos \dfrac{x}{2}$, *and* $\tan \dfrac{x}{2}$ *from the given information. Assume that* $0° < x < 360°$ *and use the facts that if* $0° < x < 180°$, *then*

$0° < \frac{x}{2} < 90°$, *and if* $180° < x < 360°$, *then* $90° < \frac{x}{2} < 180°$.

17. $\cos x = \frac{12}{13}$, $\sin x$ is positive

18. $\cot x = -\frac{8}{15}$, $\sin x$ is positive

19. $\sin x = -\frac{3}{5}$, $\tan x$ is positive

20. $\cos x = \frac{5}{13}$, $\sin x$ is negative

In Problems **21–23**, *prove the identities.*

21. $\cos^4 x - \sin^4 x = \cos 2x$

22. $\sin 2x \cot x = \cos 2x + 1$

23. $2 \sin \frac{\theta}{2} \cos \frac{\theta}{2} = \sin \theta$

24. Show that $\sin 3x = 3 \sin x - 4 \sin^3 x$. *Hint*: $\sin 3x = \sin (2x + x)$.

19-4 TRIGONOMETRIC EQUATIONS

A **trigonometric equation** is an equation involving trig functions of unknown angles. An example is $2 \sin x = 1$. Solving this equation means to find all *angles* x that make it true. We shall consider as solutions only those angles x such that $0° \leqslant x < 360°$. To solve a trig equation we use algebra (as we would with any equation) and also trig identities when they seem useful.

EXAMPLE 1

Solve $2 \sin x = 1$.

$$2 \sin x = 1$$
$$\sin x = \frac{1}{2}.$$

From our knowledge of special angles, we know that $x = 30°$ is a solution. But $\sin x$ is also positive if x is a second-quadrant angle. Thus there is a solution there, and it has $30°$ as its reference angle. It must be $180° - 30° = 150°$. The complete solution is

$$\boxed{x = 30°, 150°.}$$

Some trig equations can be solved by *factoring*, as Example 2 shows.

EXAMPLE 2

Solve $2\sin^2 x - \sin x - 1 = 0$.

$$2\sin^2 x - \sin x - 1 = 0$$

$$(2\sin x + 1)(\sin x - 1) = 0 \qquad \text{[factoring]}$$

$2\sin x + 1 = 0$	$\sin x - 1 = 0$
$2\sin x = -1$	$\sin x = 1.$
$\sin x = -\dfrac{1}{2}.$	

Thus, either $\sin x = -\frac{1}{2}$ or $\sin x = 1$. If $\sin x = -\frac{1}{2}$, then x is a third- or fourth-quadrant angle with 30° as its reference angle. Thus, $x = 210°$ or $x = 330°$. If $\sin x = 1$, then $x = 90°$. The complete solution is

$$\boxed{x = 90°, 210°, 330°.}$$

Some trig equations may be solved by writing the equation in terms of one trig function only, as Example 3 shows.

EXAMPLE 3

Solve $2\cos x - \sec x = 1$.

We can write this equation in terms of $\cos x$ only.

$$2\cos x - \sec x = 1$$

$$2\cos x - \frac{1}{\cos x} = 1. \qquad \text{[identity 2]}$$

To clear of fractions, we multiply both sides by $\cos x$.

$$2\cos^2 x - 1 = \cos x$$

$$2\cos^2 x - \cos x - 1 = 0$$

$$(2\cos x + 1)(\cos x - 1) = 0 \qquad \text{[factoring]}$$

$2\cos x + 1 = 0$	$\cos x - 1 = 0$
$2\cos x = -1$	$\cos x = 1.$
$\cos x = -\dfrac{1}{2}.$	

If $\cos x = -\frac{1}{2}$, then $x = 120°$ or $240°$. If $\cos x = 1$, then $x = 0°$. *We're not done yet.* Remember that we multiplied both sides by $\cos x$. We must check each of the angles 0°, 120°, and 240° to make sure that we didn't multiply both sides by 0.

$$\text{If } x = 0°, \text{ then } \cos x = \cos 0° = 1 \neq 0.$$

$$\text{If } x = 120°, \text{ then } \cos x = \cos 120° = -\frac{1}{2} \neq 0.$$

$$\text{If } x = 240°, \text{ then } \cos x = \cos 240° = -\frac{1}{2} \neq 0.$$

Thus the solution is

$$\boxed{x = 0°, 120°, 240°.}$$

Another method of solving some trig equations involves squaring both sides, as Example 4 shows.

EXAMPLE 4

Solve $\sin x + \cos x = 1$.

From just looking at the equation, no worthwhile substitution is obvious. One way out of this situation is to square both sides. Before squaring, a common practice is to rewrite the equation so that there is a trig function on each side.

$$\sin x + \cos x = 1$$

$$\sin x = 1 - \cos x$$

$$(\sin x)^2 = (1 - \cos x)^2$$

$$\sin^2 x = 1 - 2 \cos x + \cos^2 x$$

$$1 - \cos^2 x = 1 - 2 \cos x + \cos^2 x \quad \text{[identity 6].}$$

Now we combine terms.

$$0 = 2 \cos^2 x - 2 \cos x.$$

$$0 = 2 \cos x (\cos x - 1) \quad \text{[factoring]}$$

$2 \cos x = 0$	$\cos x - 1 = 0$
$\cos x = 0$	$\cos x = 1$
$x = 90°, 270°.$	$x = 0°.$

Since we squared both sides, we must check all values of x in the given equation. If $x = 0°$,

$$\sin 0° + \cos 0° = 0 + 1 = 1 = \text{right side.}$$

If $x = 90°$,

$$\sin 90° + \cos 90° = 1 + 0 = 1 = \text{right side.}$$

If $x = 270°$,

$$\sin 270° + \cos 270° = -1 + 0 = -1 \neq \text{right side.}$$

Thus the solution is

$$\boxed{x = 0°, 90°.}$$

In Example 4 we solved $0 = 2 \cos x(\cos x - 1)$ *by setting each factor equal to* 0. *You may be tempted to first divide both sides by* $\cos x$. *Doing this gives* $0 = 2(\cos x - 1)$, *which has* $x = 0°$ *as its only solution. However, from Example 4 we know that the original equation is true not only for* 0°, *but for* 90° *as well. Thus, by dividing by* $\cos x$ *we lose a solution. In general it is best not to divide both sides of an equation by an expression involving a variable.*

EXAMPLE 5

Solve $2 \sin x \cos x = 1$.

$$2 \sin x \cos x = 1$$
$$\sin 2x = 1 \qquad \text{[identity 16].}$$

Since $0° \leqslant x < 360°$, then by multiplying each part of this inequality by 2, we get $0° \leqslant 2x < 720°$. We want to find all angles $2x$ between 0° and 720° whose sine equals 1. Since $\sin 90° = 1$, then $2x = 90°$ and so $x = 45°$. Also, $\sin 450° = 1$. Setting $2x = 450°$ gives $x = 225°$. The complete solution is

$$\boxed{x = 45°, 225°.}$$

Problem Set 19-4

In Problems **1–28**, *solve for* x, *where* $0° \leqslant x < 360°$.

1. $\sin x = \dfrac{\sqrt{2}}{2}$

2. $\cos x = -\dfrac{\sqrt{3}}{2}$

3. $\tan x = -1$

4. $\sin x = 0$

5. $2 \cos x = 1$

6. $\sin (x + 10°) = -\frac{1}{2}$

7. $2 \cos^2 x - \cos x - 1 = 0$

8. $\sin^2 x - 2 \sin x + 1 = 0$

9. $\sin x + \sin x \cos x = 0$

10. $\sin^2 x - 1 = 0$

11. $2 \sin x - \csc x = 1$

12. $\tan x \cos x = \frac{1}{2}$

13. $\sin 2x = \dfrac{\sqrt{3}}{2}$

14. $\cos 3x = 1$

15. $\cos x - 1 - \sqrt{3} \sin x = 0$

16. $\cos x - \sin x - 1 = 0$

17. $2 \sin x \cos x = -\dfrac{\sqrt{2}}{2}$

18. $\sin 2x = \sin x$

19. $\cos x - \cos 2x = 1$

20. $\sin 2x \cos x = 0$

21. $2 \sin x - \tan x = 0$

22. $\sin x + \cos 2x = 4 \sin^2 x$

23. $2 \tan^2 x + \sec^2 x = 2$

24. $\sqrt{2 - \csc^2 x} = \csc x$

25. $\sin 2x + \cos x = 0$

26. $\tan 2x + \sec 2x = 1$

27. $\sec^2 x + \tan x = 1$

28. $\cot x - \csc^2 x = -1$

19-5 THE LAW OF SINES

In Sec. 17-5 you solved right triangles. By using certain formulas you can also solve triangles that do not contain a right angle. These triangles, called *oblique triangles*, are of two types, depending on the angles they contain. An *acute triangle* has three acute angles [Fig. 19-5(a)]. An *obtuse triangle* has one obtuse

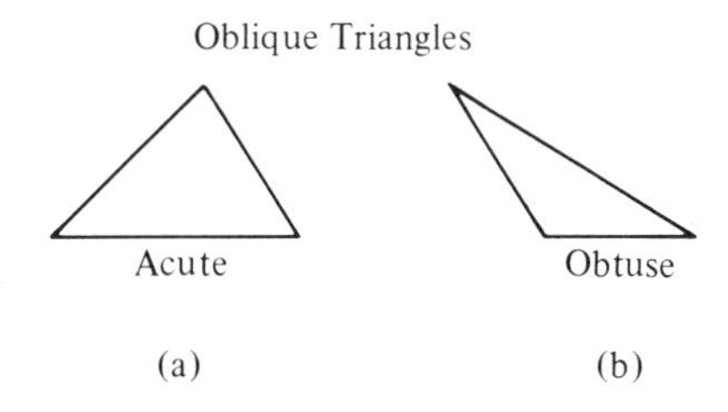

FIG. 19-5

angle, and this angle must be opposite the longest side [Fig. 19-5(b)].

The first formula we'll look at is the **law of sines**, or **sine law**. It states that for a triangle with angles A, B, and C, and opposite sides a, b, and c, respectively, then

$$\frac{a}{\sin A} = \frac{b}{\sin B} = \frac{c}{\sin C}.$$

The sine law is used to solve a triangle when you know

1. two angles and any side, or
2. two sides and the angle opposite one of them.

EXAMPLE 1

Given two angles and any side.

Solve the triangle ABC given $A = 30°$, $B = 70°$, *and* $a = 4$.

A sketch of the triangle is in Fig. 19-6. Angle C is easily found.

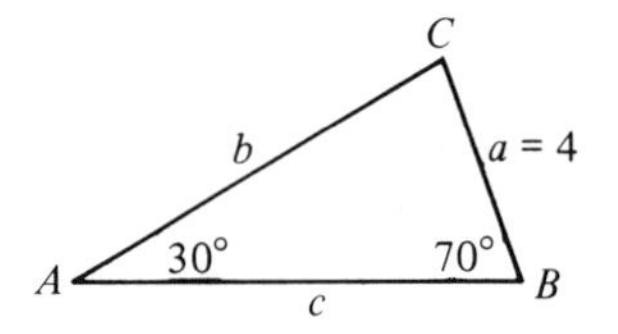

FIG. 19-6

$$C = 180° - A - B = 180° - 30° - 70° = 80°.$$

To find b, we pair the first and second expressions in the sine law and plug in our data. This leads to an equation with one unknown.

$$\frac{b}{\sin B} = \frac{a}{\sin A}$$

$$\frac{b}{\sin 70°} = \frac{4}{\sin 30°}.$$

Solving for b and using tables, we have

$$b = \frac{4 \sin 70°}{\sin 30°} = \frac{4(.9397)}{.5} = 7.518.$$

To find c, we use the sine law again.

$$\frac{c}{\sin C} = \frac{a}{\sin A}$$

$$\frac{c}{\sin 80°} = \frac{4}{\sin 30°}$$

$$c = \frac{4 \sin 80°}{\sin 30°} = \frac{4(.9848)}{.5} = 7.878.$$

Our solution is $C = 80°$, $b = 7.518$, and $c = 7.878$.

If you are given two sides and the angle opposite one of them, there may be two, one, or no triangles fitting the data. Because of these possibilities, we say that this kind of problem falls in the *ambiguous case*.

To see why these situations may occur, let's assume that the given parts are a, b, and A. It could be that side a is too short to meet the lower side of angle A (Fig. 19-7). Thus there is no triangle. On the other hand, side a may be able to meet the lower side of angle A in two places (Fig. 19-8). Thus two triangles are

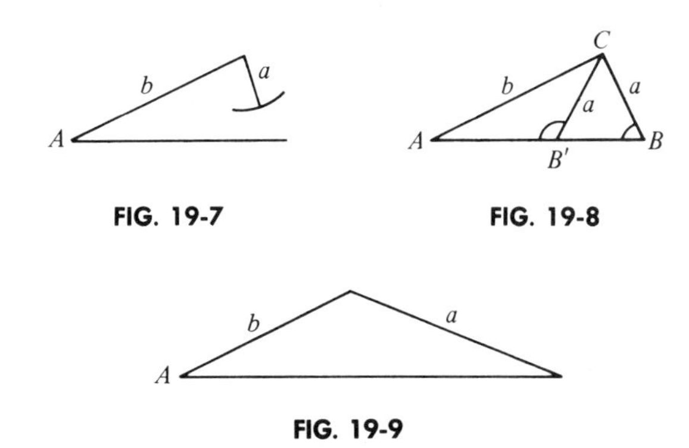

FIG. 19-7

FIG. 19-8

FIG. 19-9

determined: ABC and $AB'C$, where B' is read "B prime." Although both triangles share the same a, b, and A, notice that B is acute and B' is obtuse. It can be shown that $B + B' = 180°$; that is, B and B' are *supplementary*. Finally, a may meet the lower side of A at exactly one point (Fig. 19-9). In this case only one triangle is determined. When you have a problem of the ambiguous case, a fairly accurate sketch based on the data will often make the number of solutions obvious.

EXAMPLE 2

Given two sides and an opposite angle.

Solve the triangle ABC given that $a = 6$, $b = 10$, and $A = 30°$.

We first find B. By the sine law, $\dfrac{b}{\sin B} = \dfrac{a}{\sin A}$, which may be written

$$\frac{\sin B}{b} = \frac{\sin A}{a}$$

$$\frac{\sin B}{10} = \frac{\sin 30°}{6}$$

$$\sin B = \frac{10 \sin 30°}{6} = \frac{10(.5)}{6} = .8333.$$

There are two angles whose sines are .8333. They are $B = 56°26'$ (from tables and interpolation) and $B = 180° - 56°26' = 123°34'$. Thus two triangles are determined (Fig. 19-10).

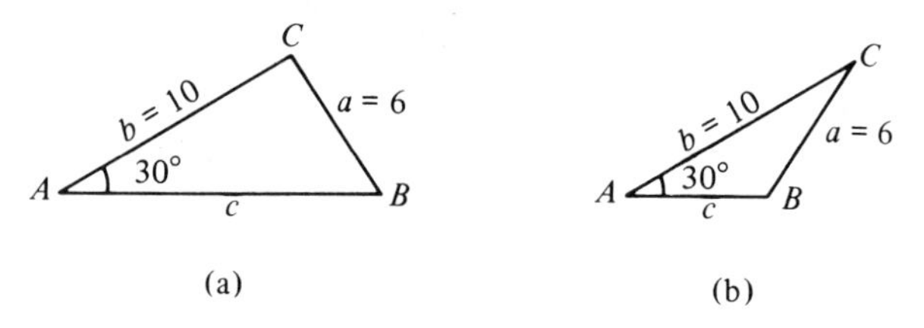

FIG. 19-10

Case 1. If $B = 56°26'$ [Fig. 19-10(a)], then

$$C = 180° - A - B = 180° - 30° - 56°26' = 93°34'.$$

Thus,

$$\frac{c}{\sin C} = \frac{a}{\sin A}$$

$$\frac{c}{\sin 93°34'} = \frac{6}{\sin 30°}$$

$$c = \frac{6(\sin 93°34')}{\sin 30°} = \frac{6(.9981)}{.5} = 11.98.$$

The solution for Fig. 19-10(a) is $c = 11.98$, $B = 56°26'$, and $C = 93°34'$.

Case 2. If $B = 123°34'$ [Fig. 19-10(b)], then

$$C = 180° - A - B = 180° - 30° - 123°34' = 26°26'.$$

$$\frac{c}{\sin C} = \frac{a}{\sin A}$$

$$c = \frac{a \sin C}{\sin A} = \frac{6(\sin 26°26')}{\sin 30°} = \frac{6(.4452)}{.5} = 5.342.$$

The solution for Fig. 19-10(b) is $c = 5.342$, $B = 123°34'$, and $C = 26°26'$.

EXAMPLE 3

Given two sides and an opposite angle.

Solve triangle ABC, given that $b = 10$, $c = 5$, *and* $B = 60°$.

By the sine law,

$$\frac{\sin C}{c} = \frac{\sin B}{b}$$

$$\frac{\sin C}{5} = \frac{\sin 60°}{10}$$

$$\sin C = \frac{5 \sin 60°}{10} = \frac{5(.8660)}{10} = .4330.$$

At this point you might be tempted to think that there are two choices for C, as in Example 2. This is *not* the case here. Since C is opposite side c, and c is *not* the longest side ($10 > 5$), then C *cannot* be obtuse. C must be acute. Thus we find that

$$C = 25°40'.$$

A sketch of the triangle is in Fig. 19-11.

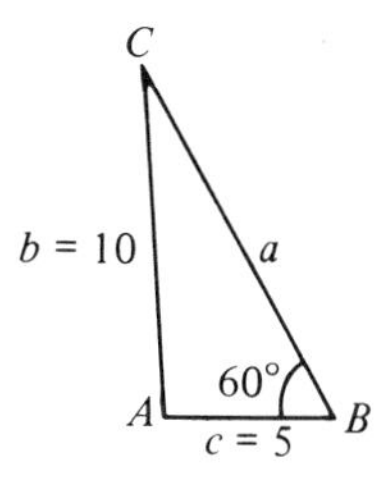

FIG. 19-11

Now,

$$A = 180° - 60° - 25°40' = 94°20'.$$

Thus, $$\frac{a}{\sin A} = \frac{b}{\sin B}$$

$$a = \frac{b \sin A}{\sin B} = \frac{10(\sin 94°20')}{\sin 60°} = \frac{10(.9971)}{.8660} = 11.51.$$

The solution is $A = 94°20'$, $C = 25°40'$, and $a = 11.51$.

EXAMPLE 4

Solve the triangle ABC, given that $a = 2$, $b = 6$, *and* $A = 20°$.

By the sine law,

$$\frac{\sin B}{b} = \frac{\sin A}{a}$$

$$\frac{\sin B}{6} = \frac{\sin 20°}{2}$$

$$\sin B = \frac{6(\sin 20°)}{2} = \frac{6(.3420)}{2} = 1.026.$$

But the sine of any angle can't be greater than 1. Thus, there is no triangle and therefore **no solution.**

Problem Set 19-5

In Problems **1–16**, *solve triangle ABC from the given information.*

1. $A = 50°$, $B = 100°$, $a = 20$

2. $A = 80°$, $C = 40°$, $c = 100$

3. $a = 10$, $b = 7$, $A = 81°$

4. $a = 20$, $b = 9$, $A = 55°$

5. $a = 7$, $b = 9$, $A = 20°$

6. $a = 67$, $b = 100$, $A = 25°$

7. $a = 50$, $b = 100$, $A = 60°$

8. $a = 30$, $b = 70$, $A = 30°$

9. $a = 7$, $c = 20$, $C = 140°$

10. $b = 110$, $c = 90$, $B = 110°$

11. $B = 60°$, $C = 72°$, $a = 80$

12. $C = 20°$, $A = 40°$, $b = 50$

13. $c = 5$, $b = 10$, $C = 30°$

14. $b = 70$, $c = 70$, $C = 68°$

15. $a = 6$, $b = 7$, $A = 104°$

16. $a = 60$, $b = 60$, $A = 100°$

17. A boat B can be seen from points A and C on the shore. A and C are 5000 ft apart. Angles BAC and BCA are found to be 65° and 75°, respectively. How far is the boat from A?

18. In order to find the distance between points A and B, another point C is marked off. The distance from B to C is known to be 200 ft. Angle ABC is measured to be 120°, and angle BCA is found to be 35°. Find the distance between A and B.

19. Two students send up a balloon with a remote-controlled camera to take a picture of the countryside. They position themselves one mile apart. When the picture is taken, the balloon is between the students and the angles of elevation from the students are 46° and 70°. Find the height of the balloon.

20. From a point on the ground, the angle of elevation of the top of a building is 60°. From another point 100 ft *farther* away from the building, the angle of elevation is 40°. Find the height of the building.

19-6 THE LAW OF COSINES

Another formula used for solving triangles is the **law of cosines**, or **cosine law**. It states that *the square of any side of a triangle equals the sum of the squares of the other two sides, minus twice the product of these sides times the cosine of their included angle*; that is,

$$a^2 = b^2 + c^2 - 2bc \cos A$$
$$b^2 = a^2 + c^2 - 2ac \cos B$$
$$c^2 = a^2 + b^2 - 2ab \cos C.$$

The cosine law is used when you know

1. two sides and the included angle, or
2. three sides.

EXAMPLE 1

Given two sides and the included angle.

Solve the triangle ABC given that $a = 10$, $b = 40$, *and* $C = 120°$ (*Fig.* 19-12).

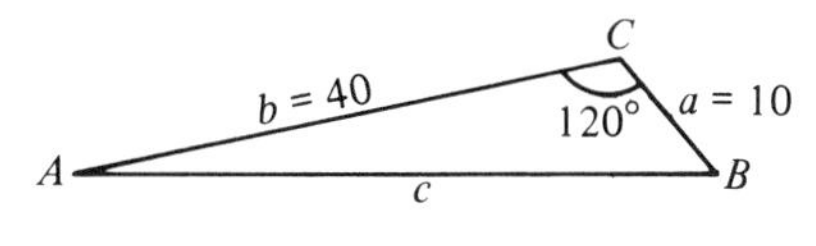

FIG. 19-12

We first use the cosine law to find side c.

$$\begin{aligned} c^2 &= a^2 + b^2 - 2ab \cos C \\ &= 10^2 + 40^2 - 2(10)(40) \cos 120° \\ &= 100 + 1600 - 800\left(-\frac{1}{2}\right) = 2100. \end{aligned}$$

Since $c^2 = 2100$, then

$$c = \sqrt{2100} = 10\sqrt{21} = 10(4.583) = 45.83.$$

Angles A and B can now be found with the cosine law. But we can simplify our work by using the sine law. To find A we have

$$\frac{\sin A}{a} = \frac{\sin C}{c}$$

$$\sin A = \frac{a \sin C}{c} = \frac{10 \sin 120°}{45.83} = \frac{10(.8660)}{45.83} = .1890.$$

From tables we find that

$$A = 10°54'.$$

Note that A must be acute, since C is obtuse. Even if C were not obtuse, A must be acute here, because a is the smallest side. Only the angle opposite the longest side has any chance of being obtuse. Finally,

$$B = 180° - A - C = 180° - 10°54' - 120° = 49°6'.$$

The solution is $c = 45.83$, $A = 10°54'$, and $B = 49°6'$.

EXAMPLE 2

Given three sides.

Solve the triangle ABC, given that $a = 7$, $b = 6$, and $c = 8$ (Fig. 19-13).

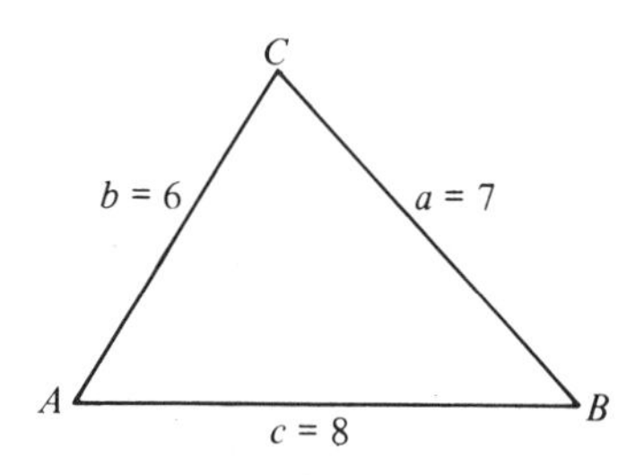

FIG. 19-13

We first find the largest angle, which is C since it is opposite the longest side. Then we'll know whether or not the triangle contains an obtuse angle. By the cosine law,

$$\begin{aligned} c^2 &= a^2 + b^2 - 2ab\cos C \\ 8^2 &= 7^2 + 6^2 - 2(7)(6)\cos C \\ 64 &= 85 - 84\cos C \\ 84\cos C &= 85 - 64 = 21 \\ \cos C &= \frac{21}{84} = \frac{1}{4} = .2500. \end{aligned}$$

Since $\cos C$ is positive, C is acute:

$$C = 75°31'.$$

To solve for A and B, we can use the cosine law again, but the work is easier if we use the sine law.

$$\frac{\sin B}{b} = \frac{\sin C}{c}$$

$$\sin B = \frac{b \sin C}{c} = \frac{6 \sin 75°31'}{8} = \frac{6(.9682)}{8} = .7262$$

$$B = 46°34'. \qquad [B \text{ must be acute}]$$

Finally, $\quad A = 180° - B - C = 180° - 46°34' - 75°31' = 57°55'.$

The solution is $A = 57°55'$, $B = 46°34'$, and $C = 75°31'$.

EXAMPLE 3

Given the triangle ABC such that $a = 9$, $b = 8$, and $c = 2$, find A.

By the cosine law,

$$a^2 = b^2 + c^2 - 2bc \cos A$$
$$9^2 = 8^2 + 2^2 - 2(8)(2) \cos A$$
$$81 = 68 - 32 \cos A$$
$$32 \cos A = 68 - 81 = -13$$
$$\cos A = -\frac{13}{32} = -.4062.$$

Since $\cos A$ is negative, angle A is obtuse. From tables, the reference angle for A is $66°2'$ and so

$$A = 180° - 66°2' = 113°58'.$$

Problem Set 19-6

Use the following approximate values in Problems **1–12**.

$\sqrt{20.83} = 4.564$	$\sqrt{587.4} = 24.24$
$\sqrt{57.90} = 7.609$	$\sqrt{1954} = 44$
$\sqrt{91} = 9.539$	$\sqrt{126{,}160} = 355$
$\sqrt{142.4} = 11.93$	

In Problems **1–8**, *solve triangle ABC from the given data.*

1. $a = 10$, $c = 9$, $B = 60°$

2. $b = 3$, $c = 4$, $A = 80°$

3. $a = 2$, $b = 3$, $c = 4$

4. $a = 4$, $b = 8$, $c = 5$

5. $a = 20$, $b = 40$, $C = 28°$

6. $b = 5$, $c = 4$, $A = 115°$

7. $a = 6$, $b = 4$, $c = 5$

8. $a = 6$, $b = 5$, $c = 7$

9. Figure 19-14 shows two forces represented by the arrows ***OA*** and ***AB***. The force

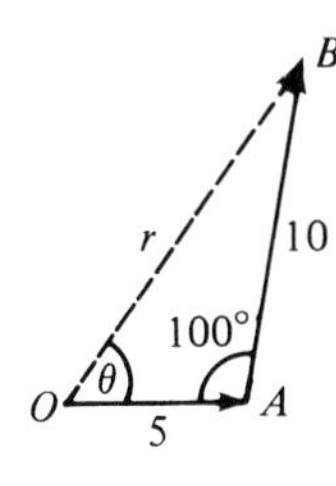

FIG. 19-14

represented by $\boldsymbol{OB}$ is called the *resultant* of $\boldsymbol{OA}$ and $\boldsymbol{AB}$. Find the length r of the resultant and find θ.

10. Repeat Problem **9** for the data in Fig. 19-15.

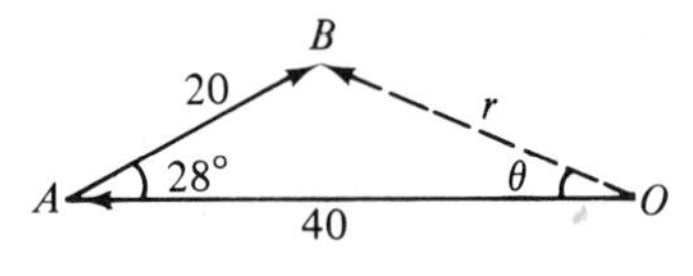

FIG. 19-15

11. Two boats leave a dock at the same time. One travels north at 10 miles per hour. The other travels northeast at 20 miles per hour. After 3 hours, how far apart are the boats? Give answer to the nearest mile.

12. Points A and B are on opposite sides of a lake. To find the distance AB, another point C on the same side of the lake as A is used. It is known that $BC = 200$ ft and $AC = 500$ ft. Angle BCA is measured to be 35°. Find AB. Give answer to the nearest foot.

19-7 REVIEW

IMPORTANT TERMS

trigonometric identity *(p. 462)*
addition formulas *(p. 469)*
subtraction formulas *(p. 471)*
double-angle formulas *(p. 475)*
half-angle formulas *(p. 476)*
oblique triangle *(p. 485)*
trigonometric equation *(p. 481)*
sine law *(p. 485)*
cosine law *(p. 490)*

REVIEW PROBLEMS

In Problems **1–8**, *prove the given identities.*

1. $\tan x + \cot x = \sec x \csc x$

2. $\cot^2 x \sin^2 x + \tan^2 x \cos^2 x = 1$

3. $\csc^2 \theta \tan^2 \theta - \sec \theta \cos \theta = \tan^2 \theta$

4. $\dfrac{\sec^2 \theta}{\cot \theta} - \tan^3 \theta = \tan \theta$

5. $\dfrac{\sin^2 x}{1 - \cos x} = 1 + \cos x$

6. $\sec^2 x + \csc^2 x = \sec^2 x \csc^2 x$

7. $\cos x + \sin x \tan x = \sec x$

8. $\sin 2x \tan x = 2 \sin^2 x$

In Problems **9–16**, *evaluate the given expressions if* α *and* β *are first-quadrant angles and* $\sin \alpha = \frac{1}{2}$ *and* $\sin \beta = \frac{4}{5}$.

9. $\sin (\alpha + \beta)$

10. $\cos (\alpha - \beta)$

11. $\tan \frac{\beta}{2}$

12. $\cos(\alpha + \beta)$

13. $\cos 2\alpha$

14. $\sin 2\alpha$

15. $\tan(\beta - 45°)$

16. $\cos \frac{\beta}{2}$

17. If $\tan x = \frac{15}{8}$ find $\sin x$, $\cos x$, $\tan x$, $\sin(-x)$, $\cos(-x)$, $\sin 2x$, $\cos 2x$, $\tan 2x$, $\sin \frac{x}{2}$, $\cos \frac{x}{2}$, and $\tan \frac{x}{2}$. Assume x is acute.

18. If $\sin x = \frac{3}{5}$, find the trig values asked for in Problem **17**. Assume x is acute.

In Problems **19–28**, *solve the triangles ABC from the given data.*

19. $a = 5$, $b = 4$, $C = 30°$. Assume $\sqrt{6.360} = 2.522$.

20. $a = 2$, $c = 3$, $B = 60°$

21. $B = 30°$, $C = 70°$, $b = 3$

22. $A = 110°$, $B = 40°$, $a = 9$

23. $a = 8$, $b = 6$, $c = 4$

24. $a = 5$, $b = 7$, $c = 9$

25. $A = 60°$, $a = 20$, $b = 10$

26. $b = 5$, $c = 8$, $C = 40°$

27. $a = 6$, $b = 8$, $A = 10°$

28. $A = 130°$, $b = 20$, $a = 10$

In Problems **29–32**, *solve the equation for x where* $0° \leqslant x < 360°$.

29. $\sin x \cos x - \cot x = 0$

30. $\sin 2x - \sqrt{2} \sin x = 0$

31. $\frac{\sqrt{3}}{2} + \tan^2 x + \sin x - \sec^2 x = -1$

32. $1 + \sin 2x = 0$

33. In Fig. 19-16 are forces ***AB*** and ***BC*** represented by arrows. These forces can be replaced by the force *AC*, called the *resultant*. Find the length r of the resultant. Leave your answer in radical form.

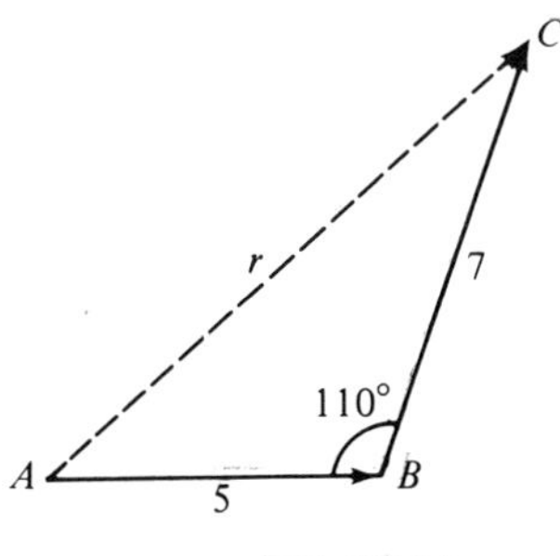

FIG. 19-16

APPENDIX A

Tables of Powers—Roots—Reciprocals

n	n^2	$\sqrt{n}$	$\sqrt{10n}$	n^3	$\sqrt[3]{n}$	$\sqrt[3]{10n}$	$\sqrt[3]{100n}$	$1/n$
1.0	1.0000	1.0000	3.1623	1.0000	1.0000	2.1544	4.6416	1.0000
1.1	1.2100	1.0488	3.3166	1.3310	1.0323	2.2240	4.7914	0.9091
1.2	1.4400	1.0954	3.4641	1.7280	1.0627	2.2894	4.9324	0.8333
1.3	1.6900	1.1402	3.6056	2.1970	1.0914	2.3513	5.0658	0.7692
1.4	1.9600	1.1832	3.7417	2.7440	1.1187	2.4101	5.1925	0.7143
1.5	2.2500	1.2247	3.8730	3.3750	1.1447	2.4662	5.3133	0.6667
1.6	2.5600	1.2649	4.0000	4.0960	1.1696	2.5198	5.4288	0.6250
1.7	2.8900	1.3038	4.1231	4.9130	1.1935	2.5713	5.5397	0.5882
1.8	3.2400	1.3416	4.2426	5.8320	1.2164	2.6207	5.6462	0.5556
1.9	3.6100	1.3784	4.3589	6.8590	1.2386	2.6684	5.7489	0.5263
2.0	4.0000	1.4142	4.4721	8.0000	1.2599	2.7144	5.8480	0.5000
2.1	4.4100	1.4491	4.5826	9.2610	1.2806	2.7589	5.9439	0.4762
2.2	4.8400	1.4832	4.6904	10.6480	1.3006	2.8020	6.0368	0.4545
2.3	5.2900	1.5166	4.7958	12.1670	1.3200	2.8439	6.1269	0.4348

n	n^2	$\sqrt{n}$	$\sqrt{10n}$	n^3	$\sqrt[3]{n}$	$\sqrt[3]{10n}$	$\sqrt[3]{100n}$	$1/n$
2.4	5.7600	1.5492	4.8990	13.8240	1.3389	2.8845	6.2145	0.4167
2.5	6.2500	1.5811	5.0000	15.6250	1.3572	2.9240	6.2996	0.4000
2.6	6.7600	1.6125	5.0990	17.5760	1.3751	2.9625	6.3825	0.3846
2.7	7.2900	1.6432	5.1962	19.6830	1.3925	3.0000	6.4633	0.3704
2.8	7.8400	1.6733	5.2915	21.9520	1.4095	3.0366	6.5421	0.3571
2.9	8.4100	1.7029	5.3852	24.3890	1.4260	3.0723	6.6191	0.3448
3.0	9.0000	1.7321	5.4772	27.0000	1.4422	3.1072	6.6943	0.3333
3.1	9.6100	1.7607	5.5678	29.7910	1.4581	3.1414	6.7679	0.3226
3.2	10.2400	1.7889	5.6569	32.7680	1.4736	3.1748	6.8399	0.3125
3.3	10.8900	1.8166	5.7446	35.9370	1.4888	3.2075	6.9104	0.3030
3.4	11.5600	1.8439	5.8310	39.3040	1.5037	3.2396	6.9795	0.2941
3.5	12.2500	1.8708	5.9161	42.8750	1.5183	3.2711	7.0473	0.2857
3.6	12.9600	1.8974	6.0000	46.6560	1.5326	3.3019	7.1138	0.2778
3.7	13.6900	1.9235	6.0828	50.6530	1.5467	3.3322	7.1791	0.2703
3.8	14.4400	1.9494	6.1644	54.8720	1.5605	3.3620	7.2432	0.2632
3.9	15.2100	1.9748	6.2450	59.3190	1.5741	3.3912	7.3061	0.2564
4.0	16.0000	2.0000	6.3246	64.0000	1.5874	3.4200	7.3681	0.2500
4.1	16.8100	2.0248	6.4031	68.9210	1.6005	3.4482	7.4290	0.2439
4.2	17.6400	2.0494	6.4807	74.0880	1.6134	3.4760	7.4889	0.2381
4.3	18.4900	2.0736	6.5574	79.5070	1.6261	3.5034	7.5478	0.2326
4.4	19.3600	2.0976	6.6333	85.1840	1.6386	3.5303	7.6059	0.2273
4.5	20.2500	2.1213	6.7082	91.1250	1.6510	3.5569	7.6631	0.2222
4.6	21.1600	2.1448	6.7823	97.3360	1.6631	3.5830	7.7194	0.2174
4.7	22.0900	2.1679	6.8557	103.823	1.6751	3.6088	7.7750	0.2128
4.8	23.0400	2.1909	6.9282	110.592	1.6869	3.6342	7.8297	0.2083
4.9	24.0100	2.2136	7.0000	117.649	1.6985	3.6593	7.8837	0.2041
5.0	25.0000	2.2361	7.0711	125.000	1.7100	3.6840	7.9370	0.2000
5.1	26.0100	2.2583	7.1414	132.651	1.7213	3.7084	7.9896	0.1961
5.2	27.0400	2.2804	7.2111	140.608	1.7325	3.7325	8.0415	0.1923
5.3	28.0900	2.3022	7.2801	148.877	1.7435	3.7563	8.0927	0.1887
5.4	29.1600	2.3238	7.3485	157.464	1.7544	3.7798	8.1433	0.1852
5.5	30.2500	2.3452	7.4162	166.375	1.7652	3.8030	8.1932	0.1818
5.6	31.3600	2.3664	7.4833	175.616	1.7758	3.8259	8.2426	0.1786
5.7	32.4900	2.3875	7.5498	185.193	1.7863	3.8485	8.2913	0.1754
5.8	33.6400	2.4083	7.6158	195.112	1.7967	3.8709	8.3396	0.1724
5.9	34.8100	2.4290	7.6811	205.379	1.8070	3.8930	8.3872	0.1695
6.0	36.0000	2.4495	7.7460	216.000	1.8171	3.9149	8.4343	0.1667
6.1	37.2100	2.4698	7.8102	226.981	1.8272	3.9365	8.4809	0.1639
6.2	38.4400	2.4900	7.8740	238.328	1.8371	3.9579	8.5270	0.1613
6.3	39.6900	2.5100	7.9372	250.047	1.8469	3.9791	8.5726	0.1587
6.4	40.9600	2.5298	8.0000	262.144	1.8566	4.0000	8.6177	0.1563
6.5	42.2500	2.5495	8.0623	274.625	1.8663	4.0207	8.6624	0.1538
6.6	43.5600	2.5690	8.1240	287.496	1.8758	4.0412	8.7066	0.1515

n	n^2	$\sqrt{n}$	$\sqrt{10n}$	n^3	$\sqrt[3]{n}$	$\sqrt[3]{10n}$	$\sqrt[3]{100n}$	$1/n$
6.7	44.8900	2.5884	8.1854	300.763	1.8852	4.0615	8.7503	0.1493
6.8	46.2400	2.6077	8.2462	314.432	1.8945	4.0817	8.7937	0.1471
6.9	47.6100	2.6268	8.3066	328.509	1.9038	4.1016	8.8366	0.1449
7.0	49.0000	2.6458	8.3666	343.000	1.9129	4.1213	8.8790	0.1429
7.1	50.4100	2.6646	8.4261	357.911	1.9220	4.1408	8.9211	0.1408
7.2	51.8400	2.6833	8.4853	373.248	1.9310	4.1602	8.9628	0.1389
7.3	53.2900	2.7019	8.5440	389.017	1.9399	4.1793	9.0041	0.1370
7.4	54.7600	2.7203	8.6023	405.224	1.9487	4.1983	9.0450	0.1351
7.5	56.2500	2.7386	8.6603	421.875	1.9574	4.2172	9.0856	0.1333
7.6	57.7600	2.7568	8.7178	438.976	1.9661	4.2358	9.1258	0.1316
7.7	59.2900	2.7749	8.7750	456.533	1.9747	4.2543	9.1657	0.1299
7.8	60.8400	2.7928	8.8318	474.552	1.9832	4.2727	9.2052	0.1282
7.9	62.4100	2.8107	8.8882	493.039	1.9916	4.2908	9.2443	0.1266
8.0	64.0000	2.8284	8.9443	512.000	2.0000	4.3089	9.2832	0.1250
8.1	65.6100	2.8460	9.0000	531.441	2.0083	4.3267	9.3217	0.1235
8.2	67.2400	2.8636	9.0554	551.368	2.0165	4.3445	9.3599	0.1220
8.3	68.8900	2.8810	9.1104	571.787	2.0247	4.3621	9.3978	0.1205
8.4	70.5600	2.8983	9.1652	592.704	2.0328	4.3795	9.4354	0.1190
8.5	72.2500	2.9155	9.2195	614.125	2.0408	4.3968	9.4727	0.1176
8.6	73.9600	2.9326	9.2736	636.056	2.0488	4.4140	9.5097	0.1163
8.7	75.6900	2.9496	9.3274	658.503	2.0567	4.4310	9.5464	0.1149
8.8	77.4400	2.9665	9.3808	681.472	2.0646	4.4480	9.5828	0.1136
8.9	79.2100	2.9833	9.4340	704.969	2.0723	4.4647	9.6190	0.1124
9.0	81.000	3.0000	9.4868	729.000	2.0801	4.4814	9.6549	0.1111
9.1	82.8100	3.0166	9.5394	753.571	2.0878	4.4979	9.6905	0.1099
9.2	84.6400	3.0332	9.5917	778.688	2.0954	4.5144	9.7259	0.1087
9.3	86.4900	3.0496	9.6436	804.357	2.1029	4.5307	9.7610	0.1075
9.4	88.3600	3.0659	9.6954	830.584	2.1105	4.5468	9.7959	0.1064
9.5	90.2500	3.0822	9.7468	857.375	2.1179	4.5629	9.8305	0.1053
9.6	92.1600	3.0984	9.7980	884.736	2.1253	4.5789	9.8648	0.1042
9.7	94.0900	3.1145	9.8489	912.673	2.1327	4.5947	9.8990	0.1031
9.8	96.0400	3.1305	9.8995	941.192	2.1400	4.6104	9.9329	0.1020
9.9	98.0100	3.1464	9.9499	970.299	2.1472	4.6261	9.9666	0.1010
10.0	100.000	3.1623	10.000	1000.00	2.1544	4.6416	10.0000	0.1000

APPENDIX B

Table of e^x and e^{-x}

x	e^x	e^{-x}	x	e^x	e^{-x}
0.00	1.0000	1.0000	0.80	2.2255	0.4493
0.05	1.0513	0.9512	0.85	2.3396	0.4274
0.10	1.1052	0.9048	0.90	2.4596	0.4066
0.15	1.1618	0.8607	0.95	2.5857	0.3867
0.20	1.2214	0.8187	1.0	2.7183	0.3679
0.25	1.2840	0.7788	1.1	3.0042	0.3329
0.30	1.3499	0.7408	1.2	3.3201	0.3012
0.35	1.4191	0.7047	1.3	3.6693	0.2725
0.40	1.4918	0.6703	1.4	4.0552	0.2466
0.45	1.5683	0.6376	1.5	4.4817	0.2231
0.50	1.6487	0.6065	1.6	4.9530	0.2019
0.55	1.7333	0.5769	1.7	5.4739	0.1827
0.60	1.8221	0.5488	1.8	6.0496	0.1653
0.65	1.9155	0.5220	1.9	6.6859	0.1496
0.70	2.0138	0.4966	2.0	7.3891	0.1353
0.75	2.1170	0.4724	2.1	8.1662	0.1225

x	e^x	e^{-x}	x	e^x	e^{-x}
2.2	9.0250	0.1108	3.9	49.402	0.0202
2.3	9.9742	0.1003	4.0	54.598	0.0183
2.4	11.023	0.0907	4.1	60.340	0.0166
2.5	12.182	0.0821	4.2	66.686	0.0150
2.6	13.464	0.0743	4.3	73.700	0.0136
2.7	14.880	0.0672	4.4	81.451	0.0123
2.8	16.445	0.0608	4.5	90.017	0.0111
2.9	18.174	0.0550	4.6	99.484	0.0101
3.0	20.086	0.0498	4.7	109.55	0.0091
3.1	22.198	0.0450	4.8	121.51	0.0082
3.2	24.533	0.0408	4.9	134.29	0.0074
3.3	27.113	0.0369	5	148.41	0.0067
3.4	29.964	0.0334	6	403.43	0.0025
3.5	33.115	0.0302	7	1096.6	0.0009
3.6	36.598	0.0273	8	2981.0	0.0003
3.7	40.447	0.0247	9	8103.1	0.0001
3.8	44.701	0.0224	10	22026	0.00005

APPENDIX C

Trigonometric Functions

Degrees	Radians	Sin	Cos	Tan	Cot	Sec	Csc		
0° 00′	.0000	.0000	1.0000	.0000	——	1.000	——	1.5708	90° 00′
10	029	029	000	029	343.8	000	343.8	679	50
20	058	058	000	058	171.9	000	171.9	650	40
30	.0087	.0087	1.0000	.0087	114.6	1.000	114.6	1.5621	30
40	116	116	.9999	116	85.94	000	85.95	592	20
50	145	145	999	145	68.75	000	68.76	563	10
1° 00′	.0175	.0175	.9998	.0175	57.29	1.000	57.30	1.5533	89° 00′
10	204	204	998	204	49.10	000	49.11	504	50
20	233	233	997	233	42.96	000	42.98	475	40
30	.0262	.0262	.9997	.0262	38.19	1.000	38.20	1.5446	30
40	291	291	996	291	34.37	000	34.38	417	20
50	320	320	995	320	31.24	001	31.26	388	10
2° 00′	.0349	.0349	.9994	.0349	28.64	1.001	28.65	1.5359	88° 00′
10	378	378	993	378	26.43	001	26.45	330	50
20	407	407	992	407	24.54	001	24.56	301	40
30	.0436	.0436	.9990	.0437	22.90	1.001	22.93	1.5272	30
40	465	465	989	466	21.47	001	21.49	243	20
50	495	494	988	495	20.21	001	20.23	213	10
3° 00′	.0524	.0523	.9986	.0524	19.08	1.001	19.11	1.5184	87° 00′
10	553	552	985	553	18.07	002	18.10	155	50
20	582	581	983	582	17.17	002	17.20	126	40
30	.0611	.0610	.9981	.0612	16.35	1.002	16.38	1.5097	30
40	640	640	980	641	15.60	002	15.64	068	20
50	669	669	978	670	14.92	002	14.96	039	10
4° 00′	.0698	.0698	.9976	.0699	14.30	1.002	14.34	1.5010	86° 00′
10	727	727	974	729	13.73	003	13.76	981	50
20	756	756	971	758	13.20	003	13.23	952	40
30	.0785	.0785	.9969	.0787	12.71	1.003	12.75	1.4923	30
40	814	814	967	816	12.25	003	12.29	893	20
50	844	843	964	846	11.83	004	11.87	864	10
5° 00′	.0873	.0872	.9962	.0875	11.43	1.004	11.47	1.4835	85° 00′
10	902	901	959	904	11.06	004	11.10	806	50
20	931	929	957	934	10.71	004	10.76	777	40
30	.0960	.0958	.9954	.0963	10.39	1.005	10.43	1.4748	30
40	989	987	951	992	10.08	005	10.13	719	20
50	.1018	.1016	948	.1022	9.788	005	9.839	690	10
6° 00′	.1047	.1045	.9945	.1051	9.514	1.006	9.567	1.4661	84° 00′
10	076	074	942	080	9.255	006	9.309	632	50
20	105	103	939	110	9.010	006	9.065	603	40
30	.1134	.1132	.9936	.1139	8.777	1.006	8.834	1.4573	30
40	164	161	932	169	8.556	007	8.614	544	20
50	193	190	929	198	8.345	007	8.405	515	10
7° 00′	.1222	.1219	.9925	.1228	8.144	1.008	8.206	1.4486	83° 00′
10	251	248	922	257	7.953	008	8.016	457	50
20	280	276	918	287	7.770	008	7.834	428	40
30	.1309	.1305	.9914	.1317	7.596	1.009	7.661	1.4399	30
40	338	334	911	346	7.429	009	7.496	370	20
50	367	363	907	376	7.269	009	7.337	341	10
8° 00′	.1396	.1392	.9903	.1405	7.115	1.010	7.185	1.4312	82° 00′
10	425	421	899	435	6.968	010	7.040	283	50
20	454	449	894	465	6.827	011	6.900	254	40
30	.1484	.1478	.9890	.1495	6.691	1.011	6.765	1.4224	30
40	513	507	886	524	6.561	012	6.636	195	20
50	542	536	881	554	6.435	012	6.512	166	10
9° 00′	.1571	.1564	.9877	.1584	6.314	1.012	6.392	1.4137	81°00′
		Cos	Sin	Cot	Tan	Csc	Sec	Radians	Degrees

Degrees	Radians	Sin	Cos	Tan	Cot	Sec	Csc		
9° 00′	.1571	.1564	.9877	.1584	6.314	1.012	6.392	1.4137	81° 00′
10	600	593	872	614	197	013	277	108	50
20	629	622	868	644	084	013	166	079	40
30	.1658	.1650	.9863	.1673	5.976	1.014	6.059	1.4050	30
40	687	679	858	703	871	014	5.955	1.4021	20
50	716	708	853	733	769	015	855	992	10
10° 00′	.1745	.1736	.9848	.1763	5.671	1.015	5.759	1.3963	80° 00′
10	774	765	843	793	576	016	665	934	50
20	804	794	838	823	485	016	575	904	40
30	.1833	.1822	.9833	.1853	5.396	1.017	5.487	1.3875	30
40	862	851	827	883	309	018	403	846	20
50	891	880	822	914	226	018	320	817	10
11° 00′	.1920	.1908	.9816	.1944	5.145	1.019	5.241	1.3788	79° 00′
10	949	937	811	974	066	019	164	759	50
20	978	965	805	.2004	4.989	020	089	730	40
30	.2007	.1994	.9799	.2035	4.915	1.020	5.016	1.3701	30
40	036	.2022	793	065	843	021	4.945	672	20
50	065	051	787	095	773	022	876	643	10
12° 00′	.2094	.2079	.9781	.2126	4.705	1.022	4.810	1.3614	78° 00′
10	123	108	775	156	638	023	745	584	50
20	153	136	769	186	574	024	682	555	40
30	.2182	.2164	.9763	.2217	4.511	1.024	4.620	1.3526	30
40	211	193	757	247	449	025	560	497	20
50	240	221	750	278	390	026	502	468	10
13° 00′	.2269	.2250	.9744	.2309	4.331	1.026	4.445	1.3439	77° 00′
10	298	278	737	339	275	027	390	410	50
20	327	306	730	370	219	028	336	381	40
30	.2356	.2334	.9724	.2401	4.165	1.028	4.284	1.3352	30
40	385	363	717	432	113	029	232	323	20
50	414	391	710	462	061	030	182	294	10
14° 00′	.2443	.2419	.9703	.2493	4.011	1.031	4.134	1.3265	76° 00′
10	473	447	696	524	3.962	031	086	235	50
20	502	476	689	555	914	032	039	206	40
30	.2531	.2504	.9681	.2586	3.867	1.033	3.994	1.3177	30
40	560	532	674	617	821	034	950	148	20
50	589	560	667	648	776	034	906	119	10
15° 00′	.2618	.2588	.9659	.2679	3.732	1.035	3.864	1.3090	75° 00′
10	647	616	652	711	689	036	822	061	50
20	676	644	644	742	647	037	782	032	40
30	.2705	.2672	.9636	.2773	3.606	1.038	3.742	1.3003	30
40	734	700	628	805	566	039	703	974	20
50	763	728	621	836	526	039	665	945	10
16° 00′	.2793	.2756	.9613	.2867	3.487	1.040	3.628	1.2915	74° 00′
10	822	784	605	899	450	041	592	886	50
20	851	812	596	931	412	042	556	857	40
30	.2880	.2840	.9588	.2962	3.376	1.043	3.521	1.2828	30
40	909	868	580	994	340	044	487	799	20
50	938	896	572	.3026	305	045	453	770	10
17° 00′	.2967	.2924	.9563	.3057	3.271	1.046	3.420	1.2741	73° 00′
10	996	952	555	089	237	047	388	712	50
20	.3025	979	546	121	204	048	356	683	40
30	.3054	.3007	.9537	.3153	3.172	1.049	3.326	1,2654	30
40	083	035	528	185	140	049	295	625	20
50	113	062	520	217	108	050	265	595	10
18° 00′	.3142	.3090	.9511	.3249	3.078	1.051	3.236	1.2566	72° 00′
		Cos	Sin	Cot	Tan	Csc	Sec	Radians	Degrees

Degrees	Radians	Sin	Cos	Tan	Cot	Sec	Csc		
18° 00′	.3142	.3090	.9511	.3249	3.078	1.051	3.236	1.2566	72° 00′
10	171	118	502	281	047	052	207	537	50
20	200	145	492	314	018	053	179	508	40
30	.3229	.3173	.9483	.3346	2.989	1.054	3.152	1.2479	30
40	258	201	474	378	960	056	124	450	20
50	287	228	465	411	932	057	098	421	10
19° 00′	.3316	.3256	.9455	.3443	2.904	1.058	3.072	1.2392	71° 00′
10	345	283	446	476	877	059	046	363	50
20	374	311	436	508	850	060	021	334	40
30	.3403	.3338	.9426	.3541	2.824	1.061	2.996	1.2305	30
40	432	365	417	574	798	062	971	275	20
50	462	393	407	607	773	063	947	246	10
20° 00′	.3491	.3420	.9397	.3640	2.747	1.064	2.924	1.2217	70° 00′
10	520	448	387	673	723	065	901	188	50
20	549	475	377	706	699	066	878	159	40
30	.3578	.3502	.9367	.3739	2.675	1.068	2.855	1.2130	30
40	607	529	356	772	651	069	833	101	20
50	636	557	346	805	628	070	812	072	10
21° 00′	.3665	.3584	.9336	.3839	2.605	1.071	2.790	1.2043	69° 00′
10	694	611	325	872	583	072	769	1.2014	50
20	723	638	315	906	560	074	749	985	40
30	.3752	.3665	.9304	.3939	2.539	1.075	2.729	1.1956	30
40	782	692	293	973	517	076	709	926	20
50	811	719	283	.4006	496	077	689	897	10
22° 00′	.3840	.3746	.9272	.4040	2.475	1.079	2.669	1.1868	68° 00′
10	869	773	261	074	455	080	650	839	50
20	898	800	250	108	434	081	632	810	40
30	.3927	.3827	.9239	.4142	2.414	1.082	2.613	1.1781	30
40	956	854	228	176	394	084	595	752	20
50	985	881	216	210	375	085	577	723	10
23° 00′	.4014	.3907	.9205	.4245	2.356	1.086	2.559	1.1694	67° 00′
10	043	934	194	279	337	088	542	665	50
20	072	961	182	314	318	089	525	636	40
30	.4102	.3987	.9171	.4348	2.300	1.090	2.508	1.1606	30
40	131	.4014	159	383	282	092	491	577	20
50	160	041	147	417	264	093	475	548	10
24° 00′	.4189	.4067	.9135	.4452	2.246	1.095	2.459	1.1519	66° 00′
10	218	094	124	487	229	096	443	490	50
20	247	120	112	522	211	097	427	461	40
30	.4276	.4147	.9100	.4557	2.194	1.099	2.411	1.1432	30
40	305	173	088	592	177	100	396	403	20
50	334	200	075	628	161	102	381	374	10
25° 00′	.4363	.4226	.9063	.4663	2.145	1.103	2.366	1.1345	65° 00′
10	392	253	051	699	128	105	352	316	50
20	422	279	038	734	112	106	337	286	40
30	.4451	.4305	.9026	.4770	2.097	1.108	2.323	1.1257	30
40	480	331	013	806	081	109	309	228	20
50	509	358	001	841	066	111	295	199	10
26° 00′	.4538	.4384	.8988	.4877	2.050	1.113	2.281	1.1170	64° 00′
10	567	410	975	913	035	114	268	141	50
20	596	436	962	950	020	116	254	112	40
30	.4625	.4462	.8949	.4986	2.006	1.117	2.241	1.1083	30
40	654	488	936	.5022	1.991	119	228	054	20
50	683	514	923	059	977	121	215	1.1025	10
27° 00′	.4712	.4540	.8910	.5095	1.963	1.122	2.203	1.0996	63° 00′
		Cos	Sin	Cot	Tan	Csc	Sec	Radians	Degrees

			s	Tan	Cot	Sec	Csc		
			10	.5095	1.963	1.122	2.203	1.0996	63° 00′
			97	132	949	124	190	966	50
			34	169	935	126	178	937	40
			70	.5206	1.921	1.127	2.166	1.0908	30
			57	243	907	129	154	879	20
			3	280	894	131	142	850	10
			9	.5317	1.881	1.133	2.130	1.0821	62° 00′
			6	354	868	134	118	792	50
			2	392	855	136	107	763	40
			8	.5430	1.842	1.138	2.096	1.0734	30
			4	467	829	140	085	705	20
				505	816	142	074	676	10
				.5543	1.804	1.143	2.063	1.0647	61° 00′
				581	792	145	052	617	50
				619	780	147	041	588	40
				.5658	1.767	1.149	2.031	1.0559	30
				696	756	151	020	530	20
				735	744	153	010	501	10
				.5774	1.732	1.155	2.000	1.0472	60° 00′
				812	720	157	1.990	443	50
				851	709	159	980	414	40
				.5890	1.698	1.161	1.970	1.0385	30
				930	686	163	961	356	20
				969	675	165	951	327	10
				.6009	1.664	1.167	1.942	1.0297	59° 00′
				048	653	169	932	268	50
				088	643	171	923	239	40
				.6128	1.632	1.173	1.914	1.0210	30
				168	621	175	905	181	20
				208	611	177	896	152	10
				.6249	1.600	1.179	1.887	1.0123	58° 00′
				289	590	181	878	094	50
				330	580	184	870	065	40
				.6371	1.570	1.186	1.861	1.0036	30
				412	560	188	853	1.0007	20
	730	422	403	453	550	190	844	977	10
33° 00′	.5760	.5446	.8387	.6494	1.540	1.192	1.836	.9948	57° 00′
10	789	471	371	536	530	195	828	919	50
20	818	495	355	577	520	197	820	890	40
30	.5847	.5519	.8339	.6619	1.511	1.199	1.812	.9861	30
40	876	544	323	661	501	202	804	832	20
50	905	568	307	703	1.492	204	796	803	10
34° 00′	.5934	.5592	.8290	.6745	1.483	1.206	1.788	.9774	56° 00′
10	963	616	274	787	473	209	781	745	50
20	992	640	258	830	464	211	773	716	40
30	.6021	.5664	.8241	.6873	1.455	1.213	1.766	.9687	30
40	050	688	225	916	446	216	758	657	20
50	080	712	208	959	437	218	751	628	10
35° 00′	.6109	.5736	.8192	.7002	1.428	1.221	1.743	.9599	55° 00′
10	138	760	175	046	419	223	736	570	50
20	167	783	158	089	411	226	729	541	40
30	.6196	.5807	.8141	.7133	1.402	1.228	1.722	.9512	30
40	225	831	124	177	393	231	715	483	20
50	254	854	107	221	385	233	708	454	10
36° 00′	.6283	.5878	.8090	.7265	1.376	1.236	1.701	.9425	54° 00′
		Cos	Sin	Cot	Tan	Csc	Sec	Radians	Degrees

Degrees	Radians	Sin	Cos	Tan	Cot	Sec	Csc		
36° 00′	.6283	.5878	.8090	.7265	1.376	1.236	1.701	.9425	54° 00′
10	312	901	073	310	368	239	695	396	50
20	341	925	056	355	360	241	688	367	40
30	.6370	.5948	.8039	.7400	1.351	1.244	1.681	.9338	30
40	400	972	021	445	343	247	675	308	20
50	429	995	004	490	335	249	668	279	10
37° 00′	.6458	.6018	.7986	.7536	1.327	1.252	1.662	.9250	53° 00′
10	487	041	969	581	319	255	655	221	50
20	516	065	951	627	311	258	649	192	40
30	.6545	.6088	.7934	.7673	1.303	1.260	1.643	.9163	30
40	574	111	916	720	295	263	636	134	20
50	603	134	898	766	288	266	630	105	10
38° 00′	.6632	.6157	.7880	.7813	1.280	1.269	1.624	.9076	52° 00′
10	661	180	862	860	272	272	618	047	50
20	690	202	844	907	265	275	612	.9018	40
30	.6720	.6225	.7826	.7954	1.257	1.278	1.606	.8988	30
40	749	248	808	.8002	250	281	601	959	20
50	778	271	790	050	242	284	595	930	10
39° 00′	.6807	.6293	.7771	.8098	1.235	1.287	1.589	.8901	51° 00′
10	836	316	753	146	228	290	583	872	50
20	865	338	735	195	220	293	578	843	40
30	.6894	.6361	.7716	.8243	1.213	1.296	1.572	.8814	30
40	923	383	698	292	206	299	567	785	20
50	952	406	679	342	199	302	561	756	10
40° 00′	.6981	.6428	.7660	.8391	1.192	1.305	1.556	.8727	50° 00′
10	.7010	450	642	441	185	309	550	698	50
20	039	472	623	491	178	312	545	668	40
30	7069	.6494	.7604	.8541	1.171	1.315	1.540	.8639	30
40	098	517	585	591	164	318	535	610	20
50	127	539	566	642	157	322	529	581	10
41° 00′	.7156	.6561	.7547	.8693	1.150	1.325	1.524	.8552	49° 00′
10	185	583	528	744	144	328	519	523	50
20	214	604	509	796	137	332	514	494	40
30	.7243	.6626	.7490	.8847	1.130	1.335	1.509	.8465	30
40	272	648	470	899	124	339	504	436	20
50	301	670	451	952	117	342	499	407	10
42° 00′	.7330	.6691	.7431	.9004	1.111	1.346	1.494	.8378	48° 00′
10	359	713	412	057	104	349	490	348	50
20	389	734	392	110	098	353	485	319	40
30	.7418	.6756	.7373	.9163	1.091	1.356	1.480	.8290	30
40	447	777	353	217	085	360	476	261	20
50	476	799	333	271	079	364	471	232	10
43° 00′	.7505	.6820	.7314	.9325	1.072	1.367	1.466	.8203	47° 00′
10	534	841	294	380	066	371	462	174	50
20	563	862	274	435	060	375	457	145	40
30	.7592	.6884	.7254	.9490	1.054	1.379	1.453	.8116	30
40	621	905	234	545	048	382	448	087	20
50	650	926	214	601	042	386	444	058	10
44° 00′	.7679	.6947	.7193	.9657	1.036	1.390	1.440	.8029	46° 00′
10	709	967	173	713	030	394	435	999	50
20	738	988	153	770	024	398	431	970	40
30	.7767	.7009	.7133	.9827	1.018	1.402	1.427	.7941	30
40	796	030	112	884	012	406	423	912	20
50	825	050	092	942	006	410	418	883	10
45° 00′	.7854	.7071	.7071	1.000	1.000	1.414	1.414	.7854	45° 00′
		Cos	Sin	Cot	Tan	Csc	Sec	Radians	Degrees

49. 25% **51.** 9.3 **53.** .2 **55.** \$736.75

57. 37.5% or $37\frac{1}{2}\%$

COMPLETION SET 2-1

1. set **2.** elements **3.** positive **4.** 4 **5.** zero

6. 0 **7.** irrational **8.** rational **9.** 4 **10.** -8

11. -3 **12.** -8 **13.** 6 **14.** -3 **15.** -12

16. 0 **17.** T **18.** F **19.** T **20.** T

21. F **22.** F

COMPLETION SET 2-2

1. greater **2.** left **3.** $4 > 3$ **4.** positive

5. absolute value **6.** $a > 6$ **7.** zero, negative **8.** $<$

9. $>$ **10.** $>$ **11.** $<$ **12.** $>$ **13.** $<$

14. 6 **15.** 270 **16.** $\frac{7}{2}$ **17.** 0 **18.** 1

19. 2 **20.** 5 **21.** -5 **22.** -4

23. 0, -3, 4, 5, -6 **24.** b, c

COMPLETION SET 2-3

1. negative **2.** 8 **3.** 12 **4.** -7 **5.** -5, $\frac{1}{5}$

6. 0 **7.** 1 **8.** $\frac{5}{3}$ **9.** 4 **10.** 0

11. $\frac{2}{x}$ **12.** 0 **13.** -5 **14.** -5

PROBLEM SET 2-3

1. 12 **3.** -9 **5.** 2 **7.** -2 **9.** -3

11. -10 **13.** -1 **15.** 0 **17.** 12 **19.** -2

21. 8 **23.** 2 **25.** $-\frac{1}{8}$ **27.** -10 **29.** 3

31. $-a$ **33.** 0 **35.** -3 **37.** $\frac{1}{4}$ **39.** 0

41. -2 43. 1 45. 63 47. -15 49. $-\frac{2}{3}$

51. $\frac{1}{6}$ 53. $-\frac{1}{2}$ 55. $-\frac{3}{4}$ 57. $\frac{1}{3}$ 59. $-\frac{4}{3}$

61. $5xy$ 63. $\frac{x}{yz}$ 65. $6y$ 67. $-\frac{x}{yz}$ 69. $\frac{6}{x}$

71. 0 73. $\frac{x}{5}$ 75. $\frac{1}{y}$

COMPLETION SET 2-4

1. $b + a$ 2. $(ab)c$ 3. $2c$ 4. $3y + 2$

5. commutative 6. associative 7. distributive

8. $+$ 9. $-$ 10. $3ab$, $6ac$, $9ad$ 11. xy

12. y 13. F 14. F 15. T 16. F

17. T 18. T

PROBLEM SET 2-4

1. $10x$ 3. $7 + x$ 5. -4 7. -8 9. 0

11. -16 13. $8 + 4x$ 15. $6x - 4$ 17. $3xz + 4yz$ 19. $xy + 3xz$

21. $4xy - 3y$ 23. $12xy$ 25. $6x - 9xy$ 27. $xy - 2xz + xw$

29. $10x - 5xy - 15xz$ 31. $3x$ 33. $\frac{9x}{4}$ 35. $24xyz$

37. $2x$ 39. $-\frac{x}{2}$ 41. $-15x$ 43. $\frac{x}{2}$ 45. $\frac{4}{3z}$

47. $\frac{9y}{2}$ 49. $\frac{3}{2p}$

PROBLEM SET 2-5

1. $-31°F$ 3. $-40°C$ 5. 8 7. $\frac{1}{13}$ 9. 3

11. $\frac{5}{4}$ 13. $\frac{19}{9}$

REVIEW PROBLEMS—CHAPTER 2

1. commutative law 3. associative law 5. commutative law

7. distributive law 9. true 11. true 13. false

15. false **17.** $-\frac{1}{2}$ **19.** $3xy + 15x$ **21.** $-10y$ **23.** -1

25. -12 **27.** 1 **29.** $20x - 12$ **31.** $x - 1$ **33.** -8

35. $3xz$ **37.** 15 **39.** 5 **41.** $28xy$ **43.** $\frac{2}{63}$

45. 12 **47.** 1 **49.** $-12x$ **51.** $8y - xy$

53. (a) -6, (b) $-\frac{13}{6}$ **55.** (a) 11, (b) 5, (c) 11, (d) -5

COMPLETION SET 3-1

1. five **2.** 6, 9 **3.** $3x$, x **4.** 1 **5.** -1

6. 1 **7.** 18 **8.** + **9.** · **10.** −

11. multiply **12.** $8x^3$

PROBLEM SET 3-1

1. 8 **3.** -16 **5.** -16 **7.** -72 **9.** $\frac{1}{3}$

11. -64 **13.** x^{11} **15.** y^9 **17.** x^7 **19.** $(x - 2)^8$

21. $\frac{x^7}{y^9}$ **23.** $14x^8$ **25.** $-12x^4$ **27.** x^{16} **29.** x^9

31. t^{2n} **33.** x^{29} **35.** x^4 **37.** $\frac{1}{x}$ **39.** $-y^6$

41. $\frac{1}{x^6}$ **43.** x^{13} **45.** x **47.** $\frac{1}{x^5}$ **49.** a^6b^6

51. $16x^4$ **53.** $45y^6$ **55.** $\frac{a^3}{b^3}$ **57.** $\frac{81}{x^4}$ **59.** x^4y^8

61. $\frac{8y^3}{z^3}$ **63.** $\frac{4}{9}a^4b^6c^{12}$ **65.** $\frac{x^6}{y^{15}}$ **67.** $\frac{x^8y^{12}}{16z^{16}}$ **69.** $-x^{13}$

71. $16x^8y^4$ **73.** $-\frac{x^5y^5}{t^4}$ **75.** x^{10} **77.** 4 **79.** $\frac{1}{32}$

81. -26

COMPLETION SET 3-2

1. fourth **2.** 2 **3.** 4 **4.** 7 **5.** index

6. radicand **7.** 0 **8.** 4, 2 **9.** 16, 4 **10.** 25, 5

PROBLEM SET 3-2

1. 7	**3.** 2	**5.** 6	**7.** -3	**9.** -4
11. 2	**13.** 0	**15.** -5	**17.** 2	**19.** .2
21. $\frac{1}{4}$	**23.** -5	**25.** 11	**27.** $\frac{1}{3}$	**29.** 5
31. 4	**33.** $5\sqrt{2}$	**35.** $2\sqrt{3}$	**37.** $2\sqrt{2}$	**39.** $3\sqrt{6}$
41. $2\sqrt[4]{3}$	**43.** $2\sqrt[5]{2}$	**45.** $-5\sqrt[3]{4}$	**47.** $\frac{1}{5}$	**49.** $\frac{\sqrt{14}}{3}$
51. $\frac{\sqrt[3]{10}}{3}$	**53.** 5	**55.** $-\frac{1}{2}$	**57.** 4	**59.** 4
61. 13	**63.** 25	**65.** 3	**67.** $\frac{3}{2}$	**69.** 2

REVIEW PROBLEMS—CHAPTER 3

In Problems **1–23**, T = true and F = false.

1. F	**3.** F	**5.** F	**7.** F	**9.** T
11. F	**13.** F	**15.** T	**17.** F	**19.** T
21. T	**23.** F	**25.** x^{13}	**27.** $\frac{x^{10}}{y^{50}}$	**29.** $-32x^5y^{20}$
31. x^{14}	**33.** $-x^5$	**35.** $\frac{1}{25}$	**37.** 3	**39.** 0
41. 13	**43.** $\frac{1}{9}$	**45.** .15	**47.** $3\sqrt{5}$	**49.** $3\sqrt[4]{10}$
51. $\frac{\sqrt{7}}{4}$	**53.** 2	**55.** 6	**57.** 6 square units	

COMPLETION SET 4-1

1. F	**2.** F	**3.** T	**4.** F	**5.** F
6. F	**7.** y	**8.** $5z$	**9.** 20, 5	**10.** 4, -2
11. 5, -5	**12.** -6, 2			

PROBLEM SET 4-1

1. $x = -3$	**3.** $x = 3$	**5.** $y = 14$	**7.** $x = -6$	**9.** $x = 3$
11. $t = -3$	**13.** $x = 15$	**15.** $x = 10$	**17.** $x = 0$	**19.** $r = -2$
21. $t = \frac{7}{4}$	**23.** $x = -3$	**25.** $y = -12$	**27.** $x = \frac{1}{2}$	**29.** $u = \frac{3}{2}$

COMPLETION SET 4-2

1. 5, 35 **2.** 3, -6 **3.** 5, $\frac{15}{4}$ **4.** 3, $-\frac{15}{2}$ **5.** 4, 16, -16

6. 10 **7.** 12 **8.** 8

PROBLEM SET 4-2

1. $x = 24$ **3.** $x = 0$ **5.** $x = \frac{8}{3}$ **7.** $y = -18$ **9.** $x = \frac{21}{2}$

11. $y = \frac{24}{5}$ **13.** $x = -\frac{15}{4}$ **15.** $t = \frac{2}{9}$ **17.** $x = \frac{1}{3}$ **19.** $x = \frac{7}{5}$

21. $u = \frac{112}{3}$ **23.** $w = \frac{6}{5}$ **25.** $x = 15$ **27.** $S = 3$ **29.** $x = -\frac{91}{9}$

31. $x = \frac{5}{2}$ **33.** $x = -\frac{1}{3}$ **35.** $x = 23$ **37.** $x = -\frac{1}{4}$ **39.** 273 cm^3

COMPLETION SET 4-3

1. a **2.** w, 84 **3.** .12, 20 **4.** 90, 15

PROBLEM SET 4-3

1. 120 m **3.** $r = \dfrac{C}{2\pi}$ **5.** $r = \dfrac{I}{Pt}$ **7.** $t = \dfrac{A - P}{Pr}$

9. $h = \dfrac{A - 2\pi r^2}{2\pi r}$ **11.** 5 ft **13.** 40 ft per sec

15. 400 **17.** 15,000 **19.** 114 **21.** $200,000 **23.** 300

25. 625 **27.** 4% **29.** 110% **31.** 3500 **33.** 42.5%

COMPLETION SET 4-4

1. 2, -1, 1 **2.** literal **3.** a, d **4.** 9, 2, 5 **5.** 7, 4

6. 5 **7.** $-$, $+$ **8.** $-$, $+$ **9.** $+$, $+$, $-$ **10.** $+$, $-$, $+$

11. (a) braces, (b) parentheses, (c) brackets **12.** 4, 4, 8, 8

13. 4, 2, 5, 4, 15, 12

PROBLEM SET 4-4

1. $12x^2$ **3.** $-7y$ **5.** $11x - 2$ **7.** $40x + 12y$

9. $-14x + 12y$ **11.** $8y - 8$ **13.** $4x - 3y$ **15.** $7a - 8b + c$

17. $5x^2 + 15$ **19.** $2xy + 5z - 7$ **21.** $2x^2 - 9x - 13$ **23.** $-10 + 4x$

25. $5x + 50$ **27.** $-a + 6$ **29.** $29x^2 - 22$

31. $8a - 11b - 13c$ **33.** $18x - 24$ **35.** $9 - 100x + 60y$ **37.** $-2x - 4$

COMPLETION SET 4-5

1. 4, 3 **2.** 2, 4 **3.** 10, 14, $\frac{14}{5}$ **4.** 6, 9, 9, 1 **5.** 4, -4, 2

PROBLEM SET 4-5

1. $x = 2$ **3.** $x = 0$ **5.** $y = 3$ **7.** $x = 0$ **9.** $x = -3$

11. $x = 2$ **13.** $y = 11$ **15.** $z = 2$ **17.** $x = -2$ **19.** $x = -\frac{5}{2}$

21. $r = -\frac{32}{11}$ **23.** $t = 4$ **25.** $x = -30$ **27.** $x = -\frac{24}{7}$ **29.** $x = -31$

31. $x = -\frac{23}{20}$ **33.** $z = \frac{1}{13}$ **35.** $x = \frac{8}{9}$ **37.** $C = \dfrac{5F - 160}{9}$

39. 105 gm of A and 175 gm of B

41. 660 Little Big Horns, 1980 Early Americans **43.** 3 mi

REVIEW PROBLEMS—CHAPTER 4

1. $11x - 2y - 3$ **3.** $-2a + 19b + 4$ **5.** $-8xy + 26$ **7.** $9x - 18$

9. $3x^2 + 18x - 36$ **11.** $x = \frac{1}{2}$ **13.** $y = \frac{3}{2}$ **15.** $x = -\frac{3}{2}$

17. $z = -\frac{20}{9}$ **19.** $u = -\frac{11}{14}$ **21.** $x = -32$ **23.** $x = \frac{37}{21}$

25. $a = \dfrac{v - v_o}{t}$ **27.** .6 **29.** 120 **31.** 12%

33. 8250 **35.** 112 tons of A, 42 tons of B, 14 tons of C

37. 4500

COMPLETION SET 5-1

1. $24x^2y^3$ **2.** 3, x, y^3, $6x^3y^4$ **3.** x, $2y$, 1 **4.** $2(x - 4)$

5. F **6.** F **7.** T

PROBLEM SET 5-1

1. $15x^3$ **3.** $6x^2y$ **5.** $3a^3b^2$ **7.** $-8x^4y^4$

9. $a^3b^3c^2$ **11.** x^3yz^4 **13.** $24x^4y^5$ **15.** $90x^{10}$

17. $ab^2c^4d^2$ **19.** $x^3 - 4x^2 + 7x$

21. $-3a^2b + a^3b^2 - a^3b$ **23.** $-5x^3y + 5xy^3 - 5x^2y^2$

25. $4x^5y^2 + 8x^4y^4 - 12x^6y^2$ **27.** $-14xy + 8x^2y + 4xy^2 - 2x^3y$

29. $x^2 + 7x + 10$ **31.** $3y^2 - 4y - 4$ **33.** $9x^2 - 6x + 1$

35. $16x^2 + 8xy + y^2$ **37.** $6x^4y + 6x^3y^2$ **39.** $t^3 - 8$

41. $x^4 - 5x^3 - x^2 + 10x - 2$ **43.** $y^5 - y^4 + 5y^3 - 3y^2 + 6y$

45. $x^2 + 2xy + y^2 - 1$ **47.** $16x^3 - 4x$ **49.** $4a^2b^2 + 4abrt + r^2t^2$

51. $2xy + y^2$ **53.** $6x$ **55.** $x^2 + x - 6$ **57.** $-15x^2y^2$

59. $-2x^2 + 3x - 1$ **61.** 0 **63.** $11\frac{1}{4}$ km **65.** $\frac{2}{3}$ hr or 40 min

67. 4 days; 56 in.

COMPLETION SET 5-2

1. $\frac{x}{5}$ **2.** $-\frac{2x^2}{z}$ **3.** $\frac{y}{2xz}$ **4.** 3, 2, $\frac{5}{3}$ **5.** 2, 1, $\frac{y}{x}$

6. 100, x **7.** T **8.** F

PROBLEM SET 5-2

1. $\frac{b}{2}$ **3.** $-\frac{2b^2}{a}$ **5.** $-c$ **7.** $\frac{x}{2yz^2}$ **9.** $\frac{3y^3z}{2x}$

11. $-\frac{ac}{b^2}$ **13.** $9x^2y$ **15.** $\frac{x^2}{9y^2}$ **17.** $2x^2 - 3x + 4$

19. $\frac{5}{x} - x + 2$ **21.** $2x^2 - 3 + \frac{2}{5x}$ **23.** $-2 - \frac{y}{x}$ **25.** $1 + \frac{x}{y}$

27. $\frac{3}{y} - x + \frac{1}{x^2y^2}$ **29.** $-2xy + \frac{2y^2}{3x} - \frac{7}{3} + \frac{4}{3x}$

31. $-10xy + \frac{5y}{2} - \frac{1}{y}$ **33.** $\frac{2}{xy} - 4y^2 - 3xy$

35. $-2x^6 - \frac{6x}{y^2} + \frac{1}{y^4}$ **37.** $x - 2 + \frac{1}{x}$

39. 20 lb

41. 420 gallons of 20% solution, 280 gallons of 30% solution

43. 32 ml

COMPLETION SET 5-3

1. a, b, d **2.** 3 **3.** -3 **4.** $x^2 - x + 1$

5. $x^2 + 0x + 1$ **6.** $-x^2 + 0x + 1$ **7.** $3x$ **8.** $-4x$

9. quotient

PROBLEM SET 5-3

1. $2x + 7 + \frac{10}{x - 2}$ **3.** $1 + \frac{1}{x + 2}$ **5.** $x - 2 - \frac{3}{4x + 1}$

7. $2x - 3 + \frac{4}{2x + 3}$ **9.** $x^2 + 3x - 2$

11. $-x^3 + x^2 - 2x + 4 - \frac{19}{3x + 5}$ **13.** $2x^2 + 1 - \frac{2}{4x - 1}$

15. $x^3 + x^2 - x - 1$ **17.** $-x^3 + 3x^2 - 9x + 27$

PROBLEM SET 5-4

1. $x^2 - x + 1, R = -2$ **3.** $3x^3 - 6x^2 + 8x - 16, R = 17$

5. $2t^3 - t^2 + 3t - 3, R = 7$ **7.** $x^3 + 3x - 2, R = 0$

9. $x^2 - 2x + 1, R = 3$ **11.** $x^2 + x + 1, R = 0$

13. $R = 0$. Thus $x - 3$ is a factor.

REVIEW PROBLEMS—CHAPTER 5

1. $2x^3y^4z^7$ **3.** $72a^5b^4$ **5.** $4x^8y^7$

7. $x^3 - 2x^2 + 4x$ **9.** $-2a^4b^2 + 2a^3b^2 - 3a^2b$

11. $x^2 - x - 12$ **13.** $x^2 - 16$ **15.** $x^2 - 5x + 6$

17. $x^2 + 4xy + 4y^2$ **19.** $x^3 - 4x^2 + 3x - 12$

21. $6x^4 - 11x^3 + 3x^2 + 15x - 5$ **23.** $2x^2 - xy + 5x - 3y^2 - 5y + 2$

25. $x^2 - 12x + 18$ **27.** $\frac{ay^2}{x}$ **29.** $\frac{ab^2}{5}$

31. $-\frac{3}{8x}$ **33.** $x - 5 + \frac{7}{x}$ **35.** $\frac{x}{y} - 5y + \frac{7}{y}$

37. $-x - 2xy^3w^2 + 2x^2y^2w$ **39.** $3x^2 + 3x - 1 - \frac{2}{2x-1}$ **41.** $x^3 - 2x^2 + 6x - 10$

43. $-x^3 + 2x^2 - x + 2 - \frac{2}{2x+1}$ **45.** $x + 3, R = 0$

47. $2x^2 - x + 1, R = 2$ **49.** $6\frac{3}{7}$ mi

51. 8000 kg of Industrial Strength and 4000 kg of Household Strength

COMPLETION SET 6-1

1. +, + **2.** −, + **3.** − **4.** 5, 5, 10, 25

5. $4x^2$, 12 **6.** 5, 25 **7.** x, 3, x^2, 9 **8.** 6

9. 6 **10.** 16, 63 **11.** 3, 10

12. 9, $2x^3$, $12x^2$, $18x$ **13.** binomial, trinomial

PROBLEM SET 6-1

1. $x^2 + 8x + 16$ **3.** $y^2 + 20y + 100$ **5.** $16x^2 + 8x + 1$

7. $x^2 + x + \frac{1}{4}$ **9.** $x^2 - 12x + 36$ **11.** $9x^2 - 12x + 4$

13. $4x^2 - 4xy + y^2$ **15.** $9x^2 + 18x + 9$ **17.** $4 - 16y + 16y^2$

19. $x^2 - 9$ **21.** $x^2 - 81$ **23.** $1 - x^2$

25. $9x^4 - 25$ **27.** $4y^2 - 9x^2$ **29.** $36 - x^2$

31. $x^2 + 11x + 24$ **33.** $x^2 + 5x + 4$ **35.** $x^2 - x - 2$

37. $t^2 + 2t - 35$ **39.** $x^2 - 5x + 6$ **41.** $x^2 - 9x + 20$

43. $y^2 + y - 6$ **45.** $2x^2 + 2x - 12$ **47.** $8x^2 - 8x + 2$

49. $6x^2 - 3x - 3$ **51.** $10x^2 - 13x + 4$ **53.** $2 + 9t - 35t^2$

55. $x^2y^4 + 2axy^2 + a^2$ **57.** $x^4 - y^4$ **59.** $x^2 - 7$

61. $4x^2 - 4x + 1$ **63.** $4x^2 + 24x + 36$ **65.** $2xy^2 - 18x$

67. $a^4b^2 - 4a^2bm^2n + 4m^4n^2$ **69.** $x^4 - 16$ **71.** 0

73. $5x^2 - 9x + 11$ **75.** $x^3 - 2x^2 - 4x + 8$ **77.** $16x^4 - 96x^3 + 144x^2$

79. $6x + 1$ **81.** $x^3 + 6x^2 + 12x + 8$

COMPLETION SET 6-2

1. factor **2.** $2x, 3y, z$ **3.** $x^3, x, 1$ **4.** 4 **5.** $2x^2$

6. 1 **7.** $-, -$ **8.** $+, -$ **9.** $-, +$ **10.** 1, 3

11. 2, 3 **12.** 2, 2 **13.** 3, 3 **14.** $2x^2, 2x^2, 3, 3$

15. $2x$ **16.** no

PROBLEM SET 6-2

1. $8(x + 1)$ **3.** $2(7y - 4)$ **5.** $5(2x - y + 5)$

7. $x(5c + 9)$ **9.** $4y(1 - 4y)$ **11.** $3x(2y + z)$

13. $x^2(2x - 1)$ **15.** $x^3y^3(2 + x^2y^2)$ **17.** $4mx^3(m - 2x)$

19. $3a^2y^3(3a^2 + y^2 - 2ayz)$ **21.** $(x + 1)(x - 1)$ **23.** $(x + 1)(x + 3)$

25. $(x + 2)(x + 5)$ **27.** $(x - 4)(x - 5)$ **29.** $(y - 4)(y + 6)$

31. $(y + \sqrt{3})(y - \sqrt{3})$ **33.** $(x + 6)^2$ **35.** $(x - 8)(x + 4)$

37. $(y - 5)^2$ **39.** $(5x + 4)(5x - 4)$ **41.** $(y + \frac{2}{3})(y - \frac{2}{3})$

43. $(3x + 1)(x + 2)$ **45.** $(2y - 1)(y - 3)$ **47.** $(4x + 1)^2$

49. $(3 + 2xy)(3 - 2xy)$ **51.** $(4y - 1)(y + 2)$ **53.** $(2x - 5)(3x + 2)$

55. $(4x + 3)(3x - 2)$ **57.** $2(x + 3)(x - 1)$ **59.** $3x(x + 3)^2$

61. $4s^2t(2t + 1)(2t - 1)$ **63.** $2(2y + 3)(y - 3)$ **65.** $2(x + 3)^2(x + 1)(x - 1)$

67. $2(x + 4)(x + 1)$ **69.** $(x^2 + 4)(x + 2)(x - 2)$

71. $(y^4 + 1)(y^2 + 1)(y + 1)(y - 1)$ **73.** $(x^2 + 2)(x + 1)(x - 1)$

75. $x(x + 1)^2(x - 1)^2$ **77.** $(x + 2)(x^2 - 2x + 4)$ **79.** $2\pi r(r + h)$

COMPLETION SET 6-3

1. x^6, y^6 **2.** nine **3.** x^6

PROBLEM SET 6-3

1. $x^3 + 12x^2 + 48x + 64$ **3.** $y^4 - 8y^3 + 24y^2 - 32y + 16$

5. $243x^5 + 405x^4 + 270x^3 + 90x^2 + 15x + 1$

7. $16z^4 - 32z^3y + 24z^2y^2 - 8zy^3 + y^4$

9. $x^{100} + 100x^{99} + 4950x^{98}$

11. $128x^7 - 1344x^6 + 6048x^5$

13. $y^{20} - 50y^{18} + 1125y^{16}$

15. $243z^{10} + 405z^8 + 270z^6$

REVIEW PROBLEMS—CHAPTER 6

1. $x^2 + 12x + 36$
3. $x^2 - 10x + 25$
5. $4x^2 + 16xy + 16y^2$
7. $x^2 - 64$
9. $9x^2 - 4$
11. $4x^2 - 16y^2$
13. $x^2 - 2x - 24$
15. $x^2 - 13x + 42$
17. $4x^2 - 14x + 12$
19. $y^4 - 16$
21. $2x^4 + 2x^3 - 24x^2$
23. $8y - 18$
25. $16x^3 + 32x^2 - 9x - 18$
27. $2xy^4(3x^2 + 2y^2)$
29. $(x - 5)(x - 6)$
31. $(4 + y)(4 - y)$
33. $x(x - 8)(x + 7)$
35. $(3x - 2)(x + 4)$
37. $2(2x + 5)(2x - 5)$
39. $(3y - 2)(5y + 4)$
41. $(x^2 + 2)(x + 2)(x - 2)$
43. $2x^3(x - 6)(x - 3)$
45. $x^4 - 16x^3 + 96x^2 - 256x + 256$
47. $32x^5 + 80x^4 + 80x^3 + 40x^2 + 10x + 1$

COMPLETION SET 7-1

1. 7
2. 3, 10
3. $x - 2$, $x + 3$, 2, -3
4. -7, 6
5. 2, -3, 4
6. x, x, 100, 0, 100
7. 36, 6
8. 4, 25, 25, 5
9. 3, 3, 4
10. 5, 3, -5, 3

PROBLEM SET 7-1

1. $x = -2, -1$
3. $x = -2, -7$
5. $t = 3, 4$
7. $z = -3, 1$
9. $x = 6$
11. $x = 0, 8$
13. $x = 0, -5$
15. $t = 0, 2$
17. $x = \pm 2$
19. $x = \pm 5$
21. $x = 0$
23. $x = \pm 2\sqrt{3}$
25. $z = \pm 3$
27. $x = \frac{1}{3}, -\frac{3}{2}$
29. $t = \pm 3$
31. $x = \pm \sqrt{7}$
33. $x = \frac{1}{2}, -4$
35. $x = 5, -2$
37. $x = -\frac{1}{2}$
39. $x = -1$
41. $t = -5, 1$
43. $x = 2, 3$
45. $x = 0, -1$
47. $x = 4, -2$
49. $y = -\frac{1}{6}, -\frac{1}{4}$
51. $x = 7, -1$
53. $x = -4 \pm 2\sqrt{2}$
55. $y = \frac{1}{2}, -\frac{3}{2}$
57. $x = 0, 1, -2$
59. $x = 2, -1$
61. $x = 0, 2, -3, 4$
63. $x = 0, \pm 1$
65. $x = 0, \pm 8$
67. $y = 0, -4, -2$
69. $x = \pm 1, \pm 3$
71. $x = 0, \pm 1$
73. $x = \pm 3, \pm 2$

COMPLETION SET 7-2

1. square
2. constant
3. 3
4. 5, 5, -5, $\sqrt{2}$
5. 8, 4, 16, 16, 16, 16, 4, 17, 4, $\sqrt{17}$, -4, $\sqrt{17}$
6. 9, 9
7. 25, 25
8. $\frac{1}{4}$, $\frac{1}{4}$

PROBLEM SET 7-2

1. $x = -3 \pm \sqrt{10}$
3. $x = 2 \pm \sqrt{3}$
5. $y = \frac{3}{2} \pm \frac{\sqrt{13}}{2}$
7. $x = -\frac{1}{2} \pm \frac{\sqrt{17}}{2}$
9. $x = \frac{7}{2} \pm \frac{\sqrt{47}}{2}$
11. $x = -\frac{1}{4} \pm \frac{\sqrt{17}}{4}$
13. $x = -1 \pm \sqrt{5}$

COMPLETION SET 7-3

1. quadratic
2. -4, 7
3. 7, -3
4. -1, 2
5. -5, 0
6. -2, -2, 1, 1, 2, 0, 1
7. $b^2 - 4ac$
8. 3, 1, 1, 5
9. zero

PROBLEM SET 7-3

1. $x = -\frac{3}{2} \pm \frac{\sqrt{5}}{2}$
3. $x = 3$
5. $y = -\frac{3}{4} \pm \frac{\sqrt{41}}{4}$
7. $x = -\frac{5}{2}$
9. $x = -1 \pm \frac{\sqrt{10}}{5}$
11. $x = -\frac{1}{3} \pm \frac{\sqrt{7}}{3}$
13. $x = \pm 6$
15. $x = 1, -7$
17. $z = 0, -\frac{3}{2}$
19. $x = -2 \pm \sqrt{2}$
21. $y = -8$
23. $x = \frac{2}{3} \pm \frac{\sqrt{10}}{3}$
25. $x = \frac{1}{5}, 4$
27. $y = 2 \pm 2\sqrt{2}$
29. $y = \frac{1}{3}$
31. $x = \frac{7}{2} \pm \frac{\sqrt{13}}{2}$
33. $x = \frac{3}{2}, -4$
35. $y = 1, -5$
37. $z = -1 \pm \sqrt{5}$
39. $x = \frac{1}{2}$
41. $x = 2 \pm \frac{\sqrt{10}}{2}$
43. $x = -\frac{1}{2}, 6$

COMPLETION SET 7-4

1. i **2.** imaginary **3.** i, $7i$ **4.** complex
5. negative **6.** -1

PROBLEM SET 7-4

1. $9i$ **3.** $\frac{1}{4}i$ or $\frac{i}{4}$ **5.** $2i\sqrt{3}$ **7.** $4i\sqrt{2}$ **9.** $i\sqrt{2}$
11. $-5i$ **13.** $x = 2 \pm i$ **15.** $x = -1 \pm i\sqrt{2}$ **17.** $x = 1 \pm i\sqrt{3}$
19. $x = \pm 2i$ **21.** $r = -\frac{2}{3} \pm \frac{i\sqrt{2}}{6}$ **23.** $r = \frac{5}{6} \pm \frac{i\sqrt{23}}{6}$
25. $x = \pm i$ **27.** $x = \frac{1}{2} \pm \frac{i\sqrt{15}}{6}$
29. 49, two different real solutions, $x = 2, -\frac{1}{3}$
31. 0, one real solution, $t = -\frac{2}{3}$
33. -20, two different imaginary solutions, $x = \frac{2}{3} \pm \frac{i\sqrt{5}}{3}$
35. 33, two different real solutions, $x = -\frac{1}{2} \pm \frac{\sqrt{33}}{6}$

PROBLEM SET 7-5

1. $3\frac{1}{2}$ sec **3.** 1.3 or 2.4 sec **5.** $x = 10$ **7.** 1 m
9. \$145, or \$155 **11.** 11 in. by 11 in. **13.** 9 cm long, 4 cm wide
15. 25 ft **17.** 5%

REVIEW PROBLEMS—CHAPTER 7

1. $x = 5$ **3.** $x = 6, -4$ **5.** $x = \frac{1}{6}, \frac{3}{2}$ **7.** $x = \pm 2\sqrt{3}$ **9.** $x = 0, \frac{1}{2}$
11. $x = 5 \pm 2\sqrt{6}$ **13.** $x = -\frac{3}{2} \pm \frac{\sqrt{11}}{2}$ **15.** $x = 3 \pm \sqrt{2}$
17. $x = -\frac{1}{2}$ **19.** $x = \frac{1}{2} \pm \frac{3i}{2}$ **21.** $x = \pm \frac{3}{4}$ **23.** $y = 4, -6$
25. $z = 2, -10$ **27.** $t = \frac{3}{4} \pm \frac{i\sqrt{15}}{4}$ **29.** $x = -\frac{3}{2} \pm \frac{\sqrt{17}}{2}$
31. $x = -\frac{5}{4}$ **33.** 4 in. **35.** 50

COMPLETION SET 8-1

1. F **2.** F **3.** T **4.** T **5.** T

6. F **7.** $3x, x^2$ **8.** $x + 3, x - 2$ **9.** $x, 10, -1$

10. x^2 **11.** $x + 3$ **12.** $x + 1$ **13.** $\frac{x - 2}{x + 1}$

PROBLEM SET 8-1

1. $\frac{x}{y}$ **3.** $\frac{a}{bc}$ **5.** $-\frac{x}{y + z}$ **7.** $\frac{x - 3}{x - 5}$ **9.** $\frac{2(2x + 1)}{x - 3}$

11. $\frac{y^2 + 1}{y^3(y - 4)}$ **13.** $\frac{1}{x + 6}$ **15.** $\frac{x + 2}{x + 1}$ **17.** $\frac{z + 1}{2z + 1}$

19. $-\frac{5 + x}{x + 3} = -\frac{x + 5}{x + 3}$ **21.** $\frac{6x^2}{(x - 1)(x + 2)}$ **23.** $-\frac{1}{x}$

25. $\frac{x}{2(x + 3)}$ **27.** $\frac{(x - 3)^2}{4}$ **29.** $x^3(x + 3)$ **31.** $\frac{x + 2}{(x - 3)(x + 1)}$

33. $\frac{x^2 + 4}{x(x - 2)}$ **35.** $-\frac{(x - 1)^2}{(x + 2)(7 + x)}$ **37.** $\frac{x}{2}$ **39.** $\frac{1}{2n}$

41. $\frac{2}{3}$ **43.** 1 **45.** $-x$ **47.** $\frac{x - 2}{x - 1}$ **49.** $\frac{2x + 3}{2x}$

51. $-\frac{(x + 2)(2 + 3x)}{9}$ **53.** $\frac{8(x + 2)}{15(x - 2)}$ **55.** $x = 10$ **57.** $y = -4$

59. $x = -2$ **61.** $x = \frac{3}{14}$ **63.** $y = \frac{10}{3}$ **65.** $x = \frac{69}{5}$ **67.** 7

COMPLETION SET 8-2

1. 4 **2.** $x - 1, 1$ **3.** x^2 **4.** $x + 1, x - 1$

5. $x - 1, x + 2, x + 2, x - 1$ **6.** F **7.** T

8. F **9.** T **10.** T

PROBLEM SET 8-2

1. $(x - 4)^5$ **3.** x^2y^3 **5.** $(x + 3)^2(x - 3)$ **7.** $2x(x + 1)$

9. $\frac{x + 5}{x - 3}$ **11.** 2 **13.** $\frac{2x - 1}{x - 1}$ **15.** $\frac{2y + 3x}{xy}$ **17.** $\frac{x - 8}{18}$

19. $\frac{3y - 4}{2xy}$ **21.** $\frac{x^2 + 4}{2x}$ **23.** $\frac{8x - 21}{(x - 2)(x - 3)}$

25. $\dfrac{5y^2 - 2x + 3x^2}{x^2y}$ **27.** $\dfrac{5x-1}{x-1}$ **29.** $\dfrac{x^2+2xy-y^2}{(x-y)(x+y)}$

31. $\dfrac{x^2+18x+9}{2(x-3)(x+3)}$ **33.** $\dfrac{x+3}{x+1}$

35. $\dfrac{-1}{(x+1)(x-1)} = -\dfrac{1}{(x+1)(x-1)}$ **37.** $\dfrac{x^2-2x+9}{(x+3)^2(x-3)}$

39. $\dfrac{-x^2-2x+1}{(x+5)(x+2)(x+1)}$ **41.** $\dfrac{x^2+4x-4}{(x-2)(x+2)}$

43. $\dfrac{2x^2+x-1}{x-1} = \dfrac{(2x-1)(x+1)}{x-1}$ **45.** $\dfrac{-x^3+3x^2-x+3}{x^2(x+1)(x+2)}$

47. $\dfrac{2y^2-y-6}{(2y+1)^2(y+3)} = \dfrac{(2y+3)(y-2)}{(2y+1)^2(y+3)}$ **49.** $\dfrac{x^2+y^2+4xy}{(x+y)^2(x-y)}$

51. $\dfrac{2x+3}{(x+y)(x-y)}$

COMPLETION SET 8-3

1. $\dfrac{3}{x}, \dfrac{x}{2}, \dfrac{x}{x-1}$ **2.** 2, 2, $\dfrac{x^2}{6}$ **3.** $(x-3)(x+2)$

PROBLEM SET 8-3

1. $\dfrac{1}{x}$ **3.** $\dfrac{2x-3}{x}$ **5.** $\dfrac{14}{6x-1}$ **7.** $\dfrac{x(3y-1)}{y(2x+1)}$ **9.** $\dfrac{x+3}{2x-3}$

11. $\dfrac{3(x+2)}{x+3}$ **13.** $\dfrac{3x-8}{3x-10}$

COMPLETION SET 8-4

1. $x+1$ **2.** x^2, 2, 1, 2, -1, 2, -1 **3.** 1

4. F **5.** F

PROBLEM SET 8-4

1. $x = \frac{1}{4}$ **3.** $x = \frac{3}{2}$ **5.** $r = \frac{2}{3}$ **7.** $x = 3, -6$ **9.** no solution

11. $y = -\frac{1}{2}$ **13.** $x = -\frac{1}{4}, -\frac{1}{2}$ **15.** $x = 0$ **17.** $x = 6, -2$

19. $x = 5, -2$ **21.** $x = -2$ **23.** $x = \frac{1}{8}$ **25.** $y = 3$

27. $x = -1, \frac{13}{10}$ **29.** $x = 1, -\frac{18}{5}$ **31.** no solution

33. $x = 2, -\frac{4}{5}$ **35.** 86.8 in. or 33.2 in. **37.** 1 microfarad

39. $\frac{12}{13}$ hr **41.** 15 mi/hr

REVIEW PROBLEMS—CHAPTER 8

1. $\dfrac{6(x-2)}{x(x-6)}$ **3.** $\dfrac{x-8}{2x}$ **5.** $\frac{3}{2}$

7. $\dfrac{-x^2+3x-5}{(x-2)(x-3)}$ **9.** $-3x(x+1)$ **11.** $\dfrac{3x^2-1}{(x-1)(x+1)}$

13. 1 **15.** -2 **17.** $\dfrac{x-2}{x+2}$ **19.** $\dfrac{2x}{(x-1)^2}$

21. $-\dfrac{3+x}{(x+4)(x+2)}$ **23.** $\dfrac{x(x+1)}{x+2}$ **25.** $-\dfrac{x+4}{x+3}$

27. $x = -15$ **29.** $x = 3$ **31.** $x = 4$ **33.** no solution

35. $a = \dfrac{bE-P}{EP}$ **37.** 50 mi/hr

COMPLETION SET 9-1

1. 0, 0 **2.** quadrants **3.** abscissa **4.** ordinate

5. second **6.** first **7.** first **8.** first

PROBLEM SET 9-1

1. (1, 1) **3.** (−4, −1) **5.** (0, 3)

7., 9., 11., 13., 15., 17. See diagram.

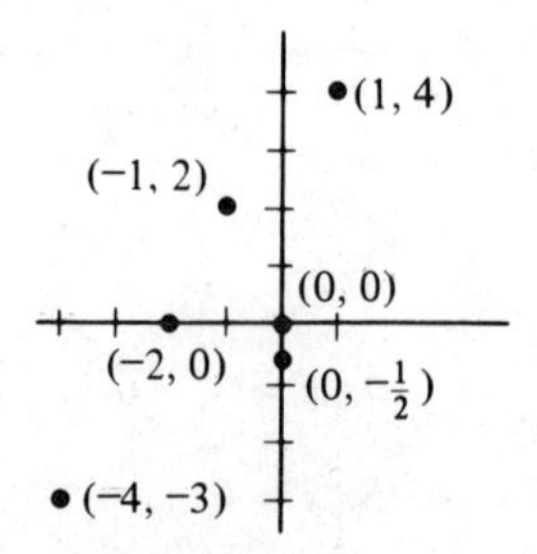

19. I　**21.** IV　**23.** II　**25.** III

27. II　**29.** I

PROBLEM SET 9-2

1.

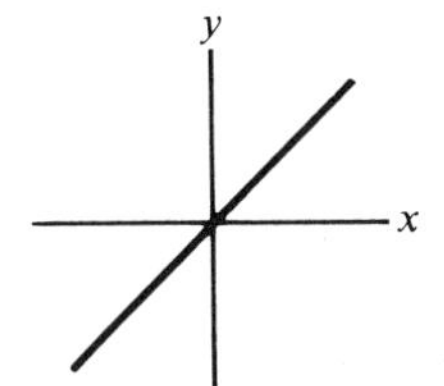

3.

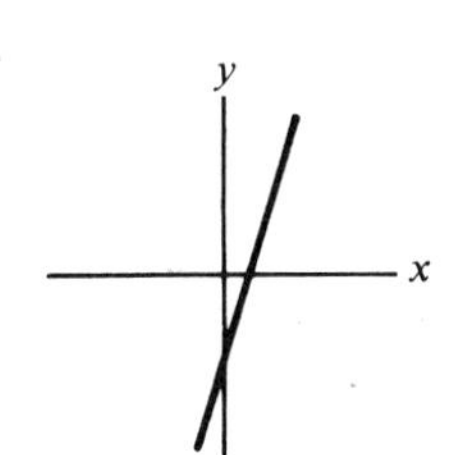

5.

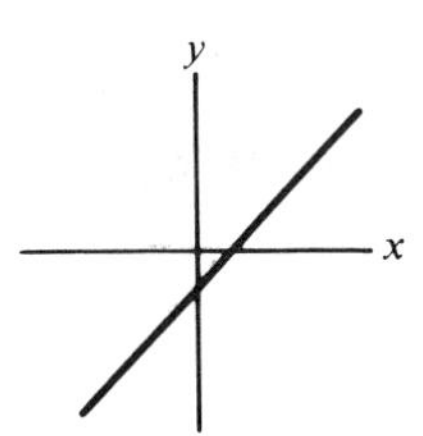

7.

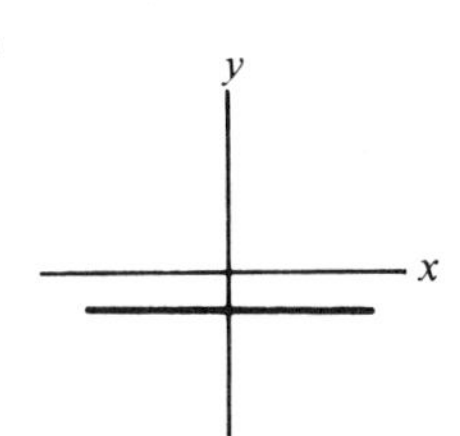

9.

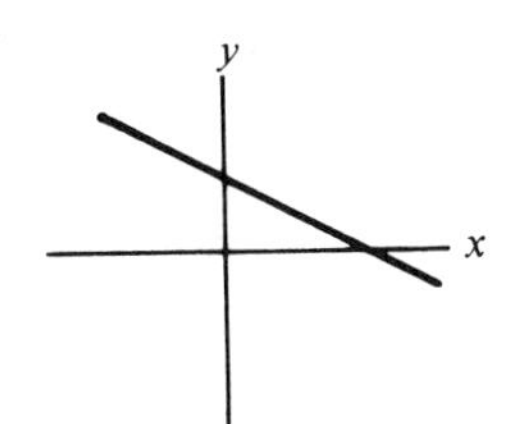

11.

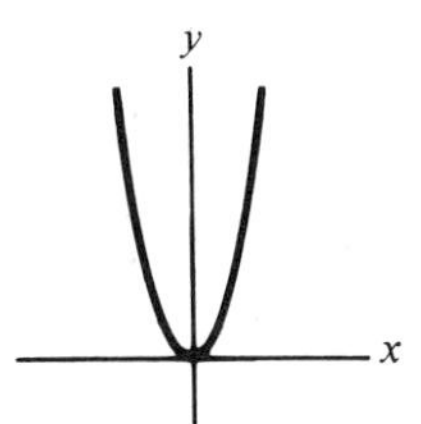

13.

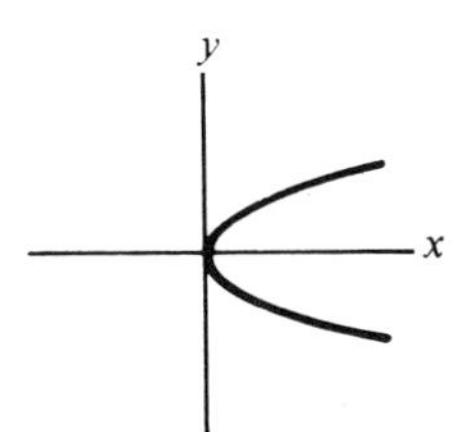

15.

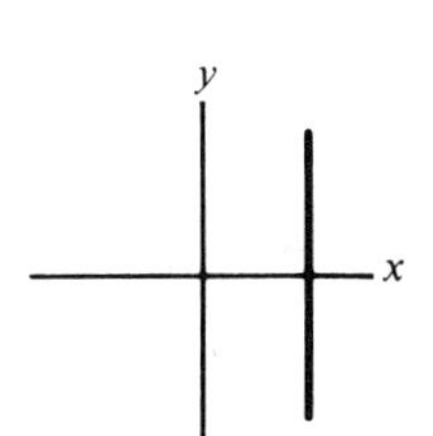

17.

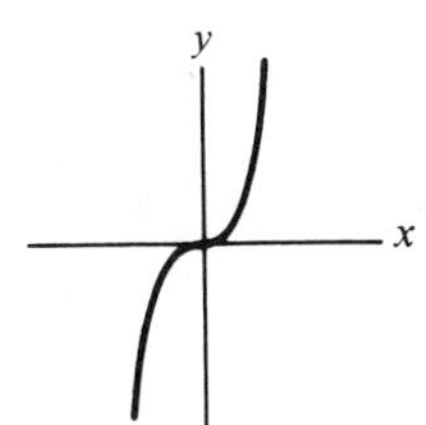

19.

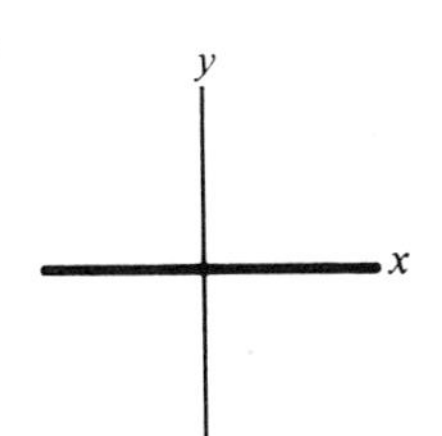

21.

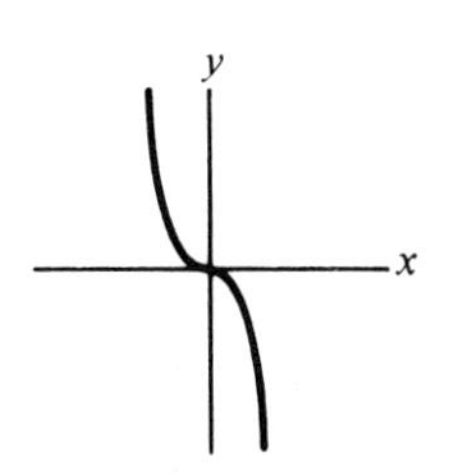

23.

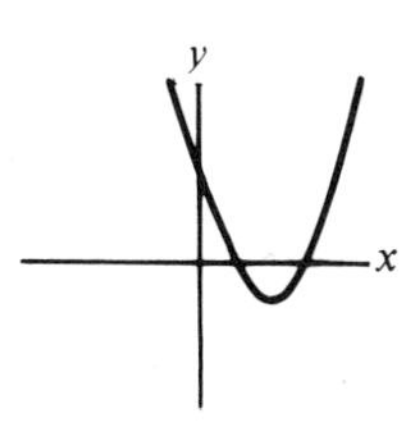

25.

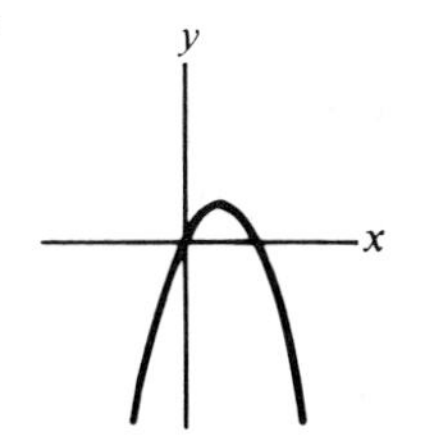

27.

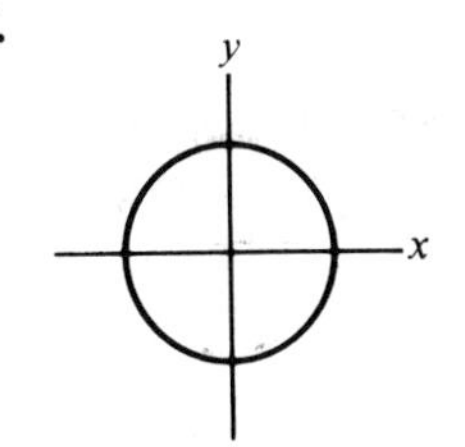

29.

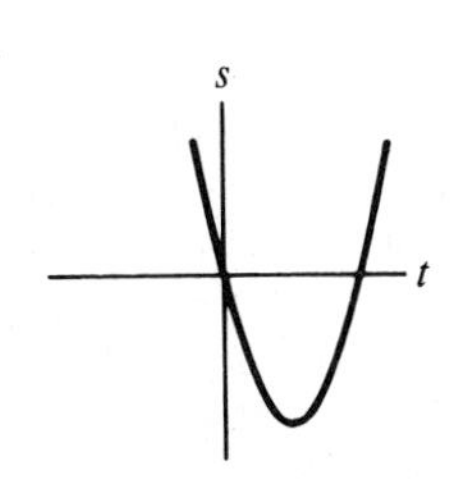

COMPLETION SET 9-3

1. one **2.** 2, 1 **3.** 5, 5, 31 **4.** $t + 1$ **5.** 5

6. constant, 3 **7.** T **8.** F **9.** T **10.** F

PROBLEM SET 9-3

1. all real numbers except 0

3. all real numbers

5. all nonnegative numbers

7. all real numbers

9. all real numbers

11. all real numbers except -2

13. all real numbers except $-\frac{5}{2}$

15. all real numbers greater than or equal to 5

17. all real numbers except 0 and 1

19. all real numbers except -1 and -5

21. all real numbers except 4 and $-\frac{1}{2}$

23. 0, 15, -20, 12

25. 1, -5, 6, $\frac{5}{2}$

27. 28, $7s$, $7(t + 1) = 7t + 7$, $7(x + 3) = 7x + 21$

29. 12, 2, $(2v)^2 + 2v = 4v^2 + 2v$, $(x^2)^2 + x^2 = x^4 + x^2$

31. 12, 12, 12, 12

33. 4, 0, $w^2 + 2w + 1$, $(x + h)^2 + 2(x + h) + 1 = x^2 + 2xh + h^2 + 2x + 2h + 1$

35. $-\frac{2}{13}$, 0, $\dfrac{3x - 5}{(3x)^2 + 4} = \dfrac{3x - 5}{9x^2 + 4}$, $\dfrac{(x + h) - 5}{(x + h)^2 + 4} = \dfrac{x + h - 5}{x^2 + 2xh + h^2 + 4}$

37. 0, 4, 3, $\sqrt{25 - p^8}$

39. 8, 3, 3, 1

41. (a) $3x + 3h - 4$; (b) 3

43. (a) 2; (b) 0

45. y is a function of x; x is a function of y

47. y is a function of x; x is not a function of y

49. 2, $\frac{1}{2}$, 3

51. 20; the price per unit at which 30 units are demanded; 30, 60, 78.

53. yes

55. 32 ft, 32 ft, no, yes

PROBLEM SET 9-4

1. Domain: all real numbers
Range: all real numbers

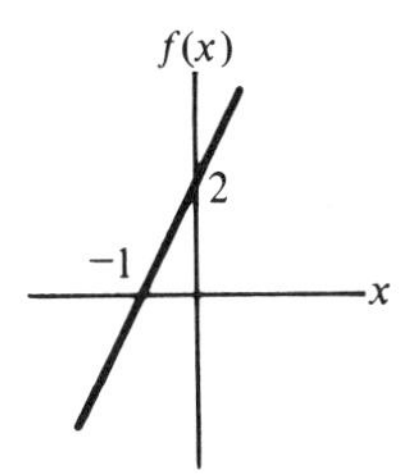

3. Domain: all real numbers
Range: all nonnegative numbers

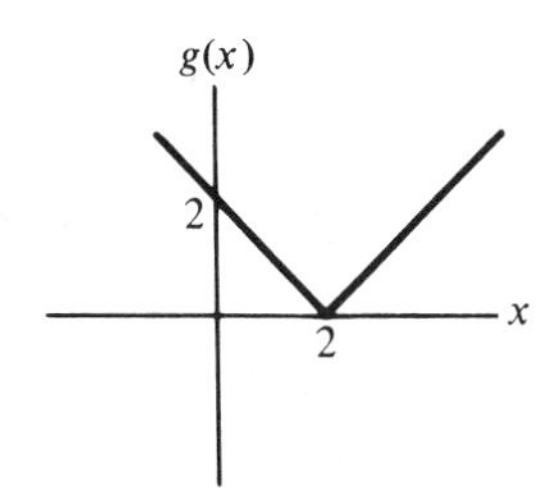

5. Domain: all nonnegative numbers
Range: all nonnegative numbers

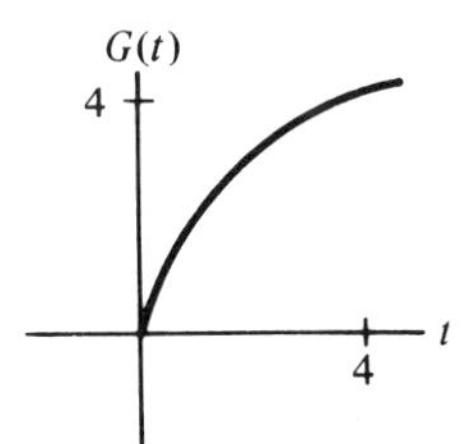

7. Domain: all real numbers
Range: 4

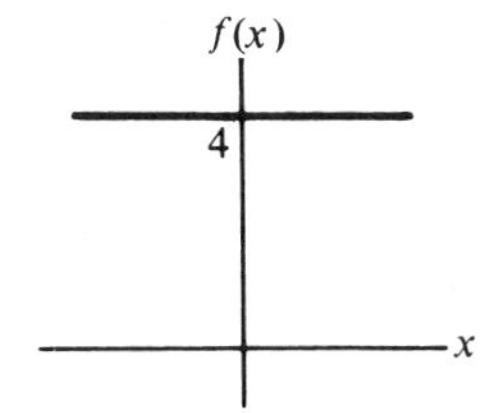

9. Domain: all real numbers
Range: all numbers less than or equal to 4

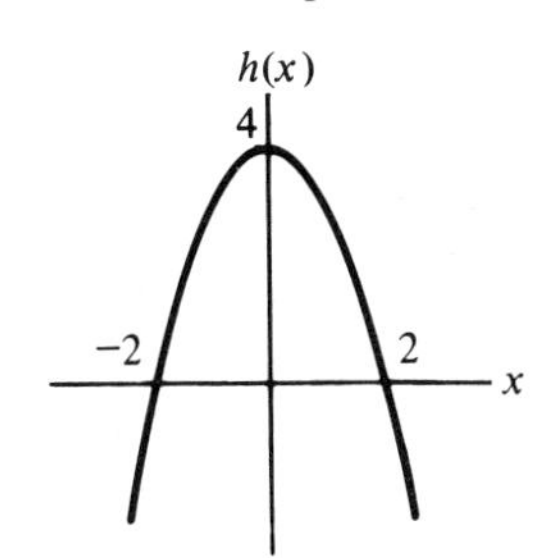

11. Domain: all real numbers except 0
Range: all real numbers except 0

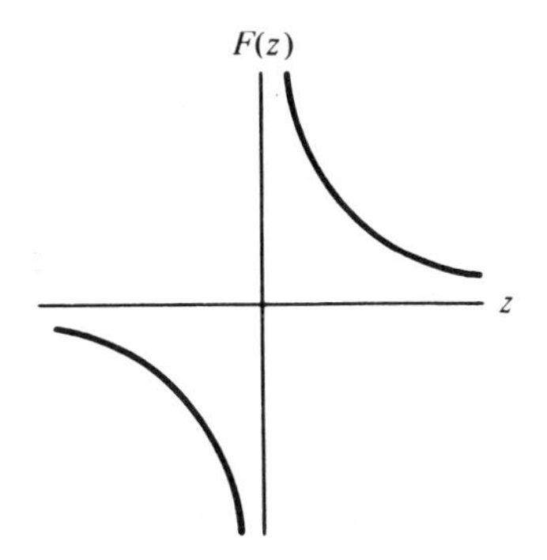

13. Domain: all real numbers except 4
Range: all real numbers except 0

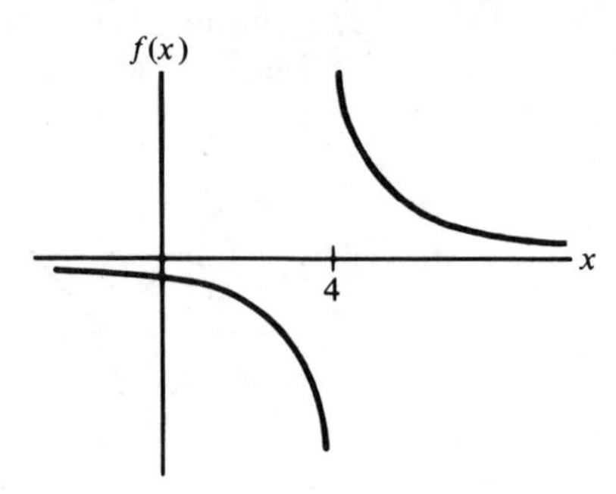

15. (a) 1, 2, 3, 0; (b) all real numbers; (c) all real numbers

17. (a) 0, -1, -1; (b) all real numbers; (c) all $y \leqslant 0$

19. (a) 20, 10, 5, 4
(b)

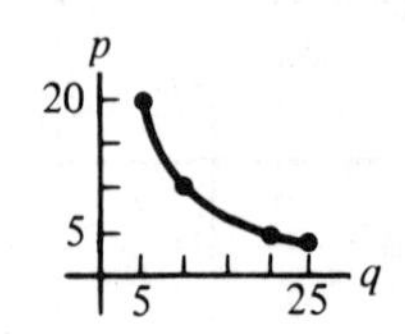

21. a, b, d

COMPLETION SET 9-5

1. 4, 13 **2.** 1, -1 **3.** $2x$, 0 **4.** $x + 1$, 1

5. $2x$ **6.** 2

PROBLEM SET 9-5

1. (a) $2x + 5$, (b) 5, (c) -3, (d) -3, (e) $x^2 + 5x + 4$, (f) -2, (g) $\dfrac{x+1}{x+4}$, (h) 0, (i) $x + 5$, (j) 8, (k) $x + 5$, (l) 10

3. (a) $2x^2 + x$, (b) 1, (c) $-x$, (d) $\frac{1}{2}$, (e) $x^4 + x^3$, (f) 0, (g) $\dfrac{x^2}{x^2 + x} = \dfrac{x}{x+1}$, (h) -1, (i) $(x^2 + x)^2 = x^4 + 2x^3 + x^2$, (j) 36, (k) $x^4 + x^2$, (l) 90

5. (a) $x^3 - 2x$, (b) $t^3 - 2t$, (c) $-2x - x^3$, (d) 12, (e) $-2x^4$, (f) $-\frac{1}{8}$, (g) $\dfrac{-2x}{x^3} = -\dfrac{2}{x^2}$, (h) $-\dfrac{2}{(r+1)^2}$, (i) $-2x^3$, (j) 16, (k) $(-2x)^3 = -8x^3$, (l) -64

7. $\sqrt{x-1}$, $\sqrt{x}-1$

9. $\dfrac{1}{x^2-1}$, $\left(\dfrac{1}{x}\right)^2 - 1 = \dfrac{1-x^2}{x^2}$

11. $3(4-3x)+4 = 16-9x$, $4-3(3x+4) = -8-9x$

REVIEW PROBLEMS—CHAPTER 9

1. all real numbers except 1 and 2

3. all real numbers

5. all nonnegative numbers except 1

7. 7, 46, 62, $3t^2-4t+7$

9. 0, 3, $\sqrt{t}$, $\sqrt{x^2-1}$

11. $\frac{3}{5}$, 0, $\dfrac{\sqrt{x+4}}{x}$, $\dfrac{\sqrt{u}}{u-4}$

13. (a) $3-7x-7h$; (b) -7

15.

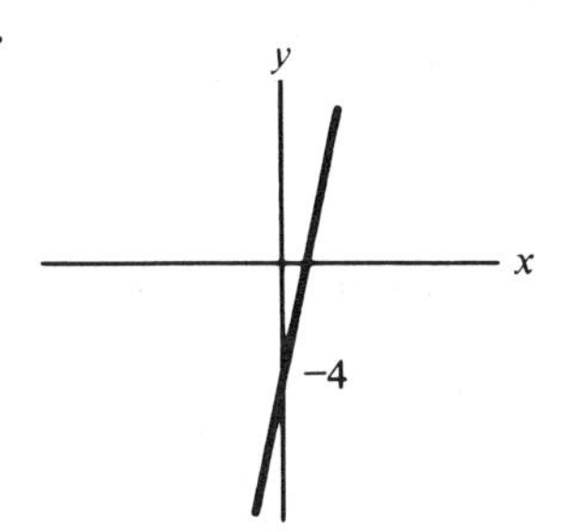

17.

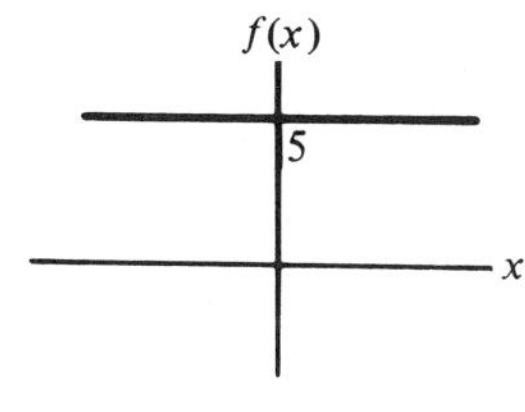

19.

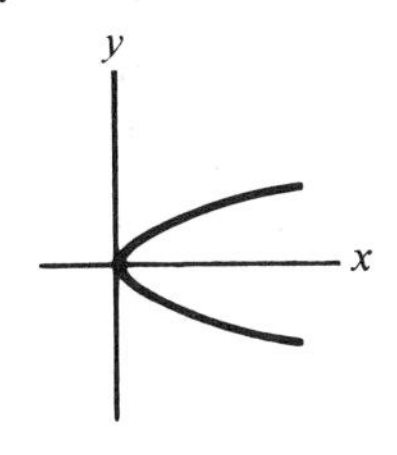

21.

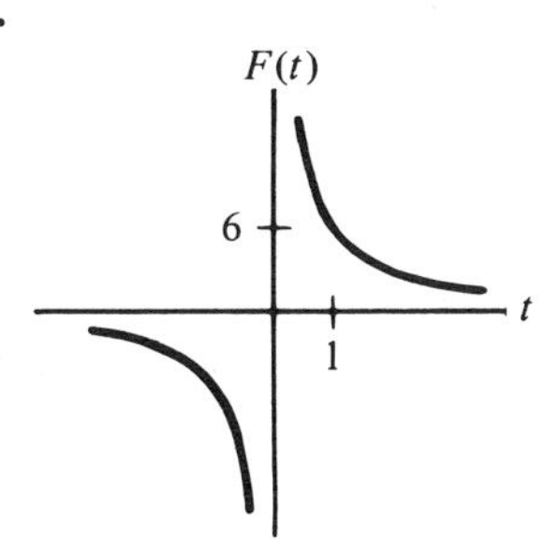

23. (a) $5x+2$, (b) 22, (c) $x-4$, (d) -6, (e) $6x^2+7x-3$, (f) 10, (g) $\dfrac{3x-1}{2x+3}$, (h) $\frac{5}{7}$, (i) $3(2x+3)-1 = 6x+8$, (j) 38, (k) $2(3x-1)+3 = 6x+1$, (l) -17

25. $\dfrac{1}{x-1}$, $\dfrac{1}{x} - 1 = \dfrac{1-x}{x}$

27. x^3+2, $(x+2)^3$

29. 15; the price per unit at which 10 units are supplied; 20, 60, 260

31. (a) 2, 2, 2, 2; (b) all real numbers; (c) 2

COMPLETION SET 10-1

1. zero **2.** 3 **3.** -4 **4.** 0 **5.** $x + 2$

6. 0, 5, -7

PROBLEM SET 10-1

1. $\frac{7}{3}$ **3.** 24 **5.** 2 **7.** ± 6 **9.** 0, -2

11. ± 2 **13.** 1, $-\frac{5}{2}$ **15.** 0, 1, -3 **17.** 0, $\dfrac{3 \pm \sqrt{5}}{2}$ **19.** 0, $\dfrac{2 \pm \sqrt{5}}{2}$

21. (a) 0, 1; (b) 0, $\frac{2}{3}$

COMPLETION SET 10-2

1. 3; 8; 1, 3; 1, 2, 4, 8 **2.** 1, integer, 1, 3

PROBLEM SET 10-2

1. yes **3.** no **5.** yes **7.** yes

9. $\pm 1, \pm 3, \pm \frac{1}{2}, \pm \frac{3}{2}, \pm \frac{1}{4}, \pm \frac{3}{4}$ **11.** $\pm 1, \pm 7$

13. $\pm 1, \pm \frac{1}{3}, \pm \frac{1}{9}$ **15.** 1 **17.** $-1, \frac{1}{2}$ **19.** $-2, 3$

21. $\frac{2}{3}, \dfrac{-1 \pm \sqrt{5}}{2}$ **23.** $-1, \pm\sqrt{5}$ **25.** $\frac{1}{2}, \dfrac{1 \pm \sqrt{13}}{2}$

27. 3, -1, -2 **29.** $\frac{1}{4}, -\frac{1}{2}$ **31.** 1, 2, -2, $-\frac{1}{2}$

33. (a) 1, -2; (b) ± 1

REVIEW PROBLEMS—CHAPTER 10

1. 2, -6, -1 **3.** 5, $-\frac{1}{2}$ **5.** $-\frac{2}{3}, \pm\sqrt{7}$ **7.** $\frac{1}{2}, \frac{3}{4}$

COMPLETION SET 11-1

1. negative slope **2.** horizontal, vertical **3.** -4, 6, 13

4. $\frac{2}{5}$ **5.** parallel **6.** rises **7.** slope **8.** $4(x - 1)$

PROBLEM SET 11-1

1. $\frac{3}{2}$ **3.** $-\frac{4}{5}$ **5.** not defined **7.** 0 **9.** $y = x$

11. $y = -\frac{1}{4}x + \frac{9}{2}$ **13.** $y = \frac{1}{3}x - 7$ **15.** $y = -5x$ **17.** $y = \frac{1}{2}x + 3$

19. $y = -x$ **21.** $y = 8x - 25$ **23.** $y = -7$ **25.** $(5, -4)$

COMPLETION SET 11-2

1. 2, 7 **2.** 7 **3.** $x = 1, y = 2$ **4.** 3, 1

5. not defined **6.** 0

PROBLEM SET 11-2

1. $y = 2x + 4$ **3.** $y = -\frac{1}{2}x - 3$ **5.** $y = -2$

7. $x = 2$ **9.** $y = 2x - 1$; 2; -1 **11.** $y = -3x + 2$; -3; 2

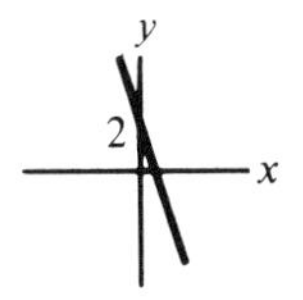

13. $y = 4x$; 4; 0 **15.** $y = -\frac{1}{3}x - 1$; $-\frac{1}{3}$; -1

17. $y = 1$; 0; 1 **19.** $x - y - 6 = 0$ **21.** $x + 2y - 2 = 0$

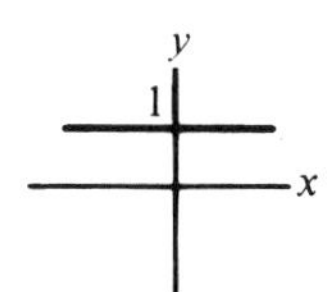

23.

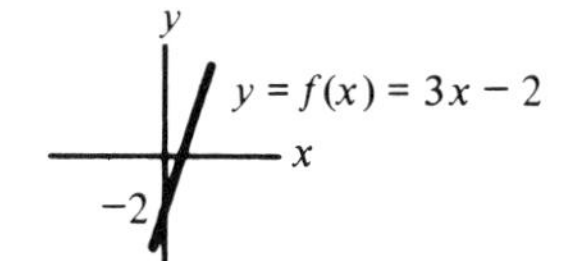

25. yes

27. $s = -\frac{5}{2}t + 70$; 10 **29.** $y = 3x + 10$; \$115 **31.**

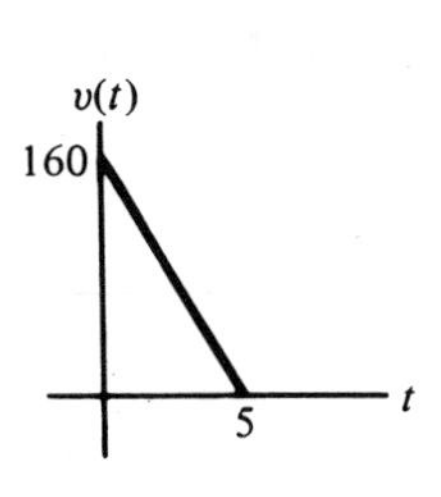

COMPLETION SET 11-3

1. 2, 2, $-\frac{1}{2}$

2. perpendicular

PROBLEM SET 11-3

1. parallel **3.** parallel **5.** perpendicular **7.** parallel

9. neither **11.** $y = 4x + 7$ **13.** $y = 1$ **15.** $y = -\frac{1}{3}x + 5$

17. $x = 7$ **19.** $y = -\frac{2}{3}x - \frac{29}{3}$

REVIEW PROBLEMS—CHAPTER 11

1. $-\frac{1}{3}$ **3.** 0 **5.** $y = -2x + 1$ **7.** $y = \frac{1}{3}x + \frac{11}{3}$

9. $y = 2$ **11.** $x = 0$ **13.** $y = 3x - 4$ **15.** $y = -\frac{3}{5}x + \frac{13}{5}$

17. $y = -5x + 7$ **19.** perpendicular **21.** neither **23.** parallel

25. $y = \frac{3}{2}x - 2$; $3x - 2y - 4 = 0$; $\frac{3}{2}$ **27.**

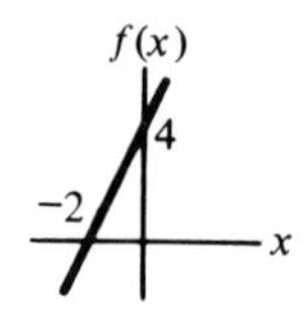

29. $a + b - 3 = 0$; 0

COMPLETION SET 12-1

1. addition, substitution

2. 4

PROBLEM SET 12-1

1. $x = -1, y = 1$ **3.** $x = 3, y = 2$ **5.** $x = 2, y = -2$ **7.** $x = \frac{1}{2}, y = \frac{3}{2}$

9. $x = 4, y = -5$ **11.** $x = 3, y = -1$ **13.** $p = 0, q = 18$

15. the coordinates of any point on the line $2x + 6y = 3$

17. no solution **19.** $x = 12, y = -12$

21. 420 gal of 20% solution, 280 gal of 30% solution

23. 60 units of Argon I, 40 units of Argon II

25. 275 mi/hr (speed of airplane in still air), 25 mi/hr (speed of wind)

27. 240 units (Early American), 200 units (Little Big Horn)

29. $x = 200, p = \$12$ **31.** 1800

PROBLEM SET 12-2

1. $x = 4, y = 2, z = 0$ **3.** $x = 13, y = 22, z = -1$

5. $x = \frac{1}{2}, y = \frac{1}{2}, z = \frac{1}{4}$ **7.** $x = 0, y = \frac{4}{3}, z = 2$

9. 100 chairs, 100 rockers, 200 chaise lounges

11. $a = 1, b = -2, c = 0$ **13.** $i_1 = -2, i_2 = 1, i_3 = -1$

REVIEW PROBLEMS—CHAPTER 12

1. $x = 2, y = -1$ **3.** no solution **5.** $x = \frac{1}{2}, y = 1$

7. the coordinates of any point on the line $6x = 3 - 9y$

9. $x = 3, y = 1, z = -2$ **11.** $r = 1, s = 3, t = \frac{1}{2}$

13. 2.5 kg of 20% copper alloy, 12.5 kg of 50% copper alloy

15. 40 semiskilled workers, 20 skilled workers, 10 shipping clerks

COMPLETION SET 13-1

1. 6, $\frac{1}{36}$ **2.** 3, 64 **3.** -6, 6, 64 **4.** 2

5. 5 **6.** 20

PROBLEM SET 13-1

1. 1 **3.** $\frac{1}{8}$ **5.** 27 **7.** 3 **9.** -4

11. -27 **13.** $\frac{1}{36}$ **15.** 1 **17.** $\frac{1}{x^6}$ **19.** x^3

21. $\frac{3}{y^4}$ **23.** $16x$ **25.** $\frac{1}{x^5y^7}$ **27.** $\frac{2a^2c^5}{b^4}$ **29.** $\frac{x^9z^4}{w^2y^{12}}$

31. x **33.** $\frac{1}{x^5}$ **35.** $\frac{6y^4}{x^3}$ **37.** $\frac{y^{20}}{x^4}$ **39.** $\frac{2y^4}{x^2}$

41. $\frac{1}{9t^2}$ **43.** $\frac{x^{15}}{y^{15}z^3}$ **45.** t^4 **47.** $\frac{y^5}{x^8}$ **49.** $\frac{xz^2w}{y^3}$

51. $\frac{8x^4y^{15}z^2}{3}$ **53.** x^2y^3 **55.** $\frac{5y^2}{8x^2}$ **57.** $-xz$ **59.** $\frac{4x^8y^{10}}{z^4}$

61. $\frac{x^5}{y^{20}z^{20}}$ **63.** $\frac{1}{x^2 - 2xy + y^2}$ **65.** 2.6×10^2 **67.** 6.3×10^{-6}

69. 4.0×10^6 **71.** 2×10^{-1} **73.** 1×10^{-2}

COMPLETION SET 13-2

1. 3 **2.** 5, 4 **3.** 2, 3 **4.** 5, 7

5. 9, 4 **6.** (a) T; (b) F

PROBLEM SET 13-2

1. 5 **3.** $\frac{1}{9}$ **5.** 9 **7.** $\frac{1}{9}$ **9.** 16

11. 8 **13.** -2 **15.** 16 **17.** $x^{1/2}$ **19.** $x^{2/3}$

21. $x^{3/4}y^{5/4}$ **23.** $x^{5/6}y^2$ **25.** $x^{9/4}$ **27.** $\frac{3}{x^{1/2}}$ **29.** $(x^2 - 5x)^{2/}$

31. x^2 **33.** x **35.** x^4 **37.** $x^{3/4}$ **39.** $3xy$

41. $x^{3/2}$ **43.** y^2 **45.** $\frac{2}{x^4}$ **47.** $\frac{8y}{x^6}$ **49.** $a^2b^{21/4}c^{9/4}$

51. $x + 3x^{1/3}$ **53.** $\frac{1}{3x^5}$ **55.** $-\frac{2}{x^2}$ **57.** $\frac{x^{1/3}}{y}$ **59.** $\frac{x^2}{y^2}$

61. $\frac{x^4}{y^2}$ **63.** $\frac{1}{x^{8/3}}$ **65.** $4x^2$ **67.** $-3x^{1/2}$ **69.** $\frac{2}{x^{1/3}}$

71. $x^{1/3} + x - x^3$ **73.** $x^{1/4} - 3x^{1/12}$ **75.** $4y^3$

COMPLETION SET 13-3

1. $x;\ z^2$ **2.** $x^2y^3,\ x^2y^3$ **3.** 16, 16, 4

4. $x^3y^3,\ x^3y^3,\ xy$ **5.** 36, 6 **6.** y^3, y^3, y^3

PROBLEM SET 13-3

1. x^2 **3.** $2x^4$ **5.** $3x^8y^9$ **7.** xy^2z^3

9. $\dfrac{x^3}{y^4}$ **11.** x^2 **13.** x **15.** $2\sqrt{3}$

17. $4\sqrt{2}$ **19.** $2\sqrt[3]{2}$ **21.** $x^3\sqrt{x}$ **23.** $2x^2\sqrt[3]{3}$

25. $x^2\sqrt[4]{xy^2}$ **27.** $x^2z\sqrt[3]{yz}$ **29.** $2ay\sqrt[3]{y^2}$ **31.** $x^4y^2z\sqrt[5]{x^3z}$

33. $9yzw^2\sqrt{xz}$ **35.** $\dfrac{\sqrt{2}}{2}$ **37.** $\dfrac{\sqrt[3]{50}}{5}$ **39.** $\dfrac{\sqrt[3]{x^2}}{y}$

41. $\dfrac{\sqrt{x}}{y^2}$ **43.** $\dfrac{\sqrt{2xy}}{y}$ **45.** $\dfrac{\sqrt[3]{2x^2y}}{xy}$ **47.** $\dfrac{\sqrt[4]{24xy^3z^2}}{2x^2yz}$

49. $\sqrt[3]{x}$ **51.** $\sqrt{3}$ **53.** $2x\sqrt{y}$ **55.** $\dfrac{\sqrt{xy}}{y}$

57. $\sqrt[6]{xyz^5}$ **59.** $x^2\sqrt{xy}$ **61.** $x^7\sqrt[3]{x^2}$ **63.** $x^2w^4\sqrt[6]{y^5w}$

65. $\dfrac{2x\sqrt[4]{x}}{y^2}$ **67.** $\dfrac{\sqrt{2}}{2}$ **69.** $\dfrac{\sqrt{xy}}{y^2}$ **71.** $\sqrt{x}$

COMPLETION SET 13-4

1. 5 **2.** $x,\ 3x$ **3.** y^3, y **4.** $x^{2/3},\ x^2$

5. $2y$ **6.** 7

PROBLEM SET 13-4

1. $3\sqrt[3]{3}$ **3.** $2x^2\sqrt{2x}$ **5.** $11\sqrt{3}$ **7.** $20\sqrt{2}$ **9.** $-y\sqrt{x}$

11. $-8\sqrt[3]{2}$ **13.** 0 **15.** $\sqrt[3]{4}$ **17.** $-2x\sqrt{2}$ **19.** 4

21. $2\sqrt{3}$ **23.** $18\sqrt{2}$ **25.** $3\sqrt[3]{12}$ **27.** $x\sqrt{6x}$ **29.** 3

31. $16x\sqrt[3]{x}$ **33.** $2\sqrt{5}$ **35.** $6\sqrt{2} - 12$ **37.** $-10 - 2\sqrt{3}$ **39.** -3

41. $9 + 4\sqrt{5}$ **43.** $2x + 3\sqrt{x} - 5$ **45.** $3xy^3\sqrt{2x}$ **47.** $31 - 10\sqrt{6}$

49. $46\sqrt{6}$ **51.** $\frac{3\sqrt{7}}{7}$ **53.** $\frac{2\sqrt{2x}}{x}$ **55.** $\frac{\sqrt[3]{4}}{2}$ **57.** $\frac{\sqrt[3]{9x^2}}{3x}$

59. 4 **61.** $a\sqrt{2}$ **63.** $\frac{2\sqrt[4]{x^3}}{x}$ **65.** $\frac{\sqrt[3]{12x^2}}{2x}$ **67.** $\frac{\sqrt[3]{50y}}{5y}$

69. $2 - \sqrt{3}$ **71.** $-\frac{\sqrt{6} + 2\sqrt{3}}{3}$ **73.** $-4 - 2\sqrt{6}$

75. $\frac{x - \sqrt{5}}{x^2 - 5}$ **77.** $3 - 2\sqrt{2}$

PROBLEM SET 13-5

1. $\sqrt[6]{x^2}$, $\sqrt[6]{x^3}$ **3.** $\sqrt[6]{4x^2}$, $\sqrt[6]{x^5}$ **5.** $\sqrt[8]{x^6y^2}$, $\sqrt[8]{y^4}$, $\sqrt[8]{xy^5}$

7. $\sqrt[4]{125}$ **9.** $x\sqrt[6]{27x}$ **11.** $3\sqrt[6]{3x^5}$ **13.** $6x\sqrt[6]{8xy^2}$

15. $y\sqrt[20]{x^{13}y^2}$ **17.** $x^2\sqrt[6]{xy^5}$ **19.** $\sqrt[4]{6}$ **21.** $\frac{\sqrt[6]{72}}{2}$

23. $\sqrt[6]{8x}$ **25.** $\frac{\sqrt[9]{2y^7}}{y}$ **27.** $2\sqrt[4]{xy}$

PROBLEM SET 13-6

1. $x = 27$ **3.** $y = \frac{41}{2}$ **5.** $x = 4$ **7.** $x = 4$ **9.** no solution

11. $x = 7$ **13.** $z = 4, 8$ **15.** $x = 2$ **17.** $x = \pm 5$

REVIEW PROBLEMS—CHAPTER 13

1. 1 **3.** $\frac{1}{5}$ **5.** 9 **7.** 10 **9.** 8

11. $\frac{1}{4}$ **13.** $\frac{1}{32}$ **15.** $4\sqrt{2}$ **17.** $x\sqrt[3]{2}$ **19.** $4x^2$

21. $3z^3$ **23.** $\frac{9t^2}{4}$ **25.** $\frac{1}{x^2y^2z^2}$ **27.** $\frac{2}{x^4}$ **29.** $\frac{1}{16t^4}$

31. $\frac{y^6}{x^6}$ **33.** $\frac{z^2}{x^3y^5}$ **35.** $2xy^2\sqrt[4]{x^3}$ **37.** $\frac{2x}{y^2}$ **39.** $-27xy^2\sqrt{x}$

41. $\frac{2\sqrt[6]{x^5}}{y^3}$ **43.** $14\sqrt{2}$ **45.** $\sqrt[3]{t^2}$ **47.** $\frac{64y^6\sqrt{x}}{x^2}$ **49.** $\frac{y^4}{z^2}$

51. $2\sqrt{2}$ **53.** $\sqrt{2} - 2\sqrt{3}$ **55.** 4 **57.** $\frac{2\sqrt{7}}{7}$ **59.** $\frac{3\sqrt[4]{x^3}}{x}$

61. $3x\sqrt{x}$ **63.** x^2 **65.** $x\sqrt{3}$ **67.** $\frac{x\sqrt{y}}{y^2}$ **69.** $\frac{3\sqrt[3]{x^2y}}{xy}$

71. $\frac{\sqrt[3]{12x}}{2}$ **73.** $\frac{\sqrt{6}+2}{2}$ **75.** $x = 10$ **77.** $x = 5$ **79.** no solution

81. $x = 10$

COMPLETION SET 14-1

1. exponential **2.** $y = 5^x$ **3.** 3

PROBLEM SET 14-1

1.

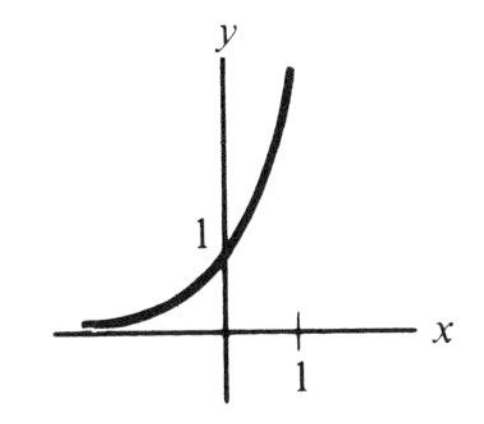

3.

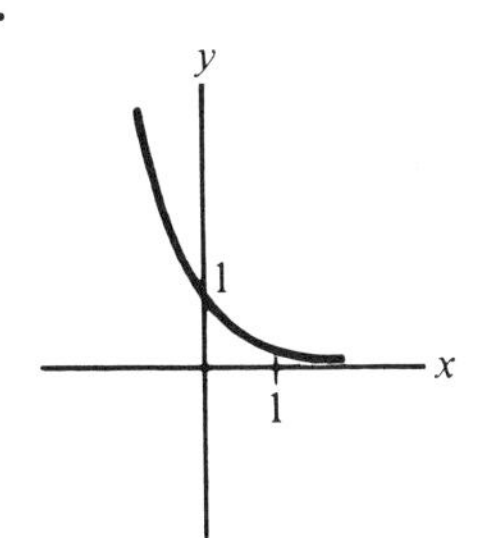

5. 20.086 **7.** 0.2231 **9.** 81, $\frac{1}{81}$, 3 **11.** 12, 6, $\frac{3}{2}$

13. 3, $\frac{33}{32}$, $\frac{3}{2}$ **15.** 140,000 **17.** .2241 **19.** 50 mg

21. \$1491.80 **23.** (a) 91 units, (b) 432 units

COMPLETION SET 14-2

1. exponential **2.** 2 **3.** 243 **4.** 128, 7

5. 3^4 **6.** x **7.** 10 **8.** e

PROBLEM SET 14-2

1. $\log_4 64 = 3$ **3.** $\log 100{,}000 = 5$ **5.** $2^6 = 64$ **7.** $8^{1/3} = 2$

9. $\log_6 1 = 0$ **11.** $2^{14} = x$ **13.** $\ln 7.3891 = 2$ **15.** $e^{1.0986} = 3$

17.

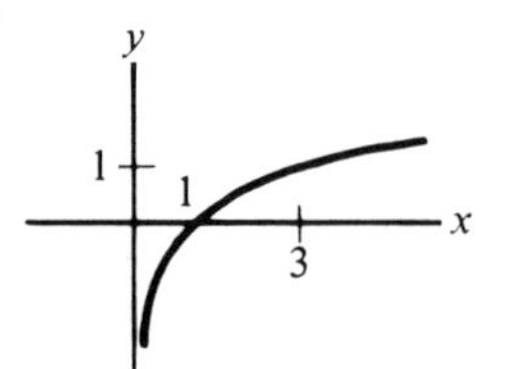

19. 2 **21.** 3

23. $\frac{1}{2}$ **25.** 1 **27.** -2 **29.** 0 **31.** -3

33. 16 **35.** 125 **37.** $\frac{1}{10}$ **39.** e^2 **41.** 2

43. 6 **45.** 2 **47.** 4 **49.** $\frac{1}{2}$ **51.** $\frac{1}{81}$

53. 9 **55.** 3 **57.** 11

COMPLETION SET 14-3

1. + **2.** − **3.** +, − **4.** 5 **5.** $\frac{1}{2}$

6. 2, 3, 4 **7.** 9, 45 **8.** 8, 2 **9.** 1 **10.** 0

PROBLEM SET 14-3

Since there are different ways to find the values in Problems **1–33**, and since the entries in Table 14-1 are only approximations, your answers may differ *slightly* from those given below.

1. .6990 **3.** 1.3222 **5.** 1.5441 **7.** 1.3980 **9.** .3521

11. −.3010 **13.** 4.5155 **15.** −2.7960 **17.** .1505 **19.** .5188

21. .4471 **23.** −.2711 **25.** 4 **27.** −2 **29.** 2.4771

31. 1.0458 **33.** −2.2677 **35.** 48 **37.** $\frac{3}{4}$ **39.** 4

41. −4 **43.** $\frac{9}{2}$ **45.** $-\frac{1}{3}$ **47.** 3 **49.** log 28

51. $\log_2 \frac{x+2}{x+1}$ **53.** $\ln \frac{25}{64}$ **55.** $\log_4 (27\sqrt{2})$

57. $\log_3 \frac{xy}{z}$ **59.** $\log_2 \frac{45}{7}$ **61.** $\ln \frac{x^2y^3}{z^4w^2}$

63. $\log x + 2 \log (x + 1)$

65. $2 \log x - 3 \log (x + 1)$

67. $3[\log x - \log (x + 1)]$

69. $\ln x - \ln (x + 1) - \ln (x + 2)$

71. $\frac{1}{2} \ln x - 2 \ln (x + 1) - 3 \ln (x + 2)$

73. $\frac{2}{5} \ln x - \frac{1}{5} \ln (x + 1) - \ln (x + 2)$

75. 9

77. 10

PROBLEM SET 14-4

1. $\dfrac{.4771}{.6990} = .6825$ **3.** $\dfrac{.9542}{.6021} = 1.585$ **5.** $\dfrac{1}{.9031} = 1.107$ **7.** $\dfrac{.3010}{.4343} = .6931$

9. $\dfrac{4.0775}{2.7080} = 1.5057$ **11.** $\dfrac{2.1826}{2} = 1.0913$

REVIEW PROBLEMS—CHAPTER 14

1. 5, 13, $\frac{13}{3}$

3. 100, $100e^4 = 5459.8$, $100e^{-3/2} = 22.31$

5.

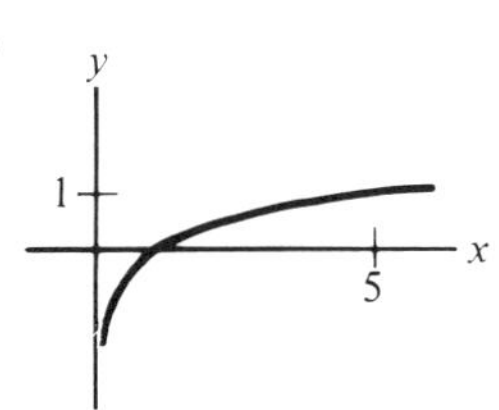

7. $\log_3 243 = 5$

9. $16^{1/4} = 2$

11. $\ln 54.598 = 4$ **13.** 3 **15.** -4 **17.** -2

19. -2 **21.** $\frac{1}{100}$ **23.** 9 **25.** 3

Since there are different ways to find the values in Problems **27–37**, and since the entries in Table 14-1 are only approximate, your answers may differ *slightly* from those given below.

27. 1.2042 **29.** .4260 **31.** .2258 **33.** 2.6990 **35.** $-.1639$

37. .1990 **39.** 0 **41.** $\frac{1}{3}$ **43.** $\frac{5}{2}$ **45.** $\log \frac{25}{27}$

47. $\ln \dfrac{x^2y}{z^3}$

49. $\log_2 \dfrac{x^{9/2}}{(x + 1)^3(x + 2)^4}$

51. $2 \ln x + \ln y - 3 \ln z$

53. $\frac{1}{3}(\ln x + \ln y + \ln z)$

55. $\frac{1}{2}(\ln y - \ln z) - \ln x$

57. (a) \$134,060; (b) \$109,760

59. $\dfrac{.6990}{.4771} = 1.465$

61. $\dfrac{2.3979}{1.3863} = 1.7297$

COMPLETION SET 15-1

1. 10, 5.1 **2.** $-8, \dfrac{15}{-2}$ **3.** $9.1, |-12|, 3^3$

4. $>$ **5.** $\geqslant$ **6.** $\geqslant$ **7.** $<$ **8.** $<$

9. 3, 4 **10.** 5, 7 **11.** 4, 1 **12.** 7, -1 **13.** b

PROBLEM SET 15-1

1. $x > 2$ **3.** $x \leqslant 3$ **5.** $x \leqslant -\frac{1}{2}$ **7.** $s < -\frac{2}{5}$

9. $x < 3$ **11.** $x \geqslant -\frac{7}{5}$ **13.** $y > 0$ **15.** $x > -\frac{2}{7}$

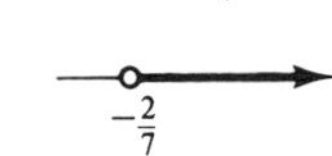

17. no solution **19.** $x < 6$ **21.** $y \leqslant -5$

23. all real numbers

25. $t > \frac{17}{9}$ **27.** $x \geqslant -\frac{14}{3}$ **29.** $r > 0$ **31.** $y < 0$

33. $x \geqslant 2$ **35.** $t \geqslant -\frac{1}{2}$

COMPLETION SET 15-2

1. 5, 6.03 **2.** 1, 2, 0, -2, -3 **3.** 3, -3, 0

4. 6 **5.** 1, 5; 1, 1, 5, 5

6. $(-)(-) = (+), (+)$; $(+)(-) = (-), (-)$; $(+)(+) = (+), (+)$

7. $4 < x < 8$; $4 \leqslant x \leqslant 8$ **8.** $x < 4$ or $x > 8$; $x \leqslant 4$ or $x \geqslant 8$

9. 0, 2 **10.** T **11.** F

PROBLEM SET 15-2

1. $1 < x < 5$ **3.** $x < -1$ or $x > 3$ **5.** $-1 < x < 1$

7. $x < -2$ or $x > 3$ **9.** $s \leqslant 0$ or $s \geqslant 5$ **11.** $-3 < x < -2$

13. $x \leqslant -5$ or $x \geqslant 4$ **15.** $-\frac{3}{2} < z < 4$ **17.** $x < -1$ or $x > -1$

19. $t \leqslant -\frac{3}{2}$ or $t \geqslant \frac{1}{2}$ **21.** $-2 < x < 1$ or $x > 4$ **23.** $x < -5$

25. $y \leqslant -1$ or $0 \leqslant y \leqslant 1$ **27.** $x = 0$ or $x \geqslant 2$ **29.** $x < -8$ or $x > 4$

31. $t \leqslant 0$ or $t > 3$ **33.** $x > -2$ **35.** $x < -3$ or $-1 < x < 1$

37. $x < -4$ or $-1 < x \leqslant 1$ or $x \geqslant 4$ **39.** $x \leqslant -6$ or $x \geqslant 2$ **41.** $t < -5$ or $t \geqslant 1$

COMPLETION SET 15-3

1. 5, 7, $\frac{37}{6}$ **2.** 6, $-\frac{8}{3}$, $|-10|$, $\frac{16}{3}$ **3.** 4, 0, 4, 4

4. 2, 0, 2, -2, 3, -1 **5.** 6, 5 **6.** 3, $x - 2$

PROBLEM SET 15-3

1. $-3 < x < 3$ **3.** $x > 6$ or $x < -6$ **5.** $-1 \leqslant x \leqslant 1$

7. $x \geqslant 3$ or $x \leqslant -3$ **9.** $-12 < x < 20$ **11.** $y \geqslant 5$ or $y \leqslant -7$

13. $-\frac{4}{3} \leqslant x \leqslant 2$ **15.** $-\frac{1}{3} < x < 1$ **17.** $t < 0$ or $t > 1$

19. $x \leqslant 0$ or $x \geqslant \frac{16}{3}$ **21.** The absolute value of any quantity is never negative.

23. (a) $|x - 7| < 3$; (b) $|x - 2| < 3$; (c) $|x + 7| \leqslant 5$; (d) $|7 - x| = 4$;
(e) $|x + 4| < 2$; (f) $|x| < 3$; (g) $|x| > 6$; (h) $|x - 6| \geqslant 4$;
(i) $|x - 105| < 3$; (j) $|x - 600| < 100$

27. $|x - .01| \leqslant .005$

REVIEW PROBLEMS—CHAPTER 15

1. $x < \frac{1}{3}$ **3.** $x \leqslant -2$ **5.** $t > 1$ **7.** $x \geqslant 40$

9. $-6 < x < 2$ **11.** $y < 0$ or $y > 6$ **13.** $x \leqslant -5$ or $x \geqslant -4$

15. $-1 < x < \frac{1}{3}$ **17.** $-4 < x < 5$ or $x > 9$

19. $p \leqslant 0$ or $0 \leqslant p \leqslant 8$; more simply $p \leqslant 8$

21. $x \leqslant -9$ or $x > -2$

23. $x < -5$ or $-2 < x < 3$ or $x > 3$ **25.** $x > 2$ or $x < -2$

27. $0 < x < \frac{1}{2}$ **29.** $x \leqslant 0$ or $x \geqslant 4$ **31.** $-2 \leqslant x \leqslant 1$ **33.** $x \geqslant -\frac{5}{2}$

35. $0 \leqslant z \leqslant 3$

PROBLEM SET 16-1

1. 5 **3.** $\sqrt{53}$ **5.** $2\sqrt{2}$ **7.** 7 **9.** $\dfrac{\sqrt{5}}{2}$

11. 13

COMPLETION SET 16-2

1. 3, -5, 49, $(3, -5)$, 7 **2.** 3, $y - 5$, 4 **3.** (1, 1)

4. (0, 0), $\sqrt{5}$ **5.** 64, 36, 64, 36, 8, 6, 10, $(-8, -6)$, 10

PROBLEM SET 16-2

1. $(x - 2)^2 + (y - 3)^2 = 36;\quad x^2 + y^2 - 4x - 6y - 23 = 0$

3. $(x + 1)^2 + (y - 6)^2 = 16;\quad x^2 + y^2 + 2x - 12y + 21 = 0$

5. $x^2 + y^2 = \frac{1}{4};\quad x^2 + y^2 - \frac{1}{4} = 0$

7. $C = (0, 0)$, $r = 3$ **9.** $C = (3, -4)$, $r = 2$ **11.** $C = (-2, 0)$, $r = 1$

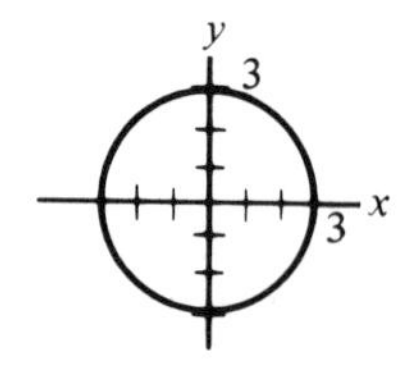

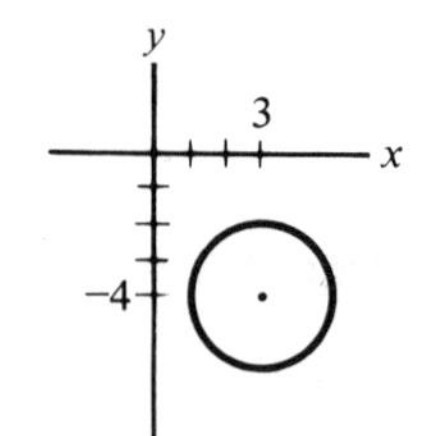

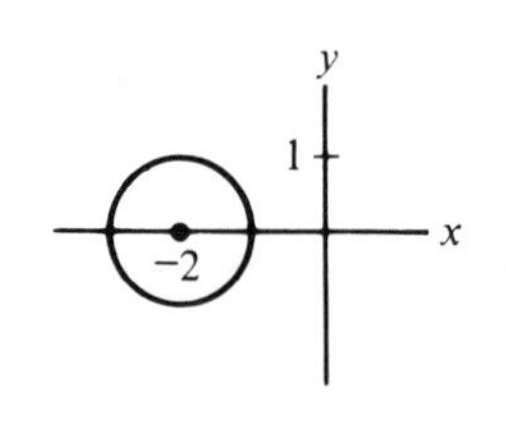

13. circle; $C = (1, 2)$, $r = 3$ **15.** circle; $C = (0, -3)$, $r = 2$

17. no graph **19.** circle; $C = (7, -2)$, $r = 4$

21. point $(\frac{3}{2}, -\frac{1}{2})$ **23.** circle; $C = (1, -\frac{7}{4})$, $r = \frac{7}{4}$

COMPLETION SET 16-3

1. parabola **2.** hyperbola **3.** hyperbola, ellipse

4. downward **5.** vertical **6.** horizontal

PROBLEM SET 16-3

1. parabola

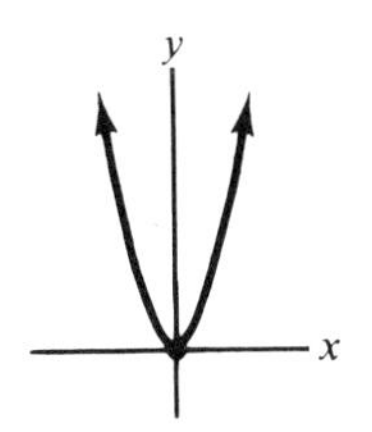

3. ellipse

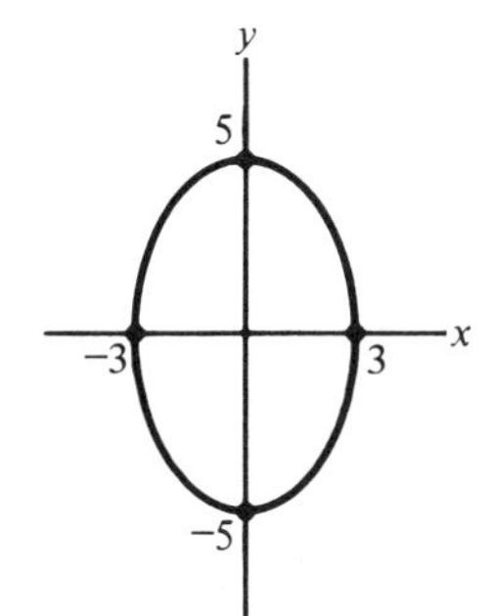

5. hyperbola

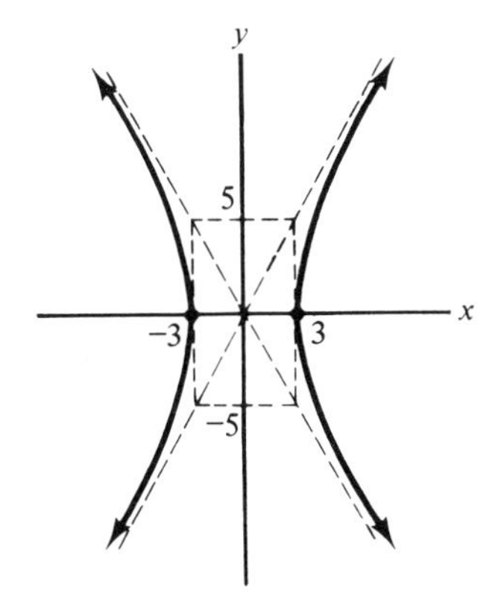

7. ellipse

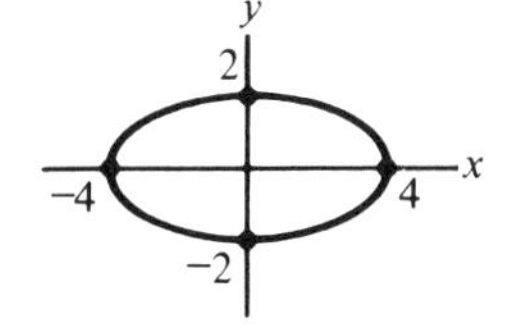

9. circle

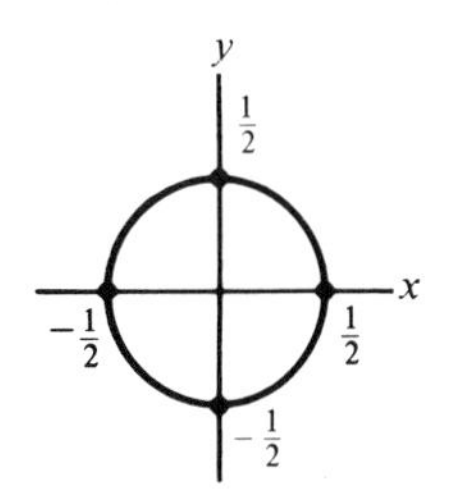

11. parabola

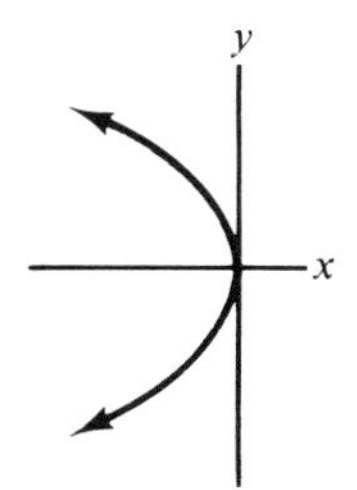

13. hyperbola

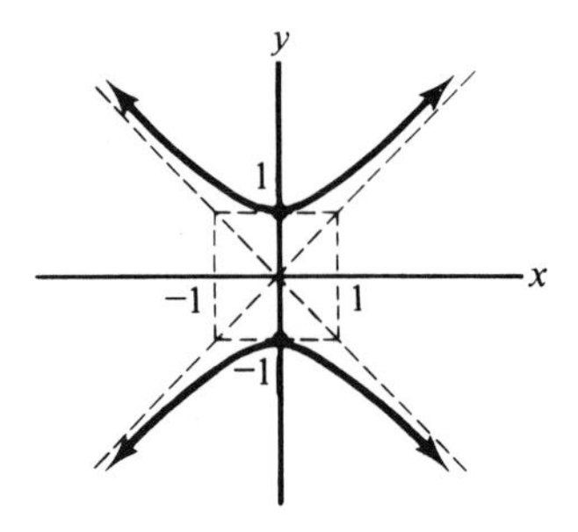

15. parabola

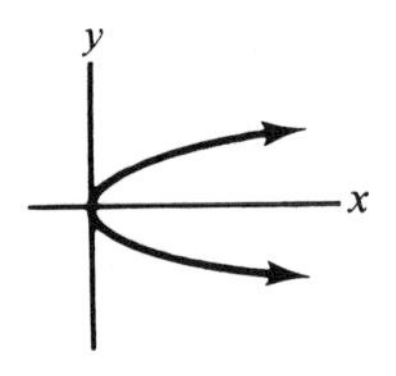

17. ellipse

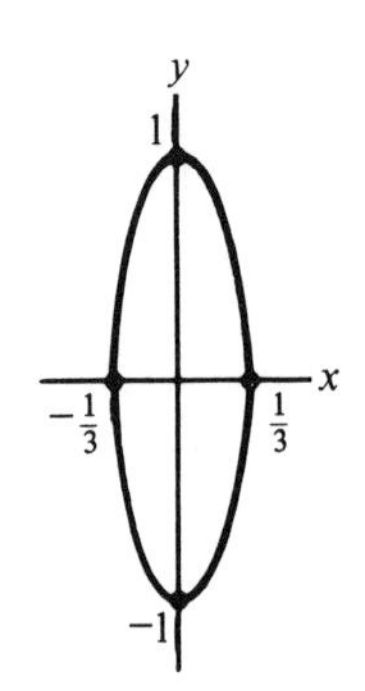

19. hyperbola

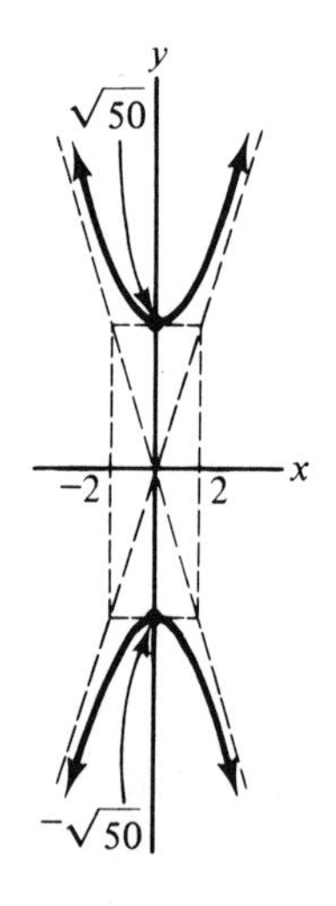

21. parabola

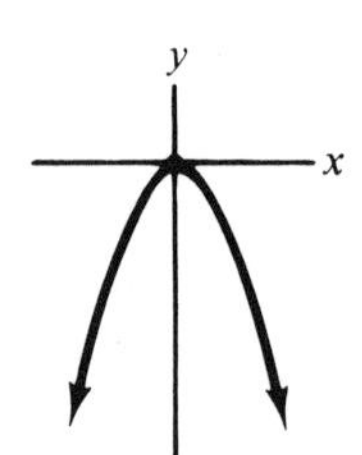

23. ellipse

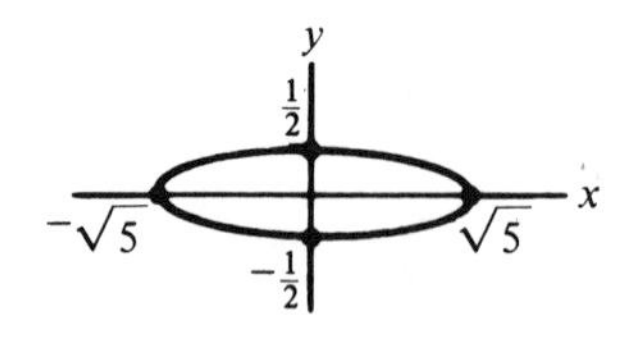

REVIEW PROBLEMS—CHAPTER 16

1. $2\sqrt{5}$ **3.** 10 **5.** $x^2 + y^2 = 25$; $x^2 + y^2 - 25 = 0$

7. $(x - 1)^2 + (y + 1)^2 = \frac{1}{4}$; $x^2 + y^2 - 2x + 2y + \frac{7}{4} = 0$

9. circle; $C = (0, 0)$, $r = 4$

11. circle; $C = (-2, 1)$, $r = \sqrt{7}$

13. circle; $C = (-\frac{5}{2}, 3)$, $r = \frac{\sqrt{53}}{2}$

15. point $(1, \frac{1}{3})$

17. hyperbola

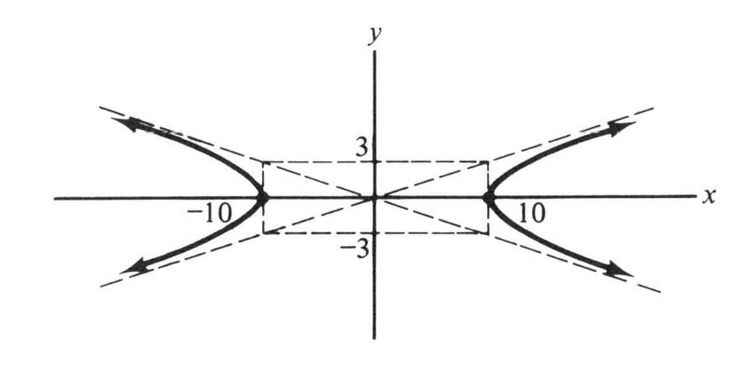

19. ellipse

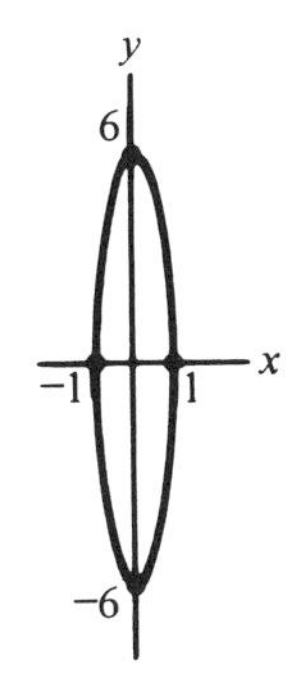

21. parabola

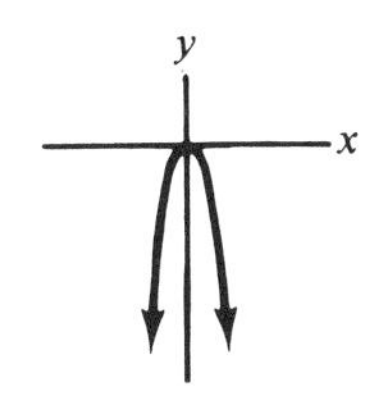

23. circle

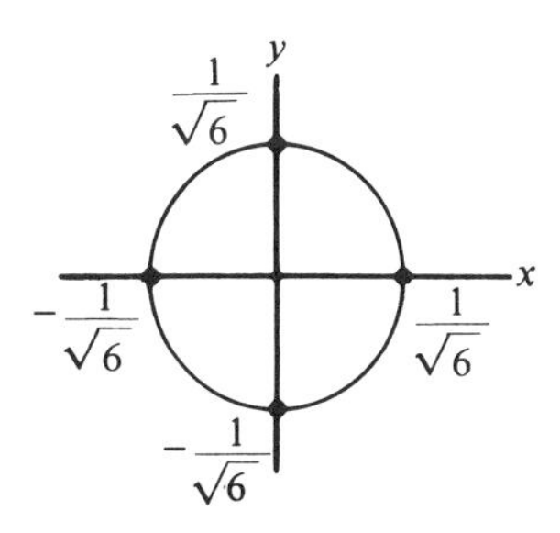

COMPLETION SET 17-1

1. (0, 0) **2.** third **3.** 90 **4.** acute

5. negative **6.** coterminal **7.** 1, 1 **8.** π

9. radians **10.** -2π **11.** $\frac{\pi \text{ rad}}{180°}$ **12.** $\frac{180°}{\pi \text{ rad}}$

PROBLEM SET 17-1

1. 0 **3.** 30° **5.** 90° **7.** $\frac{\pi}{4}$ **9.** $-\frac{2\pi}{3}$

11. 210° **13.** 300° **15.** $-\frac{3\pi}{2}$ **17.** 225° **19.** $-420°$

21. second **23.** fourth **25.** second **27.** quadrantal

29. first **31.** third **33.** quadrantal **35.** fourth

37. $C = 29° 20'$ **39.** $B = 105° 24' 45''$ **41.** $A = 66° 43' 48''$

43. $54° 42'$ **45.** $13° 14' 58''$

47. $\frac{\pi}{4}$ **49.** 2π **51.** $\frac{2\pi}{3}$ **53.** $\frac{25\pi}{3}$ **55.** $\frac{3\pi}{2}$

57. $\frac{25\pi}{2}$ **59.** 3π

In Problems **61–65**, other answers are possible.

61. 380°, 740°, −340°, −700° **63.** 675°, 1035°, −45°, −405°

65. $\frac{13\pi}{6}, \frac{25\pi}{6}, -\frac{11\pi}{6}, -\frac{23\pi}{6}$ **67.** 60° **69.** $\frac{5\pi}{4}$

71. $\frac{\pi}{2}$ **73.** 200π m **75.** $\frac{350\pi}{3}$ ft^2

COMPLETION SET 17-2

1. 1, 2, 5 **2.** y, x **3.** $\sin\theta$ **4.** $\frac{4}{3}$

5. .9848 **6.** tangent, cotangent **7.** fourth

PROBLEM SET 17-2

In Problems **1–17**, answers are in the order: $\sin\theta$, $\cos\theta$, $\tan\theta$, $\cot\theta$, $\sec\theta$, $\csc\theta$.

1. $\frac{4}{5}, \frac{3}{5}, \frac{4}{3}, \frac{3}{4}, \frac{5}{3}, \frac{5}{4}$ **3.** $\frac{4}{5}, -\frac{3}{5}, -\frac{4}{3}, -\frac{3}{4}, -\frac{5}{3}, \frac{5}{4}$

5. $-\frac{\sqrt{2}}{2}, -\frac{\sqrt{2}}{2}, 1, 1, -\sqrt{2}, -\sqrt{2}$ **7.** $\frac{\sqrt{5}}{5}, \frac{2\sqrt{5}}{5}, \frac{1}{2}, 2, \frac{\sqrt{5}}{2}, \sqrt{5}$

9. $-\frac{5\sqrt{26}}{26}, -\frac{\sqrt{26}}{26}, 5, \frac{1}{5}, -\sqrt{26}, -\frac{\sqrt{26}}{5}$

11. $-\frac{\sqrt{3}}{2}, \frac{1}{2}, -\sqrt{3}, -\frac{\sqrt{3}}{3}, 2, -\frac{2\sqrt{3}}{3}$

13. 1, 0, not defined, 0, not defined, 1

15. −1, 0, not defined, 0, not defined, −1

17. 0, −1, 0, not defined, −1, not defined

19. $\tan\theta = \frac{\sqrt{15}}{15}$, $\cot\theta = \sqrt{15}$, $\sec\theta = \frac{4\sqrt{15}}{15}$, $\csc\theta = 4$

21. $\sin\theta = -\frac{3}{5}$, $\cos\theta = -\frac{4}{5}$, $\tan\theta = \frac{3}{4}$, $\cot\theta = \frac{4}{3}$

23. $\cos\theta = -\frac{\sqrt{10}}{10}$, $\cot\theta = -\frac{1}{3}$, $\sec\theta = -\sqrt{10}$, $\csc\theta = \frac{\sqrt{10}}{3}$

25. $\frac{\sqrt{2}}{2}$ **27.** $\frac{\sqrt{3}}{2}$ **29.** $\frac{\sqrt{3}}{3}$ **31.** $\sqrt{2}$ **33.** +

35. − **37.** − **39.** + **41.** − **43.** +

45. − **47.** + **49.** −

51. $\cos\theta = -\frac{\sqrt{35}}{6}$, $\tan\theta = -\frac{\sqrt{35}}{35}$, $\cot\theta = -\sqrt{35}$, $\sec\theta = -\frac{6\sqrt{35}}{35}$, $\csc\theta = 6$

53. $\sin\theta = -\frac{\sqrt{33}}{7}$, $\cos\theta = \frac{4}{7}$, $\cot\theta = -\frac{4\sqrt{33}}{33}$, $\sec\theta = \frac{7}{4}$, $\csc\theta = -\frac{7\sqrt{33}}{33}$

55. $\sin\theta = -\frac{1}{2}$, $\cos\theta = -\frac{\sqrt{3}}{2}$, $\tan\theta = \frac{\sqrt{3}}{3}$, $\cot\theta = \sqrt{3}$, $\sec\theta = -\frac{2\sqrt{3}}{3}$

57. $\sin\theta = \frac{\sqrt{5}}{5}$, $\cos\theta = \frac{2\sqrt{5}}{5}$, $\tan\theta = \frac{1}{2}$, $\cot\theta = 2$, $\sec\theta = \frac{\sqrt{5}}{2}$, $\csc\theta = \sqrt{5}$

COMPLETION SET 17-3

1. 8, 6 **2.** 6, 8 **3.** (a) opp; (b) opp; (c) sec; (d) cot

4. third, negative, 30, −, 30, −, opp, −, 1 **5.** T

PROBLEM SET 17-3

1. $\sin A = \frac{\sqrt{7}}{4}$, $\cos A = \frac{3}{4}$, $\tan A = \frac{\sqrt{7}}{3}$, $\cot A = \frac{3\sqrt{7}}{7}$, $\sec A = \frac{4}{3}$, $\csc A = \frac{4\sqrt{7}}{7}$; $\sin B = \frac{3}{4}$, $\cos B = \frac{\sqrt{7}}{4}$, $\tan B = \frac{3\sqrt{7}}{7}$, $\cot B = \frac{\sqrt{7}}{3}$, $\sec B = \frac{4\sqrt{7}}{7}$, $\csc B = \frac{4}{3}$

3. $\frac{1}{2}$ **5.** 1 **7.** $\frac{1}{2}$ **9.** $\sqrt{3}$ **11.** $\sqrt{2}$

13. $\frac{2\sqrt{3}}{3}$ **15.** $\frac{1}{3}$ **17.** $\frac{1}{2}$ **19.** 80° **21.** 35°

23. $\frac{\pi}{3}$ **25.** $\frac{\pi}{6}$ **27.** $\frac{\sqrt{3}}{2}$ **29.** $-\frac{\sqrt{3}}{2}$ **31.** $-\frac{\sqrt{2}}{2}$

33. $-\frac{\sqrt{3}}{3}$ **35.** $\frac{2\sqrt{3}}{3}$ **37.** $-\frac{\sqrt{2}}{2}$ **39.** 2 **41.** $-\frac{1}{2}$

43. $\frac{\sqrt{3}}{3}$ **45.** $-\sqrt{2}$ **47.** −1 **49.** 2 **51.** $\frac{1}{2}$

53. $-\frac{\sqrt{3}}{3}$ **55.** $\frac{\sqrt{3}}{2}$

PROBLEM SET 17-4

1. .4067 **3.** .8796 **5.** .5150 **7.** .4108 **9.** −.9397

11. −.7679 **13.** 1.022 **15.** −1.556 **17.** .3939 **19.** 1.377

21. .0907 **23.** −.3330 **25.** .6501 **27.** 41° 40′ **29.** 74° 20′

31. 12° 14′ **33.** 26° 34′ **35.** 61° 42′ **37.** 64° 35′, 115° 25′

39. 149° 30′, 210° 30′ **41.** 319° 45′ **43.** 41° 49′ **45.** 1.638

PROBLEM SET 17-5

1. $A = 30°$, $b = 3\sqrt{3}$, $c = 6$ **3.** $B = 45°$, $a = 3\sqrt{2}$, $b = 3\sqrt{2}$

5. $A = 60°$, $B = 30°$, $a = 4\sqrt{3}$ **7.** $B = 63°$, $a = 1.8$, $b = 3.6$

9. $A = 53°\ 8'$, $B = 36°\ 52'$, $c = 10$ **11.** $A = 32°$, $a = 3.1$, $c = 5.9$

13. $A = 45°\ 34'$, $B = 44°\ 26'$, $a = \sqrt{51} = 7.1$

15. $B = 75°$, $a = 26.8$, $c = 103.5$ **17.** 336 ft

19. 358 ft **21.** 270 ft **23.** 5 min

25. $X_c = 2.5$, $R = \dfrac{5\sqrt{3}}{2} = 4.33$

REVIEW PROBLEMS—CHAPTER 17

1. $\dfrac{5\pi}{3}$ **3.** 150° **5.** $-\dfrac{5\pi}{18}$ **7.** 10°

9. third **11.** first **13.** 94° 1′ 30″, 72° 39′ 50″

15. 14π **17.** $\dfrac{3\pi}{2}$ **19.** 27π **21.** 140°

23. $\sin\theta = -\dfrac{6\sqrt{37}}{37}$, $\cos\theta = \dfrac{\sqrt{37}}{37}$, $\tan\theta = -6$, $\cot\theta = -\dfrac{1}{6}$, $\sec\theta = \sqrt{37}$, $\csc\theta = -\dfrac{\sqrt{37}}{6}$

25. $\sin\theta = -\dfrac{3\sqrt{5}}{7}$, $\cos\theta = -\dfrac{2}{7}$, $\tan\theta = \dfrac{3\sqrt{5}}{2}$, $\cot\theta = \dfrac{2\sqrt{5}}{15}$, $\sec\theta = -\dfrac{7}{2}$, $\csc\theta = -\dfrac{7\sqrt{5}}{15}$

27. $\tan\theta = -\dfrac{\sqrt{6}}{12}$, $\cot\theta = -2\sqrt{6}$, $\sec\theta = -\dfrac{5\sqrt{6}}{12}$, $\csc\theta = 5$

29. $\sin\theta = -\dfrac{2\sqrt{10}}{7}$, $\tan\theta = -\dfrac{2\sqrt{10}}{3}$, $\cot\theta = -\dfrac{3\sqrt{10}}{20}$, $\sec\theta = \dfrac{7}{3}$, $\csc\theta = -\dfrac{7\sqrt{10}}{20}$

31. $\frac{\sqrt{3}}{2}$ **33.** 0 **35.** $\sqrt{3}$ **37.** $-\frac{1}{2}$ **39.** -1

41. $\sqrt{2}$ **43.** $\frac{\sqrt{3}}{3}$ **45.** 0 **47.** $-\frac{2\sqrt{3}}{3}$ **49.** $-.9112$

51. 5.576 **53.** .8502 **55.** -1.720 **57.** 33° 23′, 146° 37′

59. 128° 53′, 308° 53′ **61.** 202° 34′ **63.** $B = 48°$, $a = 2.7$, $c = 4.0$

65. $A = 14°$, $a = 2.4$, $b = 9.7$ **67.** $A = 23° 35'$, $B = 66° 25'$, $b = \sqrt{21} = 4.6$

3. 2π **4.** 2π **5.** π

8. 1

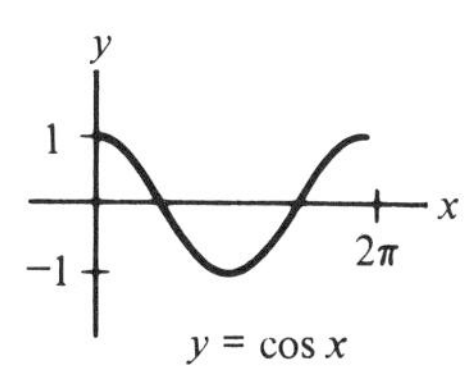

$y = \cos x$

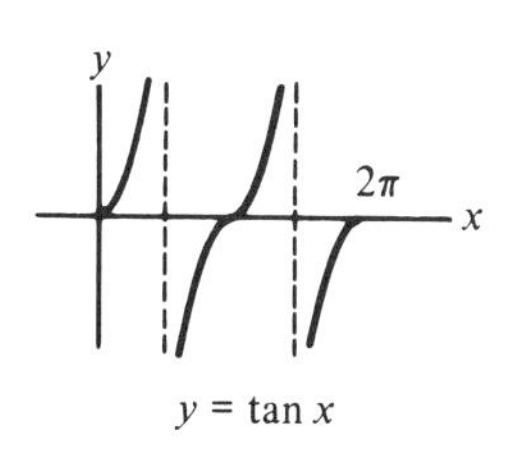

$y = \tan x$

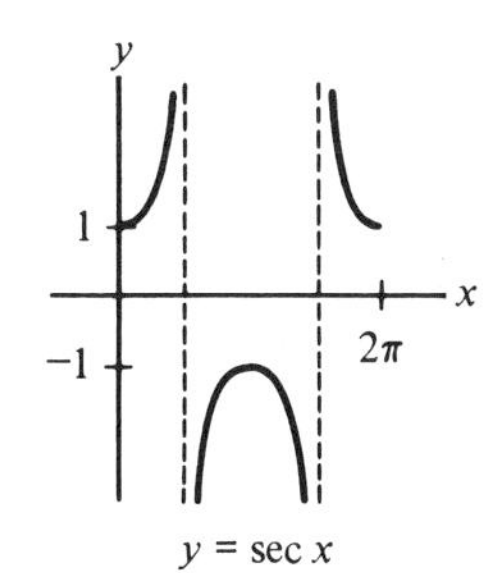

$y = \sec x$

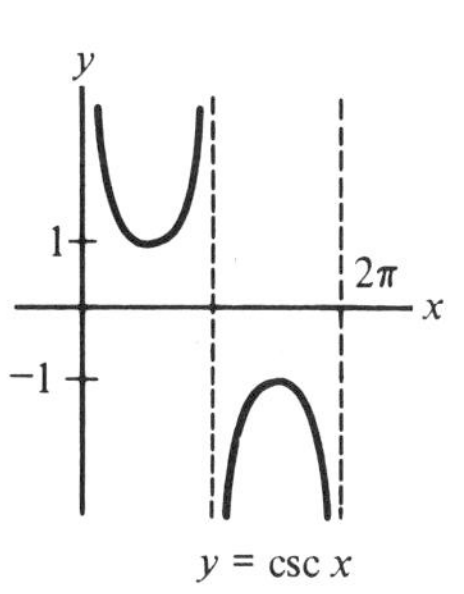

$y = \csc x$

$-\frac{}{2}, -\frac{}{2}$ **5.** $\frac{3\pi}{2}, \frac{7\pi}{2}$ **7.** $0, \pi, 2\pi$

COMPLETION SET 18-2

1. 3 **2.** -4 **3.** $2, 0, 0, 2\pi, \frac{2\pi}{3}, \frac{2\pi}{3}, 0, \frac{2\pi}{3}$

PROBLEM SET 18-2

1. $A = 3, p = 2\pi$

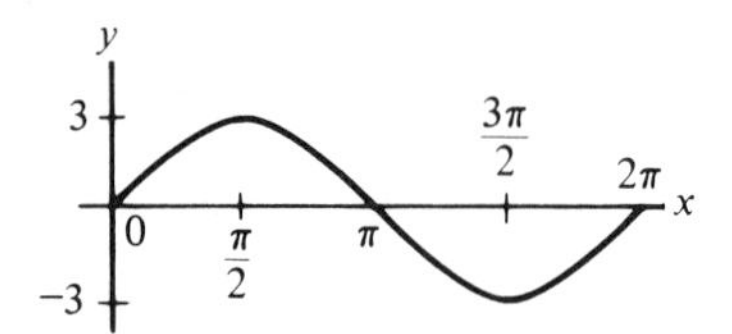

3. $A = 4, p = 2\pi$

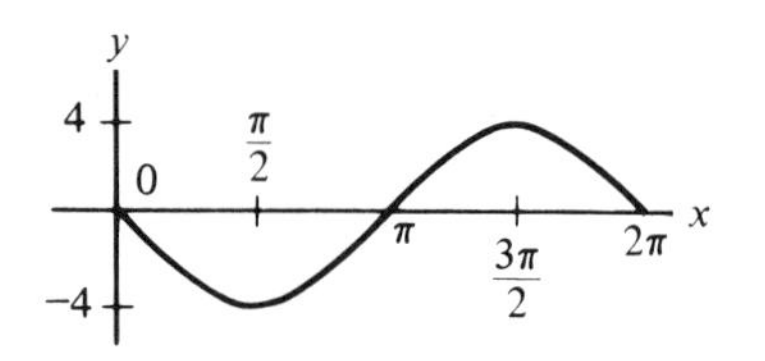

5. $A = \frac{1}{2}, p = 2\pi$

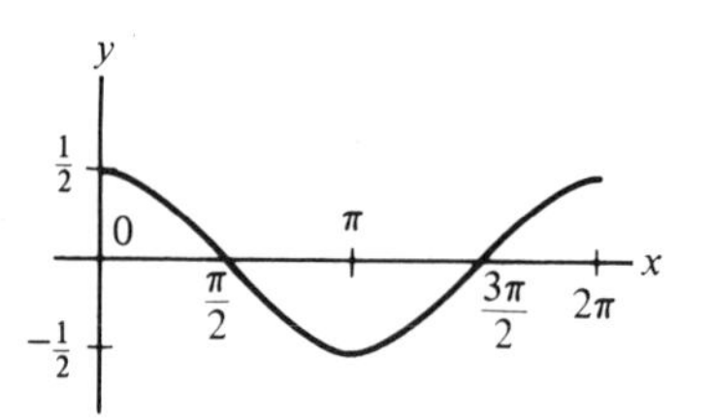

7. $A = 1, p = 2\pi$

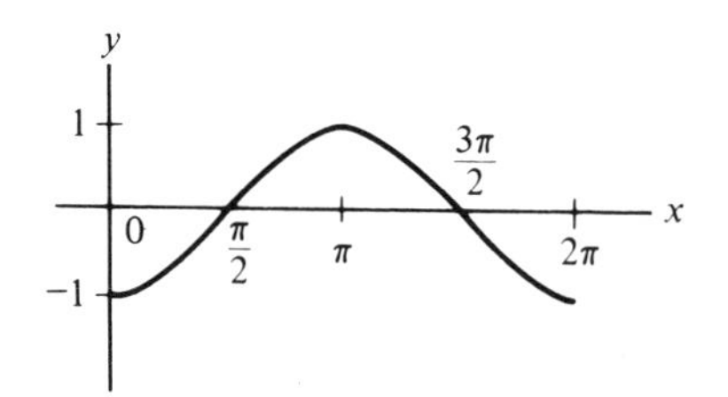

9. $A = 1, p = \pi$

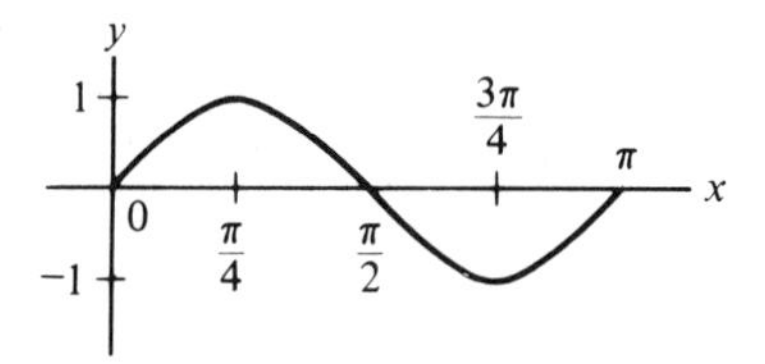

11. $A = 4, p = \frac{2\pi}{3}$

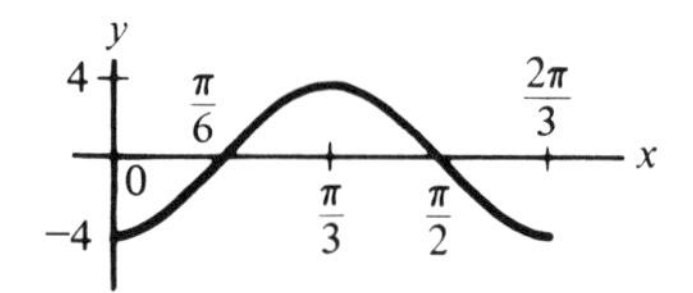

13. $A = 2, p = 6\pi$

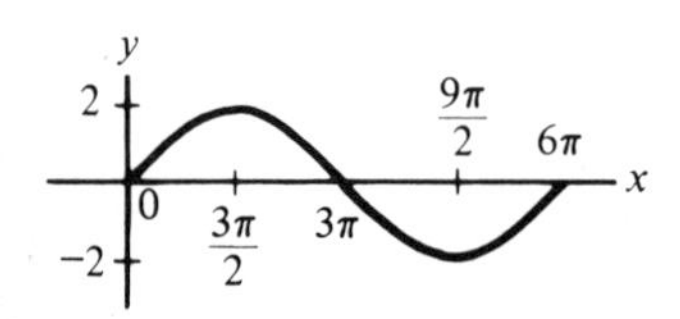

15. $A = 1, p = 2\pi$

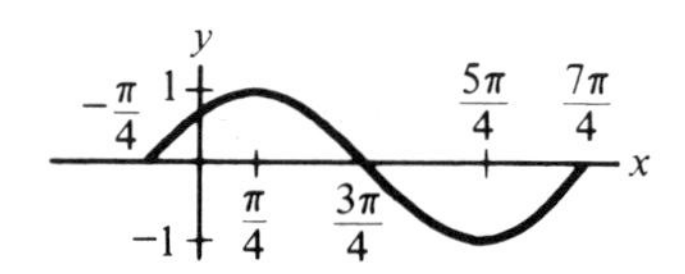

17. $A = 4, p = \pi$

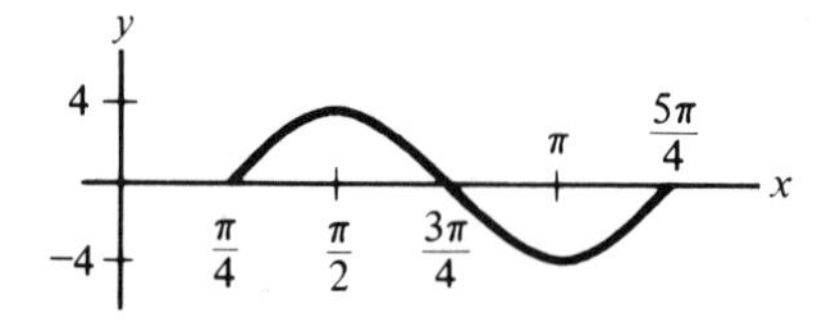

19. $A = 2, p = 2\pi$

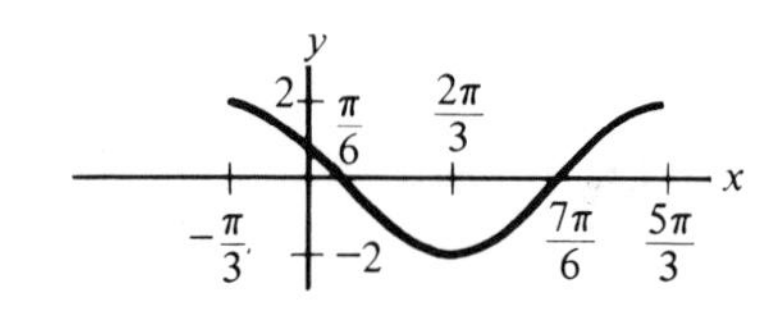

21. $A = 2, p = 4\pi$

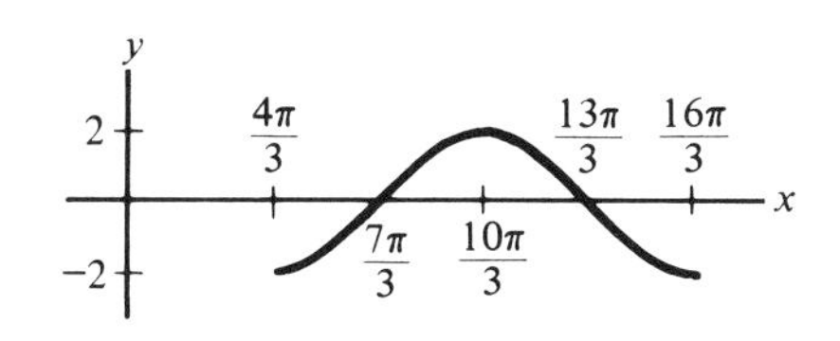

23. $A = 1, p = 2$

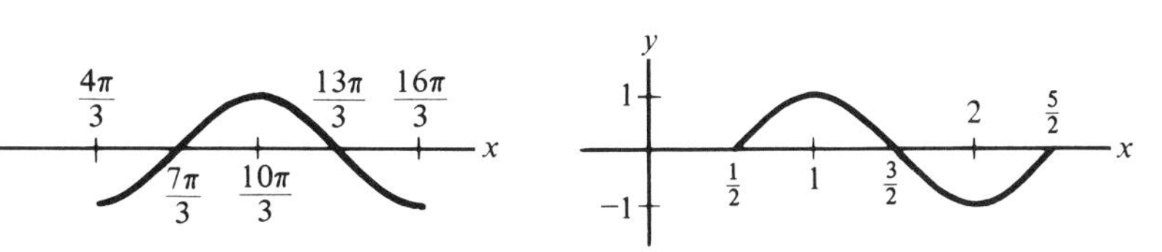

25. two

27. $\frac{4}{3}$

29.

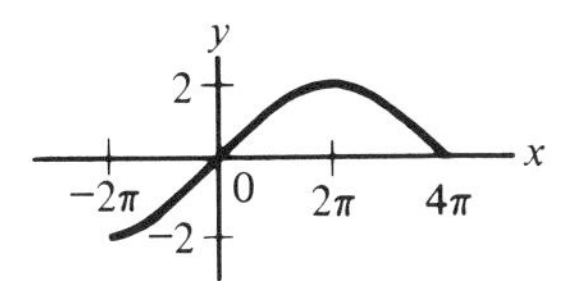

31. $\dfrac{2\pi}{b}$

COMPLETION SET 18-3

1. x

2. $0, \pi, x$

3. $-\dfrac{\pi}{2}, \dfrac{\pi}{2}$, tangent, x

4. 0

PROBLEM SET 18-3

1. $\dfrac{\pi}{2}$

3. $\dfrac{\pi}{6}$

5. 0

7. $-\dfrac{\pi}{6}$

9. $\dfrac{\pi}{4}$

11. $\dfrac{3\pi}{4}$

13. $\frac{1}{3}$

15. $\dfrac{\sqrt{3}}{2}$

17. 0

19. $\dfrac{\sqrt{2}}{2}$

21. $\dfrac{\pi}{3}$

REVIEW PROBLEMS—CHAPTER 18

1. $A = 1, p = \dfrac{2\pi}{3}$

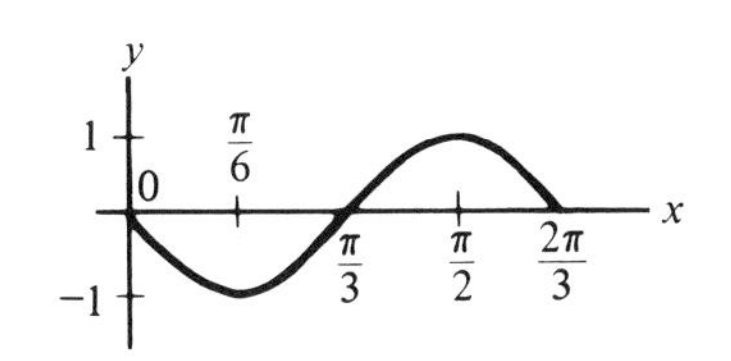

3. $A = 4, p = 2\pi$

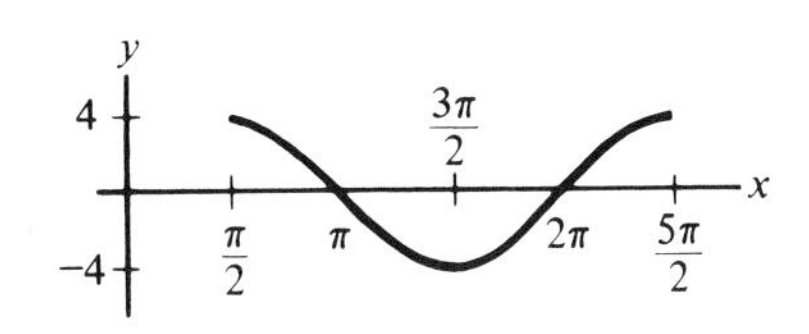

5. $A = 3, p = \pi$

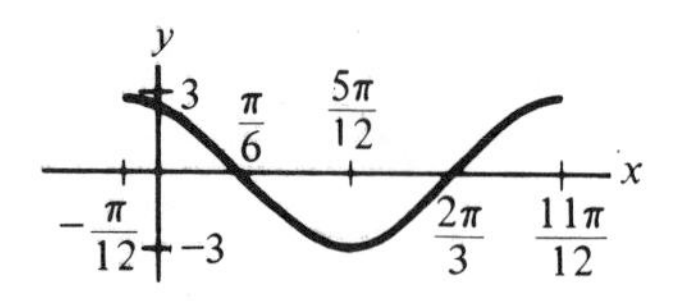

7. $-\frac{\pi}{2}$ 9. $\frac{\pi}{6}$ 11. $\frac{5\pi}{6}$

13. 0 15. $-\frac{\sqrt{3}}{2}$ 17. .8 19. π

COMPLETION SET 19-1

1. identity 2. 1 3. $\cot^2 \theta$ 4. $\tan^2 x$ 5. $\cot 2x$
6. two 7. b, c

COMPLETION SET 19-2

1. T 2. F 3. T 4. F
5. $\cos 2y, \sin 2y$ 6. $\cos \pi, \sin x, -\cos x, -\cos x$
7. 32, 4, 28 8. 60, $\frac{1}{2}$

PROBLEM SET 19-2

1. $\frac{\sqrt{6} - \sqrt{2}}{4}$ 3. $\frac{\sqrt{2} - \sqrt{6}}{4}$ 5. $2 - \sqrt{3}$ 7. $\frac{\sqrt{2} - \sqrt{6}}{4}$

9. $\frac{-\sqrt{6} - \sqrt{2}}{4}$ 11. $2 + \sqrt{3}$ 13. (a) $-\frac{7\sqrt{65}}{65}$, (b) $\frac{4\sqrt{65}}{65}$, (c) $-\frac{7}{4}$, (d) IV

15. $-\sin x$ 17. $\sin x$ 19. $\frac{1 + \tan x}{1 - \tan x}$ 21. $\cos 70°$

COMPLETION SET 19-3

1. $\cos x$ 2. $\sin^2 x$ 3. positive 4. 30 5. 80
6. cos

PROBLEM SET 19-3

1. $\frac{\sqrt{3}}{2}$ 3. $-\frac{1}{2}$ 5. $-\sqrt{3}$ 7. $\frac{\sqrt{2-\sqrt{3}}}{2}$

9. $\frac{\sqrt{2+\sqrt{2}}}{2}$ 11. $-\sqrt{2}-1$

13. $\sin x = \frac{4}{5}$, $\cos x = \frac{3}{5}$, $\tan x = \frac{4}{3}$, $\sin 2x = \frac{24}{25}$, $\cos 2x = -\frac{7}{25}$, $\tan 2x = -\frac{24}{7}$

15. $\sin x = -\frac{1}{3}$, $\cos x = -\frac{2\sqrt{2}}{3}$, $\tan x = \frac{\sqrt{2}}{4}$, $\sin 2x = \frac{4\sqrt{2}}{9}$, $\cos 2x = \frac{7}{9}$, $\tan 2x = \frac{4\sqrt{2}}{7}$

17. $\sin x = \frac{5}{13}$, $\cos x = \frac{12}{13}$, $\tan x = \frac{5}{12}$, $\sin \frac{x}{2} = \frac{\sqrt{26}}{26}$, $\cos \frac{x}{2} = \frac{5\sqrt{26}}{26}$, $\tan \frac{x}{2} = \frac{1}{5}$

19. $\sin x = -\frac{3}{5}$, $\cos x = -\frac{4}{5}$, $\tan x = \frac{3}{4}$, $\sin \frac{x}{2} = \frac{3\sqrt{10}}{10}$, $\cos \frac{x}{2} = -\frac{\sqrt{10}}{10}$, $\tan \frac{x}{2} = -3$

PROBLEM SET 19-4

1. 45°, 135° 3. 135°, 315° 5. 60°, 300° 7. 0°, 120°, 240°

9. 0°, 180° 11. 90°, 210°, 330° 13. 30°, 60°, 210°, 240°

15. 0°, 240° 17. $112\frac{1}{2}$°, $157\frac{1}{2}$°, $292\frac{1}{2}$°, $337\frac{1}{2}$°

19. 60°, 90°, 270°, 300° 21. 0°, 60°, 180°, 300° 23. 30°, 150°, 210°, 330°

25. 90°, 210°, 270°, 330° 27. 0°, 135°, 180°, 315°

PROBLEM SET 19-5

1. $C = 30°$, $b = 25.71$, $c = 13.05$

3. $B = 43° \ 44'$, $C = 55° \ 16'$, $c = 8.320$

5. (a) $B = 26° \ 5'$, $C = 133° \ 55'$, $c = 14.75$; (b) $B = 153° \ 55'$, $C = 6° \ 5'$, $c = 2.170$

7. no triangle 9. $A = 13°$, $B = 27°$, $b = 14.13$

11. $A = 48°$, $b = 93.23$, $c = 102.4$ 13. $A = 60°$, $B = 90°$, $a = 8.660$

15. no triangle 17. 7513 ft 19. .75 mi

PROBLEM SET 19-6

1. $A = 65° 12'$, $C = 54° 48'$, $b = 9.539$

3. $A = 28° 57'$, $B = 46° 34'$, $C = 104° 29'$

5. $A = 22° 47'$, $B = 129° 13'$, $c = 24.24$

7. $A = 82° 49'$, $B = 41° 25'$, $C = 55° 46'$

9. $r = 11.93$, $\theta = 55° 38'$

11. 44 mi

REVIEW PROBLEMS—CHAPTER 19

9. $\dfrac{3 + 4\sqrt{3}}{10}$

11. $\frac{1}{2}$

13. $\frac{1}{2}$

15. $\frac{1}{7}$

17. $\sin x = \dfrac{15}{17}$, $\cos x = \dfrac{8}{17}$, $\tan x = \dfrac{15}{8}$, $\sin(-x) = -\dfrac{15}{17}$, $\cos(-x) = \dfrac{8}{17}$, $\sin 2x = \dfrac{240}{289}$, $\cos 2x = -\dfrac{161}{289}$, $\tan 2x = -\dfrac{240}{161}$, $\sin\dfrac{x}{2} = \dfrac{3\sqrt{34}}{34}$, $\cos\dfrac{x}{2} = \dfrac{5\sqrt{34}}{34}$, $\tan\dfrac{x}{2} = \dfrac{3}{5}$

19. $A = 97° 32'$, $B = 52° 28'$, $c = 2.522$

21. $A = 80°$, $a = 5.909$, $c = 5.638$

23. $A = 104° 29'$, $B = 46° 34'$, $C = 28° 57'$

25. $B = 25° 40'$, $C = 94° 20'$, $c = 23.03$

27. (a) $B = 13° 23'$, $C = 156° 37'$, $c = 13.72$; (b) $B = 166° 37'$, $C = 3° 23'$, $c = 2.039$

29. $90°$, $270°$

31. $240°$, $300°$

33. $\sqrt{97.94}$

Index

A

B

C

I

L

M

N

O

P

Q

R

S

T

V

X

Y

Z